Medical Virology: From Pathogenesis to Disease Control

Series Editor

Shailendra K. Saxena, Centre for Advanced Research (CFAR), Faculty of Medicine, King George's Medical University (KGMU), Lucknow, India

This book series reviews the recent advancement in the field of medical virology including molecular epidemiology, diagnostics and therapeutic strategies for various viral infections. The individual books in this series provide a comprehensive overview of infectious diseases that are caused by emerging and re-emerging viruses including their mode of infections, immunopathology, diagnosis, treatment, epidemiology, and etiology. It also discusses the clinical recommendations in the management of infectious diseases focusing on the current practices, recent advances in diagnostic approaches and therapeutic strategies. The books also discuss progress and challenges in the development of viral vaccines and discuss the application of viruses in the translational research and human healthcare.

Jyotir Moy Chatterjee • Shailendra K. Saxena
Editors

Artificial Intelligence and Blockchain in Precision Medicine and Virology

Springer

Editors
Jyotir Moy Chatterjee
Graphic Era University
Dehradun, Uttarakhand, India

Shailendra K. Saxena
Centre for Advanced Research (CFAR)
Faculty of Medicine
King George's Medical University (KGMU)
Lucknow, India

ISSN 2662-981X ISSN 2662-9828 (electronic)
Medical Virology: From Pathogenesis to Disease Control
ISBN 978-981-96-8918-7 ISBN 978-981-96-8919-4 (eBook)
https://doi.org/10.1007/978-981-96-8919-4

This Springer imprint is published by the registered company Springer Nature Singapore Pte Ltd.
The registered company address is: 152 Beach Road, #21-01/04 Gateway East, Singapore 189721, Singapore

Preface

"Artificial Intelligence and Blockchain in Precision Medicine and Virology" stands at the intersection of medical virology, computer science engineering, and artificial intelligence. This interdisciplinary book seeks to harness the strengths of these fields to address the complex challenges of diagnosing, treating, and managing viral diseases. Medical virology, which studies viruses and virus-like agents, plays a critical role in understanding infectious diseases and developing therapeutic strategies. However, the field faces immense challenges due to the sheer volume and complexity of virological data, which demands advanced computational tools for effective analysis and application.

Computer science engineering provides the technological backbone for this initiative. Engineers in this field develop and implement sophisticated algorithms and data structures that enable the processing of large-scale virological datasets. In this context, computer science engineering is pivotal for creating the infrastructure necessary for deep learning and blockchain technologies. Deep learning, a branch of artificial intelligence, involves training neural networks to recognize patterns in vast datasets, enabling the discovery of insights that are otherwise obscured. When applied to medical virology, deep learning can significantly enhance the accuracy of viral diagnostics, predict disease outbreaks, and personalize treatment regimens based on the genetic makeup of both viruses and patients.

Unique Features of the book:

- Advanced Viral Diagnostics through Deep Learning: By employing deep learning algorithms, the book enhances the capability to analyze complex virological data. For instance, neural networks can be trained on datasets comprising viral genomic sequences, enabling them to identify viral strains rapidly and accurately. This is crucial for early detection and intervention, especially in the case of emerging viral threats.
- Blockchain for Secure Data Management: Blockchain technology ensures the secure storage and sharing of sensitive medical data. In medical virology, maintaining the integrity and confidentiality of patient data is paramount. Blockchain's decentralized ledger system prevents unauthorized access and

ensures that data is immutable, thus fostering trust among patients and healthcare providers.
- Interdisciplinary Approach to Precision Medicine: The book exemplifies a multidisciplinary approach by integrating insights from medical virology, computer science engineering, and AI. This convergence enables the development of precision medicine strategies that are tailored to individual patients based on comprehensive data analysis, leading to more effective and personalized treatments.
- Enhanced Collaboration and Data Sharing: Real-time data sharing enabled by blockchain facilitates seamless collaboration among virologists, computer scientists, and clinicians. This feature is particularly beneficial in response to viral outbreaks, where timely data sharing can accelerate research and improve public health responses.
- Patient-Centric Data Control: By giving patients control over their medical data through secure blockchain interfaces, the book enhances patient engagement and empowerment. Patients can grant or revoke access to their data as needed, ensuring transparency and fostering trust in the healthcare system.

In summary, the "Next-Generation Healthcare" book integrates deep learning and blockchain to revolutionize clinical decision-making in medical virology. The interdisciplinary collaboration between medical virology, computer science engineering, and artificial intelligence enables the development of innovative solutions for diagnosing and treating viral diseases. The unique features of this book, such as advanced diagnostics, secure data management, and enhanced collaboration, demonstrate its potential to significantly improve healthcare outcomes in the field of virology by helping the Physicians, Virologists, Researchers, etc.

Chapter 1 surveyed how artificial intelligence specifically deep learning will have a significant effect over healthcare domain in the future, particularly in the field of medical virology.

Chapter 2 presented the backbone, the implications, and new challenges of clinical virology, including human genetics, virology, diagnosis, and treatment of viral diseases.

Chapter 3 examined the relationship between deep learning and medical virology and applies deep learning to precision clinical decision making for viral diseases like human papillomavirus and human immunodeficiency virus.

Chapter 4 examined data production methodologies and datasets for various illnesses, focusing on simple deep learning algorithms, accessible tools, and visualization techniques.

Chapter 5 aimed to explore the potential of deep learning frameworks in enhancing the speed, precision, and scalability of viral detection, focusing on real-world case studies of diagnostic applications.

Chapter 6 demonstrated that the convolutional neural network is the most effective algorithm for predicting viral sequence origins of Influenza viruses.

Chapter 7 explored how medical virology can be transformed by blockchain technology and artificial intelligence, with a focus on how these technologies can ensure precise clinical decision-making.

Chapter 8 suggested a blockchain-based design for decentralized health networks.

Chapter 9 makes a speciality of key regions consisting of AI in early detection and prognosis of viral infections, customized remedy plans, and outbreak prediction, all even as safeguarding patient privacy and addressing ethical concerns.

Chapter 10 proposed a novel framework that integrates federated deep learning and blockchain technology for preserving privacy in precision medicine in medical virology

Chapter 11 proposed a methodology that integrates blockchain technology with deep learning to tackle these challenges.

Chapter 12 analyzed the impact blockchain technology has in the field of medical virology as far as safeguarding data, improving processes, and increasing efficiency are concerned.

Chapter 13 aimed to resolve kidney exchange problems optimally by utilizing various bio-inspired optimization techniques by determining the maximum groups of compatible donors/recipients from the registry of incompatible pairs submitted as transplant candidates.

Chapter 14 proposed a predictive model employing convolutional neural network, logistic regression, and support vector machine algorithms to discern the presence or absence of anemia based on anemia eye data rather than images.

Chapter 15 aimed to do extensive research about the recent techniques that are all used for the different organs.

Chapter 16 demonstrated the promise of sophisticated convolutional neural network designs for use in virology and sets the stage for further research into how to improve the efficiency and accuracy of viral classification for better disease management and response.

Chapter 17 further proposed a new deep learning model to examine clinical and genetic data for Amyotrophic Lateral Sclerosis (ALS).

Chapter 18 portrayed a thorough investigation of cutting-edge artificial intelligence paradigms that are creatively created to improve patient interaction dynamics and amplify the effectiveness of telehealth delivery.

Chapter 19 explored the possibilities for how such technologies might meet the health needs of people with disabilities while also implicating the challenges that may emerge.

Chapter 20 discussed the challenges and contributions of deep learning methods in virology, starting with an introduction to the field and discussing applications during the coronavirus pandemic.

Dehradun, Uttarakhand, India Jyotir Moy Chatterjee
Lucknow, India Shailendra K. Saxena

Acknowledgments

I extend my heartfelt gratitude to my Gurudev, Sri Sri Kalika Bhairab Brahmachari Guru Maharaj, the most significant person in my life, whose grace has transformed my long-cherished dream of this book into reality. I also express appreciation to my family and friends for their steadfast support, encouragement, and blessings, which have infused this work with unconditional love and patience.

Jyotir Moy Chatterjee
Department of CSE, Graphic Era (Deemed to be University)
Dehradun, Uttarakhand, India

Visiting Faculty, Department of IT, Lord Buddha Education Foundation
Kathmandu, Nepal

Contents

Editors and Contributors

About the Editors

Jyotir Moy Chatterjee presently holds the position of Assistant Professor in the Department of Computer Science and Engineering at Graphic Era (Deemed to be University) in Dehradun, Uttarakhand, India. Additionally, he serves as an Assistant Professor (Visiting Faculty) in the Department of Information Technology at Lord Buddha Education Foundation, affiliated with the Asia Pacific University of Technology & Innovation, Malaysia, located in Kathmandu, Nepal. Prior to these roles, he held the position of Assistant Professor and Program Leader for the B.Sc. IT program at Lord Buddha Education Foundation. He also served as an Assistant Professor in the Computer Science and Engineering department at GD-Rungta College of Engineering & Technology, affiliated with Chhattisgarh Swami Vivekananda Technical University in Bhilai, India. He earned his MTech degree in Computer Science & Engineering from Kalinga Institute of Industrial Technology (KIIT) in Bhubaneswar, Odisha, in 2016. He obtained his BTech degree in Computer Science & Engineering from Dr. MGR Educational & Research Institute in Maduravoyal, Chennai, in 2013. His research interests primarily lie in the domains of Machine Learning and Deep Learning. Specifically, he concentrates on advancing the fields of Deep Learning and Machine Learning through his research work, contributing to the ongoing developments in these areas.

Shailendra K. Saxena is the Vice Dean and distinguished Professor & Head at King George's Medical University, Lucknow, India. His research interests involve understanding the molecular mechanisms of host defense during human viral infections including Japanese encephalitis virus (JEV), HIV & neuroAIDS, COVID-19, emerging viruses, and virus-induced neurological disorders as models via stem cell/cell culture technologies and developing new predictive, preventive, and therapeutic strategies for them. Prof. Saxena has also conducted epidemiological studies that have highlighted the burden and neurological sequelae of these diseases, thereby informing public health and clinical interventions. He has significantly advanced the integration of artificial intelligence (AI) in precision medicine and virology. Through leading research, landmark publications, and developing AIdriven tools for diagnostics, treatment, and outbreak prediction, he has pioneered innovations enhancing personalized healthcare, especially in viral infections like COVID-19 and encephalitis. He plays a pivotal role in mentoring early-career scientists and fostering interdisciplinary collaborations. His research bridges virology, immunology, neurosciences and therapeutics, with the overarching goal of improving diagnostic and therapeutic outcomes in neuroinfectious and neurodegenerative diseases. His research work has been published in various high-impact-factor journals (*Science, PNAS, Nature Medicine, JMV, ACS* etc.) with a high number of citations. His work not only advances academic knowledge but also addresses real-world health challenges, particularly those affecting vulnerable populations in low- and middle-income countries. Prof. Saxena's contributions have had a transformative impact on the fields of biomedical sciences, and translational medicine, positioning him as a key figure in the international biomedical community. His ability to link cutting-edge science with practical healthcare solutions has made him a valuable contributor to global health security, especially in the face of pandemics and emerging infectious threats. With a vision rooted in scientific excellence and societal impact, Prof. Shailendra Saxena continues to shape the future of global health through innovation, collaboration, and an unwavering commitment to addressing some of the most urgent medical challenges of our time. Overall, Prof. Saxena's work embodies a blend of academic rigor and practical application, aimed at addressing pressing health issues. He has received several awards and honors in India and abroad including various Young Scientist Awards, the BBSRC India Partnering Award, and the Dr. JC Bose National Award of the Department of Biotechnology, Min. of Science and Technology, Govt. of India. He is a fellow of various international societies/academies including the Royal College of Pathologists, United Kingdom; Royal Society of Medicine, London; Royal Society of Biology, United Kingdom; Royal Society of Chemistry, London; and Academy of Translational Medicine Professionals, Austria. He was named a Global Leader in Science by The Scientist. He is also an international opinion leader/expert in vaccination for Japanese encephalitis by IPIC (UK). He has been recognized among the 'World's Top 2% Scientists' by Stanford University, California, USA; ScholarGPS™ (USA) World's Top 0.05% or better Highly Ranked Scholar-Lifetime™; and 'Clarivate's Highly Cited Researchers.' His inclusion in these prestigious databases is a testament to his exceptional contributions to global research.

Contributors

Rudraksh Agarwal Department of Computer Science and Engineering, Graphic Era Deemed to Be University, Dehradun, India

Shweta Agrawal Department of Computer Science and Engineering, Indore Institute of Science and Technology, Indore, MP, India

Sukrati Agrawal Department of Computer Science and Engineering, Indore Institute of Science and Technology, Indore, MP, India

Rehab Alaa Faculty of Science, Cairo University, Giza, Egypt

Aya Amir Faculty of Science, Cairo University, Giza, Egypt

Garvita Arora Department of Computer Science and Engineering, Graphic Era Deemed to Be University, Dehradun, India

Sameh H. Basha Faculty of Science, Cairo University, Giza, Egypt
Scientific Research School of Egypt (SRSEG), Cairo, Egypt
Faculty of Science, Galala University, Suez, Egypt

Shantanu Bhadra Department of Computational Science, Brainware University, Barasat, Kolkata, West Bengal, India

Jyotir Moy Chatterjee Department of Computer Science and Engineering, Graphic Era Deemed to Be University, Dehradun, India
CSE Department, Graphic Era University, Dehradun, India

Vanshika Chauhan Centre for Computational Biology and Bioinformatics, Amity Institute of Biotechnology, Amity University, Uttar Pradesh, Noida, India

Ashraf Darwish Scientific Research School of Egypt (SRSEG), Cairo, Egypt
Faculty of Science, Helwan University, Helwan, Egypt

Hrishikesh Desai Department of Accounting, Neil Griffin College of Business, Arkansas State University, Jonesboro, AR, USA

N. Ethiraj Department of Mechanical Engineering, Dr. M.G.R Educational and Research Institute of Technology, Chennai, Tamil Nadu, India

S. Geetha Department of Computer Science and Engineering, Dr. M.G.R Educational and Research Institute of Technology, Chennai, Tamil Nadu, India

Victo Sudha George Department of Computer Science and Engineering, Dr. M.G.R Educational and Research Institute of Technology, Chennai, Tamil Nadu, India

Nasib Singh Gill Department of Computer Science and Applications, Maharshi Dayanand University, Rohtak, Haryana, India

C. Gokulnath Department of Computing Technologies, School of Computing, SRM Institute of Science and Technology, Kattankulathur, Chengalpattu District, Tamil Nadu, India

Sanskriti Goyal Centre for Computational Biology and Bioinformatics, Amity Institute of Biotechnology, Amity University, Uttar Pradesh, Noida, India

Preeti Gulia Department of Computer Science and Applications, Maharshi Dayanand University, Rohtak, Haryana, India

Chhaya Gupta Department of Computer Science and Applications, Maharshi Dayanand University, Rohtak, Haryana, India
Vivekananda School of Information Technology, Vivekananda Institute of Professional Studies-Technical Campus, New Delhi, India

Aboul Ella Hassanien Faculty of Computers and AI, Cairo University, Cairo, Egypt
Scientific Research School of Egypt (SRSEG), Cairo, Egypt

Bala Shanmukha Sowmya Javvadhi CSE Department, Pragati Engineering College (A), Surampalem, India

Y. Jayababu CSE Department, Pragati Engineering College (A), Surampalem, India

Debasish Kar Department of Biotechnology, Faculty of Life of Allied Health Sciences, Ramaiah University of Applied Sciences, Bangalore, Karnataka, India

K. Kavitha School of Computing, Department of AI and ML, Mohan Babu University, Tirupati, India

Pasupuleti Sai Kiran Prasad V Potluri Siddhartha Institute of Technology, Vijayawada, India

Mahima K. Kishinani Department of Biotechnology, Faculty of Life of Allied Health Sciences, Ramaiah University of Applied Sciences, Bangalore, Karnataka, India

Priyan Malarvizhi Kumar Department of Information Science, College of Information Systems, University of North Texas, Denton, TX, USA

Navonita Mallick School of Law, KIIT-DU, Bhubaneswar, India

Biswadip Basu Mallik Department of Basic Science and Humanities, Institute of Engineering & Management, University of Engineering & Management, Kolkata, India

P. Mathivanan Department of Computer Science and Business Systems, Kalaignarkarunanidhi Institute of Technology, Pondicherry, India

Bhavya Mathur Centre for Computational Biology and Bioinformatics, Amity Institute of Biotechnology, Amity University, Uttar Pradesh, Noida, India

Anshit Mukerjee Department of Computer Science and Engineering, Abacus Institute of Engineering and Management, Mogra, India

Gunjan Mukherjee Department of Computational Science, Brainware University, Barasat, Kolkata, West Bengal, India

Siva Sankar Namani G Narayanamma Institute of Technology and Science, Hyderabad, India

Priyanka Narad Indian Council of Medical Research Headquarters, Ansari Nagar East, New Delhi, India

N. Nithiyanandam Department of Computing Technologies, SRM Institute of Science and Technology, Kattankulathur, India

M. Noorul Afreed Department of Information Technology, Hindustan Institute of Technology and Science, Padur, Chennai, Tamil Nadu, India

Prativa Panda School of Law, KIIT-DU, Bhubaneswar, India

Niti Patil Department of Computer Science and Engineering, Indore Institute of Science and Technology, Indore, MP, India

N. Poongavanam Department of Computer Science and Engineering, Saveetha School of Engineering, Saveetha Institute of Medical and Technical Sciences, Saveetha University, Chennai, India

G. Soniya Priyatharsini Department of Computer Science and Engineering, Dr. M.G.R Educational and Research Institute of Technology, Chennai, Tamil Nadu, India

A. Punitha Department of Information Technology, Manakula Vinayagar Institute of Technology, Pondicherry, India

M. Rajkumar Department of Smart Computing, School of Computer Science Engineering and Information Systems, Vellore Institute of Technology, Chennai, Tamil Nadu, India

R. Rajmohan Department of Computing Technologies, SRM Institute of Science and Technology, Kattankulathur, India

K. Thirupathi Rao Gandhi Institute of Technology and Management (Deemed to be University), Visakhapatnam, India

B. Tirapathi Reddy Koneru Lakshmaiah Education Foundation, Vaddeswaram, Guntur, India

V. Krishna Reddy Koneru Lakshmaiah Education Foundation, Vaddeswaram, Guntur, India

Shashwata Sahu School of Law, KIIT-DU, Bhubaneswar, India

Eman Saleh Faculty of Science, Cairo University, Giza, Egypt

G. K. Sandhia Department of Computing Technologies, SRM Institute of Science and Technology, Kattankulathur, India

Abhishek Sengupta Centre for Computational Biology and Bioinformatics, Amity Institute of Biotechnology, Amity University, Noida, India

Ramesh Chandra Sethi Department of Social Sciences and Humanities, School of Law, Kalinga Institute of Industrial Technology (KIIT), Deemed to be University, Bhubaneswar, India

Wasswa Shafik School of Digital Science, Universiti Brunei Darussalam, Gadong, Bandar Seri Begawan, Brunei Darussalam
Dig Connectivity Research Laboratory (DCRLab), Kampala, Uganda

Ankita Shome Department of Biotechnology, Vellore Institute of Technology, Vellore, Tamil Nadu, India

Arpita Shome Department of Information Technology, Haldia Institute of Technology, Haldia, West Bengal, India

N. Sivaranjani Department of Computer Science and Engineering, SCOPE, Vellore Institute of Technology, Chennai, India

R. Sujatha School of Information Technology & Engineering, Vellore Institute of Technology, Vellore, India

C. V. Suresh Babu Department of Information Technology, Hindustan Institute of Technology and Science, Padur, Chennai, Tamil Nadu, India

M. Suresh Department of Computing Technologies, SRM Institute of Science and Technology, Kattankalathur, Tamil Nadu, India

Muskan Syed Centre for Computational Biology and Bioinformatics, Amity Institute of Biotechnology, Amity University, Noida, India

Harsh Taneja School of Engineering and Technology, BML Munjal University, Kapriwas, Haryana, India

S. Uthayshangar Department of Information Technology, Manakula Vinayagar Institute of Technology, Pondicherry, India

Mobasshara Uzma Department of Biotechnology, Faculty of Life of Allied Health Sciences, Ramaiah University of Applied Sciences, Bangalore, Karnataka, India

Manas Kumar Yogi CSE Department, Pragati Engineering College (A), Surampalem, India

Part I

Fundamentals of AI and Deep Learning in Healthcare

Artificial Intelligence and Deep Learning in Healthcare 1

Harsh Taneja, Garvita Arora, Rudraksh Agarwal, and Jyotir Moy Chatterjee

Abstract

This chapter surveys how Artificial Intelligence and Deep Learning will have a significant effect on the healthcare domain in the future, particularly in the field of medical virology. As the healthcare area faces an unprecedented increase in data volume and complexity, artificial intelligence (AI) provides a track to transform patient care through effective diagnostic accuracy or systematic treatment planning and prognostic patient monitoring. By applying advanced AI and Deep Learning Models, clinical professionals can study complicated datasets and recognize and uncover nuanced illness patterns and risk factors that could otherwise remain unnoticed. This technological accuracy facilitates cautious and highly individualized care in the treatment of viral diseases, where early diagnosis and customized therapies can significantly improve the outcomes of the treatment. Moreover, this chapter deals with the advantages and difficulties of integrating AI and Deep Learning into healthcare, from making disease classification much more efficient and reducing diagnostic error rates to managing and handling ethical and resource-related issues. Healthcare becomes more accessible and scalable due to AI's ability to automate and reduce complex medical procedures by reducing time delay in diagnosis. We deal with various real-life situations, applications, and different case studies to show how AI-driven solutions are enhancing disease diagnosis, improving patient care, and enhancing the healthcare system to adjust to changing demands. This chapter highlights how

H. Taneja
School of Engineering and Technology, BML Munjal University, Kapriwas, Haryana, India

G. Arora (✉) · R. Agarwal · J. M. Chatterjee
Department of Computer Science and Engineering, Graphic Era Deemed to Be University, Dehradun, India

J. M. Chatterjee, S. K. Saxena (eds.), *Artificial Intelligence and Blockchain in Precision Medicine and Virology*, Medical Virology: From Pathogenesis to Disease Control, https://doi.org/10.1007/978-981-96-8919-4_1

AI has the ability to completely transform healthcare globally by integrating the expertise of medical professionals with AI's analytical abilities. It achieves this by providing a proactive, patient-oriented approach that increases the clinical results and sets a new point of reference for clarity in medicine. This chapter also talks about the challenges that currently exist in implementing these technologies, which include concerns with data privacy, moral dilemmas, and the processing power required to support large-scale AI applications, and how to overcome these kinds of challenges.

Keywords

Artificial Intelligence · Deep Learning Healthcare · Convolutional Neural Networks (CNNs) · Virology · Pathology · Natural Language Processing (NLP) · Graph Neural Network (GNN) · Bluedot · Healthmap

1.1 Introduction

In the field of medical healthcare, the study of disease diagnosis plays an important role. Disease is any harmful deviation from the normal structural or functional state of an organism, generally associated with certain signs and symptoms and differing in nature from physical injury. Disease may affect a person in many ways, including physical and mental affects, disrupting the normal living style of the affected person. Therefore, in order to identify the significant signs of a sickness, one must comprehend the normal state of an organism. The study of disease is called pathology. Clinical professionals interpret signs and symptoms to recognize the presence of disease. The process of determining a disease by recognizing its signs and symptoms is known as diagnosis. Another way to define diagnosis is the process of determining which disease a person has based on their symptoms and indicators. The information from a patient's medical history, physical examination, and diagnostic testing is used to make the diagnosis. There are several steps in the process, which enable doctors to collect comprehensive data. But diagnosing a disease is complex and time-consuming, and mistakes can happen that delay and prevent giving the right Medicare. The fact that not every professional or doctor has expertise in every field emphasizes the necessity of designing some automated design tools.

Managing such a huge amount of data, which is growing every day, is a major concern and makes the battle against illnesses more difficult. Artificial intelligence (AI)-driven decision support systems have the ability to integrate Deep Learning (DL) with human analysis, allowing for accurate diagnosis and better management of big datasets. Therefore, AI and DL are essential for enhancing global healthcare systems, helping to classify diseases and making the management of complex situations easier.

1.1.1 Introduction to AI and DL in Healthcare

DL and AI have become disruptive technologies in the healthcare sector, with the ability to reshape how doctors treat, diagnose, and keep records of their patients. AI refers to the ability of machines to simulate human intelligence, making decisions and solving problems based on data. Within AI, Deep Learning—powered by neural networks with numerous layers enables machines to identify complex patterns—learn from vast datasets and perform tasks that previously required human expertise (Fig. 1.1).

With their cutting-edge approaches to drug discovery, treatment planning, and disease diagnosis, AI and DL have completely transformed the healthcare industry. AI-powered solutions improve the precision and effectiveness of medical diagnostics, allowing medical practitioners to spot illnesses early and adjust treatment plans accordingly. AI-powered solutions offer a data-driven approach to improving patient outcomes in light of the exponential growth of medical data, which includes anything from patient records and epidemiological data to medical imaging and genomic sequencing.

AI has also had a significant positive impact on virology, the study of viruses and viral illnesses. Faster viral identification, epidemic prediction, and vaccine development have been made possible by the integration of AI in virology. Based on real-time data, AI models have been widely utilized to forecast disease outbreaks, examine virus–host interactions, and follow viral development. The recent COVID-19 pandemic illustrated the vital role AI plays in virology, as machine learning (ML) models aided in vaccine development, drug repurposing, and early outbreak identification.

Fig. 1.1 AI in Healthcare

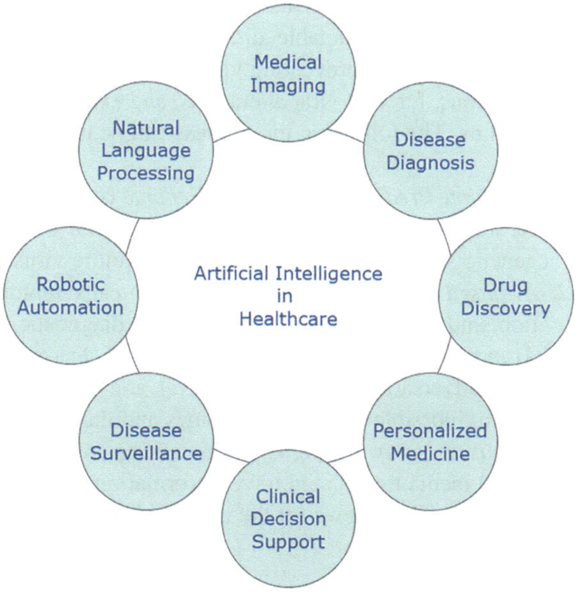

1.1.1.1 The Role of AI and DL in Healthcare

Because AI and DL may automate medical workflows, improve diagnosis accuracy, and personalize treatment, they have the potential to revolutionize conventional healthcare procedures. Conventional diagnostic techniques frequently depend on human judgment, which can be laborious, prone to mistakes, and inconsistent. AI-powered models, on the other hand, are essential in contemporary healthcare because they can analyze large volumes of data quickly, accurately, and in a scalable manner.

AI is having an impact on healthcare in a number of important areas, including:

1. *Medical Imaging and Diagnostics*: Convolutional Neural Networks (CNNs), driven by AI, analyze X-ray, MRI, CT, and ultrasound pictures to accurately identify conditions like brain tumors, pneumonia, and cancer.
2. *Pathology and Laboratory Testing*: AI helps identify anomalies in tissue biopsies and blood samples by automating the inspection of histopathological slides.
3. *Genomics and Precision Medicine*: AI systems examine genomic information to find mutations linked to inherited illnesses, supporting individualized therapeutic strategies.
4. *Clinical Decision Support Systems (CDSS)*: By evaluating patient data and suggesting evidence-based treatments, AI helps physicians with diagnosis, therapy planning, and risk assessment.
5. *AI-powered Predictive Models*: AI-powered predictive models are used in epidemiology and disease surveillance to monitor disease outbreaks, anticipate infection trends, and provide early warnings for possible pandemics.

1.1.1.2 AI in the Management of Infectious Diseases and Virology

Viruses pose serious threats to world health because of their capacity for quick mutation and unpredictable dissemination. AI has become an essential tool for comprehending and preventing viral infections. Researchers can simulate the spread of infections, forecast viral alterations, and examine genetic sequences by using DL algorithms. The following major areas can be used to classify AI's role in virology:

1. *Mutation Prediction and Viral Genome Sequencing*: AI models such as Graph Neural Networks (GNNs) and Deep Neural Networks (DNNs) examine viral genetic sequences to forecast potential future virus mutations.
2. *AI-based Disease Diagnosis*: Using clinical symptom analysis, RT-PCR automation, and medical imaging, AI-powered diagnostic tools identify viral infections.
3. *AI in Therapeutic Discovery and Vaccine Development*: AI models like Generative Adversarial Networks (GANs) and AlphaFold speed up the process of finding possible targets for vaccines and therapeutic candidates.
4. *AI-based Natural Language Processing (NLP) models* examine news articles, social media trends, and travel information to forecast and monitor epidemics in real time. This allows for the use of AI in pandemic prediction and surveillance.

During the COVID-19 pandemic, AI-driven models were utilized to monitor the virus's transmission, aid in diagnosis, and expedite the creation of vaccines, demonstrating the usefulness of AI in virology. The time needed to produce vaccines was greatly decreased by firms like DeepMind, Moderna, and Pfizer, who used AI-powered platforms to evaluate viral protein structures and refine vaccine candidates.

1.1.1.3 Obstacles and Opportunities for the Future

AI in virology and healthcare has many advantages, but it also has drawbacks. Before AI is completely incorporated into healthcare systems, there are a number of important difficulties that must be resolved, including data protection, ethical considerations, and the requirement for explainable AI (XAI). Furthermore, in order to guarantee accurate and equitable decision-making, AI models need high-quality, objective datasets.

It is anticipated that AI and DL will play an increasingly significant role in virology and healthcare as they develop. In order to provide real-time patient monitoring, early disease identification, and individualized treatments, AI in medicine will need to be integrated with wearable technologies, the Internet of Things (IoT), and cloud computing.

This chapter will examine the foundations, applications, challenges, and potential paths of AI and DL in virology and healthcare, emphasizing how AI is revolutionizing contemporary medicine.

1.2 Core Basics of AI and DL in Healthcare

By increasing diagnostic precision, improving patient care, and optimizing processes, AI and DL are essential to the transformation of healthcare. Leveraging the full potential of these technologies in the healthcare industry requires an understanding of their foundations.

1.2.1 Fundamentals of DL, ML, and AI

The simulation of human intelligence processes by machines is referred to as AI. It encompasses skills like learning, reasoning, solving problems, and comprehending natural language. AI is divided into two groups: general AI, which strives for a more comprehensive understanding similar to human cognition, and narrow AI, which concentrates on certain tasks (such as identifying diseases).

A kind of AI called machine learning allows systems to learn from data and gradually get better at what they do without explicit programming. Three primary forms of ML can be distinguished (Fig. 1.2):

1. *Supervised Learning*: Training a model with labeled data—where the intended result is known—is known as supervised learning. This method is frequently

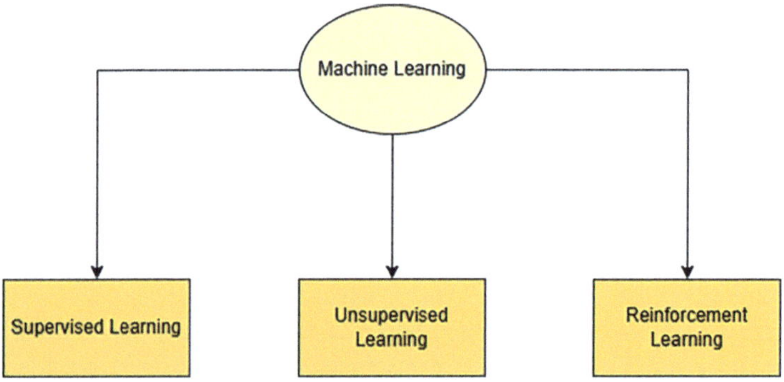

Fig. 1.2 Types of ML

applied to predictive analytics, including determining a patient's probability of contracting a disease based on their medical history.
2. *Unsupervised Learning*: Using unlabeled data, unsupervised learning seeks out underlying structures or hidden patterns. Patients might be grouped using this technique according to common medical issues.
3. *Reinforcement Learning*: By rewarding desired behavior and penalizing undesirable behavior, reinforcement learning teaches an agent to make decisions. Through gradual adaptation to patient reactions, this method can optimize treatment procedures in the healthcare industry.

A form of ML called DL uses multi-layered neural networks—thus the term "deep"—to process enormous volumes of data. DL is especially useful in healthcare applications since it performs exceptionally well at tasks involving complex datasets, such as images and natural language.

1.2.2 Important AI Methods in Healthcare

From diagnosis to treatment, AI approaches are crucial in tackling a number of healthcare issues. Healthcare practitioners can enhance patient outcomes by applying supervised, unsupervised, and reinforcement learning techniques to extract valuable insights from a variety of datasets.

1.3 Overview of DL Architectures and Neural Networks

The foundation of DL is neural networks. They are made up of layers of interconnected nodes, or neurons, that analyze input data and generate outputs. By transforming the data, each layer enables the network to recognize intricate patterns. The following DL architectures are frequently employed in the medical field:

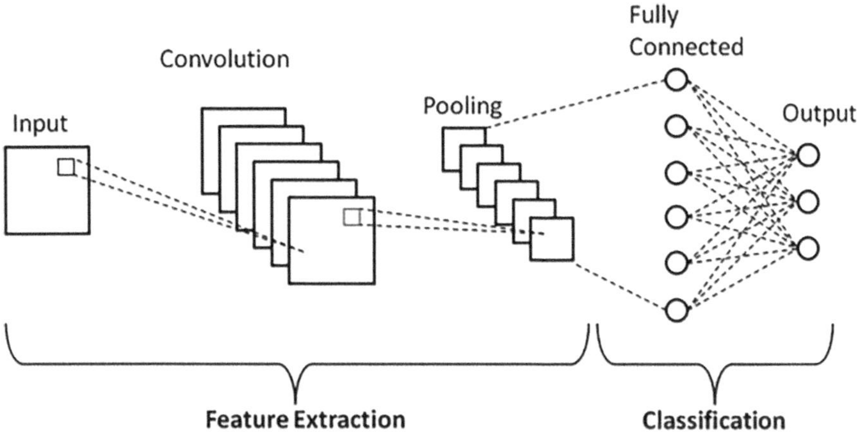

Fig. 1.3 Convolutional Neural Networks (Phung and Rhee 2019)

1. *Convolutional Neural Networks (CNNs)*: CNNs are excellent at processing visual data and are mostly utilized in medical imaging. They are able to recognize and categorize anomalies in pictures, including cancers in MRIs and X-rays. CNNs increase computational efficiency by using convolutional layers to extract features from images while lowering dimensionality (Sarvamangala and Kulkarni 2021) (Fig. 1.3).

2. *Recurrent Neural Networks (RNNs) and Long Short-Term Memory (LSTM) Networks*: RNNs (Xue and Chuah 2018) are appropriate for evaluating time-series patient data, such as tracking the evolution of an illness or monitoring vital signs, because they are built for sequential data. Long-term dependencies and the vanishing gradient issue are handled well by LSTM networks, a sophisticated variant of RNNs that enable longer-term, more accurate predictions.
3. *Transformer Models*: The use of transformer models, including BERT, GPT, and BioBERT, has transformed NLP in the medical field. Transformers are useful for therapeutic text analysis, genomics, and patient contact because of their exceptional ability to comprehend context and linkages within textual data. BioBERT, for example, is designed especially for biomedical literature, improving the extraction of pertinent data from clinical notes and research publications.
4. *Autoencoders and GANs* (Purandhar et al. 2022): Autoencoders are employed for unsupervised learning tasks, including patient record anomaly detection and data compression. When labeled data is limited, GANs can be very helpful for data augmentation and medical image synthesis, producing realistic images to improve training datasets. These methods aid in strengthening AI models' resilience, which enhances their generalizability in clinical settings.

A basis for comprehending how AI and DL in healthcare will revolutionize patient care is provided by the principles of these technologies. Healthcare practitioners can use data to enhance diagnosis, customize treatments, and streamline healthcare delivery by utilizing a variety of AI approaches, including DL architectures. The integration of AI into healthcare will be further improved by continued research and innovation as the field develops, opening the door to a more effective and efficient system.

1.4 AI in Medical Diagnostics

By improving the precision and effectiveness of disease detection and treatment, AI is transforming medical diagnostics. AI applications in CDSS, pathology, medical imaging, and NLP are revolutionizing patient diagnosis and care by utilizing cutting-edge algorithms and ML approaches (Fig. 1.4).

1.4.1 Medical Imaging Driven by AI

In the analysis and interpretation of diagnostic images, including X-rays, MRIs, CT scans, and ultrasounds, AI-powered medical imaging has advanced significantly. Since CNNs are excellent at identifying patterns and features in visual input, they are frequently utilized in these applications (IEEE Journals and Magazine | IEEE Xplore n.d.).

1. *X-rays*: AI systems are capable of analyzing chest X-rays with an accuracy level comparable to that of human radiologists in order to identify diseases including pneumonia, fractures, and malignancies. For example, AI systems have been developed to automatically determine whether pneumonia is present by spotting opacity patterns in lung images, allowing for a prompt and precise diagnosis.
2. *MRIs and CT scans*: AI technologies can help radiologists with MRI and CT imaging by automatically generating second opinions on scans, identifying problem regions, and estimating the severity of illnesses. AI models, for instance, have been used to identify brain tumors, assisting medical professionals in making better treatment choices by using quantitative imaging data.

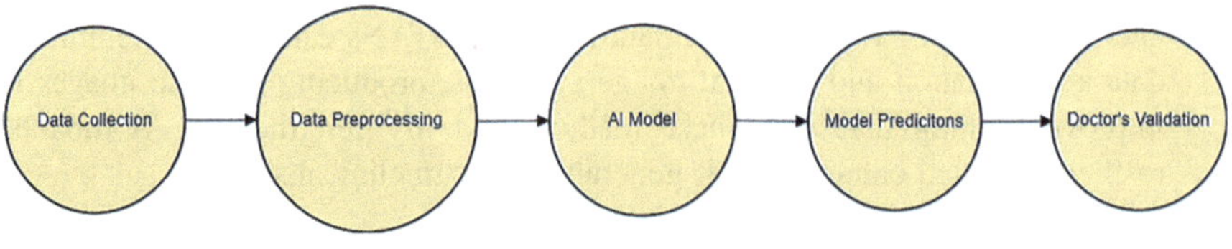

Fig. 1.4 AI-based disease diagnosis workflow

3. *Ultrasound*: AI has also enhanced ultrasound imaging, allowing for real-time image processing while treatments are being performed. By identifying characteristics in ultrasound images that indicate certain diseases, AI systems can help detect anomalies in organs such as the liver and heart.

1.4.2 Pathology and AI

AI is increasingly being used in pathology to help with automated cancer detection and histopathological sample analysis (Försch et al. 2021).

1. *Automated Cancer Detection*: Highly accurate identification of malignant cells in tissue samples is achieved by training ML algorithms. AI can differentiate between benign and malignant tissues by examining histopathology slides, giving pathologists important information. AI systems, for instance, have proven successful in detecting breast cancer from biopsy samples, resulting in more precise diagnoses and individualized treatment regimens.
2. *Histopathology*: By analyzing vast amounts of histopathological data, AI systems provide quicker and more reliable interpretations. This feature ensures that patients receive timely and appropriate care by lowering human error and diagnostic variability.

1.4.3 CDSS Powered by AI

AI-powered instruments known as clinical decision support systems are intended to help medical professionals make well-informed clinical judgments. These systems provide evidence-based recommendations for diagnosis and therapy by analyzing clinical guidelines, medical literature, and patient data.

Based on the most recent research, AI-driven CDSS can provide treatment alternatives, identify suitable diagnostic procedures, and notify practitioners of possible drug interactions. To help clinicians prioritize differential diagnoses, a CDSS, for example, can examine a patient's symptoms and medical history to indicate the possibility of specific diseases. These technologies improve patient outcomes and decision-making by integrating AI into clinical workflows.

1.4.4 NLP-Based AI Models for Analyzing Electronic Health Records

A subfield of AI called natural language processing is concerned with how computers and human language interact. Electronic health records (EHRs) (Nasarudin et al. 2024) are increasingly being analyzed using NLP-based AI models, which are able to extract useful information from patient data and unstructured clinical notes.

By analyzing massive amounts of text in EHRs, NLP algorithms are able to spot patterns, track patient outcomes, and highlight possible problems. For instance, by examining medical histories and finding pertinent risk factors, NLP can help identify patients who are at risk of unfavorable occurrences, such as readmissions to the hospital. Healthcare professionals can take proactive steps to enhance patient care and safety thanks to this capability.

1.4.5 Real-Time Usage of AI in Diagnostics

The influence of AI in identifying a range of medical disorders is demonstrated by several interesting case studies:

1. *Detecting Pneumonia*: AI systems have been trained to examine chest X-rays for indications of pneumonia in order to detect the illness. According to studies, AI systems may correctly detect cases of pneumonia, resulting in quicker diagnosis and treatment—particularly in areas with less access to radiology specialists.
2. *Tuberculosis Detection*: AI tools have been created to examine chest X-rays in order to diagnose tuberculosis (TB). These systems have proven to have excellent sensitivity and specificity, which helps fight the global tuberculosis epidemic by enabling early detection in groups that are at risk.
3. *Neurological Disorders*: By analyzing data from cognitive tests and brain scans, AI models have been used to identify neurological conditions such as Alzheimer's disease. AI can facilitate prompt intervention and improved management techniques by detecting early indicators of certain problems.

AI is revolutionizing medical diagnostics in a number of fields, including pathology, imaging, clinical decision-making, and EHR analysis. AI applications help to enhance patient outcomes and the efficiency of healthcare delivery by increasing diagnostic accuracy and efficiency. Better patient outcomes will result from the further development of AI technologies and their integration into clinical practice, which will further improve the skills of medical personnel.

1.5 AI in Virology: Understanding and Managing Viral Diseases

An emerging topic that holds great potential for improving our knowledge of and ability to treat viral infections is the nexus between AI and virology (Artificial Intelligence in Medical Virology 2023). Researchers can improve viral genome sequencing, forecast mutations, and detect and track viral outbreaks more effectively by utilizing AI technologies. By providing new tools for anticipating and reacting to viral risks, this collaboration between AI and virology is revolutionizing the field of infectious disease management.

1.5.1 AI for Predicting Mutations and Sequencing Viral Genomes

The sequencing of viral genomes has been transformed by AI technology, especially DL algorithms. Conventional sequencing techniques can be costly and time-consuming, and they frequently require a lot of laboratory effort. AI-driven methods, on the other hand, simplify the procedure and make it possible to sequence viral genomes more quickly and precisely. AI algorithms can find patterns and abnormalities in massive viral genetic material datasets, making it easier to find changes that could alter a virus's virulence or transmissibility (Fig. 1.5).

CNNs, one type of DL method, are especially useful in this field. Researchers can identify particular alterations that might appear in viral strains by using their ability to process high-dimensional genomic data. This skill is essential when dealing with viruses that are changing quickly, since prompt mutation detection can guide vaccine development and public health measures.

1.5.2 Predicting Virus–Host Interactions Using DL

Creating efficient therapies and preventative measures requires an understanding of how viruses interact with their hosts. AI-based models have demonstrated promise in forecasting these intricate relationships, especially those that make use of DL techniques. Predictive models that anticipate how viruses will interact with host cells can be created by researchers by examining a variety of biological data, such as protein structures, genetic sequences, and host immune responses.

A potent technique for simulating virus–host interactions is the GNN (Ahmedt-Aristizabal et al. 2021). The relationships between various biological entities can be represented by these networks, which makes it possible to analyze the interactions between viral proteins and host cell receptors. Researchers can anticipate possible novel interactions by training GNNs on current interaction data, opening the door to more effective vaccine formulation and tailored antiviral treatments.

1.5.3 Early Viral Outbreak Detection Assisted by AI

To prevent viral epidemics from spreading widely and lessen their impact on public health, early diagnosis is essential. AI is increasingly being used to track and examine data from a variety of sources, such as social media, travel trends, and medical reports. This data can be processed by ML algorithms to identify trends that indicate new viral threats.

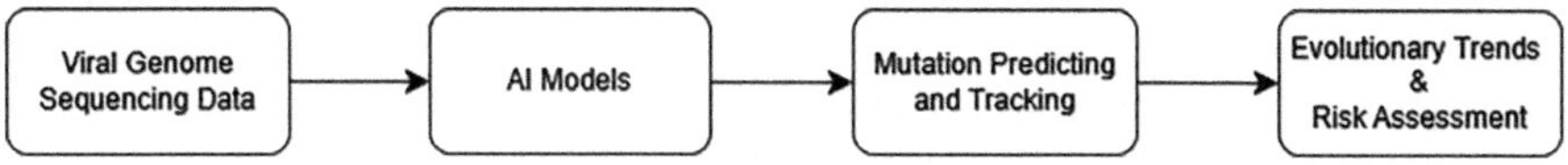

Fig. 1.5 AI in Viral Genome Sequencing and Managing Viral Diseases

For example, news articles and social media posts can be analyzed using NLP techniques to find references to viral outbreaks or symptoms. Health authorities can react quickly to possible outbreaks by receiving real-time alerts from AI systems that aggregate this data. Furthermore, by examining past outbreak data, predictive models can estimate the probability of upcoming occurrences, enabling preventative actions.

1.5.4 Models Used in Virology

1. *DNNs for Tracking Virus Mutations*: By examining genomic sequences, DNNs (LeCun et al. 2015) have demonstrated efficacy in tracking viral mutations. These models may detect possible future mutations by training on datasets of known viral mutations, which aids researchers in foreseeing shifts in viral behavior and guiding the creation of vaccines.
2. *Using GNNs to Predict Virus–Host Interaction*: GNNs are perfect for forecasting virus–host dynamics because they are excellent at simulating intricate biological interactions. These networks, which depict viruses and host cells as nodes in a graph, are able to record complex linkages and forecast potential new interactions, thereby directing treatment approaches.
3. *LSTMs for Examining Trends in Pandemic Spread*: LSTMs are extremely useful for monitoring the spread of viral pandemics because they are especially well-suited for processing time-series data. LSTMs can predict future trends by analyzing past infection rates and migration patterns, which helps public health officials with planning and resource allocation.

A new age of comprehending and treating viral illnesses is being ushered in by the integration of AI into virology. AI technologies have previously unheard-of possibilities for improving public health, from improving genome sequencing and mutation prediction to forecasting virus–host interactions and assisting in the early detection of outbreaks. The promise for better disease management and prevention becomes more apparent as researchers work to develop these models and use them in practical contexts. We can better anticipate and address the issues presented by viral infections in the future by utilizing AI.

1.6 AI-Driven Drug Discovery and Vaccine Development

The field of medication research and vaccine development is changing due to the quick growth of AI, especially when it comes to infectious diseases. Researchers can find possible antiviral drugs, speed up the development of vaccines, and improve our comprehension of drug–target interactions by utilizing advanced computational techniques. In addition to accelerating therapy development, this use of AI in pharmaceutical research is also enhancing the efficacy and safety profiles of existing therapies.

1.6.1 AI in Computational Drug Discovery: Identifying Antiviral Compounds

By allowing researchers to explore enormous chemical libraries in search of possible antiviral drugs, AI has drastically changed the drug discovery process. Long and resource-intensive laboratory tests are frequently a part of traditional drug discovery techniques. AI-driven methods, on the other hand, enable the quick screening of millions of chemicals through algorithms that can forecast their effectiveness based on known chemical and biological interactions.

Datasets with details on previously identified antiviral drugs and their biological activity can be used to train ML models, especially those that use supervised learning approaches. These models are able to spot links and trends that academics might not see right away. For instance, AI can identify new compounds with similar capabilities by examining the molecular characteristics of antiviral medications that work, making it easier to find potential candidates.

1.6.2 AI-Powered Vaccine Development: Identification of Target Problem

Finding suitable target proteins that can elicit a robust immune response is crucial to the creation of successful vaccines. By examining the viral genome and protein structures to find possible targets for vaccine development, AI technologies play a critical part in this process. This skill is especially helpful when dealing with viruses that are changing quickly, such as coronaviruses and influenza.

By precisely modeling protein folding using DL techniques, DeepMind's AlphaFold has transformed the prediction of protein structures. Researchers can find epitopes that vaccines can target by using AlphaFold to predict the three-dimensional structures of viral proteins. Since these epitopes are essential for triggering an immune response, identifying them is a crucial stage in the creation of vaccines.

1.6.3 AI Models Used in Drug Discovery and Vaccine Development

1. *AlphaFold for Protein Structure Prediction:* Our knowledge of viral proteins has been revolutionized by AlphaFold's exceptional accuracy in predicting protein structures. AlphaFold helps researchers identify possible vaccination targets by offering insights into the spatial arrangement of amino acids, which enables them to create immunogens that are more effective.
2. *GANs for the Production of Drug Molecules:* GANs (Zhang et al. 2022) are now a potent tool for creating new compounds that resemble drugs. A generator that produces new molecules and a discriminator that assesses their viability according to preset standards make up these networks. GANs can create novel

compounds that might have antiviral qualities by iteratively improving this method, which will broaden the drug discovery pipeline.

3. *Predicting Drug–Target Interactions Using Deep Reinforcement Learning (RL)*: Predicting how medications will interact with their target proteins is becoming more and more common thanks to deep reinforcement learning techniques. These algorithms are capable of identifying the most promising compounds for additional testing by simulating the intricate dynamics of drug–target interactions. This predictive ability lowers the possibility of late-stage failures and speeds up the medication development process.

1.6.4 Utility of AI in AI-Driven Vaccine Development for COVID-19 and Influenza

AI-driven methods have been essential to the quick development of influenza and COVID-19 vaccines. In order to discover important proteins that might be targets for a vaccine against COVID-19, researchers used AI to examine the genomic sequences of the SARS-CoV-2 virus. AI algorithms that evaluated the spike protein's structural characteristics and possible immunogenicity made it easier to quickly identify it as a major target.

Furthermore, data from clinical trials was analyzed using ML algorithms, which helped to improve vaccination formulations and dosage schedules. The development and approval of COVID-19 vaccinations for use in emergencies happened at a never-before-seen pace, thanks in part to this AI-driven strategy.

By examining past data on viral changes and circulating strains, AI algorithms have been used to forecast seasonal vaccination strains for influenza. Researchers can create vaccines that offer the best protection by predicting which strains will predominate in future flu seasons.

1.7 AI in Personalized and Precision Medicine

The use of AI in healthcare is transforming precision and personalized medicine, especially when it comes to treating viral infections. AI can help with patient stratification, therapy response prediction, and immune system dynamics analysis by utilizing enormous volumes of patient data. The goal of incorporating AI into clinical practice is to guarantee that patients receive individualized treatments that optimize effectiveness while reducing side effects.

1.7.1 Using AI to Identify Patients and Provide Customized Care

A key component of customized medicine is patient stratification, which enables medical professionals to pinpoint patient groups that will most benefit from particular therapies. Conventional stratification techniques frequently depend on patient demographics and clinical recommendations, which might not fully take into consideration the intricacy of each patient's reaction to treatment. A more sophisticated method is provided by AI technologies, which use ML algorithms to examine a variety of datasets, such as clinical, proteomic, and genomic data.

One of the AI models frequently used for patient classification is Support Vector Machines (SVMs) (Support Vector Machines n.d.). SVMs are capable of efficiently identifying discrete patient groupings according to their attributes by projecting high-dimensional patient data into a lower-dimensional space. By using this classification, medical professionals can customize care based on the unique characteristics of each patient, improving therapeutic results.

1.7.2 Forecasting the Reaction to Treatment: Making Use of DL Models

Predicting treatment outcomes has become more popular with DL models, especially when it comes to viral infections. By examining the intricate relationships among genetic variables, treatment plans, and patient outcomes, these models can reveal which patients are most likely to benefit from particular treatments. DL algorithms can find patterns that guide therapy choices by being trained on historical treatment data.

Neural networks, for instance, can be used to determine a patient's propensity to respond to antiviral treatment based on their genetic composition and viral traits. Clinicians might potentially avoid ineffective medications and reduce the likelihood of unpleasant responses by using this prediction power to make well-informed decisions about treatment options.

1.7.3 AI for Viral Infection Immune System Analysis

It is essential to comprehend how the immune system reacts to viral infections in order to create vaccines and therapies that work. In order to gain insight into how the immune system reacts to viral assaults, AI models are increasingly being used to examine immune system dynamics. AI can find biomarkers linked to robust immune responses by combining information from genomes, immunological testing, and patient outcomes.

Analysis of the immune system benefits greatly from the use of Bayesian networks, which describe the probabilistic correlations between variables. These networks can capture the complex interactions between various immune components, allowing researchers to predict how different factors contribute to the overall immune response. By identifying key biomarkers and understanding their relationships, healthcare providers can develop more effective therapeutic strategies and vaccine designs.

1.7.4 Case Studies: AI-Driven Approaches in HIV and Hepatitis Treatment Personalization

AI-powered methods have shown potential in treating chronic viral illnesses like hepatitis and HIV in a personalized manner. In order to find patient-specific characteristics that affect treatment response, researchers have used ML algorithms to examine genomic data and treatment history in HIV treatment. For example, using genetic profiles and past treatment outcomes, AI models have been used to forecast which antiretroviral treatment regimens will work best for certain individuals.

AI technologies have been utilized to stratify patients based on their chances of success with hepatitis therapy. AI models may categorize patients into high-risk and low-risk groups based on demographic, clinical, and viral data analysis, which helps inform treatment choices. Healthcare professionals can efficiently allocate resources and maximize treatment plans thanks to this individualized strategy.

1.8 AI in Epidemic Prediction and Disease Surveillance

One revolutionary advancement in public health is the use of AI in disease surveillance and epidemic prediction. Global health officials can now track the spread of diseases, monitor viral outbreaks, and forecast pandemic waves with previously unheard-of accuracy, thanks to AI technologies. In addition to facilitating prompt interventions, this proactive approach to disease surveillance improves the overall response to public health emergencies (Das et al. 2023).

1.8.1 AI-Powered Worldwide Viral Outbreak Monitoring

Real-time viral epidemic detection and monitoring depend on AI-powered worldwide surveillance systems. To produce a thorough picture of disease dynamics, these systems compile data from a variety of sources, such as medical facilities, labs, and public health publications. These systems can examine vast amounts of data using ML algorithms to spot new trends and possible epidemics before they get out of control.

The ability of AI to synthesize data from several platforms, such as clinical data, genomic sequencing, and environmental factors, is a significant advantage in surveillance. Health officials are able to identify outbreaks sooner and react more skillfully because of this comprehensive strategy. AI, for example, may analyze data from EHRs to spot odd sickness trends that could indicate the start of an outbreak, allowing for prompt public health measures.

1.8.2 Using AI to Monitor and Forecast Pandemic Waves

As evidenced by the COVID-19, influenza, and Ebola outbreaks, AI tools have proven crucial in monitoring and forecasting pandemic waves. ML models can estimate future disease spread by analyzing trends and patterns in historical data. Effective public health planning and resource allocation depend on this forecasting ability.

ML approaches have been used to improve the accuracy of epidemiological models, such as the Susceptible-Exposed-Infectious-Recovered (SEIR) model. Researchers can capture intricate interactions between variables that affect disease transmission by combining ML with conventional models such as Autoregressive Integrated Moving Average (ARIMA) and Long Short-Term Memory networks (LSTMs). These improved models help guide decisions on interventions such as lockdowns, vaccination campaigns, and hospital resource management by offering more precise predictions of infection rates.

1.8.3 Analyzing News Trends and Social Media with NLP-Based AI

Social media and news patterns pertaining to viral outbreaks are being analyzed using NLP tools. AI can identify public emotions and learn about new health issues by keeping an eye on news stories and internet discussions. This study helps uncover possible misconceptions that could impede public health efforts and offers insightful information about how communities are responding to epidemics.

To determine public awareness and concern regarding particular viral illnesses, for example, AI systems can examine tweets, blogs, and articles. Health professionals can improve overall public health responses by addressing community needs and misconceptions through customized public health messaging and communication methods, thanks to this real-time input.

1.8.4 Examples: AI-Powered Early Warning Systems for Pandemics

1. *BlueDot*: Known for its role in monitoring the COVID-19 pandemic, BlueDot (Wynants et al. 2021) is an AI-powered early warning system. To find possible outbreaks, BlueDot uses ML and NLP to examine a variety of data sources, such as news articles, airline ticketing information, and public health records. Days before the World Health Organization (WHO) formally issued a notice, the system was able to notify health officials about the virus's spread in Wuhan, China. This proactive strategy shows how AI may improve global health surveillance and provide timely warnings.
2. *Healthmap*: Another cutting-edge AI-powered platform that tracks and displays illness outbreaks globally is called HealthMap. HealthMap offers real-time updates on viral infections by compiling information from a variety of sources, such as news stories, social media, and official health reports. Its ML algorithms

use this data to map and identify outbreaks, keeping researchers and public health officials up to date on new hazards. The technology has been crucial in tracking the spread of illnesses like Zika and Ebola, offering insightful information for public health measures.

1.9 AI in Robotics and Automation for Healthcare

By increasing surgical accuracy, boosting diagnostic capabilities, and streamlining clinical procedures, the integration of AI into robots and automation is revolutionizing the healthcare industry. In addition to streamlining healthcare delivery, this AI and robotics synergy is essential for containing infectious diseases, which enhances patient outcomes and operational effectiveness.

1.9.1 AI-Powered Remote Diagnosis and Robotic Surgery

Robotic systems driven by AI are transforming surgery by providing increased control and precision. With the aid of AI algorithms, robotic surgeries enable surgeons to carry out intricate procedures with more precision and less invasiveness. By using real-time data processing and sophisticated imaging techniques to guide surgical tools, these devices lower the risk of problems and speed up patient recovery.

One important AI model for robotic surgery is RL (Qian and Ren 2023). Robotic systems can improve their performance by learning from previous surgical experiences through the use of RL algorithms. A robotic surgical system, for instance, can continuously improve its accuracy and efficiency by modifying its movements in response to data from prior treatments. In addition to improving surgical results, this flexibility gives surgeons access to smart instruments that can help with complex operations.

AI is enabling remote diagnostics via telemedicine systems in addition to robotic surgery. Medical imaging data, including MRIs and X-rays, can be analyzed by AI-driven diagnostic tools to find anomalies. This feature is especially helpful in underserved or distant places where it may be difficult to find professionals. Patients can obtain prompt care and referrals for additional treatment if healthcare providers are able to use AI for basic diagnoses.

1.9.2 Automation in Clinical Workflows Driven by AI

Another crucial area where AI is having a big impact is automation in clinical workflows. AI eases the workload for medical personnel by automating administrative duties, freeing them up to concentrate more on patient care. Scheduling, patient data entry, and billing procedures can be managed by AI-driven automation, which will increase operational effectiveness and lower expenses.

AI systems, for example, can evaluate patient data to optimize appointment scheduling by taking into account past attendance trends, physician availability, and patient demands. By reducing wait times and improving the overall patient experience, healthcare institutions can effectively control patient flow, thanks to this predictive capability.

1.9.3 Role of Robots in Infectious Disease Containment

The importance of robots in containing infectious diseases has become more well-known, particularly in the wake of the COVID-19 pandemic. To lessen the spread of infectious pathogens, hospitals and public areas are implementing AI-driven disinfection robots. These robots sterilize surroundings using chemical disinfectants or ultraviolet (UV) radiation, guaranteeing a secure environment for patients and medical personnel.

These disinfection robots' navigation relies heavily on computer vision models such as YOLO (Jiang et al. 2022) (You Only Look Once) and Faster R-CNN (Region-based Convolutional Neural Networks). Robots can effectively cover large areas for disinfection while detecting and avoiding impediments in real time by using computer vision algorithms. This capacity lowers the possibility of hazardous material exposure for humans while simultaneously improving the efficacy of disinfection efforts.

1.9.4 Robotics and Automation Reinforcement Using AI Models

1. *Robotic Surgery Learning (RSL)*: RSL algorithms allow robotic surgical systems to gain knowledge from past experiences and gradually enhance their performance. These technologies can improve precision, adapt to various surgical situations, and perfect their methods by analyzing data from prior surgeries.
2. *Computer Vision Models for Medical Robotics Navigation (YOLO, Faster R-CNN)*: Medical robots cannot safely explore complicated surroundings without the use of computer vision models. Real-time object identification and classification capabilities offered by YOLO and Faster R-CNN enable robots to identify and steer clear of impediments while carrying out duties like cleaning or helping with surgery.

1.10 Ethical, Regulatory, and Implementation Challenges in AI-Driven Healthcare

In order to assure its safe and equitable application, a number of ethical, legal, and implementation issues are raised by AI's growing integration into healthcare. These difficulties include issues with bias, data privacy, decision-making procedures, regulatory compliance, and the requirement for explainable AI (XAI). In order to promote trust in AI technologies and guarantee that they work in the best interests of patients and healthcare professionals, these problems must be resolved.

1.10.1 Concerns About Ethics in AI-Powered Decision-Making

There are serious ethical issues with the application of AI in healthcare, especially in relation to decision-making procedures. Diagnoses and suggested treatments are only two examples of the important medical decisions that AI algorithms might affect. But depending too much on AI systems might raise concerns about responsibility, especially if a choice has unfavorable effects. It can be difficult to determine who is at fault, whether it is the organization, the AI developers, or the healthcare practitioner.

Furthermore, the transparency of AI systems raises ethical questions. Many AI systems function as "black boxes," which means that users cannot readily see how they make decisions internally. This lack of openness may make it more difficult for medical professionals to believe in and defend AI-driven choices, which could have an effect on patient care.

1.10.2 Issues with Data Security and Privacy in AI-Powered Diagnostics

Since AI-driven healthcare systems depend on enormous volumes of sensitive patient data, data privacy and security are crucial concerns. Concerns regarding data breaches and illegal access are raised by the collection, storage, and processing of health information. Inadequate data protection procedures can have serious repercussions, such as jeopardizing patient privacy and confidence.

Regulations that set rules for data security and privacy include the Health Insurance Portability and Accountability Act (HIPAA) in the United States. But these laws frequently fail to keep up with the rapid advancement of AI technologies, leaving loopholes in protection. To reduce risks and protect patient data in AI applications, strong data governance and security procedures are crucial.

1.10.3 Fairness and Bias in AI Models

Because it can result in unjust and discriminatory outcomes when making healthcare decisions, bias in AI models is a serious concern. As AI systems are educated on past data, the models that are produced could reinforce preexisting prejudices or injustices. An AI diagnostic tool that was mostly trained on data from one demographic group, for example, would likely not function well for members of underrepresented groups, which could result in incorrect diagnoses or treatment suggestions.

It is crucial to use representative and varied training datasets to guarantee objective decision-making. Additionally, in order to detect and address biases that may develop over time, AI models must be continuously monitored and audited. Ensuring that all patients receive the right care and advancing health equity depend on the development of just and equitable AI systems.

1.10.4 The Law's Impact on AI in Healthcare

Although it is still fragmented, the regulatory environment surrounding AI in healthcare is changing. Guidelines for the authorization and supervision of AI technology in healthcare are being developed by a number of regulatory agencies, including the European Medicines Agency (EMA) in Europe and the Food and Drug Administration (FDA) in the United States. However, creating thorough legal frameworks that appropriately handle the distinctive features of AI systems is difficult due to the rapid pace of AI advancement.

Regulators have to balance encouraging innovation with protecting patient safety. This entails creating precise procedures for approving AI-powered medical equipment and algorithms while upholding strict guidelines for ethical, safety, and efficacy issues

1.10.5 XAI Is Necessary for Medical Decision-Making

In order to solve many of the moral and legal issues related to AI-driven decision-making, Explainable AI (XAI) (Giuste et al. 2022) is essential in the healthcare industry. The term "XAI" describes AI systems that are made to offer interpretable and intelligible justifications for their judgments and forecasts. XAI can improve transparency in the medical setting by enabling medical professionals to comprehend the logic behind AI-driven suggestions.

Because healthcare providers can explain the reasoning behind their decisions to patients, XAI implementation can promote confidence between patients, AI systems, and healthcare providers. Additionally (Xu et al. 2019), XAI can help healthcare workers be more accountable and see possible biases or mistakes in AI recommendations. Creating and implementing explainable AI will be essential for encouraging moral behavior and guaranteeing that patient care remains the primary focus as AI continues to play a larger role in healthcare.

1.11 Conclusion and Future Direction

Given how rapidly technology is changing the field of medical research, diagnosis, and treatment, AI in virology and healthcare has a bright future. A number of significant patterns and directions are expected to emerge as AI technologies advance, boosting virology innovation, improving pandemic preparedness, and improving healthcare delivery.

1.11.1 Future AI Trends in Virology and Healthcare

1. *Improved Predictive Analytics*: Understanding and predicting viral epidemics will be made possible by AI-driven predictive analytics. AI models can increase the precision of epidemic forecasts and enable prompt responses by utilizing large datasets, such as genetic sequencing data, epidemiological patterns, and environmental factors.
2. *Personalized Medicine*: AI integration in personalized medicine will keep growing, enabling customized treatment plans based on the unique characteristics of each patient. AI can assist in determining the best treatments for particular patient subgroups by evaluating genetic, proteomic, and clinical data, improving treatment results, and reducing adverse effects.
3. *Telemedicine and Remote Monitoring*: AI developments will help enhance telemedicine, which has grown in popularity due to the COVID-19 pandemic. Predictive models will help medical professionals remotely monitor patient conditions, while AI-powered chatbots and virtual health assistants will give patients instant access to information and advice.

1.11.2 AI's Function in Worldwide Pandemic Preparedness

The importance of AI in global pandemic preparedness is growing. By enabling real-time monitoring of viral outbreaks and trends across geographic regions, AI technology can improve surveillance systems. AI can identify potential hotspots and speed up response efforts by evaluating data from a variety of sources, such as social media, medical records, and environmental data.

Furthermore, by anticipating viral changes and assisting in the identification of target proteins, AI can help in vaccine development. In the end, this skill strengthens global health security by enabling researchers to stay ahead of new viral dangers and react quickly with efficient vaccinations and therapies.

1.11.3 New Computational Methods and AI Models

It is anticipated that new computational methods and AI models will enhance AI's potential in healthcare. Advanced NLP (Zhou et al. 2022), GANs, and deep RL are just a few of the techniques that will continue to develop and provide creative answers to challenging healthcare problems.

For example, by modeling different treatment scenarios and outcomes, deep RL might optimize therapy methods. In a similar vein, GANs can help in drug discovery by producing new drug candidates with particular characteristics. Researchers and medical professionals will be able to develop more potent interventions and enhance patient care, thanks to these new methods.

1.11.4 Combining AI with Wearable Technology, the IoT, and Cloud-Based Healthcare Solutions

Patient monitoring and care delivery will be significantly improved by the combination of AI with wearable technology, the IoT, and cloud-based healthcare solutions. AI-enabled wearables can continuously monitor health indicators and vital signs, enabling real-time analysis and warnings of possible problems.

Cloud-based healthcare solutions will make it easier for healthcare professionals to share data, allowing for informed decision-making and collaborative care (Chauhan and Kumar 2013). Healthcare systems can improve patient outcomes and operational efficiency by leveraging AI's analytical powers to extract relevant insights from the data gathered by wearables and IoT devices.

Conclusion: AI as a Facilitator, Not a Substitute for Medical Experts
It's critical to understand that as AI develops in virology and healthcare, it supports medical experts rather than replaces them. The human element is still indispensable in healthcare, even though AI technology can improve diagnostic precision, expedite processes, and aid in decision-making. AI cannot replace the empathy, ethical considerations, and sophisticated comprehension of patient requirements that healthcare practitioners possess.

Collaboration between technology and human skills will be necessary for the successful integration of AI into healthcare. AI and medical professionals can collaborate to improve patient outcomes, raise the standard of care, and handle the challenges of contemporary medicine. In the end, AI in healthcare aims to enhance human skills, empower healthcare professionals, and enhance patients' overall healthcare experiences.

1.12 Conclusion

With new trends and technologies poised to transform medical research, pandemic preparedness, and patient care, AI in virology and healthcare is expected to make major strides in the future. Stakeholders can fully utilize AI's promise to improve public health outcomes and healthcare delivery by addressing the challenges and seizing the opportunities it presents. To ensure that technology is a strong ally in the pursuit of improved health for everyone, it is critical that we prioritize ethical issues, data protection, and the cooperative connection between AI and medical professionals as we proceed.

References

Ahmedt-Aristizabal D, Armin MA, Denman S, Fookes C, Petersson L (2021) Graph-Based Deep Learning for medical diagnosis and analysis: past, present and future. Sensors 21(14):4758. https://doi.org/10.3390/s21144758

Artificial intelligence in medical virology (2023) Med Virol. https://doi.org/10.1007/978-981-99-0369-6

Chauhan R, Kumar A (2013) Cloud computing for improved healthcare: techniques, potential and challenges. E-Health Bioeng Conf (EHB) *2022*:1–4. https://doi.org/10.1109/ehb.2013.6707234

Das S, Mahanta P, Brahma A, Sharma S, Baruah R, Singh AK (eds) (2023) INDUSTRY 4.0 (First). https://www.researchgate.net/profile/Syed-Munawar-5/publication/378568212_Smart_Factories/links/65e07071e7670d36abe625c1/Smart-Factories.pdf#page=59

Försch S, Klauschen F, Hufnagl P, Roth W (2021) Artificial intelligence in pathology. Deutsches Arzteblatt Int 118(12):194–204. https://doi.org/10.3238/arztebl.m2021.0011

Giuste F, Shi W, Zhu Y, Naren T, Isgut M, Sha Y, Wang MD (2022) Explainable artificial intelligence methods in combating pandemics: a systematic review. IEEE Rev Biomed Eng 16:5–21. https://ieeexplore.ieee.org/document/9804787

Imaging Intelligence: AI is transforming medical imaging across the imaging spectrum (n.d.) IEEE Journals & Magazine | IEEE Xplore.. https://ieeexplore.ieee.org/abstract/document/8471151

Jiang P, Ergu D, Liu F, Cai Y, Ma B (2022) A review of Yolo algorithm developments. Proc Comput Sci *199*:1066–1073. https://doi.org/10.1016/j.procs.2022.01.135

LeCun Y, Bengio Y, Hinton G (2015) Deep learning. Nature 521(7553):436–444. https://doi.org/10.1038/nature14539

Nasarudin NA, Al Jasmi F, Sinnott RO et al (2024) A review of deep learning models and online healthcare databases for electronic health records and their use for health prediction. Artif Intell Rev 57:249. https://doi.org/10.1007/s10462-024-10876-2

Phung VH, Rhee EJ (2019) A high-accuracy model average ensemble of convolutional neural networks for classification of cloud image patches on small datasets. Appl Sci 9(21):4500

Purandhar N, Ayyasamy S, Kumar PS (2022) Classification of clustered health care data analysis using generative adversarial networks (GAN). Soft Comput 26(12):5511–5521. https://doi.org/10.1007/s00500-022-07026-7

Qian C, Ren H (2023) Deep reinforcement learning in surgical robotics: enhancing the automation level. arXiv (Cornell University). https://doi.org/10.48550/arxiv.2309.00773

Sarvamangala DR, Kulkarni RV (2021) Convolutional neural networks in medical image understanding: a survey. Evol Intel 15(1):1–22. https://doi.org/10.1007/s12065-020-00540-3

Support Vector Machines (n.d.) Scikit-learn. https://scikitlearn.org/stable/modules/svm.html

Wynants L, Van Calster B, Collins GS, Riley RD, Heinze G, Schuit E, Van Smeden M (2021) Leveraging artificial intelligence for pandemic preparedness and response: a scoping review to identify key use cases. Npj Digit Med 4:1–10. https://www.nature.com/articles/s41746-021-00459-8

Xu F, Uszkoreit H, Du Y, Fan W, Zhao D, Zhu J (2019) Explainable AI: A brief survey on history, research areas, approaches and challenges. Lect Notes Comput Sci:563–574. https://doi.org/10.1007/978-3-030-32236-6_51

Xue Q, Chuah MC (2018) New attacks on RNN based healthcare learning system and their detections. Smart Health 9–10:144–157. https://doi.org/10.1016/j.smhl.2018.07.015

Zhang Z, Li F, Guan J, Kong Z, Shi L, Zhou S (2022) GANS for molecule generation in drug design and discovery. Intell Syst Ref Libr:233–273. https://doi.org/10.1007/978-3-030-91390-8_11

Zhou B, Yang G, Shi Z, Ma S (2022) Natural language processing for smart healthcare. IEEE Rev Biomed Eng 17:4–18. https://doi.org/10.1109/rbme.2022.3210270

Human-Centric Artificial Intelligence (HCAI) for Precision Clinical and Medical Virology

2

Wasswa Shafik

Abstract

Artificial intelligence (AI) has become the key driver of innovations that can transform clinical practice and virology research. Although AI in virology can range from basic research to public health and medicine, human challenges and patient care put human needs and human values in the spotlight of any AI development. Herein, Human-Centric Artificial Intelligence (HCAI), or the approach of placing human values and human-centric considerations at the core and the beginning of AI and its development, is of particular relevance in clinical virology. Being guided by respectful and participative, patient-friendly AI, this kind of AI technology in virology can support the clinical virologist in pivotal decisions. To virtually cover this technology and the human-virology interface, this paper presents the backbone, the implications, and new challenges of clinical virology, including human genetics, virology, diagnosis, and treatment of viral diseases. The multidisciplinary CARER model, or comprehensive assessment and referral of environments and roles, addresses environmental and individual factors. The motivation for this chapter is driven by the ethical choices that are necessary for virology research when it comes to viral infections and opposing patient needs. Moreover, recent outbreaks of viral infections have supported the necessity for paradigm shifts and novel approaches. The Human-Centric AI for Virology and Precision Clinical Decision-Making Workshop was held to bring together AI, virology, informatics, and medicine specialists for optimal discussions of topics involving AI and clinical virology. The medical parts

W. Shafik (✉)
School of Digital Science, Universiti Brunei Darussalam, Gadong, Bandar Seri Begawan, Brunei Darussalam

Dig Connectivity Research Laboratory (DCRLab), Kampala, Uganda

J. M. Chatterjee, S. K. Saxena (eds.), *Artificial Intelligence and Blockchain in Precision Medicine and Virology*, Medical Virology: From Pathogenesis to Disease Control, https://doi.org/10.1007/978-981-96-8919-4_2

cover continuous problems around the importance of personalized medicine. In addition, precision and personalized medicine must consider human genetics and individual variabilities as well as viral mutations. Traditionally, these are addressed by various diagnostic methods on the molecular, cellular, and functional levels to provide a particular precedent in clinical decision-making. With patients, such steps are always technology-mediated, a kind of technological triangulation. Physicians focus on educating and informing each patient in the context of the available evidence-based information and sharing the decision-making process with them. More and more complex HCAI is proposed in the literature and proves to be not just technology-mediated. It encloses a judgment driven by human values and interests: strictly speaking, because HCAI uses inductive solid relationships, it needs an understanding of the backward-looking notion of causation.

Keywords

Artificial intelligence · Medical virology · Privacy and data security · Precision clinical procedures · Machine learning in healthcare · Precision clinical virology

2.1 Introduction

Artificial intelligence (AI) has become the key driver of innovations that can transform clinical practice and virology research. Although AI in virology can range from basic research to public health and medicine, human challenges and patient care put human needs and human values in the spotlight of any AI development. Herein, Human-Centric Artificial Intelligence (HCAI), or the approach of placing human values and human-centric considerations at the core and the beginning of AI and its development, is of particular relevance in clinical virology. Being guided by respectful and participative patient-friendly AI, this kind of AI technology in virology can support the clinical virologist in pivotal decisions (Zhang 2024). To virtually cover this technology and the human-virology interface, this paper presents the backbone, the implications, and new challenges of clinical virology, including human genetics, virology, diagnosis, and treatment of viral diseases. The multidisciplinary CARER model, or comprehensive assessment and referral of environments and roles, addresses environmental and individual factors (Lee and Zhang 2024).

HCAI aims to integrate AI into medical practice to enhance diagnostic and therapeutic processes. This seems reasonable since AI has grown to surpass human expert skills in some areas, yielding findings of high accuracy. A gain in efficiency is also expected through the use of AI, as it has the potential to reduce marginal human errors by 80%. This is highly significant, as the diagnosis of diseases would be conducted at an earlier, asymptomatic stage. One of the emerging trends in HCAI is the application of AI in virology. Clinical virology laboratories are often faced with increasing workloads, new technology and novel tests, and a more complex patient population, making errors or diagnostic delays more likely. Because of these constraints, HCAI can play a significant role in the successful diagnosis of a viral infection in the clinical virology laboratory (Tanos et al. 2024). The technology prevents the development of specific individual or combined abilities of human beings that are needed to enable professional activity. In the diagnostic and therapeutic process, the action of AI will tend to execute all or most of the associated capacities, reducing the exercise of such abilities. This discussion discusses the importance of human-centric implementation of AI in the virology domain, a domain that supports the meme metaphor based on competition between information strands and the capture and domestication of high-yield sources in the form of creative-constructive chains (Raisch and Fomina 2024). It is not yet clear how systems such as those discussed here are positioned in structuring advanced medical practices. Therefore, medical practice must anticipate, apply, and incorporate set mechanisms and tools into the established habitual medical practice or at least be prepared to do so.

Research into AI that focuses on usage, application, and outcomes has been making ground of late as we shift our focus to the user in the drive to build interactive systems. Special consideration is being given to adjusting systems that we develop and the methodology we apply to do so to meet the individual needs of the people we know will use them. Given the pervasiveness of complex technology, it should come as no surprise that the healthcare machine is involved in managing large quantities of data. Tapping into the capabilities of HCAI will enable health system stakeholders to detect issues that suggest damaged standard-of-care situations and weak adherence to evidence-based best practices, investigate possible obligations further in collaboration with both technologists and healthcare professionals, and find explanations that use that same electronic health record data (Leng et al. 2024). There can be only one demand to keep AI alive and widely accepted. A human-centric approach that creates an ethical, technical solution by maintaining human needs remains the primary driver of technological innovation.

The human-centered AI approach provides a human-like effect (greater acceptance by the end user of the system and additional ethical guarantees) supported by the abilities of expert systems, as shown in Fig. 2.1. Regardless of the discipline, the positive side of employing this technique is quite clear. The human-centered method has begun to gain attention, particularly in several areas where people's problems need to be solved. Another technique that employs social media data mining and topic modeling has been successfully orchestrated and has proved to be outstanding in both pharmaceutical companies and the health system. The plan will further aid both individuals and organizations in making quick, informed decisions in the healthcare sector using the future monitoring system (Partarakis and Zabulis 2024). The people-centered approach has also been identified as the most suitable paradigm for managing individual information, as it is particularly tailored to people's preferences and meets their demands.

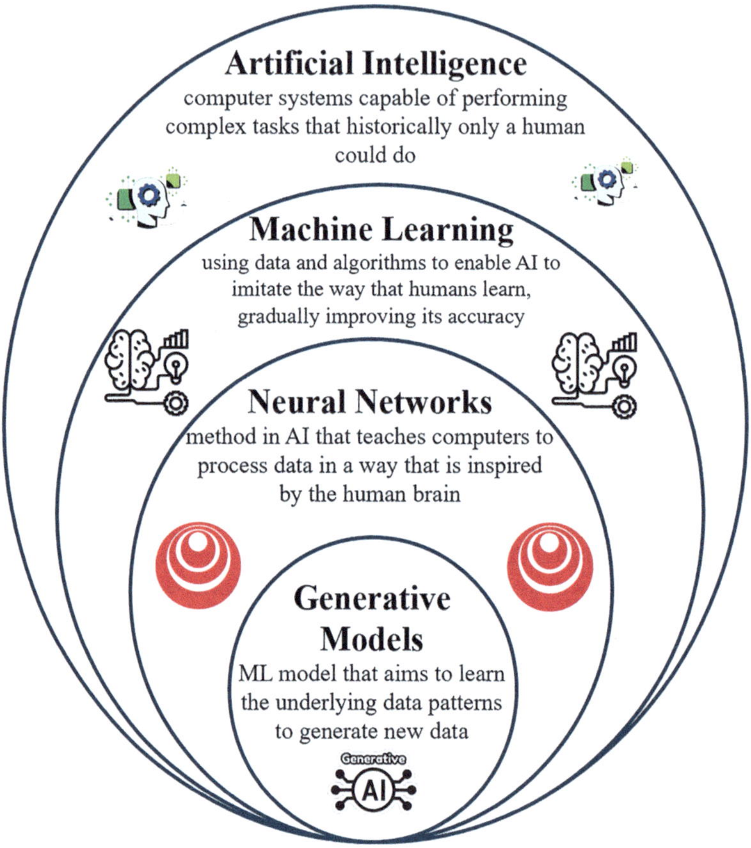

Fig. 2.1 HCAI capacities

2.1.1 Chapter Motivation

This is driven by the ethical choices that are necessary for virology research when it comes to viral infections and opposing patient needs. Moreover, recent outbreaks of viral infections have supported the necessity for paradigm shifts and novel approaches. The Human-Centric AI for Virology and Precision Clinical Decision-Making Workshop was held to bring together HCAI, virology, informatics, and medicine specialists for optimal discussions of topics involving HCAI and clinical virology. The medical parts cover continuous problems around the importance of personalized medicine. In addition, precision and personalized medicine must consider human genetics and individual variabilities as well as viral mutations. Traditionally, these are addressed by various diagnostic methods on the molecular, cellular, and functional levels to provide a specific precedent in clinical decision-making. With patients, such steps are always technology-mediated, a kind of technological triangulation. The focus of physicians is educating and informing each patient in the context of the available evidence-based information and sharing the decision-making process with them. More and more complex HCAI is proposed in the literature and proves to be not just technology-mediated. It encloses a judgment driven by human values and interests: strictly speaking, because HCAI uses inductive solid relationships, it needs an understanding of the backward-looking notion of causation.

2.2 Fundamentals of Medical Virology and Clinical Practice

Medical virology is a branch of medicine that deals with human viral pathogens, viral cytogenetics, viral genetics, virological diagnostics, molecular virology of the cell, current antiviral treatment, preparation, long-term prevention from infection, and symptoms in virus-affected people. It has a broad and fundamental theoretical, practical, and functional correlation with the clinical setting and helps in identifying the etiological diagnosis, genus, and species of the viruses revealed in the affected patient, in molecular analysis of the viral genome, in symptomatic diagnostic prediction, in prevention, and in promoting proper management (Sung et al. 2024). Virology has been integrated into the clinical setting and today is known as medical virology. Viral pathogens cause a variety of diseases, such as infectious exanthems, acute respiratory infections, hepatitis B, C, D, and E, viral gastroenteritis, genitourinary infections, ocular infections, infections of the central nervous system like meningoencephalitis, spongiform encephalopathy, myelitis, polyradiculitis, viral sexually transmitted diseases, congenital disabilities, skin and mucosal infections, acute viral hemorrhagic fever, and human immunodeficiency virus. The basics of virology and the management of a viral outbreak will be discussed here. A healthcare provider wishes to review virology and its correlation with clinical, laboratory, and therapeutic aspects of human infections. The patient's recovery clinic is based on virology (Pistolesi et al. 2024). When a physician reads an admission request list, they are offered a differential diagnosis based on patient history, geography, occupation, investigations, and clinical suspicion, and they initiate antivirals using primary care management of HCV, NDV, Ebola, etc.

Recent advancements in the fields of molecular biology, single-cell biology, and nanomedicine have laid the foundation for human-centric virology and AI. AI will play a growing role in precision virology, shaping the daily research of clinical and medical virology. Only recently significant changes have occurred in the discipline of virology, leading to shifts in clinical and medical practice. These shifts have been caused by developing medical paradigms constituting the basis of the new concept in medicine, known under different names but especially as precision medicine. Precision medicine can be defined as a treatment model considering, in the diagnosis and therapy processes, individual patient characteristics, mainly including a patient's genes, environment, and lifestyle (Sung et al. 2024). As both authors are involved in the field of virology, mainly clinical, we have tried to explore the significance of virology in this unique and promising field of precision medicine. In recent decades, applying individualized therapeutic strategies has been a very intense area of interest among physicians and scientists. This trend has also included combating pathogen infections, especially viral infections, which cannot be treated with classical chemotherapeutic agents, such as antibiotic drugs. The identification of viral and host biomarkers is critical in the characterization of viral behavior and can help to predict the outcome of infection in a patient and can contribute to their further classification. Consequently, it appears that this study fills a gap in the area of virology. This viewpoint describes the definition and importance of precision medicine and gives examples of major viral diseases in which this therapeutic strategy is currently applied (Partarakis and Zabulis 2024). Due to the importance and progress of precision medicine, virologists are morally obliged to fulfill the precision medicine requirements to be certified as "truly valuable" virologists in the precision medicine era.

When developing medical AI applications, it is paramount to adopt a user-centered or human-centric approach. While AI can interoperate with stored medical knowledge rapidly, support inferences across multiple domains, and aid inpatient care throughout the continuum, technology-mediated practice in inpatient care may foster transformative clinical–HCAI interactions. In instances where healthcare practitioners, including administrators, work alongside medical AI tools, they provide perspectives on the use-case requirements that drive AI design. The importance of human oversight is amplified in medical applications of HCAI as a safety device because software errors can cause complications ranging from suboptimal performance to life-threatening risks. Efforts to design AI tools that prioritize human experience, improve user interfaces, enhance usability, and foster acceptance are critical factors influencing the final clinical or workflow outcomes (Leng et al. 2024). In December 2020, the FDA issued an EUA for the first AI-powered readmissions predictor in an attempt to optimize patient outcomes and their transition to mainstream clinical facilities while simultaneously reducing care costs. Notwithstanding these possibilities, barriers may exist in the integration of AI into clinical workflow, which ultimately determines its adoption. In this review, we will present a cutting-edge status that showcases how HCAI and other clinical automation advance the practice of virology, first providing guidelines for harnessing virology-generated bioinformatics information and policy implications. Suppose such guidelines and policy implications are given serious consideration. In that case, it will prepare the necessary innovations or research and development pipelines required at the intersection of virology, informatics, and data science. Forming an appreciation of the human factor in AI is crucial, particularly concerning technologies that have the potential to impact clinical decision-making (Zhang 2024). Careful application of HCAI, in light of societal needs, drives adoption and the desired outcomes of enhanced patient care.

2.3 Overview of Expert Systems in Healthcare

Expert systems such as AI and machine learning have become ubiquitous tools used in the healthcare sector, driven mainly by their ability to make sense of health and medical data and predict future outcomes. The scope of AI and associated technologies is vast; they play crucial roles in precision medicine efforts, treatment development, and workflow improvements. However, the meaning and scope of "AI" encompass a diverse range of techniques and methodologies, including data mining, machine learning, and neural networks; furthermore, each of these techniques can be used for classification, clustering, prediction, outlier detection, and association discovery. It is essential to break down these categories, or genres, of AI when discussing their application to virology. It can be said, with some certainty, that practically, "real-world" AI spans two categories: natural language processing and image recognition (Raisch and Fomina 2024; Zhang 2024). The advancement of health technology positioning natural language processing in AI and AI-led decision support will support clinicians with evidence-based decisions for virology services. Moreover, AI could be utilized in cost-effectiveness assessments of virological diagnostic interventions.

In machine learning, given a dataset, algorithms can learn to automatically discover functions and find complex patterns after performing a "training" or "learning" process. It comes in unsupervised, supervised, and semi-supervised flavors. Building on machine learning, deep learning was invented. It is more akin to AI than machine learning in that "intelligence" is used to create neural networks because it is a complex and multi-layered affair. Deep learning is integral to a range of AI applications, including machine learning. The adoption of AI technologies in healthcare settings is increasing. The potential for these technologies to improve diagnostic accuracy, support the personalization of treatment options, and enhance patient outcomes has become increasingly apparent. The evolution and commercialization of HCAI technologies and their ability to facilitate clinical decision-making, as well as operational efficiencies, have been discussed in the academic literature (Lee and Zhang 2024). However, there are challenges associated with the adoption and deployment of HCAI within the healthcare ecosystem, including technical, regulatory, behavioral, and ethical issues. This underscores the need to enable interdisciplinary research.

2.4 Applications of AI in Clinical Virology and Medical Virology Research

AI is widely applied in clinical virology and medical virology research. Since their outbreak in 2020, coronavirus disease (COVID-19) and its causative pathogen, SARS-CoV-2, have threatened humanity and stalled nearly every facet of socio-economic and human activities.

2.4.1 Monitoring the Epidemiological Trends

AI tools ranging from classical machine learning models to the most recent deep learning with promising results have been applied in SARS-CoV-2 diagnosis and allied viral infections. Functional genomics and integrative AI are helpful in identifying virus-host interactions, pathogenesis of SARS-CoV-2, and host-targeted therapeutics. In drug repurposing and biomarker exploration, AI is broadly used for COVID-19 identification and prediction and medical decision support systems. AI is an accelerating partner in designing epitope-based vaccines against SARS-CoV-2 and also in monitoring the epidemiological trends for preventing new viral outbreaks. AI in clinical microbiology and virology, especially for molecular diagnosis, is tracking the rate of infection in a given period and predicting the future trajectory of the pandemic spread post-vaccination (Raisch and Fomina 2024; Tanos et al. 2024). These AI tools usually have options to track and represent the data in refined spatial and temporal aspects, including data disaggregation. The study of AI tools and the application of AI in designing therapeutics bridge precision medicine by identifying specific gene variations in a given population for carrying out treatment. Consortia of virologists and data scientists have developed AI-powered tools to manage chronic and emerging viral infections. Such an integrated approach aids in diagnosis, identifying new therapeutic compounds and designing AI therapeutics. Consortia of virologists and data scientists collaborated to develop AI tools that can be trained with the known knowledge of a virus and can predict potential drug candidates, whereas others may sequence the viral genome to study the virus's mutations and evolution (Zhang 2024). In silico AI models can help predict the outcome of vaccine design.

2.4.2 Disease Diagnosis and Treatment

AI in Disease Diagnosis and Treatment. Medical data regarding diseases, drugs, and patients have remarkably accumulated since the turn of the century. Such big data, however, often creates information overload and makes it challenging for physicians to read thousands of publications of new knowledge. AI has great potential to help in disease diagnosis and treatment, since it can analyze big medical data and identify patterns in medical images. With respect to disease diagnosis, AI applications can identify details in medical images and make them available to humans. For some tasks, AI has even been shown to outperform practicing radiologists. Furthermore, the need for new and better tools for medical diagnosis is critical, given that many deadly diseases, such as pancreatic cancer, are not currently treated on time. Unchecked, this problem has the potential to wipe out many lives due to the sheer overwhelming magnitude of medical image inflow (Tanos et al. 2024; Zhang 2024). AI can speed up the process of studying these images, and markets such as imaging are especially ripe for AI work.

2.4.3 Unpredictable Therapeutic Interventions

AI can capture a very complex, multidimensional description of a patient and create a trajectory. This is very important to predict progression or response to therapy. High dimensionality captures the complexity of the human patient in disease, including epigenetics, environmental factors, metabolites, and other layers, not only the phenotype. Capturing this complexity allows for extracting hidden disease subtypes, essential genes, and critical pathways, resulting in more accurate disease diagnosis and sub-diagnosis and, in some fields, previously unpredictable therapeutic interventions. Since humans are not accustomed to processing big data, these models tend to outperform manual curation. Hence, AI in medicine can be used to create a digitized patient and a digitized disease, and this can be used in translational medicine and drug and vaccine discovery (Leng et al. 2024; Tanos et al. 2024). AI-based models can also help in disease treatment because they demonstrate rapid hypothesis discovery or treatment pathways and provide personalized therapy. These models forecast results through simulations and show a reduced time for clinical trials and drug discovery. In addition, some models show that the onboarding of AI was highly likely to circumvent other complications, such as avoiding liver toxicity, revealing novel vaccine targets, and expediting the understanding of diseases. Hence, AI will pave the way for patients to receive the best level of therapy. HCAI is expected to have a significant effect on complex conditions such as chronic diseases, where a group of genes codes for a disease characterized by various symptoms and having a mutual etiology, often determined by exogenous and environmental factors (Sung et al. 2024).

2.4.4 Cancer Diagnosis and Precision Medicine

The success of genome-wide association studies and publicly available large-scale omics big data have become instrumental in aiding the rapid rise of precision medicine in recent years. AI, deep learning-based networks, can efficiently mine multi-omics data. The results could be utilized to typify treatment outcomes for cancer patients and discover novel biomarkers. This innovative approach considers a holistic overview of the knowledge about cancer that is focused on personalized medicine. AI could efficiently analyze patient multimodal data, such as electronic health records, medical imaging, genetics, and pathology data. Such findings could corroborate or help clinicians plan diagnosis or propose efficient risk-stratification therapy (Lee and Zhang 2024). The ultimate aim of AI is to aid the precise treatment for each patient based on their unique, personalized multimodal data. The technical advances in image capture and the consequential data tsunami produced a fertile ground for numerous deep learning AI algorithms to branch out from computer science and enter the world of applications in biomedicine. HCAI is also destined to position the field of computer-aided imaging as a powerful actor in healthcare. AI was recently used to diagnose pediatric patients with brain tumors accurately, demonstrating diagnostic accuracy, inter-annotator disagreement, and the capability of outperforming radionics. This method provides clinicians with local diagnostic information as well as a "choro-map" of the tissues. These findings show that AI interpretability is on the rise since understanding region-based brain assemblies is essential when it concerns underlying neural dysfunction (Pagani and Champion 2021; Zhang 2024). Therefore, AI has the potential to enhance diagnostic power and alleviate the risk stratification of the tumor as well as improve treatment planning options.

2.4.5 Infectious Disease Surveillance

One of the most commonly cited AI applications in healthcare involves the surveillance of infectious diseases. After an outbreak, it takes time for epidemiologists to identify and report the case counts, often leading to considerable delays. AI is well-suited to the task of handling large volumes of data in real-time from multiple sources, such as labs and healthcare providers, and can also draw on unstructured public communication data. In addition, some AI models are predictive and can be trained on historical data to learn the features of outbreaks and guide decision-making about the public health response and resource allocation. These predictive models use various data sources, including climate and satellite data, and can model animal populations as an early warning system in the case of zoonoses (Castellano 2020). There already is a growing body of evidence for the utility of AI approaches to infectious disease surveillance. Surveillance with AI brings with it several ethical, technical, and operational concerns. From a legal perspective, using people's data for surveillance and to prevent outbreaks may provide a legitimate and necessary purpose. Ethical considerations must take into account the danger of violating individuals' privacy and the potential for abuse of surveillance mechanisms. Additionally, arguments for conduct that would facilitate data sharing between nations have been made, as greater access to regional or global data is needed, particularly in the case of pandemics that are widespread or cross boundaries (Lee and Zhang 2024).

2.4.6 Artificial Intelligence for Biological Preparedness

The recent success of global collaborations must also be leveraged in order to strengthen capacities and systems in-country. The use of AI to address infectious diseases, particularly pandemics, will impact national biosafety and biosecurity measures and international information sharing and collaborative mechanisms. Meeting global health challenges such as the newly identified variant urgently demands international solidarity and cooperation, borne in mind for policymaking and to facilitate coordinated global communications. Worldwide, there are marked disparities and weaknesses in the surveillance systems of both LMICs and some HICs. The use of AI has the potential to provide a technology platform to strengthen nationwide disease event detection, event-based surveillance mechanisms, and the information and analysis required to provide a capability to assess and respond to infectious disease incidents rapidly (Partarakis and Zabulis 2024). The shift to leveraging AI for biological preparedness with new warehouses of virus-sampling robots, prototypes of automated self-cleaning buses, smart systems for contact tracing, and drones is part of a multi-country effort to place machine learning, robotics, and advanced environmental sensing at the heart of global efforts to track and counter new biological threats. Coordination and standard-setting mechanisms can be used to overcome the "data-rich and data-poor" conundrum in the sphere of surveillance. The most significant power of AI in outbreak preparedness comes during an outbreak and its tertiary uses, including the monitoring of drug resistance and vaccine coverage and use in tailored infection control and surge supply chain modeling. As the world is entering its second year of the largest pandemic in living memory, the interest in HCAI for medicine has become mainstream, but the lessons of the need for broader availability and consistent user engagement underpinned by trust and acceptance are as valid today as with the advent of the virus (Raisch and Fomina 2024).

2.4.7 Enhancing Patient Experience with AI

While still at a nascent stage, AI technologies have the potential to lift the patient experience significantly. The personalization of patient interactions is a central promise of AI tools like virtual assistants or chatbots. By leveraging both historical and real-time patient data, an organization can offer patients more targeted and convenient care plans as well as provide a high degree of service. Often, this type of service can lead to high interest and increased satisfaction from the serviced population. Frequent use cases include appointment booking, pre-visit assessment technology, educational material dissemination, and communication solutions while patients are in a hospital, clinic, or home (AI Commission 2019). The ability to use AI to streamline administrative processes, such as appointment booking, resource navigation, and real-time care coordination, can be a crucial differentiator when it comes to patient satisfaction.

Increasing ease and convenient care will lead to reduced wait times, more straightforward access to care, and a new level of knowledge and control. For example, legacy forms of communication between doctors, nurses, and patients aren't always efficient. Utilizing a virtual assistant to increase awareness and guide patients through simple tests and upcoming treatments decreases anxiety and stress for families. An interactive AI tool can assist in promoting health education and self-management. Taking mere minutes, it captures the essence of a patient's symptoms, provides a potential diagnosis, and assists in booking an appointment. Then, if the diagnosis is accurate, the chatbot endorses an over-the-counter medication and suggests home remedies. Once patients have visited their healthcare provider, the chatbot follows up to see how the patient is progressing (Fuglsang 2019). Of course, the patient also has the opportunity to have their questions answered and go over their concerns.

2.5 Challenges and Opportunities in Implementing AI in Precision Clinical Virology

The application of AI in precision clinical virology faces several challenges. Firstly, hospitals, research institutes, and the general public have privacy concerns when using these technologies. Second, model reliability and fairness concerns stem from biases in the input data used to train the algorithm. Finally, healthcare systems have procedures they follow, which AI algorithms must accommodate. Precision medicine is driving the healthcare field in giving the right treatment to the right person at the right dose and at the right time. However, the need for large, high-quality datasets is becoming a challenging hurdle. The integration of AI to cater to biological data, clinical data, point-of-care device data, and patients' behaviors are further challenges (Nishimura et al. 2021).

One of the challenges to implementing AI technologies into the clinic is making big data accessible, such as scripts via programmed programming interfaces, and transferring data. We want to access multiple institutions' and partners' data to conduct more thorough analyses. However, we recognize that data sharing is a complex issue relating to data value, data privacy, safety concerns, competitive gains, institutional interests, and academic practices. A major challenge in the use of AI models for analysis in functional genomics, including virology, is transparency, interpretability, and trust. A functional AI model must be regulated so that it will not only serve the purpose of accurate prediction but will also generate highly transparent outcomes supporting these considerations leading to the prediction (AI Commission 2019; Rathnayaka et al. 2023). These issues need to be addressed since predictive outcomes could have serious consequences for patient management; with the application of HCAI, mobile medication is now possible, as shown in Fig. 2.2.

One of the means to reach this is through open source and open data access, scientific reporting, and verification built from extensive validation using data not used to trigger model development and model assessment in the biomedical AI industry. Another challenge concerning model trustworthiness is avoiding overfitting of the training data by testing the performance of the models in target populations or in real-world situations in which the technology will work. Thus, an ethical pathway to ensure data trustworthiness and certification for predictive risks should be established to prevent the misuse of the designed AI model. When used in the real world, AI models are like new medications that have to go through a transparent development and evaluation system (AI Commission 2019; Nishimura et al. 2021). A cut-to-come in silico drug, machine, or AI robot testing might modify the standard. Decisively, if used in the analysis of viruses and AI patient management, these ethical considerations must be extended to encompass the risks of invading other individuals' privacy. Confronting these issues could lead to novel technologies that analyze deeply and accurately and allow people to provide

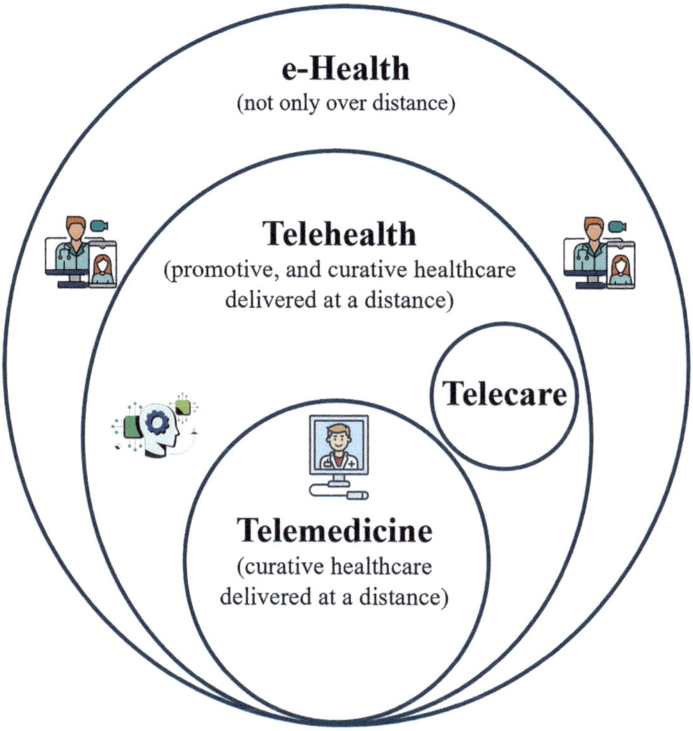

Fig. 2.2 Categories of HCAI-assisted mobile medication

significant opportunities. Providing an accessible source of information and ethical knowledge is key to this. Understanding data fabrication controls will help hinder incorrect bias or unjust assumptions, regardless of the use of artificial intelligence for deep factual insights or correct privacy (AI Commission 2019; Rathnayaka et al. 2023). Using AI in clinical virology has surfaced with several promising prospects, including the elimination of empiric anti-infective treatment and the establishment of appropriate antimicrobial management. When new theories underlying viral infectious diseases unfold, technology systems can help to comprehend the genomic information of the disease. Indeed, the predictive role of AI is changing physical research in a synergistic way and permits the modification of the treatment of diseases through precision medicine. The degree to which the model of viral behavior is limited by biological know-how could have significant future ramifications for the virtually supervised suitable growth of the treatment disease care field (Li and Liu 2023). Interdisciplinary communication is essential for the advancement of various domains and for the development of cutting-edge human data opportunities.

2.6 Artificial Intelligence Applications Considerations for Medical Virology

We are faced with many ethical and regulatory considerations as we begin to integrate HCAI into applications of medical virology. A foundational principle of "ethical AI" has been developed that is based on fairness, accountability, and transparency. Fairness calls for a critical examination of the performance of predictive models in different user groups, while accountability prioritizes the ethical implications at a decision-making level within AI algorithms. The principle of transparency concerns the importance of transparency to the user. The primary purpose of ethical guidance in AI may be understood as a focus on consequences that align broadly with the good of the individual. Developing the healthiest and best-informed decisions and practices in this area may entail revisiting our understanding (Adikari et al. 2021).

2.6.1 Regulatory Considerations

One random algorithmic decision may not appear especially meaningful when contemplated in isolation by an external observer, yet these decisions can communicate significant meanings about their relevance to the affected inhabitants. In the absence of carefully considered and purposeful application, AI procedures may unintentionally erode trust, foster fear, and compromise optimally delivered patient care. This is because patient care represents not a quantifiable notion of an outcome but an experience. Hospitable care environments invest in the soft skills suitable for engagement, emotional intelligence, and social grace demanded by the practice of rendering a service as unique and profound as the medical act of embracing the individual (Rožanec et al. 2022b). Rather than focusing backward on comforting an indeterminate population of individuals, regulatory focus within AI may become future-oriented and begin to safeguard continued investment in building an AI healthcare future that is replete with a character of trust, not fear. It must be acknowledged, rather than resented, that such expansive dialogue will require time and energy for stakeholders in the medical-ethical fields. To undeceive ourselves of the fallacious comfort that the hard questions have been answered; to re-evaluate; to reflect again is to engage in a fundamental contextualizing of guidelines (Rezaev and Tregubova 2023). It is a process that should draw us nearer to a shared philosophy. If we find ourselves momentarily marooned from an extremely rigid concept of the "ethical," we may find not faint-heartedness but the opportunity for enlightened reinvigoration. At the very least, we may take comfort in an excellent example of a return to reason in ethics-led discourse; Fig. 2.3 illustrates available sustainable devices that can be used to ensure easy medical and social assess.

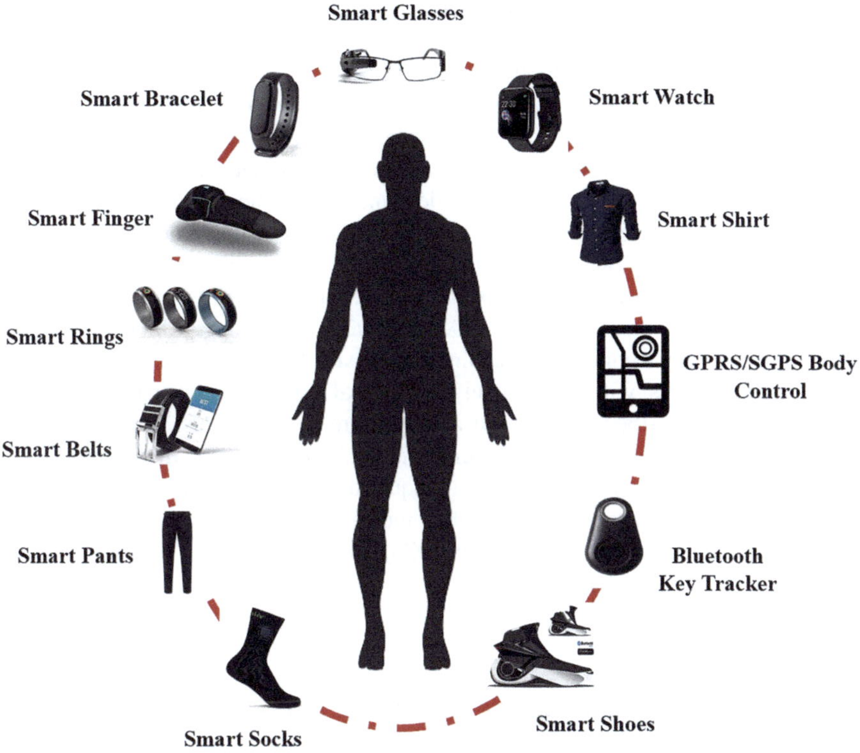

Fig. 2.3 HCAI-sustainable devices

2.6.2 Ethical Considerations

There are emerging and unresolved ethical implications of AI-based approaches to clinical and medical virology. If used by healthcare professionals, AI tools for virology may or may not successfully improve patient care but could create new ethical and regulatory challenges. Commonly named risks associated with AI include biases in the algorithms and their outputs, the extent of which in medical applications is not yet fully known, and privacy concerns associated with large-scale data. Some AI for virology risks may also be unique or specific to virology; for example, if AI algorithms to identify acute human viral infections lead to a lack of differential diagnoses, increasing the risk of misdiagnosis (Pagani and Champion 2021). In virology, responsible development and clinical implementation of AI in alignment with patient care advocates for careful ethical and professional guidance of AI developers and healthcare providers. AI technologies in healthcare are heavily regulated, with regulation balancing health and safety-related outcomes with the increasing importance of data in healthcare innovation. Ethical AI is founded on multifaceted ethical landscapes, both with guiding principles developed in-house and in the scrutiny of pre-existing laws and arguments for why they do or do not apply. Nonetheless, these ethical guidelines provide normative values rather than a legal framework (Fuglsang 2019).

In the case of healthcare, the loss of privacy and personal data in favor of public safety and healthcare outcomes argues critically in favor of a robust, albeit negotiable, legal framework around AI. AI manufacturers should comply not just with existing laws but also with public and professional expectations and recommendations, likely against a backdrop of pre-existing knowledge of how existing laws and regulations interface with existing practices and technologies. This immediately creates a problem in the case of AI: there is currently no clear consensus on new laws for patient AI nor standards against which to hold the developers of the technology accountable. In situations where there is no clear law, AI developers will have to rely on ethical guidelines, instead only acting where clearly described actions and activities are defined. This creates a chilling effect, slowing the rollout of new technologies and possibilities to ill patients. That is to say, the default, even if it is not legally required, must be to negotiate consensus on acceptable violations or face a dangerous increase in patient morbidity and immorality (Lee and Zhang 2024; Tanos et al. 2024). This highlights the need for proactive engagement with the ethics of AI in virology, even if the outcomes of such work do not become enshrined in law.

2.6.3 Data Collection and Management in Precision Virology

Precision virology often integrates an array of heterogeneous data, including both clinical data, such as transcriptomic and proteomic measurements, as well as individual host genomics. To better model the viral life cycle, viral genotypes can be obtained through high-throughput sequencing of host or non-environmental samples, as well as from historical and often more plentiful sequence repositories. While our present efforts aim to capture all relevant data to increase precision in clinical virology and medical virology, unexpected data sources, such as air quality measurements, will continue to surprise us. The development of practical, high-throughput methods to collect and store observational data from electronic health records, medical imaging, omics measurements, and various environmental data in a way that is accessible for machine learning techniques, therefore, represents an optimal setting for modeling in virology (Sung et al. 2024; Zhang 2024). The use of machine learning in virology demands high-quality data. These data are often generated within the clinic, laboratory, and environmental settings and are almost universally linked to patient IDs and patient record data.

The collection of individual genomic data and other big data linked to patients and potentially identifiable healthcare settings is subject to data governance and privacy regulations. Even under assumed informed consent, developers of systems incorporating HCAI will need robust ethical frameworks with oversight to ensure that data standardization and data storage are handled in a way that patients, payers, providers, policies, and innovators might all recognize. Now this has the ability to

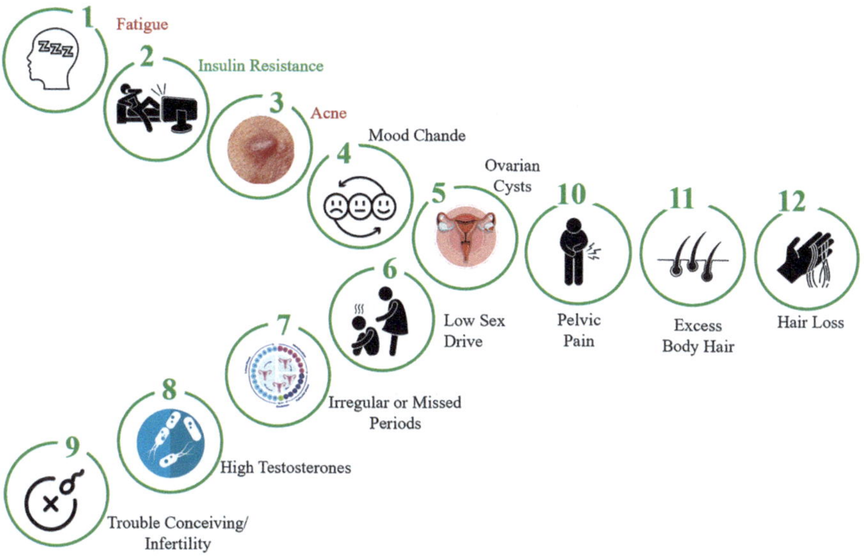

Fig. 2.4 HCAI detection and classification abilities

detect or classify aspects of humans, as shown in Fig. 2.4. This is true for hospital settings, where clinicopathological data is captured primarily in electronic health records and where imaging and biostatistics data are typically managed in picture archiving and communication systems, but it is also true for research groups attempting to share and analyze data more broadly. Data stewardship is facilitated by efficient and wide-ranging data management (Rožanec et al. 2022a). At a broad level, data management promotes efficient and reliable data integration in terms of aggregated data, augmented data, and model results. Data management, therefore, codifies the orderly development of AI-driven insights based on the foundational integration of these data across management contexts, such as formalized data and metadata models, master data management, and knowledge graph data models. Consequently, efficient data management is also required to capitalize on the insights derived from AI and efficiently execute biologically productive experiments. While defined outputs of an AI-driven health data stack may differ by stakeholder group, integrated data held in a robust data management system will be preconditions for most applications (Ball Dunlap et al. 2024).

2.6.4 Privacy and Data Security

Healthcare information privacy awareness has grown with increasing concerns about measuring and addressing sensitive issues in healthcare data security. Healthcare data contain information regarding patients' physical and mental health, family history, drug prescriptions, genetic information, and details of diseases and personal lifestyles. After social security numbers, personal health information and payments are considered valuable, which makes health and personal information sensitive and preferred targets for hacking in the healthcare insurance industry. It is easier to test the logic for safety and digital verification in different healthcare insurance personal details because names and dates might be the same, and digital surveys might correlate what the hospitals have to deliver and how many middlemen are involved (Ramírez-Gordillo et al. 2024).

In the US, the Health Insurance Portability and Accountability Act implements rules for tagging healthcare data that links personal data with health care data, including guidelines for data privacy and security. Many IT solutions protect PHI and medical information, such as firewalls, security systems, data scraping, malware detection, encryption, and anonymization. The process of anonymization seems simple, but it is difficult; re-identification during data sharing is injurious. Anonymization and encryption pose certain challenges in big data; hence, AI techniques could be implemented. The most important point is to wait to see which AI model is traceable and generates the most informative results if the model can be traced to the correct patient. Healthcare is cost-sensitive, including the accumulation of point contracts and electronic healthcare records (Ng et al. 2024). Various safeguards are taken to secure firewalls, intruder security systems, data scrubbing, malware detection, and data anonymization over IT healthcare information. Data processed and shared through IT healthcare and health insurance companies is helpful in keeping a record of personal patient data linked to their healthcare data, tracking healthcare facility usage, tracing insurance claims, and further transferring to hospital management separately. Customer data from healthcare insurance, such as names and addresses for digital confirmation, gets traced when the same person uses a different healthcare policy across different cities or states or when children come for nursery check-ups at their relatives' clinics (Alam et al. 2024). Sensitive personal and health insurance policy details can help in gaining easy access by probing into different health insurance details where the person might be investing.

2.6.5 Bias and Fairness in Artificial Intelligence Algorithms

Substantial attention has been devoted to the issues of bias and fairness in designing AI applications. This attention and discussion have been particularly intense in the field of healthcare, as the consequences of biased decisions can be particularly harmful. The influence of discrimination caused by biases in the data is one of the commonly argued critical aspects. It has been estimated that machine learning models constructed using patient data can produce risk estimations that are greater for specific individuals compared to others, which contradicts the fact that the disease incidence of these individuals is actually lower because of the age difference in the patient cohorts. When the decision support systems (DSSs) are biased, situations may change from a potentially neutral classification process to an actively unfair one (Restrepo Amariles and Baquero 2023). To achieve fairness, it has been strongly emphasized that the training data that an AI system uses should cover the diversity of society so that the public tends to believe that a truly unbiased performance can be obtained.

This conception demands diversity; namely, a model should have similar performance for different protected groups. This is not as simple as it seems, as it already includes a number of different connotations, such as variations in a sliding scale or variations under nominal conditions. The imbalance of classes, or the imbalance of outcomes that are desired to be constructed within the design process, may lead to fair or unbiased assessments of the system. There are different methodologies proposed to detect and mitigate biases in AI algorithms. The implied perspective suggests three guidelines for fair AI, which include AI safety, value-sensitive integration, and individual fair treatment. The ethics assessment of DSSs has to consider social, demographic, contextual, and historical features, including potential effects, politics, and socio-economic factors of using the DSS. If the system may contribute to health disparities, it may be considered an ethical issue (Bedenkov et al. 2021). Therefore, it is argued that only considering the fairness of a specific classification step is not enough; in principle, fairness evaluation should take into account the evaluation of causal effects of the whole system. It emphasizes that instead of fairness at a system level or fairness of models, the value of individual fair treatment is required, which is the third guideline in this argument. According to this perspective, the implications of AI beyond the measurement of the quality of the algorithms have to be taken into consideration. Moreover, it is suggested that the auditing of fairness in AI should be conducted continuously instead of using fairness indicators as an internal pre-determination tool in model training (Rolf et al. 2022). It is also suggested that the application of prediction models may standardly include a model-specific fairness report as part of the audit procedures.

2.6.6 User-Centered Design Principles for Healthcare AI

For AI in healthcare—like any technology deployed in a professional setting—to be effective and used, it is pretty clear that it needs to be designed with the end-users and other stakeholders in mind. This document describes a set of user-centered design principles, known as the User-Centered AI Design Checklist, which can be used to help build developer communities and design hubs focused on human-centric AI in healthcare. They were developed as a result of a broad consultation with researchers, clinicians, industry representatives, and other stakeholders. In particular, the checklist is centered on the idea that involving user communities (whether clinicians and allied healthcare professionals, patients, or other stakeholders) in the design process is a fundamental requirement (Trotsek and Setiyawan 2013). By doing so, it should ensure that AI tools and technologies developed in response to this process are genuinely responsive to the needs and requirements of these "real-world" communities; these AI can remind people of different behaviors of life, as shown in Fig. 2.5.

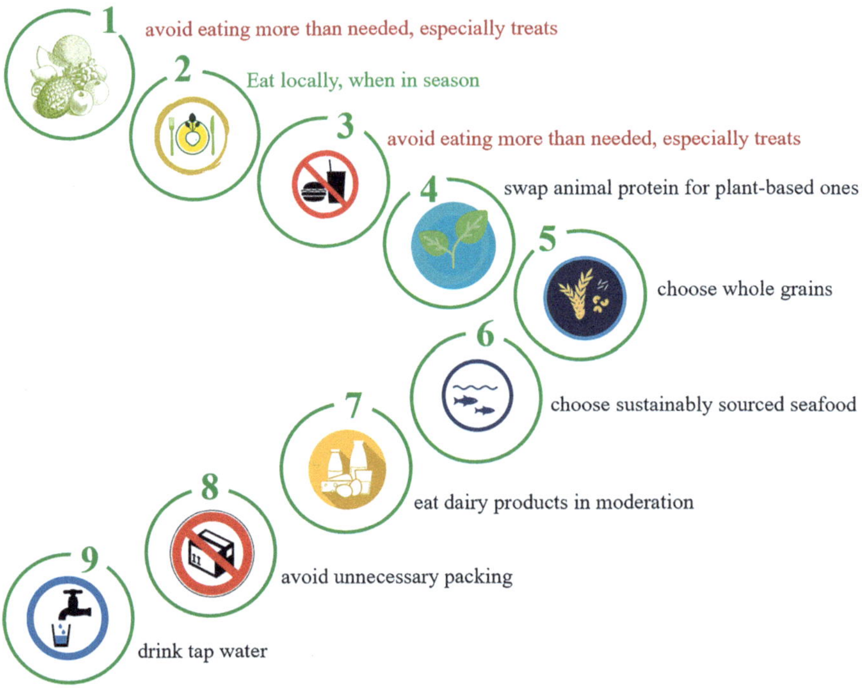

Fig. 2.5 HCAI behavioral alerts

Guidelines and literature universally describe the importance of iterative design processes and continuous feedback loops so that emerging technology tools are able to be tested against real-world professional use and to respond to feedback, updating products in response to changes in practice and real-world usability. These ensure that AI is use-informing, context-aware, and created to accommodate current capability. They were shaped by those who will use them day-to-day for their work, and they also help invest in product comprehension and acceptability. User-centered criteria, such as accessibility and inclusiveness, can also generate feedback from marginalized groups that benefit the central design process and highlight potentially discriminatory system design strategies (Adikari et al. 2021). However, in the field of medical diagnostics and the process of care and health, as of today, little regard is given to what effective technology design could be and how it would be put to support practice and end-users in their objectives. The following real-world examples show how essential it is to adopt human factors in the design field of HCAI and illustrate how their application to AI has enabled a gap between technology concepts and practice to be filled in several particular cases.

2.6.7 Frameworks and Compliance in Healthcare Artificial Intelligence

The absence of clear regulations is an obstacle when it comes to patient safety and the threat of bias and discrimination. HCAI in healthcare is at a stage of transformation from less human-supervised machine learning systems to more complex and self-learning AI systems. Examples of guidance can be found in a cooperation paper on the use of AI in healthcare systems. This paper provides recommendations for decision-makers. AI-based systems that qualify as medical devices are subject to the FDA and CE Mark pathway. The twenty-first Century Cures Act was designed to cover ongoing digital health measures in the US broadly. Low- and medium-risk categories are under discretion and have not been authorized, cleared, or recognized by the FDA (Rolf et al. 2022). Still, data-driven technologies are under continuing rapid adoption, and the government agency is attempting to catch up with policy writing and guidance.

Several stakeholders may play an essential role in shaping the policy regarding AI in healthcare, including patients, healthcare professionals, companies and professional organizations, clinical research organizations, the pharmaceutical industry, and regulation and oversight agencies. Harassment, discrimination, or unconscious bias might find their way into the algorithms learning from structured and unstructured biomedical data. Standardization of decision-making helps to explain why a decision has been taken, which becomes important in a court of law. However, although technological innovation tends to steer guidance and policymaking toward standardization, such as in quality assurance guidelines, this should be balanced. This could result in highly structured AI algorithms that prioritize mechanical reproducibility, potentially at the expense of development and creativity. Because the use of technology in healthcare decisions might be life-threatening, any new technologies, including AI applications, must provide adequate clinical evidence showing that patient safety and data integrity are assured (Bedenkov et al. 2021). In areas where there have already been established regulatory frameworks and compliance guidelines, these regulations must be followed.

2.6.8 Limitations of AI in Precision Virology

HCAI, especially machine learning, has seen explosive growth, with significant technological advancements in application across many diverse fields, including healthcare and virology. Many challenges and limitations associated with integrating AI in precision virology relate to the composition and quality of data. Perhaps the most relevant question for virology is whether our current datasets and knowledge bases are sufficiently diverse to represent the evolutionary history, host-pathogen-environment interactions, and viral diversity of SARS-CoV-2 and other human and veterinary viral pathogens. Moreover, many AI methods can be subject to bias if the data provided to train them is skewed (Ng et al. 2024). The systems under investigation in virology are biochemically, immunologically, genetically, and phylogenetically quite complex. Predicting the outcome of viral pathology throughout evolution requires a comprehensive systems approach, along with highly complex, robust, and non-specific AI predictive software and algorithms.

Of course, in some instances, even while recognizing these limitations, AI cannot be trained and validated due to the unavailability of large and diverse datasets. Based on these considerations, it is likely that AI will be most helpful in virology by providing "data-derived hypotheses." To achieve immediate and longer-term success, therefore, requires a blending of virologists, epidemiologists, other scientists, and computational mathematicians and engineers who study AI to develop robust methods for the validation of AI software outputs and performance in infectious disease and veterinary and human virology and to aid in the setup of massive data repositories to ensure that the correct questionnaires with required demographic and clinical data and imaging required to understand a new or established pathogen and its clinical impact are developed and in use. Moreover, given the ethical considerations and regulatory control of AI and how AI-derived data-driven hypotheses will be validated, these aspects of AI in virology need to be critically studied and defined (Ramírez-Gordillo et al. 2024). In this section, the key challenges and limitations on the application of AI in virology and possible future directions for research in AI in virology are discussed.

2.7 Case Studies and Success Stories in AI-Driven Precision Clinical Virology

In this section, we present a collection of case studies and success stories that illustrate the various ways AI impacts the clinical decision-making process in virology. This demonstration of AI technology serves as a showcase for the implementation of AI in the specific case study, illustrating the logical flow of the process. Each development introduces the task or problem, followed by the methods and implementation tailored to highlight a specific AI technology (Rezaev and Tregubova 2023; Rolf et al. 2022). We then present the remainder of the case study step-by-step: The results demonstrate the unique way of impacting human-driven data analytics in advanced analytics and machine learning typically developed in other research domains, except in the specific reported clinical virology case study in the form of case validation, and finally, some conclusions and implications in terms of how the methodology or AI technology learned here can guide future collection of case studies are discussed (Leng et al. 2024).

2.7.1 AI-Driven Drug Discovery and Development in Virology

Conventional drug discovery and development is known for its high costs, labor intensity, and high likelihood of successful development failure. The current era of information technology, specifically AI, enhances the opportunity to analyze valuable data that accumulates over decades. Enhanced capabilities for full-spectrum computational modeling complement this as a means of identifying potential candidates for new or repositioned drug treatment development. This can involve brute-force data mining, modeling of molecular interactions to identify potential new targets for drug development, predictions for potential lead compounds, and modeling of their ability to disrupt interactions in advanced formulations that can directly inform their uptake in potential future clinical use. When concerning antiviral research, AI-driven drug discovery on viral infections can also go through a steady pipeline (Nishimura et al. 2021). For instance, the phase of AI-driven high-throughput virtual screening can identify potential lead compounds for one or more new antiviral targets. This will then be followed by advanced drug combination formulations leading to new synergy results, thereby discovering new development entities in addition to new antiviral hits.

Various case studies (like the example in Fig. 2.6) have already shown that using AI-driven approaches results in valid desired antiviral activity, more rapid development, reduced costs, and reduced human errors since several approaches are automated. Approaches like these exemplify the potential to predict the groundbreaking field accurately. One of the AI-driven technologies highlighted in these case studies is based on molecular modeling, which supports the evolution of potentially viable drug development by reducing development time, costs, and failure rates while enhancing the synergy of drug property profiles. The primary source of drug development failures is attributed to target selection or target

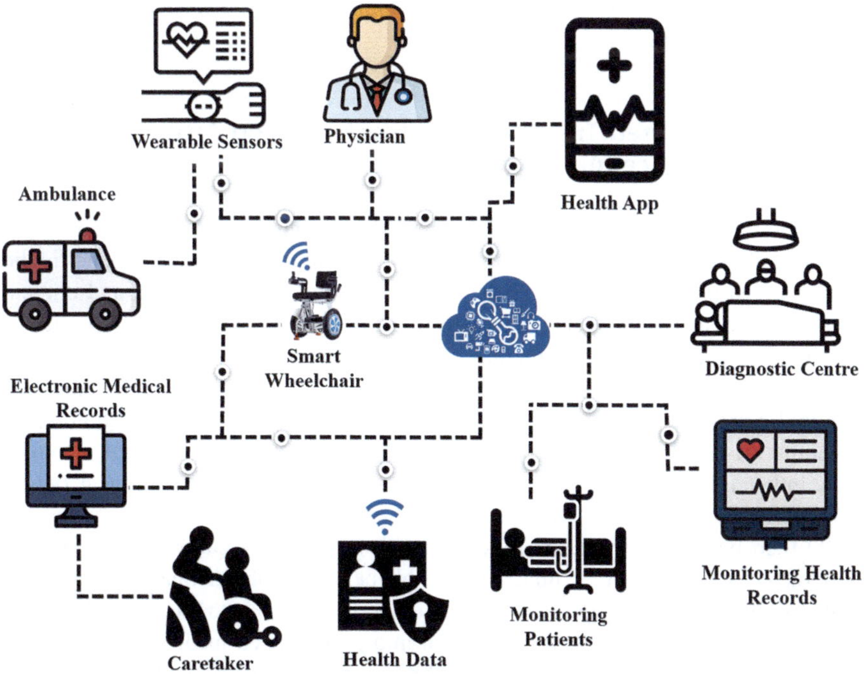

Fig. 2.6 HCAI-medical application scenario

characterization (Rezaev and Tregubova 2023). For projects to be successful, multidisciplinary teams are required with expertise in virology, epidemiology, data science, informatics, regulatory drug development, and technology.

Drug development typically requires three phases of clinical trials, which are often time-consuming and expensive with a small likelihood of success. To start human Phase I clinical studies, detailed preclinical safety and toxicology evaluations will be needed, for example, converting the computational results for human data and toxicology evaluations to preclinical evaluations, often requiring systems beyond just the virology results. Manufacture and regulation will also be important prior to entry into the clinic. Trials may need to combine single and multiple ascending doses to assess safety and tolerability, conduct pharmacokinetic studies to determine plasma half-life metrics and exalidons, evaluate pharmacodynamic effects including human PD and virological target engagement, and perform transmission studies involving human volunteers. The limitations include the need to comply with the rigors of the clinical trial process for novel molecule development, which could still be a gating factor for speeding up treatments of future pandemics out of compounds into human use. Trials used to take years, but now trials can start, and patient recruitment in the many thousands can occur in days to weeks (AI Commission 2019; Fuglsang 2019). The use of AI may also result in the emergence of new potential innovative products such as devices that plug into mobile devices to support rapid analyses of novel infection agents with a single injection in the skin. Additional opportunities predicted are AI-supported insights into current treatment to allow speedier tweaking in order to enhance recovery and reduce hospital stays.

2.7.2 Personalized Treatment Approaches Using Artificial Intelligence (AI) in Virology

Continuously evolving antivirals and a better understanding of pathogenesis and individual host responses offer the opportunity to tailor therapeutic approaches in the setting of viral infections even more. Though several host-associated and clinical factors impact treatment strategies in virology, host genetics are mainly attributed to helping individualize antiviral therapy. In the context of chronic HCV and HBV, a 48-week treatment schedule for a patient with the CC genotype of IL-28B was required compared to a 72-week schedule in patients with CT and TT genotypes. Potential AI-tailored approaches may include analyzing various biological omics data, either infectious agents or host-related, to identify the best treatment option (Castellano 2020; Rathnayaka et al. 2023). The treatment plan, based on the results, could then be guided by AI-informed input from clinical treatment guidelines. In outlook, various challenges hinder the implementation of personalized treatment with specific antiviral drugs. AI offers a number of methods that could help tackle these obstacles and facilitate the transition into personalized medicine. There is an infinite possibility that AI could change the healthcare industry, and there are real-life scenarios when AI met precision medicine. AI is used to detect false results in a commercial genetic home testing kit, which can cause misinformed decisions to be made by patients about their treatment. AI analysis of large virology data sets is needed to help doctors consider treatment options in patients who might respond to the drugs available in new ways, whereas such a level of personalization under normal clinical practice is not possible (Rezaev and Tregubova 2023). AI could predict disease progression based on patients' genomics, therefore avoiding unnecessary chemotherapy for non-advantageous patients.

2.7.3 AI-Enhanced Public Health Surveillance and Response Systems

Automated and real-time public health surveillance systems improved by AI represent powerful tools to manage viral outbreaks. Such systems continually collect and analyze information from multiple data sources worldwide, including electronic health records, demographic data, social media, laboratory results, purchase data, data from wearables, and news reports. AI-based technologies make it feasible to search, read, analyze, and prioritize requests for up-to-date epidemiological intelligence from diverse and even disparate sources. Thus, a spokes-council of the world's public health stakeholders can more quickly look through a larger part of a smaller mountain of data than is currently possible to identify epidemiological trends faster than is currently achievable. This supports real-time public health decision-making, including recommendations for the cancellation or continuation of mass gatherings and the special protective measures that might be taken in hotspots of reintroduced community disease that were incubated in another part of the world and imported in the bodies of air passengers (Bedenkov et al. 2021; Rožanec et al. 2022b).

AI, as an additional tool, can forecast if and when viruses might suddenly belong where they do not belong, at what times of year, and in which geographies. This can also contribute to richer data-led pandemic preparedness planning and a faster start to epidemic control. Several research groups have shown that AI-supported surveillance systems for pandemic and epidemic forecasting are feasible. They have even generated AI systems with sufficient predictive accuracy against a number of previously recorded viruses in diverse populations and environments. These include the global and retrospective prediction of infectious disease outbreaks of avian influenza in humans, MERS, and EBOV, as well as retroactive recognition of the progression of the H1N1 pandemic. Accurate predictive accuracy in such public surveillance is not translated into real-world practical value until even larger public surveillance systems with the same level of sophistication have also been integrated into international public health policymaking. AI holds great promise for global health and, in particular, epidemic preparedness. Public health and the global economy have much to gain by investing in their development. However, obstacles remain to the introduction of AI in global health due to concerns about data privacy, the precision of analyses, and the adequacy of protection against falsification through desensitization and centralized infection management, as is currently practiced (Partarakis and Zabulis 2024; Restrepo Amariles and Baquero 2023). All stakeholders in the global health sector are needed to help define these obstacles and assuage the concerns they raise.

2.7.4 Artificial Intelligence for Rapid Diagnostics

AI-driven label-free optofluidic screening for biochemical detection of pathogenic viruses in point-of-care settings and nanoparticles could be easily achieved at extremely rapid assay times. Opportunities for further breakthroughs in diagnostics for virology are clearly on the horizon. AI in Drug Prescription and Personalized Treatment Development: Our multiscale mathematical model and AI-informed knowledge graphs as predictive medicine can inform the appropriate personalized treatment combinations, dosages, and timing of drug initiation in COVID-19 patients with confirmed delta variant. AI in Improved Management of Pandemic Risk: Here, we describe how various new technologies developed in our COVID-19 Integrated Diagnostic Radar Program can mitigate the public health and economic havoc caused by ongoing and future pandemics (Castellano 2020). By understanding the origin, spread, and potential for both diagnosis and personalized treatment with precision diagnostics with up to 97% diagnostic sensitivity and potentially zero false positive test errors, we take critical steps forward.

The challenges posed by the COVID-19 pandemic cannot be understated; however, through collaborative public-private partnerships, great strides have been made in the international community on how to confront the challenges posed by the present effort. Outlining the AI tool development process to impact the real world, i.e., going from the AI method to the case study, is the basis of our approach in this paper. The most attractive qualities that extend each case study are, in part, its basis in pulling together collective action and coordination and cooperation among various groups. Our examples support communities coming together across research, industry, clinic, and government settings. These efforts need a shared vision and goals, and they need vigorous, transparent leadership to convene and drive momentum. In the case of this reported success, this coalition recognized and acted on these synergies, demonstrating advancement in the field of medicine (Pagani and Champion 2021). We believe that these lessons have shown value and success, and we hope to foster further dedication to these principles across fields where AI and healthcare intersect. Table 2.1 summarizes the HCAI technologies in precision clinical and medical virology, including aspects, descriptions, and their impact.

2.8 Future Directions and Emerging Trends in Human-Centric AI for Medical Virology

This chapter introduces and discusses HCAI applications for medical virology, primarily focusing on clinical virology, data interpretation, and precision virology. Despite significant developments discussed in the individual chapters, the discussion of future directions and emerging trends in human-centric AI explicitly targeted to the field of medical virology is missing.

Table 2.1 Human-centric artificial intelligence technologies in precision clinical and medical virology

Technology	Aspect	Aspect	Effect on medical virology	Effect on precision clinical virology
AI-enhanced imaging	Personalized diagnostics	AI-enhanced imaging techniques (for instance, CT scans) detect viral impacts in organs, particularly in identifying viral pneumonia or related complications.	Improves precision in diagnosing complications related to viral infections.	Allows for accurate evaluation of viral impacts on patients' organs.
Collaborative AI tools	Multidisciplinary collaboration	AI platforms facilitate collaboration between clinicians, AI experts, virologists, and ethicists to make informed decisions.	Encourages shared knowledge and research in the field of medical virology.	Enhances the coordination of precision treatments across disciplines.
Virtual reality	Patient education and support	VR technology educates patients on virology-related procedures, symptoms, and treatment in an immersive format.	Provides better patient engagement in virology-focused education.	Reduces patient anxiety and improves understanding of their condition.
Social media analytics	Early outbreak detection	AI scans social media and news for mentions of symptoms or viral activity, detecting early signs of outbreaks.	Helps in rapid identification and response to localized viral outbreaks.	Aids in tracking the spread and public perception of viruses globally.
Deep learning models	Continuous learning systems	Deep learning models adapt to new viral data, enabling continuous updates in treatment protocols.	Facilitates rapid adaptation to emerging viral threats in clinical settings.	Ensures virology models are up to date with the latest research data.
Digital twins	AI-guided treatment plans	Digital twin technology	Personalizes treatments	Enhances the testing of

(continued)

Table 2.1 (continued)

Technology	Aspect	Aspect	Effect on medical virology	Effect on precision clinical virology
		creates virtual patient models to simulate responses to different antiviral treatments.	based on simulated outcomes in virtual patients.	personalized antiviral strategies without patient risk.
Blockchain	Data privacy and security	Blockchain secures patient health records, ensuring privacy and transparency in data transactions.	Ensures secure and compliant handling of patient data in virology research.	Blockchain secures patient health records, ensuring privacy and transparency in data transactions.
Explainable artificial intelligence (EAI)		EAI provides human-interpretable decisions, ensuring clinicians and patients understand AI-recommended treatments.	Ensures AI decisions can be scrutinized and ethically validated.	Builds trust in AI-driven virology diagnostics and treatments.
Connected Internet of Things devices	Real-time monitoring	Internet of Things (IoT) devices continuously collect and transmit patient health data for analysis.	Provides clinicians with real-time data to make timely interventions.	Improves proactive management of patients with viral infections.
AI-driven drug discovery	Drug development and testing	AI models simulate drug interactions and propose potential antiviral compounds faster than traditional methods.	Enables identification of effective drug combinations for viral treatment.	Accelerates the discovery of drugs for specific viral strains.
Personalized AI chatbots	Emotional and social support	AI chatbots are used to offer emotional support, answer questions, and remind patients of	Reduces anxiety among patients by providing timely information.	Provides consistent patient engagement and emotional support during treatment.

(continued)

Table 2.1 (continued)

Technology	Aspect	Aspect	Effect on medical virology	Effect on precision clinical virology
		medication schedules.		
Natural language processing	Patient education	NLP tools provide patient-friendly information regarding viral infections, symptoms, and treatments.	Enhances communication between healthcare providers and patients.	Empowers patients with personalized information to make informed decisions.
Predictive analytics	Predictive modeling	Machine learning models predict the spread of viral outbreaks by analyzing environmental, social, and patient-specific factors.	Improves public health responses and planning for epidemic control.	Helps clinicians prepare for potential outbreaks and allocate resources efficiently.
AI-based diagnostics	Personalized diagnostics	AI algorithms analyze patient health records, genetic data, and symptoms to provide targeted diagnoses for viral infections.	Enhances accuracy and speed of diagnosing specific viral infections.	Supports precision in determining subtypes and strains of viruses.
Wearable health devices	Patient-centered data collection	Wearable devices collect patient data like vital signs, which are used to monitor health and detect early signs of viral infections.	Improves patient adherence to monitoring for chronic viral conditions.	Enables early detection and continuous monitoring of viral symptoms.

The speculation about developments and technological trends in AI and machine learning that can be used to bring precision medicine to a deeper level and hope to stimulate a constructive debate on future innovative practices for collaboration among virologists, clinicians, and AI practitioners to accelerate personalized medicine, especially in the field of viral diseases. In conclusion, we foresee a significant change in virology and healthcare achieved by a combination of AI techniques and genomic data analysis. The biggest challenge at present is to facilitate the discussion between experts in virology, computer science, medicine, pharmacology, and regulatory science on this topic. Although there are limits to practical human-centric AI applications at present, we would like to accelerate the application of AI and machine learning to research against COVID-19 by integrating viruses and healthcare to promote innovation (Rožanec et al. 2022b). Our purpose is to provide some meaningful thoughts and scientific contributions while engaging in an ethic of care in which respectful dialogue is sought in order to generate positive social impacts for a better future for humanity.

The development of AI has been ongoing for the past few decades and shows no sign of slowing down. Applications of human-centric AI in various sectors continue to evolve, and virology is no exception. Integrating AI into the study of viral genomes and catalyzing the movement toward precision medicine, telemedicine, national health, genomic medicine, and systems biology, among other things, is the focus of future research. Researchers should observe recent research trends in these fields to identify the path of development, and such literature should be equally related to virology. The next decade is expected to be the golden age of genomics, where science, large data, and omics will integrate research (Castellano 2020). In particular, customized health delivery specific to local epidemics is one of the latest focuses of viral research. Research and medical practices are moving toward full data sharing, creating a database that forges a new paradigm to meet the number of effective therapies for disease.

The study, the virosome, which, when related to other viruses with increasing frequency over time, is more nominal than each study and anything beneficial to the virologist, specializes in this field alone. Urgent needs for scientifically integrated AI to solve complex challenges include the need for public policy and practices to build public confidence in AI and address moral, ethical, and security issues. This is an ongoing process, especially in light of current examples of HCAI applications and facial recognition technologies. Close cooperation and collaboration between virological experts and HCAI researchers is a must to integrate this progress in the scientific transformation of the introductory provisions of medicine (Fuglsang 2019). In particular, to scale HCAI across different units, it is a challenging task to ensure that it can be shared and processed in multiple healthcare infrastructures, meet the standards, and provide the necessary information. We are confident that the study of cutting-edge research will guide the development of AI.

2.9 Conclusion

The tremendous potential of HCAI technologies to advance the frontiers of virology from "average" to "precision" clinical and medical virology was outlined in this chapter. These AI-based tools will eventually go on to transform clinical decision-making that is human-centric by focusing on enhanced patient care, minimizing the time to result, and ensuring better patient health outcomes. In the broader context, they will make the healthcare ecosystem efficient and secure in AI governance and, therefore, human-centric. However, for the benefits to accrue, there is a requirement for a collaborative approach among the healthcare ecosystem, the technocrats who are developing these AI algorithms and frameworks, and the policymakers and regulatory agencies who will eventually endorse these developed AI algorithms and frameworks for public use. This chapter discusses the potential and the need for human-centric AI development in medical virology in detail. We have also discussed the challenges and limitations that the community is already facing in order to develop and pass these AI-based and data-driven models, especially in the medical virology domain. The AI platforms also have a detailed discussion on the regulatory apparatus around the medical practice of AI. The chapter finally summarizes and suggests that although these challenges and limitations exist during the initial developmental phase of AI development in healthcare and diagnostics, a careful, human-centric, and sustained research innovation ecosystem leveraging AI technologies is undoubtedly needed. This will enable exploration in the area of virology and interface with pathogens, their virulence, and host responses to develop application products that will have the capability to respond to emerging viral pandemics, ensuring global societal benefits. Nevertheless, it is critical that the technology solution is provided with a "human-in-the-loop" and is used responsibly.

References

Adikari A, Nawaratne R, de Silva D, Ranasinghe S, Alahakoon O, Alahakoon D (2021) Emotions of COVID-19: content analysis of self-reported information using artificial intelligence. J Med Internet Res 23(4). https://doi.org/10.2196/27341

AI Commission (2019) European Commission – PRESS RELEASES – Press release – artificial intelligence: commission takes forward its work on ethics guidelines. European Commission

Alam MS, Rashid MM, Faizabadi AR, Zaki HFM (2024) Power of alignment: exploring the effect of face alignment on ASD diagnosis using facial images. IIUM Eng J 25(1). https://doi.org/10.31436/iiumej.v25i1.2838

Ball Dunlap PA, Nahm E-S, Umberfield EE (2024) Data-centric machine learning in nursing. CIN: Comput Inform Nurs 42(5). https://doi.org/10.1097/cin.0000000000001102

Bedenkov A, Moreno C, Agustin L, Jain N, Newman A, Feng L, Kostello G (2021) Customer centricity in medical affairs needs human-centric artificial intelligence. Pharm Med 35(1). https://doi.org/10.1007/s40290-020-00378-1

Castellano G (2020) What kind of human-centric robotics do we need? Investigations from human-robot interactions in socially assistive scenarios. In: Proceedings of the 8th international conference on human-agent interaction

Fuglsang L (2019) Human-centric service co-innovation in public services from a practice-based perspective: a case of elderly care. In: Human-centered digitalization and services. https://doi.org/10.1007/978-981-13-7725-9_2

Lee MJ, Zhang R (2024) Human-centric artificial intelligence of things–based indoor environment quality modeling framework for supporting student well-being in educational facilities. J Comput Civ Eng 38(2). https://doi.org/10.1061/jccee5.cpeng-5632

Leng J, Zhu X, Huang Z, Li X, Zheng P, Zhou X, Mourtzis D, Wang B, Qi Q, Shao H, Wan J, Chen X, Wang L, Liu Q (2024) Unlocking the power of industrial artificial intelligence towards Industry 5.0: insights, pathways, and challenges. J Manuf Syst 73. https://doi.org/10.1016/j.jmsy.2024.02.010

Li S, Liu Y (2023) Human-centric artificial intelligence enabled digital images and videos forensic triage. In: Proceedings – 2023 human-centered cognitive systems, HCCS 2023. https://doi.org/10.1109/HCCS59561.2023.10452651

Ng JY, Cramer H, Lee MS, Moher D (2024) Traditional, complementary, and integrative medicine and artificial intelligence: novel opportunities in healthcare. Integr Med Res 13(1). https://doi.org/10.1016/j.imr.2024.101024
Nishimura S, Egami S, Ugai T, Oono M, Kitamura K, Fukuda K (2021) Ontologies of action and object in home environment towards injury prevention. In: ACM international conference proceeding series. https://doi.org/10.1145/3502223.3502239
Pagani M, Champion R (2021) Creating business value through human-centric AI. In: Artificial intelligence for sustainable value creation. https://doi.org/10.4337/9781839104398.00010
Partarakis N, Zabulis X (2024) A review of immersive technologies, knowledge representation, and AI for human-centered digital experiences. Electronics (Switzerland) 13(2). https://doi.org/10.3390/electronics13020269
Pistolesi F, Baldassini M, Lazzerini B (2024) A human-centric system combining smartwatch and LiDAR data to assess the risk of musculoskeletal disorders and improve ergonomics of Industry 5.0 manufacturing workers. Comput Ind 155. https://doi.org/10.1016/j.compind.2023.104042
Raisch S, Fomina K (2024) Combining human and artificial intelligence: hybrid problem-solving in organizations. Acad Manag Rev. https://doi.org/10.5465/amr.2021.0421
Ramírez-Gordillo T, Mora H, Pujol-Lopez FA, Jimeno-Morenilla A, Maciá-Lillo A (2024) Industry 5.0: towards human centered design in human machine interaction. In: Springer proceedings in complexity. https://doi.org/10.1007/978-3-031-44721-1_50
Rathnayaka P, Gamage G, De Silva D, Alahakoon D, Manic M (2023) EmoZen: a robust word embedding for implicit and explicit expressions of emotion. In: IECON proceedings (industrial electronics conference). https://doi.org/10.1109/IECON51785.2023.10312470
Restrepo Amariles D, Baquero PM (2023) Promises and limits of law for a human-centric artificial intelligence. Comput Law Secur Rev 48. https://doi.org/10.1016/j.clsr.2023.105795
Rezaev AV, Tregubova ND (2023) Looking at human-centered artificial intelligence as a problem and prospect for sociology: an analytic review. Curr Sociol. https://doi.org/10.1177/00113921231211580
Rolf B, Mebarki N, Lang S, Reggelin T, Cardin O, Mouchère H, Dolgui A (2022) Using knowledge graphs and human-centric artificial intelligence for reconfigurable supply chains: a research framework. IFAC-PapersOnLine 55(10). https://doi.org/10.1016/j.ifacol.2022.09.641
Rožanec JM, Novalija I, Zajec P, Kenda K, Tavakoli Ghinani H, Suh S, Veliou E, Papamartzivanos D, Giannetsos T, Menesidou SA, Alonso R, Cauli N, Meloni A, Recupero DR, Kyriazis D, Sofianidis G, Theodoropoulos S, Fortuna B, Mladenić D, Soldatos J (2022a) Human-centric artificial intelligence architecture for Industry 5.0 applications. Int J Prod Res. https://doi.org/10.1080/00207543.2022.2138611
Rožanec J, Trajkova E, Novalija I, Zajec P, Kenda K, Fortuna B, Mladenić D (2022b) Enriching artificial intelligence explanations with knowledge fragments. Future Internet 14(5). https://doi.org/10.3390/fi14050134
Sung SH, Suh JM, Hwang YJ, Jang HW, Park JG, Jun SC (2024) Data-centric artificial olfactory system based on the eigengraph. Nat Commun 15(1). https://doi.org/10.1038/s41467-024-45430-9
Tanos P, Yiangou I, Prokopiou G, Kakas A, Tanos V (2024) Gynaecological Artificial Intelligence Diagnostics (GAID) GAID and its performance as a tool for the specialist doctor. Healthcare (Switzerland) 12(2). https://doi.org/10.3390/healthcare12020223
Trotsek D, Setiyawan (2013) Building trust in human-centric artificial intelligence. J Chem Inf Model 53(9). https://ec.europa.eu/futurium/en/ai-alliance-consultation/guidelines.1.html
Zhang C (2024) Human-centric artificial intelligence. Appl Comput Eng 33(1). https://doi.org/10.54254/2755-2721/33/20230273

Deep Learning Algorithms and Techniques 3

Muskan Syed, Vanshika Chauhan, Bhavya Mathur, Sanskriti Goyal, Priyanka Narad, and Abhishek Sengupta

Abstract

Deep learning is a subdiscipline of machine learning that has gained a lot of attention in the recent past due to its effectiveness in applying artificial neural networks for pattern recognition and information extraction from large datasets, thus enabling automation of decision-making processes. Backpropagation, autoencoders, restricted Boltzmann machines, convolutional neural networks, recurrent neural networks, generative adversarial networks, capsule networks, Transformers, and other significant deep learning algorithms and their architectures will all be covered in this chapter. The effectiveness of these algorithms has been shown in many research domains, but the healthcare sector has experienced some of the most success. This chapter examines the relationship between deep learning and medical virology and applies deep learning to precision clinical decision-making for viral diseases like human immunodeficiency virus (HIV) and HPV. The development of targeted therapies, the analysis of viral genomic data to look for strain differences, and diagnostic imaging of viral-induced abnormalities are some of the main applications discussed. The role of deep learning techniques in enhancing the detection and management of cervical abnormalities associated with HPV and HIV-related lung diseases is also explored. These technologies also raise significant practical challenges, e.g., in terms of data confidentiality, model interpretability, and neural architecture search. Possible outcomes of these issues are presented in the recent works on

M. Syed · V. Chauhan · B. Mathur · S. Goyal · A. Sengupta (✉)
Centre for Computational Biology and Bioinformatics, Amity Institute of Biotechnology, Amity University, Uttar Pradesh, Noida, India
e-mail: asengupta@amity.edu

P. Narad
Indian Council of Medical Research Headquarters, Ansari Nagar East, New Delhi, India

J. M. Chatterjee, S. K. Saxena (eds.), *Artificial Intelligence and Blockchain in Precision Medicine and Virology*, Medical Virology: From Pathogenesis to Disease Control, https://doi.org/10.1007/978-981-96-8919-4_3

AutoML-Zero, federated learning, and evolutionary deep learning, as well as their solutions. The contribution of this chapter is to provide a systematic review of the current state of play in applying deep learning to precision medicine for viral conditions, with a particular focus on its relationship with medical virology. Researchers and practitioners can benefit from this chapter, which offers a systematic account of deep learning algorithms in healthcare contexts, including their pros and cons. By providing crucial insights into the future trajectory of deep learning in medical virology, it fosters continuous innovation in this transformative field. Ethical content on the use of AI in healthcare implementation is also addressed to ensure equitable access to these developments.

Keywords

Artificial intelligence · HPV · HIV · Deep learning · Blockchain · Medical virology

3.1 Introduction

Deep learning has transformed the possibilities of technology across various domains, particularly in medicine, by enabling machines to examine extensive datasets, identify patterns, and produce insights that were once unattainable (Li et al. 2023). In contrast to traditional algorithms, deep learning models autonomously acquire the ability to recognize and comprehend patterns directly from unprocessed data, reshaping the frameworks of operations, image recognition, natural language processing (NLP), and predictive analytics. Furthermore, in the field of medicine, this technology addresses the complexities and volume of medical data, which includes genomic sequences and medical imaging, by providing tools for precision medicine, early diagnosis, and error reduction (Quazi 2022). Deep learning's neural network framework is intended to replicate brain functions, with methods such as backpropagation, activation functions, and optimization algorithms used to improve efficiency. Its role in medical virology (e.g., HIV and HPV detection, genomic analysis, and diagnostic imaging) highlights its transformative potential (Kufel et al. 2023). Notwithstanding concerns regarding data privacy, model interpretability, and ethical considerations, novel solutions are being developed through advancements in machine learning, such as AutoML-Zero, federated learning, and evolutionary algorithms (Nguyen et al. 2025).

The key feature of DL is its capability for analyzing heterogeneous information, integrating the structured and unstructured information resources into a single framework for the extraction of deep insights (Yang et al. 2022). As an example, convolutional neural networks (CNNs) are essential in medical picture analysis, detecting anomalies that may be overlooked by human observation, whereas RNNs have a significant strength in sequential data handling like patient histories (Muñoz-Aseguinolaza et al. 2023). These models have been modified to take on complex medical problems such as disease progression modeling to inform treatment strategies and personalize plans to the individual profile of a patient (Brice et al. 2023). Deep learning is very useful in medical virology for the rapid identification and reaction to outbreak diseases by viral mutation and variation analysis at a genomic level, which contributes to the identification of mutations and variations that affect the efficiency of drugs/vaccines (Lebatteux et al. 2024).

Beyond its technical applications, deep learning also addresses the global need for scalable healthcare solutions (Tobón et al. 2022). The COVID-19 pandemic highlighted the critical need for rapid diagnostic devices, and deep learning models emerged to meet this need and paved the way for the rapid identification of viral infection in radioscopic images as well as other biomarkers (Rafique et al. 2023). These developments not only increased response times but also reduced the strain on the health system by automating tedious chores (Alotaibi and Federico 2017). Likewise, epidemiological patterns are being tracked and predicted using deep learning models, which offer vital information for public health planning and resource distribution (Tariq and Ismail 2024).

3.2 Key Deep Learning Algorithms and Architectures

The merging of AI, BT, and virology speaks to a noteworthy headway in present-day healthcare (Tagde et al. 2021). As viral issues ended up more complex, such as quickly advancing HIV strains and constant HPV contaminations, advanced expository instruments are fundamental (Hewavisenti et al. 2023). The aim of this chapter is to analyze the considerable effect of deep learning techniques, particularly backpropagation ones, in our ability to understand, forecast, and treat all kinds of viral diseases while maintaining data safety through blockchain technology (Kim and Huh 2020).

3.2.1 Backpropagation

Backpropagation is a crucial technique in artificial intelligence, particularly for analyzing and evaluating viruses (Choubey et al. 2022). The approach that virologists have traditionally used to deepen their insights into viral behavior—careful observation and learning from mistakes—closely resembles our mathematical model (Schiffer et al. 2018). Like how a scientist hones their hypothesis through experiments, backpropagation allows artificial neural networks to improve from mistakes by methodically modifying their internal parameters (Taye 2023). Blockchain technology ensures the privacy and accuracy of these assessments, and this educational approach has shown significant worth in viral studies, such as identifying high-risk HPV strains and forecasting HIV mutations (Bennis et al. 2024). The algorithm serves as an essential resource in contemporary virology and medicine due to its capacity to manage complex, nonlinear data connections, which enables it to adeptly clarify the subtle patterns of viral behavior (Padhi et al. 2023).

3.2.1.1 Concept and Functioning

The key method that transformed deep learning was backpropagation, especially when it came to pathogen analysis and prediction (Taherdoost 2023a). Essentially, backpropagation mirrors the iterative method by which virologists refine their understanding of viral activity via continuous observation and hypothesis adjustment (Prussin et al. 2020). The approach operates by retroactively propagating error signals across a neural network, like tracking the origin of a virus transmission (Shi et al. 2022). A neural network assessing viral protein sequences modifies its internal settings based on prediction mistakes about viral mutations, hence improving future predictive accuracy (Sokhansanj et al. 2022).

Backpropagation in neural networks resembles the traditional approach used by virologists to study viral behavior—characterized by careful observation and iterative improvement of understanding (Linardatos et al. 2021). Analogous to the methods used by HIV researchers to track the virus's development to evade the immune system, backpropagation helps neural networks in learning from mistakes and improving their predictive accuracy over time (Li and Li 2020).

The procedure begins when a neural network analyzes viral data, including HIV genetic sequences (Ivanov et al. 2020). The network assesses its predictions on viral activity against verified outcomes. Discrepancies between expected and actual results serve as an error indication (Goergen et al. 2022). This error then disseminates retrogressively across the network, altering the strength of connections among artificial neurons, much to the progression of our understanding of HIV's resistance mechanisms over decades of research (Steiner et al. 2020).

Analyze the case of HIV mutation assessment: A neural network may initially achieve only 60% accuracy in forecasting the evolution of HIV in response to a specific antiviral drug (Gandhi et al. 2023). The network adjusts its internal settings via backpropagation in reaction to its faults (Sarker 2022). After failing to forecast a specific mutation in the virus's envelope protein, the algorithm strengthens the connections that may have led to the correct prediction while reducing those that caused the error (Alizadehmohajer et al. 2023). The backpropagation technique in viral analysis adheres to a methodical methodology akin to a virologist's refinement of knowledge via laboratory experimentation (Chen et al. 2020a). The procedure adheres to these essential stages:

(a) *Forward Pass*

The network analyzes viral data (including protein sequences or structural information) across its layers, generating preliminary predictions on viral activity, mutation patterns, or protein folding (Mall et al. 2021). During the forward pass, viral data traverses consecutive layers of the neural network, analogous to a sample progressing through various phases of laboratory investigation (Kumar and Srivastava 2024). Every layer of neurons analyzes the information it receives and transmits its outcomes to the subsequent layer (Montesinos López et al. 2022).

For example, while examining an HIV protease sequence, the network evaluates the amino acid sequence across many layers:

- The first layer recognizes fundamental amino acid sequences.
- The intermediate layers identify possible cleavage sites.
- Deeper layers identify structural features that may signify medication resistance.
- The output layer ultimately predicts the virus's probable resistance to certain protease inhibitors.

(b) *Error Assessment: Evaluating Predictive Accuracy*

The error calculating process is comparable to juxtaposing laboratory data with established outcomes and its predictions with established viral results, quantifying the discrepancy or "loss." The network evaluates its predictions against confirmed viral behavior data, quantifying the extent of deviation from actual outcomes (Elmes et al. 2020).

In the context of HIV, if the network forecasts an 85% likelihood of resistance to a certain treatment, however, laboratory analyses reveal that the virus is really totally resistant, the system quantifies this inconsistency (Deeks et al. 2021). The error measurement evaluates not only the accuracy of the forecast but also the magnitude of the deviation (Jierula et al. 2021). The quantification of error, sometimes referred to as the "loss function," may consider (Alnmr et al. 2024):

(i) The extent of predictive inaccuracy.
(ii) The predictive confidence level.
(iii) The therapeutic implications of the errors (overlooking a resistant strain may be more critical than incorrectly identifying a susceptible one).

(c) *The Backward Pass: Error Signal Propagation*

It involves the propagation of the error gradient across the network, modifying connection weights in accordance with their respective contributions to the error (Biró et al. 2024). The backward pass signifies the network's learning phase, much like how scientists may retrace their experimental methods to pinpoint errors in their assumptions (Singh et al. 2020). The error gradient propagates backward across the network, ascertaining the contribution of each link to the ultimate error (Jo et al. 2019). This approach is essential in viral analysis, as it aids in identifying the aspects of the viral sequence that contributed most significantly to erroneous predictions (Wohl et al. 2016).

For instance, in the analysis of HPV (Zhong et al. 2024):

- The network may ascertain that it assigned excessive significance to some genetic markers.
- It recognized that it had underestimated the significance of certain protein structures.
- The system may identify trends in the impact of various sequence combinations on viral behavior. The backward pass computes partial derivatives at each link, thereby addressing the inquiry: "To what extent did this specific connection contribute to our error?"

(d) *Weight Updates*

The network adjusts its parameters using the gradient descent technique, enhancing its capacity to identify infectious patterns (Zhu et al. 2025). The last stage is modifying the network's parameters according to the computed gradients, much like how a researcher could refine their experimental techniques based on prior findings (Mehmood et al. 2023). The gradient descent method calculates the necessary adjustments to each connection weight to reduce future mistakes (Alzubaidi et al. 2021).

This adjustment procedure is complex and considers several factors:

(i) *Learning Rate*: The magnitude of modifications required (excessive values may overshoot ideal parameters, while insufficient values may impede the learning process) (Ribeiro and Moniz 2020).
(ii) *Momentum*: The extent to which prior updates should affect current ones, aiding in the avoidance of local minima (Katona et al. 2024).
(iii) *Regularization*: Avoiding excessive specialization of the network to the training data, hence neglecting larger patterns (Ming et al. 2023).

These upgrades in viral analysis could include the following:

- Enhancing associations that accurately detected resistance mutations.
- Deteriorating associations that resulted in inaccurate forecasts about viral behavior.
- Modifying the network's sensitivity to certain sequence patterns.

The neural network gradually improves its ability to understand viral data by repeating these four stages one after the other (Seguritan et al. 2012). A steady and reliable technique for viral analysis is established using blockchain, which ensures the secure documentation of these learning outcomes and permits verification by other researchers (Ayub Khan et al. 2021).

3.2.1.2 Applications in Healthcare

The integration of backpropagation-based deep learning and BT has entirely altered our approaches to virology research and healthcare management (Razzak et al. 2020). The integration has significantly propelled viral research and treatment forward, establishing a robust ecosystem that improves data security and analytical capabilities (Alahi et al. 2023).

(a) *Secure Viral Data Examination*

Modern viral research generates substantial amounts of sensitive data, requiring stringent security protocols and sophisticated analysis (Tariq et al. 2023). By utilizing a comprehensive system architecture, backpropagation and blockchain work together to fulfill these two needs (Sadri et al. 2023). In studies of HIV, this method enables secure management of patient viral load information, genetic sequences, and treatment results (Falkenhagen and Joshi 2018). When examining a new HIV sample, the genetic sequence is initially encoded and logged on the blockchain with a distinctive identifier (Andrew et al. 2023). This ensures that, despite having access to data, advanced cryptographic techniques protect patient confidentiality (Basil et al. 2022). The deep learning models, trained through backpropagation, subsequently analyze this encrypted data to identify patterns that signal medication resistance or the potential for mutation (Askr et al. 2023). For instance, when analyzing HIV protease sequences, the system might identify new mutation patterns that confer resistance to specific antiviral drugs (Pikalyova et al. 2022). The outcomes are subsequently recorded on the blockchain, creating an unchangeable log that can be verified by other researchers while maintaining patient privacy (Singh et al. 2024).

(b) *Real-Time Mutation Surveillance*

Real-time tracking of viral evolution was a major step in virus surveillance and epidemiology. The system works in an uninterrupted manner without ceasing by observing the development of viruses in the different geographical areas and patient groups. Evaluate the situation concerning HPV surveillance:

- When a new HPV variant is identified in a clinical sample, the neural network rapidly assesses its genomic sequence, comparing it with known variants (Campos et al. 2022). When significant mutations are detected, particularly those that may affect vaccination efficacy or cancer susceptibility, the system triggers a range of automatic actions (Fan et al. 2023).
- Mutation information is recorded on the blockchain, providing a timestamp and an unchangeable record of when the variation was first identified (Worley and Skjellum 2019). Smart contracts evaluate the importance of the mutation autonomously using established criteria (Alowais et al. 2023). On the off chance that the change outperforms particular peril limits, the framework rapidly informs pertinent healthcare suppliers and inquires about organizations (Giancotti et al. 2024). This incite checking has illustrated impressive significance amid viral episodes, as instantly distinguishing modern changes can illuminate open well-being measures and treatment procedures (Baker et al. 2022). The system's capability to memorize and adjust through backpropagation progresses its accuracy in anticipating changes that seem to lead to expanded destructiveness or sedate resistance (Whittington and Bogacz 2019).

(c) *Distributed Learning Systems*

The integration of blockchain verification and federated learning has revolutionized cooperation between healthcare organizations while ensuring data confidentiality (Ngoupayou Limbepe et al. 2025). This varied approach is especially important in viral research, since different patient groups and regional variations can significantly influence viral dynamics and treatment results (Glingston et al. 2019). In research on HIV treatment, various hospitals from different regions participate in this distributed learning network (Bouabida et al. 2023). Every institution maintains its regional dataset and trains its neural network using backpropagation (Wright et al. 2022). Once a significant pattern is recognized by a local model, like a new drug resistance mechanism, this finding is shared throughout the network securely and anonymously (Truman et al. 2014).

The blockchain plays a crucial role in this procedure by (Kasyapa and Vanmathi 2024):

- Authenticating the veracity of disseminated model updates
- Documenting the origin of each contribution to the global knowledge repository
- Guaranteeing the preservation of patient confidentiality throughout the procedure
- Administering access rights and data sharing agreements

The worldwide model consistently enhances via this collaborative methodology, integrating varied patient demographics and treatment results (Abutaleb et al. 2023). For instance, if a hospital in a specific area finds an HIV mutation that reacts well to a specific combination of treatments, this information is securely shared and verified through blockchain technology, allowing other institutions to utilize it while maintaining patient privacy (Hernandez et al. 2024). This distributed approach has shown significant efficacy in expediting research and enhancing treatment results. HPV research has facilitated the swift discovery of geographical disparities in virus strains and their reactions to various therapies, resulting in more precise and successful intervention tactics (Johnson et al. 2022).

The system is always evolving, with new applications often arising. Recent advancements encompass:

- Incorporation of genetic databases for enhanced viral analysis
- Sophisticated predictive models for therapeutic outcomes
- Pre-emptive alert systems for prospective viral epidemics
- Automated algorithms for matching clinical trial participants
- Concurrent assessment of therapeutic effectiveness across diverse populations

These applications illustrate that the integration of backpropagation and blockchain technology is enhancing our comprehension of viral infections while also revolutionizing treatment and preventative tactics in contemporary healthcare (Burmeister et al. 2022).

3.3 Autoencoders

In the intricate domain of viral research, where extensive genomic and clinical data need advanced analysis, autoencoders serve as effective instruments that connect basic data to significant findings (Maleki Varnosfaderani and Forouzanfar 2024).

These neural network topologies have revolutionized the processing and comprehension of viral information, especially concerning HIV and HPV studies. Researchers may now evaluate sensitive viral data by integrating autoencoder technology with blockchain security, so protecting patient privacy and data integrity (Aljohani 2023).

3.3.1 Working Principles

Autoencoders fundamentally operate as advanced data compression systems, with a notable capacity to autonomously identify the most significant aspects of viral material (Dias and Torkamani 2019). Envision a virologist endeavoring to comprehend the fundamental attributes of myriad viral samples; autoencoders execute this process methodically and effectively (Bolukbasi et al. 2016).

The design comprises two principal components: an encoding system and a system for decoding.

While the decoding procedure aims to restore the original data from this compressed format, the encoding method compresses the input data into a compact representation known as the latent space (Sudlow et al. 2015).

This method holds considerable importance in virology for various reasons: Analyze a situation pertaining to the assessment of an HIV genome. In the study of several HIV genetic sequences, the encoder may compress a complicated genetic sequence, including thousands of base pairs, into a concise representation that captures critical mutation trends (Hayawi et al. 2024). The tool for decoding thereafter endeavors to recover the original sequence from the compressed version. The automatic encoding, through training, discerns the genomic patterns essential for understanding viral activity (Lee 2023a).

The training procedure includes many essential phases:

- *Data Input Phase*: The autoencoder acquires viral data, such as HPV genomic sequences or HIV protein structures. The data has been standardized and pre-processed to provide uniform analysis (Kumar et al. 2023).
- *The encoding process*: It involves the network compressing input data while discerning the most essential aspects (Shafay et al. 2023). In HIV analysis, this may include identifying patterns in the virus's envelope protein that signify medication resistance. The compressed data resides in the latent space representation. This depiction succinctly encapsulates fundamental viral characteristics (Ali et al. 2023).
- *The decoding process*: This involves the decoder's effort to recreate the original data. Discrepancies between the input and output facilitate the network's understanding of which traits are crucial to retain (Chen and Guo 2023).

In viral research, this procedure is safeguarded by blockchain integration, which guarantees that both the original data and the acquired patterns are maintained immutably and securely (Berahmand et al. 2024).

3.3.2 Use Cases in Medical Data Compression and Anomaly Detection

The use of autoencoders in virology has resulted in several significant advancements, especially in data reduction and anomaly detection (Lapique and Benenson 2018).

(a) *Compression of Medical Data*
In viral research, effective data reduction is essential for handling extensive genomic and clinical data (Chen et al. 2021). Autoencoders demonstrate proficiency in this domain via several applications:
   (i) *Viral Genome Compression*: In the analysis of many HIV genomic sequences, autoencoders may effectively compress this data while preserving essential information on mutation trends (Mühr et al. 2021). A research institution investigating HIV drug resistance may consolidate hundreds of genomic sequences to discern prevalent patterns indicative of resistance to certain antiretroviral medications (Lee et al. 2020).
   (ii) *Clinical Data Integration*: In HPV research, autoencoders may condense and amalgamate diverse data sets, including the following (Jia et al. 2016):
      - Genetic sequencing of several HPV strains
      - Diagnostic photographs of the patient
      - Data on treatment response
      - Clinical history data

      This condensed form enables researchers to see trends that may remain obscured when analyzing each data type in isolation (Prabakaran et al. 2007).
(b) *Anomaly Detection*
Autoencoders have shown significant efficacy in detecting atypical virus patterns or activities (Papadopoulos and Karalis 2023).
   (i) *HIV Mutation Detection*
   By analyzing the standard patterns of HIV genomic sequences, autoencoders may swiftly detect atypical mutations that may signify (Albshaier et al. 2024):
      - Emergence of pharmacological resistance
      - Alterations in viral virulence
      - Innovative transmission patterns

      Upon the detection of an anomalous pattern, the information is documented on the blockchain, establishing an immutable record that can be safely disseminated to other research organizations (Torabi et al. 2023).
   (ii) *Classification of HPV Strains*: Autoencoders facilitate the identification of atypical HPV variations by (Foxley-Marrable et al. 2023):
      - Acquiring knowledge of the standard characteristics of recognized HPV strains
      - Identifying sequences that diverge from anticipated patterns
      - Identifying prospective high-risk variations

(c) *Treatment Optimization*
By examining compressed viral data in conjunction with treatment results, autoencoders facilitate (Ahuja et al. 2023):
- Determine the most effective therapy combinations.
- Anticipate possible pharmacological resistance.
- Tailor therapeutic strategies according to viral characteristics.

The integration of autoencoder technology with blockchain security has established a robust environment for viral research, enabling advanced analysis and safe data storage (Ahakonye et al. 2024). This integration perpetuates advancements in viral research and therapeutic development (Clipman et al. 2022).

3.4 Restricted Boltzmann Machines (RBMs)

They are a type of generative stochastic network frequently employed in unsupervised machine learning (Meiring et al. 2012). These two-layer networks have a visible layer (denoting input characteristics) and a hidden layer (indicating learned feature representations), characterized by symmetric connections between the layers and an absence of intralayer connections (Thudumu et al. 2020). RBMs are proficient in feature extraction, dimensionality reduction, and uncovering hidden patterns in intricate datasets, rendering them very pertinent for virology applications, especially when combined with sophisticated technologies (Chabria et al. 2014).

AI-based RBMs may establish new approaches to virology analysis that embrace large and complicated datasets, ranging from genetic sequences to mutations in patients. Their secure and distributed nature will therefore improve the ability of RBMs to ensure traceability and data integrity (Atlam et al. 2024). Together, these technologies may enhance predictive modeling, individualized treatment, and the development of diagnostic tools for viral diseases (Combes and Franceschi 2018).

3.4.1 Structure and Learning Process

The makeup of RBMs comprises the following:

- *Visible Units*: Depicting the input data, such as genetic sequences, clinical features, or diagnostic test results.
- *Hidden Units*: Grasping the underlying traits that distinguish higher patterns or representations within the data.
- *Weights*: Balanced connections between observable and latent units that are gradually improved throughout the training process (Junaid et al. 2022).

These networks typically utilize an energy-based model, wherein the energy function defines the joint probability distribution of observable and hidden components (Wong et al. 2011). To identify the most likely input data setups, one aims to reduce this energy (Ahakonye et al. 2024).

Techniques such as Contrastive Divergence (CD) are utilized in training RBMs since they can minimize the gap between the distribution produced by the model and the actual data distribution by modifying the weights (Khan et al. 1825).

- *Fine-tuning*: A series of pre-trained Restricted Boltzmann Machines (RBMs) can be arranged in layers to form Deep Belief Networks (DBNs), facilitating further fine-tuning for regression or classification purposes (Vrábel et al. 2020).

Example RBMs can analyze HPV genetic data in virology to detect mutations linked to cancer-related strains. RBMs can be used to analyze trends in HIV medication resistance data, which helps forecast how well antiretroviral therapy will work.

3.4.2 Healthcare Applications

RBMs hold great potential in the medical field, particularly in the areas of pharmaceutical discovery, patient classification, and disease modeling (Miotto et al. 2018). Regarding virology, RBMs could offer:

(a) *Genomic Examination*

In the case of HPV, genome-wide association studies (GWAS) data can be analyzed using RBMs to find biomarkers linked to cervical cancer susceptibility (Ngufor et al. 2020). By examining latent patterns in genomic data, RBMs may distinguish between benign variants and high-risk HPV strains (Davids and Ashrafian 2022). RBMs may analyze HIV viral RNA sequences to find mutations associated with medication resistance to help create more effective antiretroviral regimens (Fröhlich et al. 2018).

(b) *Predictive Modeling*

By evaluating the patient's clinical characteristics and genetic composition, RBMs can ascertain the likelihood of HPV existence. RBMs can assist in modifying treatment strategies by modeling patient reactions to HIV therapy (Koel et al. 2023).

(c) *Incorporation with Blockchain*

Blockchain preserves patient confidentiality by allowing safe information sharing between researchers and healthcare providers (Takeuchi et al. 2019). Genomic data examined by RBMs for HPV or HIV may be securely stored and accessible over blockchain, enabling collaborative study. The traceability attributes of blockchain can authenticate the data used for training RBMs, hence guaranteeing dependable AI-generated predictions (Adebamowo et al. 2020).

(d) *Pharmaceutical Development*

RBMs may model interactions between antiviral agents and viral proteins, accelerating the identification of innovative therapies for HPV-related malignancies or HIV infections (Espinoza et al. 2021).

Recent research on HIV used RBMs to forecast the efficacy of combination antiretroviral treatment (cART) by analyzing patient demographic, clinical, and genetic data (Lee et al. 2022). By revealing latent factors, RBMs found subpopulations with unique treatment responses, facilitating individualized medicine strategies (Pavone et al. 2024). Utilizing the convergence of RBMs, AI, and blockchain, virology research may surpass conventional limitations, promoting creative methods for addressing viral infections such as HPV and HIV (Liu et al. 2018).

3.5 Convolutional Neural Networks (CNNs)

These networks are a unique type of deep learning model that is specifically designed to analyze inputs that resemble structured grids, like photographs. Since CNNs can autonomously learn spatial hierarchies of characteristics from input data, they have transformed domains like computer vision and medical diagnostics (Brickman and Palefsky 2015). In virology, convolutional neural networks (CNNs) are essential for the analysis of imaging data, such as histopathology slides, medical scans, and viral imagery. When integrated with blockchain technology, convolutional neural networks (CNNs) provide safe and transparent management of medical picture collections, promoting trust and cooperation in virology research (Elvas et al. 2023).

The amalgamation of AI-driven convolutional neural networks with blockchain technology enables the development of novel diagnostic instruments for viral diseases (Gürsoy et al. 2022). These instruments facilitate early identification, individualized therapy, and the formulation of data-informed healthcare initiatives (Jin et al. 2019).

3.5.1 Architecture and Key Components

CNNs' architecture was created especially for analyzing images, characterized by its hierarchical structure, which comprises:

- *Convolutional Layers*: Using filters (kernels), these layers carry out convolutional operations that retrieve information from input images, such as edges, textures, and patterns. Every filter gains unique characteristics relevant to the work (Park et al. 2021).

- *Pooling Layers*: By decreasing the spatial dimensions associated with feature maps, these layers improve the network's computational efficiency while maintaining crucial data. Pooling further mitigates overfitting by consolidating characteristics (Zhao and Zhang 2024).
- *Activation Functions*: Non-linearity is incorporated into the network with nonlinear activation functions, like ReLU (Rectified Linear Unit), which makes it easier to learn intricate patterns (Nirthika et al. 2022).
- *Fully Connected Layers*: By connecting every neuron in one layer to every other layer, these layers combine the properties learned from convolutional and pooling layers to produce predictions (Zafar et al. 2024).
- *Dropout and Regularization*: Techniques like dropout randomly deactivate neurons during training to mitigate overfitting, thus improving the model's capacity to extrapolate to unidentified data (Wang 2025).

In virology, CNNs may analyze cervical cytology images, such as Pap smears, to detect pre-cancerous lesions induced by high-risk HPV strains. Convolutional neural networks (CNNs) can evaluate medical imaging data for HIV to determine the virus's effects on organ systems, including brain scans for identifying HIV-associated neurocognitive disorders (HAND) (Mohammed et al. 2021).

3.5.2 Image Analysis in Medical Diagnostics

Convolutional neural networks (CNNs) are essential in medical diagnostics because of their proficiency in accurately recognizing patterns in medical imagery (Tan et al. 2021). Their applications in virology encompass:

(a) *Histopathological Examination*: Convolutional neural networks (CNNs) can identify dysplasia and neoplastic alterations in cervical tissue samples with superior accuracy compared to conventional manual testing for HPV (Wu et al. 2024a). Trained on extensive datasets of labeled pictures, CNNs can differentiate between benign and malignant tumors, enhancing diagnostic efficiency. CNNs may examine histopathology slides to detect opportunistic infections or malignancies, such as Kaposi's sarcoma, related to HIV (Allahqoli et al. 2022).

(b) *Medical Imaging*: Convolutional Neural Networks (CNNs) may be used to evaluate 3D imaging modalities, such as magnetic resonance imaging (MRI) or computed tomography (CT) scans, for tumor identification and therapy planning in patients with cervical cancer related to HPV (Hou et al. 2022). CNNs may evaluate the advancement of HIV-related problems via imaging techniques such as PET scans for visualizing metabolic activity and CT scans for detecting lung infections like *Pneumocystis pneumonia* (Yamashita et al. 2018).

(c) *Blockchain Integration*: Blockchain guarantees safe, immutable storage and dissemination of medical picture datasets used for training convolutional neural networks. Anonymized cervical cancer imaging data processed for HPV detection may be safely accessible by international research teams. Blockchain facilitates traceability, guaranteeing the verifiability of the origin of training data for CNNs, thereby augmenting the trustworthiness of AI-driven diagnoses (Sarvamangala and Kulkarni 2022).
(d) *Prognostic and Customized Healthcare*: Convolutional neural networks trained on HPV imaging data can forecast patient outcomes, assisting doctors in formulating individualized treatment strategies. By examining cervical imaging data, CNNs can forecast the probability of recurrence after therapy. CNNs can evaluate the efficacy of antiretroviral medication for HIV by examining sequential imaging data and facilitating therapy modification. In a recent application, CNNs were used to examine Pap smear pictures for automated diagnosis of HPV, attaining sensitivity and specificity rates equivalent to those of professional pathologists. In HIV research, CNNs have been used to analyze brain MRI images for the detection of early indicators of HAND, facilitating prompt intervention (Kourounis et al. 2023).

The integration of CNNs, AI, and blockchain is revolutionizing virology, especially by providing reliable, scalable, and transparent options for medical diagnosis and personalized treatment. Such innovations lead to better health outcomes in the treatment of viral infections, such as HIV and HPV (Yadav and Jadhav 2019).

3.6 Recurrent Neural Networks (RNNs)

A type of artificial neural network specifically structured to retain a memory of previous inputs, which is useful in the processing of sequential data, is known as a recurrent neural network (RNN). By keeping a memory of the info they store, RNNs can show sequences or temporal details, which makes them quite good at processing time series or sequential information (Prasad et al. 2024). In the case of virology, the evolving nature of diseases, mutations in viruses, and patient responses to therapies often yield sequential data. Combining RNNs, AI, and blockchain technology can help researchers to extract insights from data streams to ensure confidentiality, transparency, and data integrity (Mohammad Shahin et al. 2024).

Recurrent neural networks (RNNs) play a pivotal role in providing genomic sequence analysis, patient outcome predictions, and personalized treatment strategies for viral diseases like HPV and HIV by interpreting sequential medical data (Sarhangi et al. 2024).

3.6.1 Structure and Types (LSTM, GRU)

RNNs are characterized by their structure, which specifically caters to the sequential nature of data, and their recurrent connections, which allow them to retain information across time steps. However, traditional RNNs suffer from vanishing or exploding gradient problems that limit their ability to learn long-term dependencies; advanced architectures such as Gated Recurrent Units (GRUs) and Long Short-Term Memory (LSTM) networks help to overcome these challenges (Wu et al. 2024b).

(a) *Long Short-Term Memory (LSTM)*

LSTMs add gating mechanisms (input gate, forget gate, output gate, etc.) to minimize the information flows. This structure of LSTMs is capable of learning long-term dependencies in sequential data (Mustafa et al. 1763). In virology examples, LSTMs can review patient vaccination records and later data that predicts the risk of HPV-related conditions. In HIV, the temporal dynamics of viral load can be captured using LSTMs, allowing therapy to be optimized (Klontzas et al. 2024).

(b) *Gated Recurrent Unit (GRU)*

The GRU simplifies the LSTM structure by merging the input and the forget gates into one update gate, thus decreasing the computational complexity while holding on to the efficiency (Srikantamurthy et al. 2023). In the areas of virology, GRUs may analyze gene expression data for time-varying patterns associated with viral status during HPV and HIV infections (Araújo et al. 2017).

3.6.2 Applications in Sequential Medical Data

RNNs are particularly useful for analyzing sequential data in medical diagnostics, treatment planning, and epidemiological modeling. Their applications in virology include the following (Dörrich et al. 2023):

(a) *Genomic Sequence Analysis*

For HPV, RNNs can process DNA sequences to identify patterns associated with high-risk oncogenic strains, enabling targeted screening programs. For HIV, RNNs can analyze viral RNA sequences to detect mutations conferring drug resistance, helping clinicians adapt therapy plans in real time (Mehedi et al. 2024).

(b) *Disease Progression Modeling*

RNNs can track the progression of viral infections using longitudinal patient data. For instance, in HPV, RNNs can analyze cervical lesion progression over time to predict cancer risk. For HIV, RNNs can model the temporal relationship between viral load, CD4 counts, and patient adherence to antiretroviral therapy (ART) (An et al. 2024).

(c) *Integration with Blockchain*

Blockchain ensures secure storage and sharing of sequential patient data analyzed by RNNs. Blockchain technology, for instance, can be used to store HPV vaccination schedules or HIV treatment adherence data, improving data integrity and provider collaboration. Additionally, blockchain offers an unchangeable audit trail that confirms the legitimacy of the sequential datasets used to train RNNs (Ali et al. 2023).

(d) *Predictive Analytics and Personalized Medicine*

RNNs can review patients' histories in order to forecast future health outcomes. For HPV, RNNs could predict the effectiveness of vaccination programs by modeling the trends of infections. With respect to HIV, they can predict patient responses to ART from treatment history and genetic data. As a result, the utilization of this variable enjoyed immense success (Shi et al. 2020).

(e) *Epidemiological Studies*

Public health interventions to control the epidemic spread of infections in populations often rely on models that interpret time-series data in the form of recurrent neural networks (RNNs). For instance, HPV vaccination campaigns impact the rates of infection, and this can be predicted using RNNs. For instance, in the case of HIV, RNNs can help monitor the effectiveness of preventive interventions, such as pre-exposure prophylaxis (PrEP) (Taherdoost 2023b).

Example from the Real World: LSTM networks were applied in analyzing the influence of sequential treatment data on the viral load in HIV, where sequential treatment data was utilized to predict changes in viral load, and high accuracy for determining treatment outcome was reached (Li et al. 2020a). Similarly, GRUs have been used to analyze HPV vaccination data in order to produce recommendations for booster shots and duration of immunity. AI, Blockchain, and RNNs Together: Combining AI and blockchain with RNNs encourages innovation in virology research by enabling researchers to make data-supported decisions while preserving security in joint initiatives to fight viral diseases like HIV and HPV (Liu et al. 2022).

3.7 Generative Adversarial Networks (GANs)

Generative adversarial networks (GANs) are a category of machine learning algorithms designed to generate new, artificial data that mimics real data. Introduced by Ian Goodfellow in 2014, generative adversarial networks (GANs) include two adversarial neural networks, named a generator and a discriminator, which participate in a zero-sum game to enhance the quality of data production (Kwak and Hwang 2024). In virology, GANs provide disruptive potential by producing synthetic datasets for research, diagnostics, and AI model training while addressing issues like data scarcity and privacy concerns. The use of blockchain technology guarantees the safe and ethical management of synthetic datasets, particularly in critical medical fields (Şandru et al. 2024).

Generative adversarial networks (GANs) provide novel approaches to producing synthetic medical pictures for HPV and HIV, enhancing datasets for diagnostic artificial intelligence models, and modeling viral evolution for predictive analysis (Arora and Arora 2022).

3.7.1 Functionality and Adversarial Training

The fundamental operation of GANs is based on its adversarial training process, which comprises two networks (Beltrán et al. 2024):

(a) *Generator*: This network produces synthetic data by converting random noise into data samples that mimic the target dataset. For instance, it may produce synthetic cytology pictures for HPV studies or replicate RNA sequences for HIV investigations.
(b) *Discriminator*: This network assesses the veracity of the data, differentiating between genuine and manufactured samples. Providing input to the generator facilitates the enhancement of synthetic data quality (Lim et al. 2024).

The adversarial training approach involves iterative optimization:

- The generator produces artificial data.
- The discriminator assesses the data and generates a loss function that indicates whether the data is authentic or fabricated.
- The generator adjusts its settings to reduce the discriminator's loss, enhancing its capacity to provide realistic data.
- The discriminator modifies its settings to enhance the discriminator's ability to differentiate between the authentic and the fabricated data (Ramzan et al. 2024).

Example In virology, GANs may provide high-quality synthetic cervical histopathology pictures for HPV, thereby enhancing datasets used in AI-driven diagnosis algorithms. For HIV, GANs may replicate patient-specific viral mutation patterns, facilitating individualized therapy approaches (Goyal and Mahmoud 2024).

3.7.2 Use in Synthetic Data Generation and Augmentation

Generative adversarial networks (GANs) are especially beneficial for generating synthetic data and enhancing current datasets, tackling issues like the scarcity of annotated medical data and privacy problems (Rashidi et al. 2024).

(a) *Generation of Synthetic Medical Images*

Generative adversarial networks (GANs) may produce synthetic medical pictures (Chai et al. 2024). These pictures closely resemble authentic samples, facilitating the effective training of diagnostic AI models without the need for extensive real-world datasets. In HPV diagnostics, GANs may provide a variety of cervical cell pictures that illustrate different phases of dysplasia, hence enhancing the robustness of AI models designed to identify pre-cancerous lesions (Lacan et al. 2023).

(b) *Data Augmentation*

GANs augment the variety of current datasets by producing variants of authentic samples. For example, GANs may enhance HIV genomic data by modeling mutations or changes in many environmental contexts. Enhancing constrained datasets of HIV viral load trajectories helps optimize AI models intended for predicting treatment results (Lee 2023b).

(c) *Confidentiality-Respecting Data Exchange*

GANs provide synthetic data that replicates authentic patient information while safeguarding sensitive details. Sharing HIV and HPV data for cooperative research while adhering to privacy regulations like the General Data Protection Regulation (GDPR) requires this kind of ability. While maintaining the anonymity of real patients, GANs may produce artificial patient cohorts that mimic HPV vaccine responses, promoting international research collaboration (Yue and Dutta 2022).

(d) *Modeling Viral Evolution*

Generative adversarial networks (GANs) may simulate the progression of viruses such as HPV and HIV, producing synthetic sequences that forecast future mutations. These findings guide vaccine research and antiviral medication formulation. For HIV, GANs may model resistance mutations resulting from antiretroviral treatment, facilitating the development of next-generation pharmaceuticals (Elste et al. 2024).

(e) *Blockchain Integration*

Blockchain improves the moral and safe use of GAN-generated artificial data by offering immutable records of data provenance. Researchers may authenticate the provenance and alterations of synthetic datasets, so assuring accountability and trust.

Synthetic HPV cytology datasets recorded on a blockchain may be accessible and scrutinized by many stakeholders, facilitating transparent AI model training (Michelutti et al. 2024).

(f) *Training and Validating Artificial Intelligence Models*

Datasets created by GANs may be used to train AI models for uncommon ailments or marginalized groups, promoting equitable healthcare solutions. For HPV, GANs may provide synthetic datasets that depict uncommonly high-risk strains, thereby enhancing the efficacy of diagnostic models. Generative adversarial networks (GANs) can produce artificial patient data for training AI models aimed at predicting drug adherence in marginalized areas affected by HIV. Utilizing GANs, AI, and blockchain technologies, virology research may surmount data constraints, guarantee ethical standards, and improve the efficacy of AI-based diagnoses and therapies for HPV and HIV (Coyner et al. 2022).

3.8 Transformers

Transformers have transformed artificial intelligence by presenting a highly adaptable and robust model architecture that processes sequential input without dependence on recurrence (Cao et al. 2024). The fundamental aspect of Transformers is their self-attention mechanism, which allows for the recording of long-range dependencies and contextual linkages within data. Initially designed for natural language processing (NLP), Transformers have proliferated into other fields, including medical data analysis and virology research (Zhang et al. 2023). When combined with AI and blockchain, Transformers has the potential to revolutionize the secure and efficient processing, analysis, and trade of medical data. In virology, Transformers are essential for analyzing intricate information, including genetic sequences and medical literature, while ensuring data security and provenance via blockchain integration. Transformers provide innovative solutions for enhancing diagnosis, treatment planning, and public health initiatives for illnesses such as HPV and HIV (Patwardhan et al. 2023).

3.8.1 Architecture and Self-Attention Mechanism

The Transformer design has encoder and decoder layers, each constructed using multi-head self-attention mechanisms and feedforward neural networks. Essential elements comprise (Steurer et al. 2024):

(a) *Self-Attention Mechanism*

This approach calculates attention ratings for each token (e.g., word, nucleotide) in a sequence in relation to others, enabling the structure to concentrate on significant segments of the input (Kayikci and Khoshgoftaar 2024). In HPV research, the self-attention mechanism may discern pivotal genomic areas linked to carcinogenic activity via the analysis of sequencing data. In HIV research, self-attention may identify areas of viral RNA sequences associated with medication-resistance mutations (Li et al. 2020b).

(b) *Positional Encoding*

Transformers use positional encodings to preserve the order of input, allowing the model to properly understand temporal and spatial linkages (Choi and Lee 2023). For HPV, positional encoding facilitates the examination of temporal patterns in patient vaccination data. It assists in simulating the temporal evolution of viral loads in HIV (Alharbi and Vakanski 2023).

(c) *Multi-Head Attention*

This approach improves the model's capacity to identify varied arrangements in data by calculating attention across several subspaces. Multi-head attention may concurrently evaluate many HPV-related genomic characteristics, including mutations and expression levels, for precise diagnosis (Nie et al. 2023). In HIV, it may analyze patient demographics, viral mutations, and treatment history to provide individualized therapeutic recommendations (Deeks et al. 2015).

(d) *Feedforward Networks*

These networks implement transformations on the attention outputs, augmenting the model's expressive capacity (van Zyl et al. 2018).

3.9 Applications in Natural Language Processing for Medical Data

Transformers are proficient in natural language processing tasks, which include text categorization, summarization, and translation, making them essential for the analysis of medical data. Principal applications encompass (Tucudean et al. 2024):

(a) *Literature Extraction*

Transformers may get insights from extensive collections of virology research publications. For HPV, relationships between vaccination rates and cervical cancer incidence may be discerned from the scientific literature. Transformers may encapsulate facts about the effectiveness of antiretroviral medication for HIV (Cho et al. 2024).

(b) *Analysis of Electronic Health Records (EHR)*

Transformers may analyze unstructured language in electronic health records to discern illness patterns and suggest therapies (Nassiri and Akhloufi 2024).

Examining HPV-associated symptoms and patient outcomes in clinical documentation. Analyzing changes in antiretroviral therapy adherence and their impact on viral suppression in HIV (Byrd et al. 2020).

(c) *Annotation of Genomic Data*

Transformers can evaluate nucleotide sequences to identify functional areas, including promoters and genes (Clauwaert et al. 2021). Annotating genomic areas of HPV associated with high-risk phenotypes. Identifying conserved areas for the creation of an HIV vaccine (Ji et al. 2021).

(d) *Predictive Modeling*

Transformers are capable of forecasting illness outcomes with medical history and demographic information. Assessing the likelihood of cervical cancer in HPV-positive individuals. Forecasting treatment response for HIV based on viral genotype and patient adherence (Mohammed et al. 2024).

(e) *Applications in Public Health*

Transformers may assess social media and public health information to monitor the spread of viral diseases. Monitoring public mood and disinformation pertaining to HPV vaccination initiatives. Utilizing NLP-driven analysis of news items to identify developing hotspots for HIV (Mapaona et al. 2024).

(f) *Blockchain Integration for Secure Data Exchange*

Transformers are capable of generating and validating metadata for blockchain systems, hence assuring the safe exchange of medical data. Verifying the authenticity of HPV vaccination efficacy research disseminated throughout international health organizations. Facilitating secure cooperation on patient cohort studies for HIV (Ortiz et al. 2019).

The Bidirectional Encoder Representations from Transformers (BERT) model has been used in medical NLP projects to extract pertinent clinical information from unstructured text. In virology, BERT has been refined to detect biomarkers for HPV-related malignancies and HIV-associated comorbidities, offering practical insights for researchers and physicians. Utilizing Transformers, virology researchers may tackle significant issues in data analysis and sharing, enhancing the comprehension and control of HPV and HIV via the combination of AI and blockchain technology (Biswas and Shah 2023).

3.10 Deep Learning in Precision Clinical Decision-Making

3.10.1 Introduction to Precision Medicine

Precision medicine is a cutting-edge approach to the healthcare industry which personalizes medical care to each patient's unique traits, taking account of lifestyle, ecological, and biological variables to inform decisions about illness prevention, diagnosis, and treatment (Zhang et al. 2020). This methodology aims to enhance patient outcomes by focusing on three primary objectives:

- *Personalization*: Creating targeted therapies that work better for specific groups of patients.
- *Proactivity*: Predicting health issues before they manifest through predictive analytics.
- *Precision*: Improving the accuracy of diagnoses and treatments by considering individual variability.

Precision medicine offers transformative potential in the context of viral infections, particularly HIV and HPV. Traditional diagnostic methods often face challenges in distinguishing between different viral strains and tailoring treatment responses, complicating therapeutic decision-making. Precision medicine seeks to improve clinical outcomes by identifying altered biological markers early and directing treatment based on the response of the individual patient (Tao et al. 2024).

Drug resistance along with pharmacogenomic profiles has shown that precision medicine works wonderfully in the context of HIV treatment. In fact, these methods allow the tailoring of antiretroviral therapy (ART) according to the genetic makeup of the patient, with higher efficiency and fewer side effects as well (Bekbolatova et al. 2024). Machine learning, along with improving efficacy improvements in diagnostics, identifies patterns in large biological datasets that tie to specific viral infections. Another important aspect of precision medicine is reducing unnecessary treatments for well-characterized infections. This is quite important because misdiagnosis leads to either a delay in the effective therapy or misapplication of medication, which does not help (Pushkaran and Arabi 2024).

For HPV, precision medicine can focus on the mechanisms of infection persistence for this virus and its role in cervical carcinogenesis. As new discoveries in the fields of epigenetics and transcriptomics search for the mechanisms by which HPV integrates into the human genome, some research efforts focus on finding biomarkers that can contribute to early detection and intervention methods (Mu et al. 2018). In summary, the treatment of infectious diseases like HIV and HPV has changed because of precision medicine. As shown in Fig. 3.1, using the patients' varied characteristics and cutting-edge diagnostic techniques can significantly improve clinical outcomes and customize the treatment plan for viral infections.

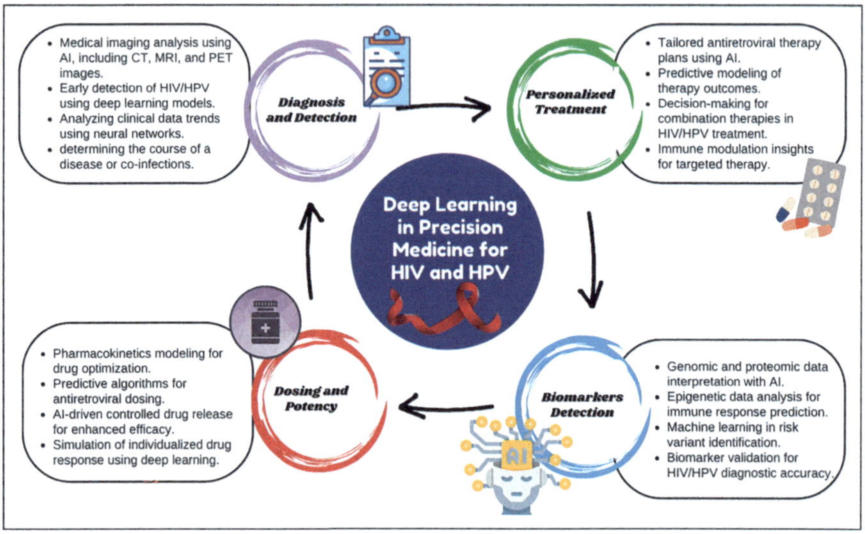

Fig. 3.1 Deep learning applications in precision medicine, specifically in biomarker profiling, dosage schemes, HIV and HPV diagnostics, and personalized healthcare parameterization

3.10.2 Case Studies in Medical Virology

The use of DL algorithms in precision medicine has revolutionized both the diagnosis and management of viral infections. Advanced technologies have demonstrated promising capability to respond to important challenges in the detection, treatment, and monitoring of infections like HIV and HPV. Deep learning, through the analysis of complex biological and clinical data, improves early detection, optimizes treatment plans, and reduces the risk of misdiagnosis. The two examples given below have been highlighted as the exemplary case studies in medical virology:

- *HIV*: This case highlights advancements in early detection and treatment optimization through pharmacogenomics and imaging technologies.
- *HPV*: This case focuses on the early identification of cervical abnormalities using deep learning techniques in diagnostic imaging and biomarker discovery. Through these examples, we illustrate how deep learning-driven precision medicine is revolutionizing clinical decision-making and improving patient outcomes in virology.

3.10.2.1 HIV: Early Detection and Treatment Optimization

The human immunodeficiency virus (HIV) remains a global public health issue, and timely diagnosis and individualized therapeutic approaches are essential to achieve optimal outcomes.

(a) *Deep Learning for Early Detection*

Deep learning models can identify pneumonia and other common HIV patient complications from chest X-ray images. Medical professionals can be immediately alerted because some studies have shown that convolutional neural networks (CNNs) can accurately identify pneumonia from such images (Jayawickrama et al. 2024). Moreover, deep learning algorithms were able to classify images of the rapid HIV tests with a sensitivity of 97.8% and 100% specificity—an improvement over diagnostics that occur in the field (Carini and Seyhan 2024).

(b) *Pharmacogenetic Profile for Personalized Pharmacotherapy*

Precision medicine for customizing antiretroviral therapy to its unique genetic variants is pharmacogenetic profiling. As has been clearly shown, personalizing treatment in this way ultimately also leads to better patient adherence and outcomes by ensuring improved efficacy of treatment and lesser adverse drug events (Yammouri and Lahcen 2024).

3.10.2.2 HPV: Early Identification of Cervical Abnormalities

Cervical cancer is primarily caused by human papillomavirus (HPV) infection, which is the most common sexually transmitted infection worldwide (Langford et al. 2024). Timely detection and intervention are essential to reduce the risk of progression to malignancy. Recent advancements in diagnostic methods, bolstered by deep learning technologies and precision medicine, have significantly transformed the management of cervical abnormalities associated with HPV (Wang et al. 2024).

(a) *Improving Detection Through AI-Powered Cytology Screening*

AI for cytology screening has considerably improved the speed and accuracy of the identification of cervical abnormalities. The impressive capabilities of AI systems, particularly deep learning models, to analyze Pap smear images are described in research published in Cytopathology (Jiang et al. 2023a). Compared with traditional methods, these systems exhibit an enhanced capacity for unearthing abnormal cells, as evidenced by markedly lower false-negative rates, leading to expedited responses within the treatment framework.

(b) *Increasing Accessibility Through the Use of Self-Sampling Techniques*

When appropriate medical care is not available, HPV DNA testing combined with self-sampling methods has emerged as a practical alternative. A comprehensive study found that the use of self-collected samples along with molecular diagnostic tools successfully increased the number of people who participated in cervical screening programs (Dellino et al. 2024). This method, which is especially helpful in settings with limited resources, allows women to conduct on-screenings at home and submit their samples for lab-based HPV testing.

(c) *Risk Assessment and Biomarker Discovery*

The risk of stratification has significantly increased thanks to HPV biomarkers. E6/E7 mRNA expression levels are one of several biomarkers that have been used to differentiate between infections that are temporary and those that have a higher chance of becoming cancer or high-grade lesions. Together with AI algorithms, these biomarkers can then be used to precisely classify patients according to their risk, allowing physicians to create far more individualized and customized treatment regimens (Hays 2024).

3.10.3 Targeted Therapies Development

3.10.3.1 Personalization Based on Genomic Data

Experts in viral genomics rely on this genomic approach to tailor treatment for viral infection according to individual characteristics. By analyzing the genomes of the virus and the host, precision medicine can pinpoint genetic differences that shape the severity of the disease and the success of treatment. This enables medical professionals to choose the most effective treatment plans based on each patient's unique genetic traits, thereby increasing the effectiveness of therapy.

3.10.3.2 Genomic Profiling for Antiviral Therapy

Genomic data has been tremendously useful in optimizing antiretroviral therapy (ART) for the human immunodeficiency virus (HIV). For example, hypersensitivity to one of the most prescribed antiretrovirals, abacavir, is linked with genetic determinants, including HLA-B*57 (Letchumanan and Theva Das 2025). Pharmacogenomic profiling enhances patient safety and therapy efficacy, allowing clinicians to anticipate potential adverse reactions and choose the most suitable course of action.

3.10.3.3 Identifying Viral Strain Differences

Understanding strain differences in the human papillomavirus (HPV) is essential for assessing the course of the disease and the efficacy of treatment. Cervical cancer is strongly linked to high-risk strains of HPV, including HPV-16 and HPV-18 (Kolou et al. 2021). Deep learning algorithms that examine viral genome data to detect these strains and aid in the creation of focused vaccinations and therapeutic approaches.

3.10.4 Diagnostic Imaging

Modern healthcare diagnostics includes the imaging of anatomic structures of the body that is important for diagnosing and treating a patient. Thus, medical images belong to a large number of specialized diagnostic imaging techniques, including X-rays, computed tomography (CT), magnetic resonance imaging (MRI), ultrasound, and nuclear medicine. In addition, each of them has its own strengths and applications in diagnosing various diseases; thus, clinicians can accurately help the patient with a precise diagnosis and proper care (Hussain et al. 2022).

3.10.4.1 Techniques for Detecting Viruses-Induced Anomalies

Emerging imaging technologies powered by deep learning are changing the way we detect the aberrations produced by viruses. Recognizing intricate patterns in medical images, such systems boost diagnostic accuracy and lower the rate of false-negative results. Imaging modalities including computed tomography (CT), magnetic resonance imaging (MRI), and ultrasound are being used to identify those tissue changes caused by viral infections such as those caused by HIV and HPV.

(a) *Computed Tomography (CT) Scans*:

CT scans are often employed to diagnose HIV-associated lung problems such as pneumonia and tuberculosis. Deep learning algorithms are very precise when assessing CT images which enables early diagnosis and appropriate treatment ultimately resulting in a better prognosis for the patient (Nabrawi and Alanazi 2023).

(b) *Magnetic Resonance Imaging*:

It is an essential technique for the early identification of neurological complications in HIV patients. When integrated with deep learning approaches, it even reveals more subtle structural brain changes that have been notoriously validated for distinguishing even uninfected-related abnormal brain anatomy.

(c) *Ultrasound*:

HPV patients are usually subjected to ultrasound scans to assess their abdominal and pelvic regions. An AI-made analysis of ultrasound images increases the chances of detecting any lesions and other relevant changes indicative of cervical cancer.

3.10.5 Examples and Case Studies

3.10.5.1 Case Study 1: HIV Imaging with Deep Learning

A study involving HIV-positive patients employed deep learning algorithms to analyze chest X-rays for detecting pulmonary tuberculosis. The results demonstrated an accuracy rate of over 90%, highlighting the ability of AI to enhance diagnostic accuracy, especially in resource-limited environments. Early detection of opportunistic infections like tuberculosis is essential for effective patient management and treatment outcomes (Nijiati et al. 2022).

3.10.5.2 Case Study 2: AI-Enhanced HPV Screening

In a case study focused on cervical cancer screening, a combination of self-sampling and AI-driven HPV DNA testing significantly improved participation rates and diagnostic precision in underserved communities (Ibrahim et al. 2024). This innovative approach allowed women to conduct initial screenings at home, facilitating earlier identification of cervical abnormalities related to HPV.

3.10.5.3 Case Study 3: Diagnosing Cerebral Toxoplasmosis in HIV Patients

A retrospective study conducted between 2015 and 2018 in Russia examined cases of cerebral toxoplasmosis in HIV-infected patients (Devasia et al. 2023). The study emphasized the crucial role of MRI in diagnosing this condition, as it revealed characteristic lesions indicative of cerebral toxoplasmosis. Early diagnosis using advanced imaging techniques was key to improving patient outcomes and reducing mortality rates.

3.11 Challenges in Implementing Deep Learning in Healthcare

3.11.1 Data Volume

Deep learning involves computational models that require significant processing power, exemplified by fully connected multi-layer neural networks, which need accurate estimation of numerous parameters. Success hinges on large datasets, with a common guideline suggesting around ten times as many samples as parameters. This is crucial in fields like computer vision and natural language processing, where large datasets are available.

In contrast, the healthcare sector faces challenges due to limited patient access to essential services, restricting the data pool for training deep learning models. Additionally, understanding diseases is more complex than tasks like image or speech recognition, necessitating a much larger volume of medical data to develop effective deep learning models (Kumar et al. 2024).

In the case of HIV, it is essential to gather data from a wide range of patient populations, as the disease progresses differently, and individuals respond to treatment in various ways.

Likewise, HPV poses difficulties because of its numerous strains and differing oncogenic capabilities, which require extensive and diverse datasets to develop effective diagnostic models (Yenikaya et al. 2024).

3.11.2 Data Quality

Healthcare data is different from data in other fields as it is often highly heterogeneous, ambiguous, noisy, and incomplete. The development of a deep learning model working with such big and heterogeneous data sets entails numerous challenges. We need to address a number of issues, including but not limited to missing values, redundancy, and even data sparsity.

However, challenges in generalizing models rise due to heterogeneity in HIV data containing virologic load measurements, CD4 counts, or demographic characteristics. The heterogeneity of patient-reported outcomes or diagnostic imaging may also make HPV datasets difficult to analyze (Javed et al. 2025).

3.11.3 Temporality

Disease progression is a dynamic process, with unpredictable evolution over time. This heterogeneity is due to a myriad of factors, including genetic differences, environmental factors, and the response of individual patients to a given treatment. For proper diagnosis and management, it is important to understand the temporal dynamics of diseases, as symptomatology, severity, or overall patient health status often change as the disease progresses. While this dynamic behavior is a fundamental property, many currently popular deep learning models, especially for medical applications, assume static vector-based inputs. These models make use of data as fixed snapshots, neglecting the sequences and variations over time. This static paradigm creates a substantial vulnerability to their ability to represent disease progression adequately. A model is no more intelligent than the datasets it is trained on, so if a model only takes a patient's health status at the moment when they walk in through the door and feeds it to an algorithm without being compensated for patients' prior medical history or the trajectory of their symptoms, it would blind itself to any important insights that could lead to informing treatment decisions for the patient. Likewise, when it comes to chronic diseases, whose health status can be naturally volatile, fixed data used for eventual prediction and interventions can lead to less performant results (Ezechi et al. 2014).

However, they all have their shortcomings, and there is an increasing demand for deep learning models that account for the temporal nature of data and are able to adjust to the dynamic characteristics of diseases. RNNs, LSTMs, or other architectures can also be employed to learn from this sequence data.

More relevant is the need to monitor temporal data on HIV itself, including viral load and CD4 counts, over time in order to see how a disease progresses and how effective treatment may be. In contrast, HPV requires longitudinal follow-up to monitor the progression from infection to potential cancer development (Ye et al. 2020).

3.11.4 Domain Complexity

Compared to fields like image and speech analysis, biomedicine and healthcare face more complicated problems because of a number of interrelated, medically specific factors.A major contributor to this complexity is the significant heterogeneity of diseases. Unlike more homogeneous datasets used for image or speech analysis, diseases are highly heterogeneous and are affected by lifestyle, environment, genetics, and other health problems. The fact that two patients suffering from the same diagnosis might have different symptoms and responses to therapy complicates normalizing diagnostic and treatment protocols. Additionally, many medical conditions show variability not only between individuals but also within the same individual over time. Chronic illnesses can fluctuate in intensity and symptoms, making it difficult to predict outcomes and tailor treatment strategies. Moreover, the integration of diverse data types adds to the complexity in biomedicine. Healthcare data includes clinical documentation, imaging results, lab findings, and genomic data, each with its own structure and format, complicating the development of comprehensive models to capture the nuances of patient health.

For example, HIV involves a wide spectrum of clinical manifestations and treatment responses, varying by genetic, environmental, and socio-economic factors. HPV has over 200 strains, with certain high-risk types like HPV-16 and HPV-18 being linked to cancers, adding to the complexity of modeling the disease (Shimbre et al. 2024).

3.11.5 Interpretability

Deep learning models have demonstrated considerable success across various application domains; however, they are frequently regarded as black boxes. This lack of transparency may not pose significant issues in more deterministic fields, such as image annotation, where end users can objectively verify the accuracy of the assigned tags. In the healthcare sector, however, it is essential not only to assess the quantitative performance of algorithms but also to understand the underlying reasons for their functioning.

Indeed, model interpretability—specifically, identifying the phenotypes that influence predictions—is vital for persuading medical professionals regarding the recommended actions from predictive systems, such as the prescription of specific medications or the identification of individuals at high risk for certain diseases (Dessie et al. 2020).

Effective HIV management involves understanding viral load, CD4 counts, therapy adherence, and co-infections to predict treatment failure and customize interventions. Patients with low CD4 counts or high viral loads may need closer monitoring and treatment adjustments, while considering social determinants of health aids in personalized care. Similarly, managing HPV requires analyzing biomarkers and imaging data to assess cervical cancer risk. By evaluating HPV types and cytological results, healthcare providers can categorize patients by risk and implement targeted surveillance (Ennab and Mcheick 2022).

3.11.6 Medical Data Representation and Transformation

Medical data representation and transformation are able to generate the most precise observations and predictions when given the right kind and amount of data. Real-world medical data currently exists in unstructured formats, such as text data (such as descriptions of symptoms, tumors, and medical records), trees (such as XML documents, parse trees, and RNA), sequences (such as time series, audio and video signals, and DNA), or various combinations of these formats (Rahman et al. 2024). The fundamental aspect of deep learning techniques is that they are limited to processing numeric input data, as all information is ultimately converted into binary strings for computational purposes. Qualitative data often presents challenges in terms of conversion into a usable format, which can complicate the processing.

In contrast, humans can readily interpret and derive meaning from such data, particularly when there are simultaneous variations in factors like intensity and quantity; for instance, changes in temperature and light can be easily understood and adjusted accordingly. In deep learning, representing analogous processes and conditions necessitates extensive encoding and intricate mathematical formulations in dimensional reduction and transformation. An algorithm for cross-domain feature learning that is based on stacked denoising autoencoders has been proposed to effectively represent features that describe data with multimodal characteristics, such as signals, images, videos, and audio (Ahmed et al. 2023). To manage certain real-world scenarios, a DL architecture that can integrate several data types at once is required (Tobore et al. 2019).

3.11.7 Hardware Requirements for Medical Big Data

Effective deep learning solutions require substantial training data, which is abundant and continuously growing in the medical field. Adequate processing power is essential for executing tasks and developing models, leading to the creation of multicore, high-performance GPUs, as traditional CPUs are insufficient for large-scale deep learning. However, hospitals often lack easy access to these graphics processing units, and they are costly and power-hungry. Companies like Nervana Systems and Wolfram Mathematica offer cloud-based services to help researchers accelerate the training process (Kalinin et al. 2018). Standard deep learning systems have the drawback of requiring costly data centers, which are frequently unavailable in medical settings. After training, deployment occurs on smart devices with limited processing power, like laptops and wearables. As deep learning architectures become more complex, the computational demands increase, making real-world applications costly and resource intensive. Without effective deep learning methods to analyze mHealth data, valuable insights remain untapped. For example, genomic analysis in HIV research and high-resolution imaging for HPV diagnosis require substantial computational resources, which may not be accessible in low-resource settings (Wu et al. 2023).

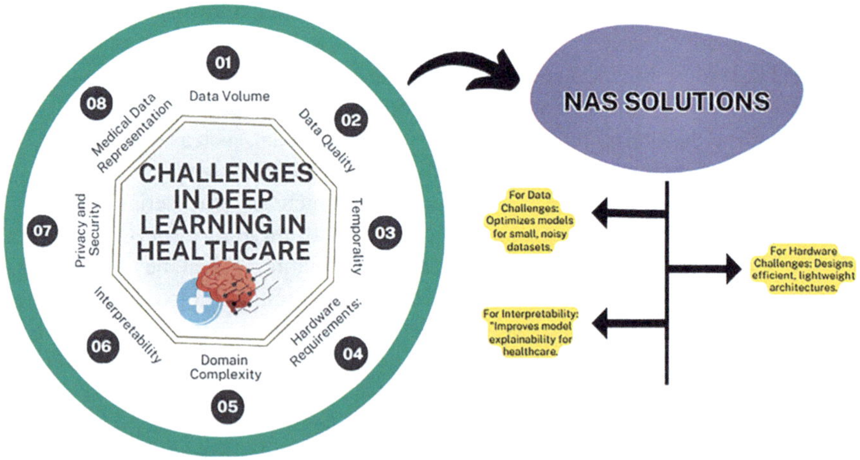

Fig. 3.2 Challenges in implementing deep learning in healthcare

3.11.8 Privacy and Data Security

Healthcare systems handle enormous volumes of private patient data, such as financial and medical histories, so data security and privacy are crucial. Protecting this information from unauthorized access is crucial for legal compliance and maintaining patient trust. Compromised patient data can damage relationships between patients and providers, leading to hesitance in sharing vital health information and affecting care quality. Healthcare professionals have an ethical duty to safeguard this data, as breaches can harm reputations and result in disciplinary actions. There are also legal aspects, as organizations need to adhere to numerous regulations (e.g., HIPAA, CCPA, and GDPR) in the United States, and not complying can lead to severe penalties and more scrutiny on these organizations. So, the mechanisms of strong data privacy and data security are very essential for better and faster management of sensitive patient data, as shown in Fig. 3.2.

3.11.9 Addressing Healthcare Challenges with NAS

Neural Architecture Search (NAS) is an automated process of designing neural network architecture, targeting to uncover optimal models adapted to custom tasks with limited human intervention. NAS holds great potential in the healthcare field, combating numerous obstacles associated with the implementation of DL solutions.

(a) *Data Quality and Volume*:

All healthcare datasets are small and have missing values or noise. To overcome these limitations, systems should be optimized to perform well, and NAS can improve the accuracy and robustness of such models. For several datasets, a lightweight NAS model outperformed the medical picture categorization, illustrating its robustness against variations in data volume and quality (Arora et al. 2023).

(b) *Hardware Constraints*:

When implementing deep learning models across healthcare settings, it is important that practical hardware limitations be considered, especially in resource-constrained environments. This procedure is known as hardware awareness and allows each neural network design to be tailored to suit the capabilities of its hardware while allowing it to function effectively without any compromise.

(c) *Interpretability and Compliance*:

In the case of medical applications, model interpretability is crucial to win medical practitioners' trust and comply with regulatory guidelines. NAS frameworks can be utilized to design architectures with similarities to healthcare standards in both extremely effective and understandable manners.

(d) *Disease-Specific Model Optimization*:

NAS has been utilized in developing specialized models in different healthcare applications and fields, including health monitoring, analysis of medical images, and categorization of disease diagnosis. Until now, evolutionary NAS methods have not only proved capable of those tasks but also been adopted into specialized domains such as medicine that differ tremendously, customizing systems tailored around their specific medical workloads (Zhou et al. 2021).

Neural architecture search (NAS) promises to automate the design of neural network architectures, reducing reliance on domain-specific deep learning expertise, accelerating the development speed, and producing models that are better suited to the particular challenges of healthcare data and environments. Automating this key procedure can encourage even more integration of deep learning into the healthcare industry, potentially leading to better diagnostic tools, personalized therapies, and overall patient health outcomes.

3.12 Recent Developments and Solutions

Deep learning is evolving at an unprecedented pace, spurring novel approaches and advancements that navigate the complexities of healthcare data, privacy, and model performance. Tools like AutoML-Zero, federated learning, and evolutionary deep learning are sophisticated solutions that could revolutionize healthcare through improved model training, data securement, and system-friendly construction.

3.12.1 AutoML-Zero

3.12.1.1 Overview and Significance

AutoML-Zero, Another novel approach in automated machine learning (AutoML) streamlines the process by removing any dependencies on prior knowledge of machine learning algorithms and architectures (Chen et al. 2020b). Instead, it constructs models from scratch with primitive operations and evolutionary search methods to ascertain the most appropriate algorithm for a specific task. Armed with the data for such a method, one that greatly reduces the need for unnecessary human intervention and expertise, we can build machine learning models you truly have no preconceived notions or biases about.

3.12.1.2 Applications in Healthcare

In healthcare, the definition of which is to find ailments, AutoML-Zero can present a wide range of possibilities. AutoML-Zero can hence be used to create better tools to predict patients' outcomes, diagnose diseases, and choose a treatment automatically. To illustrate, it can help build optimized models for examining medical imaging data, genetic sequences, or electronic health records, thus enhancing personalized healthcare and treatment precision (Habehh and Gohel 2021).AutoML-Zero also represents progress in several important dimensions:

(a) *Drug Discovery*: AutoML-Zero can accelerate the identification of potential drug candidates by optimizing molecular docking simulations and analyzing extensive chemical libraries (Zhuhadar and Lytras 2023). This capability allows researchers to efficiently explore vast datasets, leading to faster discovery of viable compounds for further development.
(b) *Early Disease Detection*: The approach is instrumental in designing models that integrate proteomic, transcriptomic, and genomic data to find subtle disease indicators. By leveraging diverse biological data, AutoML-Zero can enhance early detection capabilities, potentially identifying diseases at stages where intervention is most effective (Koh et al. 2022; Walter et al. 2023).
(c) *Treatment Response Prediction*: AutoML-Zero facilitates the development of adaptive models that predict patient-specific responses to therapies. This personalization enables real-time adjustments to treatment plans based on individual patient profiles, optimizing therapeutic outcomes.
(d) *Future Potential*: With the increasing complexity and size of healthcare datasets, AutoML-Zero's capacity to generate highly specialized models without human involvement will only increase in value. Future advancements may include the integration of quantum computing, which could lead to even faster model generation. Additionally, extending its application to global health initiatives could address resource constraints by providing efficient, automated solutions. Collaborative efforts between research institutions and healthcare providers could further enhance its utility, promoting widespread adoption in both clinical and research settings.

3.12.2 Federated Learning

3.12.2.1 Concept and Benefits for Data Privacy

Federated learning is a decentralized machine learning method that protects sensitive data while enabling collaborative model training across several institutions (Romero et al. 2022). Each institution utilizes its own data to wire its model on its infrastructure and sends only model updates (weights or gradients) to a central server. This approach helps to resolve key challenges of data security and privacy, as it ensures that the raw data remains within the concerned institutions. In certain industries—healthcare, for example—patient data must comply with rigorous confidentiality laws, like GDPR in Europe and HIPAA in the United States.

Key Benefits of Federated Learning:

(a) *Enhanced Privacy*: Sensitive data never crosses the organizational boundary, drastically reducing the risk of data leaks (Chakraborty et al. 2024).
(b) *Collaboration without Compromise*: Institutions may synergize and learn from each other without sacrificing their proprietary or sensitive information.
(c) *Compliance with Regulations*: When data is kept on-premise, federated learning can comply with stringent data protection legislation.

3.12.2.2 Applications in Healthcare Collaborations

Federated learning has changed the face of healthcare by allowing hospitals, research institutions, and pharmaceutical companies to work together. Notable applications include the following:

(a) *Medical Imaging*: A prominent application of federated learning is the area of medical imaging, where models can be trained to improve diagnostic accuracy in radiology across diverse datasets from various hospitals. For example, when identifying abnormalities in chest X-rays, such features ensure that models are strong and applicable across different populations and imaging equipment.
(b) *Drug Discovery and Genomics*: Federated learning is used by pharmaceutical firms to glean insights from genomic data while maintaining patient identity. This technique speeds up drug generation by utilizing various genetic datasets to discover treatment targets (Liu et al. 2024).
(c) *Disease Prediction Models*: Using federated learning allows for the creation of predictive models for chronic diseases, such as diabetes, CVDs, and cancer, using information obtained from multiple health providers. Models are less homogenous, both demographically and clinically.
(d) *Multi-Center Clinical Trials*: Federated learning allows combining trial data from multiple organizations, increasing the validity and generalization of trial results, while protecting the privacy of study participants (Hasin et al. 2017).

3.12.3 Evolutionary Deep Learning

Evolutionary deep learning merges evolutionary algorithms with deep learning techniques to optimize neural network architectures and hyperparameters. By mimicking natural evolutionary processes—such as selection, mutation, crossover, and inheritance—this approach allows for an efficient exploration of a vast search space. This capability is particularly advantageous for discovering innovative architectures and configurations that traditional optimization methods, like grid or random search, may struggle to identify (Ivanisevic and Sewduth 2023).

3.12.3.1 Key Principles and Applications of Evolutionary Deep Learning

(a) *Selection*: This involves retaining the most promising neural network configurations based on performance metrics, such as accuracy or loss.
(b) *Mutation*: Small random changes are introduced in network parameters or structures to explore new solutions.
(c) *Crossover*: Attributes from multiple high-performing networks are combined to create new architectures.
(d) *Inheritance*: Successful traits are propagated across generations to iteratively refine models.

Some of the applications cited for evolutionary deep learning are given below:

(i) *Automated Neural Architecture Search (NAS)*: Evolutionary algorithms automate the design of neural networks, resulting in architectures specifically optimized for tasks like image recognition, NLPs, and medical diagnostics (Sanches et al. 2024; Mohr et al. 2024).
(ii) *Hyperparameter Optimization*: Parameters like learning rates, activation functions, and dropout rates are fine-tuned to enhance performance (Kiliçarslan and Celik 2024).
(iii) *Model Compression and Efficiency*: Evolutionary techniques are employed to create lightweight models that are deployable on devices with limited resources, like wearables and edge devices.

3.12.3.2 Potential in Optimizing Healthcare Models

Evolutionary deep learning holds significant promise in healthcare, particularly for addressing complex and high-stakes challenges. Key applications include the following:

(a) *Personalized Treatment Planning*: By analyzing extensive and diverse patient data, evolutionary deep learning can create models that suggest personalized treatment regimens depending on each patient's unique genetic, environmental, and lifestyle characteristics (Fatima et al. 2025).

(b) *Drug Discovery*: Evolutionary algorithms expedite the identification of drug candidates by optimizing models that predict drug efficacy, toxicity, and interactions based on molecular data (Paul et al. 2021).
(c) *Medical Imaging*: Complex tasks such as detecting rare diseases, segmenting tumors, or classifying disease stages are enhanced by models designed through evolutionary techniques. For instance, evolutionary NAS has been utilized to optimize CNNs for analyzing MRI or CT scans (Blanco-González et al. 2023).
(d) *Multimodal Diagnostics*: Evolutionary deep learning integrates data from various modalities (e.g., imaging, genomics, and clinical records) to create comprehensive diagnostic models (Serrano et al. 2024).

3.12.4 Case Studies and Examples

(a) *Cancer Detection*: In breast cancer detection, evolutionary NAS has been employed to develop CNNs that excel in identifying malignancies in mammograms while minimizing false positives (Mak et al. 2024).
(b) *Rare Disease Prediction*: Evolutionary techniques help create specialized models for detecting rare conditions with limited training data by efficiently exploring feature combinations and architectures (Visan and Negut 2024).
(c) *Drug-Target Interaction Models*: Evolutionary deep learning has optimized recurrent neural networks (RNNs) for predicting interactions between drugs and proteins, which is crucial in drug repurposing efforts (Qi et al. 2024).

3.13 Ethical Considerations in AI for Healthcare

AI is transforming healthcare by improving patient outcomes, streamlining processes, and advancing science. However, its integration raises ethical challenges, including equitable access, biases, and the need for ethical frameworks. The debate continues on whether AI should fit within existing legal frameworks or require a new classification. Four major ethical concerns need to be resolved in order to fully utilize AI's potential in healthcare: data privacy, algorithmic fairness, safety and transparency, and informed consent for data use (Dara et al. 2022).

3.13.1 Privacy and Data Security

Protecting sensitive medical information and patient privacy poses significant ethical challenges in implementing AI and machine learning in healthcare. Organizations collect vast amounts of patient data, which is essential for training AI algorithms. However, this need raises concerns about security breaches and data privacy (Yi et al. 2021). In the context of artificial intelligence and machine learning, privacy concerns extend beyond standard data security procedures and cover the proper handling and application of private medical data. Healthcare institutions are required to make sure that patient data is gathered, kept, and handled in accordance with ethical guidelines and privacy laws. Since unauthorized access to patient records can result in identity theft, confidentiality violations, or misuse of private medical information, it poses major risks to patient autonomy and trust in the healthcare system (Murdoch 2021).

Throughout the entire AI and ML lifespan, including data collection, storage, algorithm development, and deployment, strong data security protocols are essential. Strict authentication, access controls, and encryption are put in place to guard against data breaches and illegal access. Prioritizing data minimization and anonymization will also help healthcare organizations address privacy issues with AI and ML and lower risks to sensitive patient data (Yadav et al. 2023). Furthermore, to protect patient confidentiality and guarantee adherence to legal and ethical standards, compliance with privacy regulations such as the General Data Protection Regulation (GDPR) in the European Union (Wang et al. 2022), the Health Insurance Portability and Accountability Act (HIPAA) in the United States (Williamson and Prybutok 2024), and comparable laws and regulations around the world is essential.

3.13.2 Bias and Fairness

Algorithmic bias is one of the greatest challenges to fairness and justice in healthcare, presenting ethical challenges not just with respect to AI, but also with respect to machine learning. Biases related to race, gender, or socio-economic status may produce discriminatory outcomes due to imbalances in either the training data sets or the algorithms themselves (Chen et al. 2023a; Ferrara 2024). In health care, an ethical issue called algorithmic bias leads to differences in diagnosis, treatment, and outcomes among different populations. To illustrate, a diagnostic algorithm that has been trained using data from a certain demographic can provide bad outcomes in terms of false diagnoses or delays in treatment for underrepresented groups. In addition, prediction models based on socio-economic factors may worsen already existing health disparities by perpetuating inequitable care access and resource allocation (Cross et al. 2024; Belenguer 2022).

Once neglected, algorithmic bias can threaten the values of justice, equity, and fairness in healthcare, leading to systemic discrimination and a loss of public trust in the medical establishment (Ueda et al. 2024). In addition to harming individual patients, bias in AI and ML algorithms has broader societal consequences, which, in turn, hinder efforts to attain health equity and exacerbate existing inequalities (Chen et al. 2023b).

Preventing bias in AI and machine learning requires coordination in data collection, business logic, and model evaluation. Healthcare organizations should track diversity and representation in their training datasets. To reduce biases, algorithm developers should apply the following fairness strategies: fairness-aware training algorithms and post-processing techniques. Additionally, ensuring that algorithms are both transparent and interpretable is crucial for recognizing biases and helping stakeholders understand their origins and take effective action (Nazer et al. 2023).

3.13.3 Transparency and Explainability

The black box nature of AI and machine learning models, for example, presents challenges to the explainability that is necessary for accountability and trust in healthcare. Acceptance of these technologies may not be universal, as skepticism over previously unknown mechanisms for AI-generated clinical predictions can lead to diminished trust among users and the public (Norori et al. 2021). Technical advancement has resulted in mistrust and decreased trust among patients and healthcare providers. Transparency in healthcare is the degree to which information about AI and ML algorithms is accessible and intelligible to the relevant stakeholders, while explainability is the extent to which the relevant stakeholders can understand and interpret the outcomes of those algorithms. Explainability and transparency are paramount to trustworthiness in AI and ML recommendations and the ability to assist in qualified patient care decisions (Albaroudi et al. 2024). Healthcare organizations can promote transparency by providing algorithmic transparency reports, in clear and simple terms, outlining how AI and ML models are created, verified, and function, which may provide information such as algorithmic architecture and parameters, performance metrics, training data used, and use restrictions or warnings related to algorithmic predictions.

3.13.4 Clinical Validation and Regulation

The rapid growth of AI and ML technologies in the field of healthcare requires stringent regulatory oversight and comprehensive clinical validation studies to ensure patient safety and efficacy of these technologies (Li et al. 2024). AI and ML algorithms are challenged with validation and regulation—very different from how conventional medical interventions are subjected to rigorous clinical trials and government approval. The ever-changing nature of healthcare data and the iterative processes involved during each stage of the algorithms' development led to challenges in assessing the clinical validity and reliability of AI-powered tools (Bajwa et al. 2021). Clinical validation is critical to ensuring AI and ML algorithms provide accurate and relevant results in healthcare. However, classic validation techniques might be insufficient because of the pace with which these algorithms evolve. In contrast to drugs or medical devices, which are subjected to comprehensive assessments, AI algorithms must undergo continuous validation to maintain performance, as they may face updates as frequent as every 2 days for some applications (Miller et al. 2023).

Standardized frameworks and methodologies for validating AI and ML algorithms in healthcare settings are sorely needed to address these challenges. Such frameworks should encompass rigorous testing protocols, standardized performance metrics, and guidance for collecting data, building models, and designing validation studies. In addition, validation studies should reflect multiple patient populations and practical clinical settings to ensure scalability and generalizability of AI-based technologies across different healthcare environments and patient populations (Tsopra et al. 2021).

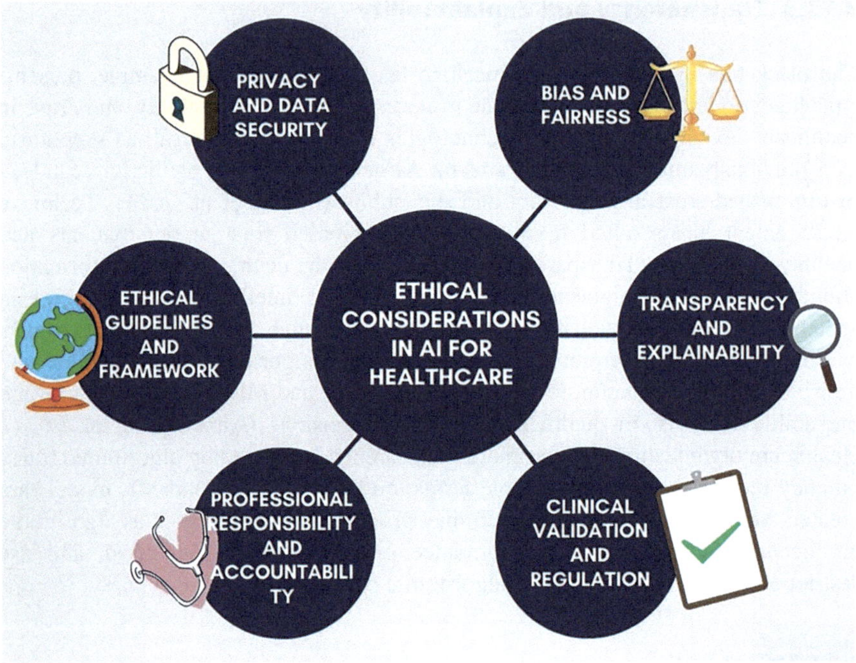

Fig. 3.3 Ethical considerations in AI for healthcare

3.13.5 Professional Responsibility and Accountability

It is important not to let AI and ML technologies be integrated into clinical practice without scrutiny, which is the ethical obligation of all healthcare professionals for patients' highest quality care and safety. Healthcare professionals, being the custodians of the patients' health and well-being, should cautiously, honorably, and ethically navigate through the complex landscape of AI and ML in healthcare (Aldoseri et al. 2023). First and foremost, it is the responsibility of healthcare professionals to be vigilant in evaluating the reliability and validity of AI-driven recommendations. Despite their potential to enhance patient outcomes, treatment planning, and diagnostic precision, AI and ML technologies are not perfect (Rajpurkar et al. 2022), as shown in Fig. 3.3.

3.13.6 Existing Frameworks and Standards of Ethics for AI in Healthcare

In the virtue of encouraging the responsible development and application of AI systems in healthcare, numerous institutions and organizations have set ethical standards, guaranteeing that these systems reduce risks while also benefiting society.

Fairness, accountability, transparency, and privacy are just a few of the ethical issues that can be addressed with these principles (Singhal et al. 2024). The World Health Organization's (WHO) Ethics and Governance of AI for Health, the IEEE's Ethically Aligned Design, and the European Commission's Ethics Guidelines for Trustworthy AI are a few of the most significant ethical frameworks (Ueda et al. 2024).

3.13.6.1 The European Commission's Ethics Guidelines for Trustworthy AI

The European Commission's *Ethics Guidelines for Trustworthy AI* provide a comprehensive framework that aligns AI development with fundamental European values, including human rights, justice, and democracy. These guidelines accentuate seven essential necessities for AI systems in healthcare (Bouderhem 2024):

- *Human agency and oversight*: AI should augment, and not replace, human decision-making, and AI systems should be subject to appropriate human oversight.
- *Technical robustness and safety*: AI systems should be secure to use, reliable, and resilient, minimizing the risks of failure or misuse.
- *Meticulousness and management of data*: AI systems should be mindful of the privacy of individuals and secure data as per the requirement of the GDPR.
- *Transparency*: AI systems should be transparent by communicating clearly about their capabilities and limitations.
- *Equal treatment, diversity, and equity*: AI systems should be fairness-promoting or avoiding bias and should use fair principles in providing design and opportunities to all.
- *Societal and Environmental well-being*: AI should be beneficial to society and the environment and not detrimental.
- *Accountability*: The developers and operators should be held accountable for the outputs generated (Solanki et al. 2023).

3.13.6.2 The World Health Organization's Ethics and Governance of Artificial Intelligence for Health

The World Health Organization (WHO) has set forth guidelines that emphasize the ethical use of AI in healthcare to promote health equity and human dignity. WHO's guidelines stress the following principles (Al Kuwaiti et al. 2023):

- *Patient-Centered Care*: AI should be used to enhance patient care and improve health outcomes without compromising patient rights or interests.
- *Global Health Equity*: AI should be leveraged to reduce health disparities globally, especially in low-resource settings.
- *Transparency and Trust*: Building public trust in AI through transparency in the design, deployment, and usage of AI systems (Maleki Varnosfaderani and Forouzanfar 2024).

3.13.6.3 IEEE's Ethically Aligned Design

The Institute of Electrical and Electronics Engineers (IEEE) has developed the *Ethically Aligned Design* framework, a set of principles aimed at ensuring the ethical development of AI technologies (Adus et al. 2023). This framework emphasizes the following principles for AI in healthcare:

- *Human Rights and Well-Being*: Artificial intelligence should be developed in a manner that prioritizes the well-being of the individuals and the communities (Bernal and Mazo 2022).
- *Bias Reduction*: Mitigating biases in AI models to ensure that decisions made by AI do not perpetuate injustice or inequality (Novelli et al. 2024).
- *Accountability*: Clear responsibility should be defined for the outcomes of AI decisions, ensuring accountability for developers, implementers, and users of AI systems (Guillaume et al. 2024).

3.14 Emerging Trends and Advancements in Medical Virology Technology

Current global health issues, which include viruses like HIV and HPV, highlight the critical need to develop effective solutions, as existing methods are often inefficient. Traditional approaches are results-oriented; however, they tend to be slow, inaccurate, and hard to switch to other tasks with never-ending resources and goals (Baryakova et al. 2023). Working as a potent subfield of AI, deep learning is revolutionizing this entire area. As we are data-rich, deep learning algorithms are able to spot intricate associations that are frequently beyond human comprehension. Such an ability not only streamlines the diagnosis but also allows for real-time personalized medicine (Huang et al. 2024). At the same time, blockchain technology establishes a secure and distributed system for protecting sensitive health information, thus guaranteeing confidentiality and integrity in an environment where data breaches are on the rise. Collectively, these technologies are likely to greatly impact the field of medical virology by enabling a proactive and efficient approach toward viral infections (Ghadi et al. 2024a).

3.14.1 Enhanced Diagnostic Accuracy

In the area of medical imaging, deep learning models, especially CNNs, have proved to be extremely competent. For example (Jiang et al. 2023b):

(a) *Detecting Viral Pneumonia*: Viral pneumonia detection has proved to be extremely accurate on DL models, especially CNNs that have been designed to interpret chest X-rays. This means that much shorter time will be spent on diagnosing, and treatment will start at shorter time (Yousef et al. 2022).

(b) *HPV Screening*: Deep learning algorithms are changing the landscape of HPV detection and screening, which is a significant cause of burden in cervical malignancy. Modern advancements in artificial intelligence have enabled various metrics to be utilized, such as the colposcopy interpretation, as they are superior in sensitivity and selectivity in comparison to cytology. These models observe these minor cellular changes to evaluate high-risk HPV types, which can help in reducing the risk of cancer (Salehi et al. 2023). Cancer development could be avoided if these tools could be paired together with telemedicine to allow patients in remote areas to assist with proper diagnostic assessment and treatment (Zhang and Qie 2023).

3.14.2 Predictive Analytics

Drawing from electronic health records, genomic data, or even social trends, machine learning algorithms can forecast viral outbreaks (Santangelo et al. 2023). For example:

(a) *Simulation of HIV Transmission*: Complex mechanisms of transmission often demand sophisticated modeling for prediction and intervention. Such AI models are developed for simulating as well as predicting transmission patterns, based on behavioral, demographic, and epidemiological data. By analyzing sexual behaviors, access to preventive methods like pre-exposure prophylaxis (PrEP), and available health structures, models give insights concerning high-risk populations (Khan and Ramsahai 2021). Predictive models can also be valuable for resource allocation by policymakers, as they could send mobile clinics or community outreach programs to areas considered very high risk of new infection. Moreover, real-time data from electronic health records enhances these predictions and creates timeliness for public health responses (Heidari et al. 2022).

(b) *Seasonal Flu Predictions*: Such queries on search engines, along with mentions across social media, have demonstrated that AI models can accurately forecast influenza trends, thus proving the worth of real-time data to control outbreaks (Shih et al. 2024).

3.14.3 Accelerated Drug Discovery

The traditional way of discovering antiviral drugs is often slow and resource heavy. However, deep learning presents a promising alternative through (Yang et al. 2021):

(a) *Drug Repurposing*: AI models sift through databases of existing medications to pinpoint candidates that might work against specific viral strains, which helps cut down the time and expenses linked to drug development (Woo et al. 2016). For example, medications that have already been approved for conditions like rheumatoid arthritis or hepatitis could potentially be effective in treating new viral diseases. This approach to repurposing also helps to navigate the regulatory challenges typically encountered in drug approval, speeding up the process of making treatments available to patients (El Morr et al. 2024).

(b) *Compound Screening*: Algorithms can forecast how effective chemical compounds will be against viral targets, making preclinical research more efficient and significantly shortening the time needed to find viable candidates (Yang et al. 2015). With AI's integration, researchers are now able to examine extensive libraries of chemical compounds, anticipate their interactions with viral proteins, and prioritize those most likely to succeed in clinical trials. This data-driven strategy not only accelerates the drug discovery process but also reduces costs, leading to more accessible treatment options (You et al. 2022).

3.14.4 Genomic Analysis

Using deep learning models, viral genomes can be analyzed to identify mutations that predispose against antiviral drugs or highlight virulence factors, giving the scientific community a stronger toolkit against virulent HIV strains (Bonsall et al. 2020). Some examples include the following:

(a) *HIV Mutation Tracking*: AI tools are continuously monitoring refractive SARS-CoV-2 mutations, enabling clinicians to customize treatment regimens in real time. Such tools analyze the genomic sequences of multiple HIV strains and identify drug resistance-associated mutations, delivering actionable recommendations for patient-specific treatment plans (Meriki et al. 2019). In this way, these models help to calibrate antiretroviral treatments to the virus's specific genetic signature to minimize the risk of treatment failure and the recurrence of resistant variants (Anstett et al. 2017).

(b) *SARS-CoV-2 Sequencing*: Complex algorithms examined large genomic datasets to identify new variants, evaluate their effects on transmissibility and immune escape, and direct changes to vaccines (Avihingsanon et al. 2022). By detecting changes in spike proteins and other important sites, scientists would be able to anticipate changes in virus activity and guide public health strategies. These results also led to the quick production of updated vaccines and monoclonal antibodies to combat emerging variants (Lindsay DeVane 2024).

(c) *Pan-Viral Genomic Analysis*: In addition to focusing on specific viruses, deep learning models are being employed to investigate genetic mutations across a wide range of viruses. These models are used to identify conserved genetic regions that might serve as universal targets for antiviral drugs or vaccines (Puccetti et al. 2024). By integrating data from various viral genomes, researchers are revealing patterns that provide insights into virus evolution, zoonotic spillover events, and mechanisms of immune evasion (Gregson et al. 2016).

3.14.5 Innovative Technologies on the Horizon

Advances in technology continue to pave the way for groundbreaking innovations in medical virology, transforming the landscape of disease prevention, diagnosis, and treatment. Here are some of the most promising technological breakthroughs and their applications (Mahara et al. 2023):

3.14.5.1 Natural Language Processing (NLP)

NLP algorithms are used to analyze unstructured data, turning it into actionable insights. Applications include the following:

- *Research Summaries*: NLP tools can comb through thousands of research articles, synthesizing findings to guide future studies and highlight trends in virology (Zuo et al. 2024).
- *Clinical Documentation*: Automating the extraction of critical information from patient records ensures more accurate diagnostics and streamlined workflows (Mermin-Bunnell et al. 2023).
- *Real-Time Clinical Decision Support*: NLP can flag potential issues in patient records, such as interactions between prescribed antiviral medications (Zhou et al. 2020).

3.14.5.2 Wearable Health Devices and Biosensors

The integration of biosensors in wearable devices provides a continuous stream of patient data, enabling proactive care (Vo and Trinh 2024):

- *Monitoring Vital Signs*: Devices track indicators like body temperature, oxygen levels, and heart rate to detect early signs of viral infections (Masoumian Hosseini et al. 2023).
- *Medication Adherence*: Alerts and reminders help ensure patients take prescribed antiviral medications on time.
- *Real-Time Immune Response Tracking*: Advanced sensors analyze biomarkers to gauge treatment effectiveness.
- *Telemedicine and Remote Care*: Telemedicine has become a vital tool for providing healthcare in the digital age.
- *AI-Enhanced Symptom Checkers*: These platforms assess symptoms and recommend actions, such as scheduling an in-person consultation or initiating antiviral therapy (Al-Halhouli et al. 2021).
- *Remote Diagnosis and Treatment*: AI tools analyze patient-submitted data, such as home testing results or imaging scans, to guide treatment plans (Islam et al. 2020).
- *Augmented Reality (AR) in Medical Training and Diagnostics*: AR technology creates immersive learning and diagnostic experiences for healthcare professionals (Ellebrecht et al. 2022).

- *Interactive Tutorials*: Training modules simulate real-world diagnostic challenges, enhancing skill development (Lv and Li 2022).
- *Enhanced Visualization*: Some overlays provide detailed views of viral structures, aiding in the interpretation of diagnostic images (Kim et al. 2024).

3.14.5.3 Blockchain for Health Data Security

The secure and decentralized nature of blockchain makes it ideal for managing sensitive health data (Ajakwe et al. 2024):

- *Data Integrity*: Blockchain is a technique that makes sure that diagnostic records are tamper-proof and only accessible to authorized personnel (Rajput et al. 2021).
- *Decentralized Patient Records*: Patients can securely share their data with healthcare providers without compromising privacy.
- *Facilitating Research Collaboration*: Blockchain enables the sharing of anonymized data between institutions, accelerating virology research (Jabarulla and Lee 2021).

3.14.5.4 Advanced Computational Models and Simulations

Simulation tools, powered by supercomputers and advanced algorithms, help predict viral behaviors and treatment outcomes (Marbouh et al. 2020):

(a) *Outbreak Modeling*: Simulations based on real-time data can forecast the trajectory of viral outbreaks, informing public health strategies (de Haro-Olmo et al. 2020).
(b) *Drug Resistance Analysis*: Computational models assess how new antiviral drugs may interact with rapidly mutating viruses, providing foresight into potential resistance mechanisms (Mongia et al. 2021).

3.14.5.5 Integration of Multi-Omics Data

The combination of proteomics, genomics, and metabolomics data is revealing intricate details about viral mechanisms (Chen et al. 2020c):

(a) *Comprehensive Analysis*: Multi-omics approaches uncover biomarkers that distinguish viral infections from other illnesses (Fortuin and Soares 2022).
(b) *Personalized Therapeutics*: Integrating diverse datasets enables the development of customized treatment plans that can be customized based on an individual's genetic and metabolic history and profile. By implementing these technologies, medical virology is on the verge of a revolution that looks quite promising toward improved diagnostics, efficient treatments, and a deeper understanding of viral diseases (Smelik et al. 2024).

3.15 Conclusion

This chapter has noted the significant repercussions that advanced computation methods may have in the future of the healthcare industry. As one of the most important components of AI, deep learning is improving processes in medical virology and virome studies by enabling researchers to analyze enormous, complex datasets at an unprecedented speed and accuracy. Such algorithms are essential, for instance, in identifying the viral genome, predicting viral and host interactions, and modeling the dynamics of viral outbreaks. In addition, the fusion of deep learning and blockchain helps to boost confidence and security because the data sharing confidentiality concerns are properly managed. With the patient blockchain approach, his or her data is secure but available for collaborative research and clinical data applications due to the immutability and transparency of deep blockchain technology.

In the field of medical virology, this technological integration can significantly enhance efforts focused on understanding viral evolution, identifying therapeutic targets, and improving vaccine development methods. Using transcriptomic and proteomic information, deep learning algorithms can identify certain patterns of viral activity, whereas blockchain ensures secure and decentralized access to datasets from multiple institutions. By deploying such systems, researchers can rise to the challenge of predicting viral resistance mechanisms and monitoring pandemic spread in real time. In this chapter, a plea is made for the acceptability and investment in these transformational technologies by scholars, practitioners, and policymakers. Interdisciplinary collaboration, maximized by equitable access to these tools, will assist in bringing the healthcare and scientific communities to bear upon their full potential toward driving meaningful advancement in virology and related fields. Thus, it responds to pressing calls in health crises and inspires hope for a safe, innovative, and resilient future in the healthcare domain.

References

Abutaleb RA, Alqahtany SS, Syed TA (2023) Integrity and privacy-aware, patient-centric health record access control framework using a blockchain. Appl Sci 13:1028. https://doi.org/10.3390/app13021028

Adebamowo SN, Adeyemo AA, Rotimi CN et al (2020) Genome-wide association study of prevalent and persistent cervical high-risk human papillomavirus (HPV) infection. BMC Med Genet 21:231. https://doi.org/10.1186/s12881-020-01156-1

Adus S, Macklin J, Pinto A (2023) Exploring patient perspectives on how they can and should be engaged in the development of artificial intelligence (AI) applications in health care. BMC Health Serv Res 23:1163. https://doi.org/10.1186/s12913-023-10098-2

Ahakonye LAC, Nwakanma CI, Kim DS (2024) Tides of blockchain in IoT cybersecurity. Sensors (Basel) 24(10):3111. https://doi.org/10.3390/s24103111

Ahmed SF, Alam MSB, Hassan M et al (2023) Deep learning modelling techniques: current progress, applications, advantages, and challenges. Artif Intell Rev 56:13521–13617. https://doi.org/10.1007/s10462-023-10466-8

Ahuja SK, Shrimankar DD, Durge AR (2023) A study and analysis of disease identification using genomic sequence processing models: an empirical review. Curr Genom 24(4):207–235. https://doi.org/10.2174/0113892029269523231101051455

Ajakwe SO, Saviour II, Ihekoronye VU, Nwankwo OU, Dini MA, Uchechi IU, Kim D-S, Lee JM (2024) Medical IoT record security and blockchain: systematic review of milieu, milestones, and momentum. Big Data Cogn Comput 8:121. https://doi.org/10.3390/bdcc8090121

Al Kuwaiti A, Nazer K, Al-Reedy A, Al-Shehri S, Al-Muhanna A, Subbarayalu AV, Al Muhanna D, Al-Muhanna FA (2023) A review of the role of artificial intelligence in healthcare. J Pers Med 13(6):951. https://doi.org/10.3390/jpm13060951

Alahi MEE, Sukkuea A, Tina FW, Nag A, Kurdthongmee W, Suwannarat K, Mukhopadhyay SC (2023) Integration of IoT-enabled technologies and artificial intelligence (AI) for Smart City scenario: recent advancements and future trends. Sensors 23:5206. https://doi.org/10.3390/s23115206

Albaroudi E, Mansouri T, Alameer A (2024) A comprehensive review of AI techniques for addressing algorithmic bias in job hiring. AI 5:383–404. https://doi.org/10.3390/ai5010019

Albshaier L, Almarri S, Hafizur Rahman MM (2024) A review of blockchain's role in E-commerce transactions: open challenges, and future research directions. Computers 13:27. https://doi.org/10.3390/computers13010027

Aldoseri A, Al-Khalifa KN, Hamouda AM (2023) Re-thinking data strategy and integration for artificial intelligence: concepts, opportunities, and challenges. Appl Sci 13:7082. https://doi.org/10.3390/app13127082

Al-Halhouli A, Albagdady A, Alawadi J, Abeeleh MA (2021) Monitoring symptoms of infectious diseases: perspectives for printed wearable sensors. Micromachines 12:620. https://doi.org/10.3390/mi12060620

Alharbi F, Vakanski A (2023) Machine learning methods for cancer classification using gene expression data: a review. Bioengineering (Basel) 10(2):173. https://doi.org/10.3390/bioengineering10020173

Ali A, Ali H, Saeed A, Ahmed Khan A, Tin TT, Assam M, Ghadi YY, Mohamed HG (2023) Blockchain-powered healthcare systems: enhancing scalability and security with hybrid deep learning. Sensors 23:7740. https://doi.org/10.3390/s23187740

Alizadehmohajer N, Zahedifar S, Sohrabi E, Shaddel Basir S, Nourigheimasi S, Falak R, Nedaeinia RA, Ferns G, Emami Nejad A, Manian M (2023) Using in Silico bioinformatics algorithms for the accurate prediction of the impact of spike protein mutations on the pathogenicity, stability, and functionality of the SARS-CoV-2 virus and analysis of potential therapeutic targets. Biochem Genet 61(2):778–808. https://doi.org/10.1007/s10528-022-10282-9

Aljohani A (2023) Predictive analytics and machine learning for real-time supply chain risk mitigation and agility. Sustain For 15:15088. https://doi.org/10.3390/su152015088

Allahqoli L, Laganà AS, Mazidimoradi A, Salehiniya H, Günther V, Chiantera V, Karimi Goghari S, Ghiasvand MM, Rahmani A, Momenimovahed Z, Alkatout I (2022) Diagnosis of cervical cancer and pre-cancerous lesions by artificial intelligence: a systematic review. Diagnostics (Basel). 12(11):2771. https://doi.org/10.3390/diagnostics12112771

Alnmr A, Ray R, Alzawi MO (2024) A novel approach to swell mitigation: machine-learning-powered optimal unit weight and stress prediction in expansive soils. Appl Sci 14:1411. https://doi.org/10.3390/app14041411

Alotaibi YK, Federico F (2017) The impact of health information technology on patient safety. Saudi Med J 38(12):1173–1180. https://doi.org/10.15537/smj.2017.12.20631

Alowais SA, Alghamdi SS, Alsuhebany N et al (2023) Revolutionizing healthcare: the role of artificial intelligence in clinical practice. BMC Med Educ 23:689. https://doi.org/10.1186/s12909-023-04698-z

Alzubaidi L, Zhang J, Humaidi AJ et al (2021) Review of deep learning: concepts, CNN architectures, challenges, applications, future directions. J Big Data 8:53. https://doi.org/10.1186/s40537-021-00444-8

An Q, Chen W, Shao W (2024) A deep convolutional neural network for pneumonia detection in X-ray images with attention ensemble. Diagnostics (Basel). 14(4):390. https://doi.org/10.3390/diagnostics14040390

Andrew J, Isravel DP, Sagayam KM, Bhushan B, Sei Y, Eunice J (2023) Blockchain for healthcare systems: architecture, security challenges, trends and future directions. J Netw Comput Appl 215:103633. ISSN 1084-8045. https://doi.org/10.1016/j.jnca.2023.103633

Anstett K, Brenner B, Mesplede T, Wainberg MA (2017) HIV drug resistance against strand transfer integrase inhibitors. Retrovirology 14:36. https://doi.org/10.1186/s12977-017-0360-7

Araújo T, Aresta G, Castro E, Rouco J, Aguiar P et al (2017) Classification of breast cancer histology images using convolutional neural networks. PLoS One 12(6):e0177544. https://doi.org/10.1371/journal.pone.0177544

Arora A, Arora A (2022) Generative adversarial networks and synthetic patient data: current challenges and future perspectives. Future Healthc J 9(2):190–193. https://doi.org/10.7861/fhj.2022-0013

Arora A, Alderman JE, Palmer J, Ganapathi S, Laws E, McCradden MD, Oakden-Rayner L, Pfohl SR, Ghassemi M, McKay F, Treanor D, Rostamzadeh N, Mateen B, Gath J, Adebajo AO, Kuku S, Matin R, Heller K, Sapey E, Sebire NJ, Cole-Lewis H, Calvert M, Denniston A, Liu X (2023) The value of standards for health datasets in artificial intelligence-based applications. Nat Med 29(11):2929–2938. https://doi.org/10.1038/s41591-023-02608-w

Askr H, Elgeldawi E, Aboul Ella H et al (2023) Deep learning in drug discovery: an integrative review and future challenges. Artif Intell Rev 56:5975–6037. https://doi.org/10.1007/s10462-022-10306-1

Atlam HF, Ekuri N, Azad MA, Lallie HS (2024) Blockchain forensics: a systematic literature review of techniques, applications, challenges, and future directions. Electronics 13:3568. https://doi.org/10.3390/electronics13173568

Avihingsanon A, Hughes MD, Salata R, Godfrey C, McCarthy C, Mugyenyi P et al (2022) Third-line antiretroviral therapy, including raltegravir (RAL), darunavir (DRV/r) and/or etravirine (ETR), is well tolerated and achieves durable virologic suppression over 144 weeks in resource-limited settings: ACTG A5288 strategy trial. J Int AIDS Soc 25:e25905. https://doi.org/10.1002/jia2.25905

Ayub Khan A, Laghari AA, Shaikh AA, Bourouis S, Mamlouk AM, Alshazly H (2021) Educational blockchain: a secure degree attestation and verification traceability architecture for higher education commission. Appl Sci 11:10917. https://doi.org/10.3390/app112210917

Bajwa J, Munir U, Nori A, Williams B (2021) Artificial intelligence in healthcare: transforming the practice of medicine. Future Healthc J 8(2):e188–e194. https://doi.org/10.7861/fhj.2021-0095

Baker RE, Mahmud AS, Miller IF et al (2022) Infectious disease in an era of global change. Nat Rev Microbiol 20:193–205. https://doi.org/10.1038/s41579-021-00639-z

Baryakova TH, Pogostin BH, Langer R et al (2023) Overcoming barriers to patient adherence: the case for developing innovative drug delivery systems. Nat Rev Drug Discov 22:387–409. https://doi.org/10.1038/s41573-023-00670-0

Basil NN, Ambe S, Ekhator C, Fonkem E (2022) Health records database and inherent security concerns: a review of literature. Cureus 14(10):e30168. https://doi.org/10.7759/cureus.30168

Bekbolatova M, Mayer J, Ong CW, Toma M (2024) Transformative potential of AI in healthcare: definitions, applications, and navigating the ethical landscape and public perspectives. Healthcare 12:125. https://doi.org/10.3390/healthcare12020125

Belenguer L (2022) AI bias: exploring discriminatory algorithmic decision-making models and the application of possible machine-centric solutions adapted from the pharmaceutical industry. AI Ethics 2:771–787. https://doi.org/10.1007/s43681-022-00138-8

Beltrán JF, Herrera-Belén L, Yáñez AJ et al (2024) Prediction of viral oncoproteins through the combination of generative adversarial networks and machine learning techniques. Sci Rep 14: 27108. https://doi.org/10.1038/s41598-024-77028-y

Bennis SL, Yared NF, Horvath KJ, Baker JV, Waterboer T, Thyagarajan B, Kulasingam S (2024) HPV vaccination status in HIV-negative MSM and its association with high-risk HPV detection using HPV serology and anorectal swabs. Vaccines (Basel) 12(10):1154. https://doi.org/10.3390/vaccines12101154

Berahmand K, Daneshfar F, Salehi ES et al (2024) Autoencoders and their applications in machine learning: a survey. Artif Intell Rev 57:28. https://doi.org/10.1007/s10462-023-10662-6

Bernal J, Mazo C (2022) Transparency of artificial intelligence in healthcare: insights from professionals in computing and healthcare worldwide. Appl Sci 12:10228. https://doi.org/10.3390/app122010228

Biró A, Cuesta-Vargas AI, Szilágyi L (2024) AI-assisted fatigue and stamina control for performance sports on IMU-generated multivariate times series datasets. Sensors 24:132. https://doi.org/10.3390/s24010132

Biswas MR, Shah Z (2023) Extracting factors associated with vaccination from Twitter data and mapping to behavioral models. Hum Vaccin Immunother 19(3):2281729. https://doi.org/10.1080/21645515.2023.2281729

Blanco-González A, Cabezón A, Seco-González A, Conde-Torres D, Antelo-Riveiro P, Piñeiro Á, Garcia-Fandino R (2023) The role of AI in drug discovery: challenges, opportunities, and strategies. Pharmaceuticals (Basel) 16(6):891. https://doi.org/10.3390/ph16060891

Bolukbasi T, Chang K-W, Zou JY, Saligrama V, Kalai AT (2016) Man is to computer programmer as woman is to homemaker? Debiasing word embeddings. In: Lee DD, Sugiyama M, Luxburg UV, Guyon I, Garnett R (eds) Advances in neural information processing systems 29. Proceedings of the 30th Conference on Neural Information Processing Systems (NIPS 2016), Barcelona, pp 4349–4357. https://papers.nips.cc/paper/6228-man-is-to-computer-programmer-as-woman-is-to-homemaker-debiasing-word-embeddings.pdf. Accessed 31 Oct 2019

Bonsall D, Golubchik T, de Cesare M, Limbada M, Kosloff B, MacIntyre-Cockett G, Hall M, Wymant C, Ansari MA, Abeler-Dörner L, Schaap A, Brown A, Barnes E, Piwowar-Manning E, Eshleman S, Wilson E, Emel L, Hayes R, Fidler S, Ayles H, Bowden R, Fraser C (2020) HPTN 071 (PopART) team. A comprehensive genomics solution for HIV surveillance and clinical monitoring in low-income settings. J Clin Microbiol 58(10):e00382–e00320. https://doi.org/10.1128/JCM.00382-20

Bouabida K, Chaves BG, Anane E (2023) Challenges and barriers to HIV care engagement and care cascade: viewpoint. Front Reprod Health 5:1201087. https://doi.org/10.3389/frph.2023.1201087

Bouderhem R (2024) Shaping the future of AI in healthcare through ethics and governance. Humanit Soc Sci Commun 11:416. https://doi.org/10.1057/s41599-024-02894-w

Brice SN, Harper PR, Gartner D, Behrens DA (2023) Modeling disease progression and treatment pathways for depression jointly using agent-based modeling and system dynamics. Front Public Health 10:1011104. https://doi.org/10.3389/fpubh.2022.1011104

Brickman C, Palefsky JM (2015) Human papillomavirus in the HIV-infected host: epidemiology and pathogenesis in the antiretroviral era. Curr HIV/AIDS Rep 12(1):6–15. https://doi.org/10.1007/s11904-014-0254-4

Burmeister CA, Khan SF, Schäfer G, Mbatani N, Adams T, Moodley J, Prince S (2022) Cervical cancer therapies: current challenges and future perspectives. Tumour Virus Res 13:200238. https://doi.org/10.1016/j.tvr.2022.200238

Byrd KK, Hou JG, Bush T, Hazen R, Kirkham H, Delpino A, Weidle PJ, Shankle MD, Camp NM, Suzuki S, Clay PG, Patient-centered HIV Care Model Team (2020) Adherence and viral suppression among participants of the patient-centered human immunodeficiency virus (HIV) care model project: a collaboration between community-based pharmacists and HIV clinical providers. Clin Infect Dis 70(5):789–797. https://doi.org/10.1093/cid/ciz276

Campos FS, de Arruda LB, da Fonseca FG (2022) Special issue "viral infections in developing countries". Viruses 14(2):405. https://doi.org/10.3390/v14020405

Cao K, Zhang T, Huang J (2024) Advanced hybrid LSTM-transformer architecture for real-time multi-task prediction in engineering systems. Sci Rep 14:4890. https://doi.org/10.1038/s41598-024-55483-x

Carini C, Seyhan AA (2024) Tribulations and future opportunities for artificial intelligence in precision medicine. J Transl Med 22:411. https://doi.org/10.1186/s12967-024-05067-0

Chabria SB, Gupta S, Kozal MJ (2014) Deep sequencing of HIV: clinical and research applications. Annu Rev Genomics Hum Genet 15:295–325. https://doi.org/10.1146/annurev-genom-091212-153406

Chai P, Hou L, Zhang G, Tushar Q, Yang Z (2024) Generative adversarial networks in construction applications. Autom Constr 159:105265. ISSN 0926-5805. https://doi.org/10.1016/j.autcon.2024.105265

Chakraborty C, Bhattacharya M, Pal S, Lee S-S (2024) From machine learning to deep learning: advances of the recent data-driven paradigm shift in medicine and healthcare. Curr Res Biotechnol 7:100164. ISSN 2590-2628. https://doi.org/10.1016/j.crbiot.2023.100164

Chen S, Guo W (2023) Auto-encoders in deep learning—a review with new perspectives. Mathematics 11:1777. https://doi.org/10.3390/math11081777

Chen J, Wang R, Wang M, Wei GW (2020a) Mutations strengthened SARS-CoV-2 infectivity. J Mol Biol 432(19):5212–5226. https://doi.org/10.1016/j.jmb.2020.07.009

Chen W, Chen Y, Tsangaratos P, Ilia I, Wang X (2020b) Combining evolutionary algorithms and machine learning models in landslide susceptibility assessments. Remote Sens 12:3854. https://doi.org/10.3390/rs12233854

Chen C, Wang J, Pan D, Wang X, Xu Y, Yan J, Wang L, Yang X, Yang M, Liu GP (2020c) Applications of multi-omics analysis in human diseases. MedComm 4(4):e315. https://doi.org/10.1002/mco2.315

Chen Z, Zhang J, Liu J, Dai Y, Lee D, Min MR, Xu M, Gerstein M (2021) DECODE: a deep-learning framework for condensing enhancers and refining boundaries with large-scale functional assays. Bioinformatics 37(Suppl_1):i280–i288. https://doi.org/10.1093/bioinformatics/btab283

Chen RJ, Wang JJ, Williamson DFK, Chen TY, Lipkova J, Lu MY, Sahai S, Mahmood F (2023a) Algorithmic fairness in artificial intelligence for medicine and healthcare. Nat Biomed Eng 7(6): 719–742. https://doi.org/10.1038/s41551-023-01056-8

Chen Y, Clayton EW, Novak LL, Anders S, Malin B (2023b) Human-centered design to address biases in artificial intelligence. J Med Internet Res 25:e43251. https://doi.org/10.2196/43251

Cho HN, Jun TJ, Kim YH, Kang H, Ahn I, Gwon H, Kim Y, Seo J, Choi H, Kim M, Han J, Kee G, Park S, Ko S (2024) Task-specific transformer-based language models in health care: scoping review. JMIR Med Inform 12:e49724. https://doi.org/10.2196/49724

Choi SR, Lee M (2023) Transformer architecture and attention mechanisms in genome data analysis: a comprehensive review. Biology (Basel) 12(7):1033. https://doi.org/10.3390/biology12071033

Choubey S, Barde S, Badholia A (2022) Enhancing the prediction efficiency of virus borne diseases using enhanced backpropagation with an artificial neural network. Meas Sens 24:100505. ISSN 2665-9174. https://doi.org/10.1016/j.measen.2022.100505

Clauwaert J, Menschaert G, Waegeman W (2021) Explainability in transformer models for functional genomics. Brief Bioinform 22(5):bbab060. https://doi.org/10.1093/bib/bbab060

Clipman SJ, Solomon SS, Srikrishnan AK, McFall AM, Gomathi S, Saravanan S, Anand S, Vasudevan CK, Kumar MS, Celentano DD, Mehta SH, Lucas GM (2022) Antiretroviral drug resistance in HIV sequences from people who inject drugs and men who have sex with men across 21 cities in India. Open Forum Infect Dis 9(10):ofac481. https://doi.org/10.1093/ofid/ofac481

Combes JD, Franceschi S (2018) Human papillomavirus genome variants and head and neck cancers: a perspective. Infect Agents Cancer 13:13. https://doi.org/10.1186/s13027-018-0185-6

Coyner AS, Chen JS, Chang K, Singh P, Ostmo S, RVP C, Chiang MF, Kalpathy-Cramer J, Campbell JP (2022) Imaging and informatics in retinopathy of prematurity consortium. synthetic medical images for robust, privacy-preserving training of artificial intelligence: application to retinopathy of prematurity diagnosis. Ophthalmol Sci 2(2):100126. https://doi.org/10.1016/j.xops

Cross JL, Choma MA, Onofrey JA (2024) Bias in medical AI: implications for clinical decision-making. PLOS Digit Health 3(11):e0000651. https://doi.org/10.1371/journal.pdig.0000651

Dara S, Dhamercherla S, Jadav SS et al (2022) Machine learning in drug discovery: a review. Artif Intell Rev 55:1947–1999. https://doi.org/10.1007/s10462-021-10058-4

Davids J, Ashrafian H (2022) AIM in neurodegenerative diseases: Parkinson and Alzheimer. In: Lidströmer N, Ashrafian H (eds) Artificial intelligence in medicine. Springer, Cham. https://doi.org/10.1007/978-3-030-64573-1_190

de Haro-Olmo FJ, Varela-Vaca ÁJ, Álvarez-Bermejo JA (2020) Blockchain from the perspective of privacy and anonymisation: a systematic literature review. Sensors 20:7171. https://doi.org/10.3390/s20247171

Deeks S, Overbaugh J, Phillips A et al (2015) HIV infection. Nat Rev Dis Primers 1:15035. https://doi.org/10.1038/nrdp.2015.35

Deeks SG, Archin N, Cannon P et al (2021) Research priorities for an HIV cure: international AIDS society global scientific strategy 2021. Nat Med 27:2085–2098. https://doi.org/10.1038/s41591-021-01590-5

Dellino M, Cerbone M, d'Amati A, Bochicchio M, Laganà AS, Etrusco A, Malvasi A, Vitagliano A, Pinto V, Cicinelli E et al (2024) Artificial intelligence in cervical cancer screening: opportunities and challenges. AI 5:2984–3000. https://doi.org/10.3390/ai5040144

Dessie ZG, Zewotir T, Mwambi H et al (2020) Modelling of viral load dynamics and CD4 cell count progression in an antiretroviral naive cohort: using a joint linear mixed and multistate Markov model. BMC Infect Dis 20:246. https://doi.org/10.1186/s12879-020-04972-1

Devasia J, Goswami H, Lakshminarayanan S et al (2023) Deep learning classification of active tuberculosis lung zones wise manifestations using chest X-rays: a multi label approach. Sci Rep 13:887. https://doi.org/10.1038/s41598-023-28079-0

Dias R, Torkamani A (2019) Artificial intelligence in clinical and genomic diagnostics. Genome Med 11:70. https://doi.org/10.1186/s13073-019-0689-8

Dörrich M, Hecht M, Fietkau R et al (2023) Explainable convolutional neural networks for assessing head and neck cancer histopathology. Diagn Pathol 18:121. https://doi.org/10.1186/s13000-023-01407-8

El Morr C, Ozdemir D, Asdaah Y, Saab A, El-Lahib Y, Sokhn ES (2024) AI-based epidemic and pandemic early warning systems: a systematic scoping review. Health Informatics J 30(3). https://doi.org/10.1177/14604582241275844

Ellebrecht DB, Gola D, Kaschwich M (2022) Evaluation of a wearable in-ear sensor for temperature and heart rate monitoring: a pilot study. J Med Syst 46:91. https://doi.org/10.1007/s10916-022-01872-6

Elmes A, Alemohammad H, Avery R, Caylor K, Eastman JR, Fishgold L, Friedl MA, Jain M, Kohli D, Laso Bayas JC et al (2020) Accounting for training data error in machine learning applied to earth observations. Remote Sens 12:1034. https://doi.org/10.3390/rs12061034

Elste J, Saini A, Mejia-Alvarez R, Mejía A, Millán-Pacheco C, Swanson-Mungerson M, Tiwari V (2024) Significance of artificial intelligence in the study of virus–host cell interactions. Biomol Ther 14:911. https://doi.org/10.3390/biom14080911

Elvas LB, Serrão C, Ferreira JC (2023) Sharing health information using a blockchain. Healthcare (Basel) 11(2):170. https://doi.org/10.3390/healthcare11020170

Ennab M, Mcheick H (2022) Designing an interpretability-based model to explain the artificial intelligence algorithms in healthcare. Diagnostics (Basel). 12(7):1557. https://doi.org/10.3390/diagnostics12071557

Espinoza H, Ha KT, Pham TT, Espinoza JL (2021) Genetic predisposition to persistent human papillomavirus-infection and virus-induced cancers. Microorganisms 2092:9. https://doi.org/10.3390/microorganisms9102092

Ezechi OC, Ostergren PO, Nwaokorie FO et al (2014) The burden, distribution and risk factors for cervical oncogenic human papilloma virus infection in HIV positive Nigerian women. Virol J 11:5. https://doi.org/10.1186/1743-422X-11-5

Falkenhagen A, Joshi S (2018) Genetic strategies for HIV treatment and prevention. Mol Ther Nucleic Acids 13:514–533. https://doi.org/10.1016/j.omtn.2018.09.018

Fan T, Zhang M, Yang J et al (2023) Therapeutic cancer vaccines: advancements, challenges and prospects. Sig Transduct Target Ther 8:450. https://doi.org/10.1038/s41392-023-01674-3

Fatima S, Reddy CKK, Hanafiah MM, Mohana RM (2025) Bio-inspired algorithms-based machine learning and deep learning models in healthcare 6.0. In: Gunjan VK, Senatore S, Kumar A (eds) Cybernetics, human cognition, and machine learning in communicative applications, Cognitive science and technology. Springer, Singapore. https://doi.org/10.1007/978-981-97-8533-9_9

Ferrara E (2024) Fairness and bias in artificial intelligence: a brief survey of sources, impacts, and mitigation strategies. Sci 6:3. https://doi.org/10.3390/sci6010003

Fortuin S, Soares NC (2022) The integration of proteomics and metabolomics data paving the way for a better understanding of the mechanisms underlying microbial acquired drug resistance. Front Med (Lausanne) 9:849838. https://doi.org/10.3389/fmed.2022.849838

Foxley-Marrable M, D'Cruz L, Meredith P, Glaysher S, Beckett AH, Goudarzi S, Fearn C, Cook KF, Loveson KF, Dent H, Paul H, Elliott S, Wyllie S, Lloyd A, Bicknell K, Lumley S, McNicholas J, Prytherch D, COVID-19 Genomics UK (COG-UK) consortium, Lundgren A, Graur O, Chauhan AJ, Robson SC (2023) Combining viral genomics and clinical data to assess risk factors for severe COVID-19 (mortality, ICU admission, or intubation) amongst hospital patients in a large acute UK NHS hospital Trust. PLoS One 18(3):e0283447. https://doi.org/10.1371/journal.pone.0283447

Fröhlich H, Balling R, Beerenwinkel N et al (2018) From hype to reality: data science enabling personalized medicine. BMC Med 16:150. https://doi.org/10.1186/s12916-018-1122-7

Gandhi RT, Bedimo R, Hoy JF et al (2023) Antiretroviral drugs for treatment and prevention of HIV infection in adults: 2022 recommendations of the international antiviral society–USA panel. JAMA 329(1):63–84. https://doi.org/10.1001/jama.2022.22246

Ghadi YY, Mazhar T, Shahzad T, Amir Khan M, Abd-Alrazaq A, Ahmed A, Hamam H (2024a) The role of blockchain to secure internet of medical things. Sci Rep 14(1):18422. https://doi.org/10.1038/s41598-024-68529-x. Erratum in: Sci Rep 2024 Sep 12;14(1):21337. https://doi.org/10.1038/s41598-024-71990-3

Giancotti M, Lopreite M, Mauro M, Puliga M (2024) Innovating health prevention models in detecting infectious disease outbreaks through social media data: an umbrella review of the evidence. Front Public Health 12:1435724. https://doi.org/10.3389/fpubh.2024.1435724

Glingston RS, Deb R, Kumar S, Nagotu S (2019) Organelle dynamics and viral infections: at crossroads. Microbes Infect 21(1):20–32. https://doi.org/10.1016/j.micinf.2018.06.002

Goergen CJ, Tweardy MJ, Steinhubl SR, Wegerich SW, Singh K, Mieloszyk RJ, Dunn J (2022) Detection and monitoring of viral infections via wearable devices and biometric data. Annu Rev Biomed Eng 24:1–27. https://doi.org/10.1146/annurev-bioeng-103020-040136

Goyal M, Mahmoud QH (2024) A systematic review of synthetic data generation techniques using generative AI. Electronics 13:3509. https://doi.org/10.3390/electronics13173509

Gregson J, Tang M, Ndembi N, Hamers RL, Rhee S-Y, Marconi VC, Diero L et al (2016) Global epidemiology of drug resistance after failure of WHO recommended first-line regimens for adult HIV-1 infection: a multicentre retrospective cohort study. Lancet Infect Dis 16(5):565–575

Guillaume D, Waheed DE, Schleiff M, Muralidharan KK, Vorsters A, Limaye RJ (2024) Global perspectives of determinants influencing HPV vaccine introduction and scale-up in low- and middle-income countries. PLoS One 19(1):e0291990. https://doi.org/10.1371/journal.pone.0291990

Gürsoy G, Brannon CM, Ni E et al (2022) Storing and analyzing a genome on a blockchain. Genome Biol 23:134. https://doi.org/10.1186/s13059-022-02699-7
Habehh H, Gohel S (2021) Machine learning in healthcare. Curr Genomics 22(4):291–300. https://doi.org/10.2174/1389202922666210705124359
Hasin Y, Seldin M, Lusis A (2017) Multi-omics approaches to disease. Genome Biol 18:83. https://doi.org/10.1186/s13059-017-1215-1
Hayawi K, Shahriar S, Alashwal H, Serhani MA (2024) Generative AI and large language models: a new frontier in reverse vaccinology. Inform Med Unlocked 48:101533. ISSN 2352-9148. https://doi.org/10.1016/j.imu.2024.101533
Hays P (2024) Artificial intelligence in cytopathological applications for cancer: a review of accuracy and analytic validity. Eur J Med Res 29:553. https://doi.org/10.1186/s40001-024-02138-2
Heidari A, Jafari Navimipour N, Unal M et al (2022) Machine learning applications for COVID-19 outbreak management. Neural Comput & Applic 34:15313–15348. https://doi.org/10.1007/s00521-022-07424-w
Hernandez V, Nasser L, Do C, Lee W-C (2024) Healing the whole: an international review of the collaborative care model between primary care and psychiatry. Healthcare 12:1679. https://doi.org/10.3390/healthcare12161679
Hewavisenti RV, Arena J, Ahlenstiel CL, Sasson SC (2023) Human papillomavirus in the setting of immunodeficiency: pathogenesis and the emergence of next-generation therapies to reduce the high associated cancer risk. Front Immunol 14:1112513. https://doi.org/10.3389/fimmu.2023.1112513
Hou X, Shen G, Zhou L, Li Y, Wang T, Ma X (2022) Artificial intelligence in cervical cancer screening and diagnosis. Front Oncol 12:851367. https://doi.org/10.3389/fonc.2022
Huang Y, Guo X, Wu Y et al (2024) Nanotechnology's frontier in combatting infectious and inflammatory diseases: prevention and treatment. Sig Transduct Target Ther 9:34. https://doi.org/10.1038/s41392-024-01745-z
Hussain S, Mubeen I, Ullah N, Shah SSUD, Khan BA, Zahoor M, Ullah R, Khan FA, Sultan MA (2022) Modern diagnostic imaging technique applications and risk factors in the medical field: a review. Biomed Res Int 2022:5164970. https://doi.org/10.1155/2022/5164970
Ibrahim AU, Ozsoz M, Serte S et al (2024) Pneumonia classification using deep learning from chest X-ray images during COVID-19. Cogn Comput 16:1589–1601. https://doi.org/10.1007/s12559-020-09787-5
Islam M, Mahmud S, Muhammad LJ et al (2020) Wearable technology to assist the patients infected with novel coronavirus (COVID-19). SN Comput Sci 1:320. https://doi.org/10.1007/s42979-020-00335-4
Ivanisevic T, Sewduth RN (2023) Multi-omics integration for the design of novel therapies and the identification of novel biomarkers. Proteomes 11(4):34. https://doi.org/10.3390/proteomes11040034
Ivanov S, Lagunin A, Filimonov D, Tarasova O (2020) Network-based analysis of OMICs data to understand HIV-host interaction. Front Microbiol 11:1314. https://doi.org/10.3389/fmicb.2020.01314
Jabarulla MY, Lee HN (2021) A blockchain and artificial intelligence-based, patient-centric healthcare system for combating the COVID-19 pandemic: opportunities and applications. Healthcare (Basel) 9(8):1019. https://doi.org/10.3390/healthcare9081019
Javed H, El-Sappagh S, Abuhmed T (2025) Robustness in deep learning models for medical diagnostics: security and adversarial challenges towards robust AI applications. Artif Intell Rev 58:12. https://doi.org/10.1007/s10462-024-11005-9
Jayawickrama SM, Ranaweera PM, Pradeep RGGR, Jayasinghe YA, Senevirathna K, Hilmi AJ, Rajapakse RMG, Kanmodi KK, Jayasinghe RD (2024) Developments and future prospects of personalized medicine in head and neck squamous cell carcinoma diagnoses and treatments. Cancer Rep 7(3):e2045. https://doi.org/10.1002/cnr2.2045

Ji Y, Zhou Z, Liu H, Davuluri RV (2021) DNABERT: pre-trained bidirectional encoder representations from transformers model for DNA-language in genome. Bioinformatics 37: 2112–2120

Jia L, Li L, Gui T et al (2016) Analysis of HIV-1 intersubtype recombination breakpoints suggests region with high pairing probability may be a more fundamental factor than sequence similarity affecting HIV-1 recombination. Virol J 13:156. https://doi.org/10.1186/s12985-016-0616-1

Jiang P, Li X, Shen H et al (2023a) A systematic review of deep learning-based cervical cytology screening: from cell identification to whole slide image analysis. Artif Intell Rev 56(Suppl 2): 2687–2758. https://doi.org/10.1007/s10462-023-10588-z

Jiang X, Hu Z, Wang S, Zhang Y (2023b) Deep learning for medical image-based cancer diagnosis. Cancers (Basel) 15(14):3608. https://doi.org/10.3390/cancers15143608

Jierula A, Wang S, Oh T-M, Wang P (2021) Study on accuracy metrics for evaluating the predictions of damage locations in deep piles using artificial neural networks with acoustic emission data. Appl Sci 11:2314. https://doi.org/10.3390/app11052314

Jin XL, Zhang M, Zhou Z, Yu X (2019) Application of a blockchain platform to manage and secure personal genomic data: a case study of LifeCODE.ai in China. J Med Internet Res 21(9):e13587. https://doi.org/10.2196/13587

Jo T, Nho K, Saykin AJ (2019) Deep learning in Alzheimer's disease: diagnostic classification and prognostic prediction using neuroimaging data. Front Aging Neurosci 11:220. https://doi.org/10.3389/fnagi.2019.00220

Johnson MM, Jones CE, Clark DN (2022) The effect of treatment-associated mutations on HIV replication and transmission cycles. Viruses 15(1):107. https://doi.org/10.3390/v15010107

Junaid SB, Imam AA, Balogun AO, De Silva LC, Surakat YA, Kumar G, Abdulkarim M, Shuaibu AN, Garba A, Sahalu Y, Mohammed A, Mohammed TY, Abdulkadir BA, Abba AA, Kakumi NAI, Mahamad S (2022) Recent advancements in emerging technologies for healthcare management systems: a survey. Healthcare (Basel) 10(10):1940. https://doi.org/10.3390/healthcare10101940

Kalinin AA, Higgins GA, Reamaroon N, Soroushmehr S, Allyn-Feuer A, Dinov ID, Najarian K, Athey BD (2018) Deep learning in pharmacogenomics: from gene regulation to patient stratification. Pharmacogenomics 19(7):629–650. https://doi.org/10.2217/pgs-2018-0008

Kasyapa MSB, Vanmathi C (2024) Blockchain integration in healthcare: a comprehensive investigation of use cases, performance issues, and mitigation strategies. Front Digit Health 6: 1359858. https://doi.org/10.3389/fdgth.2024

Katona K, Neamah HA, Korondi P (2024) Obstacle avoidance and path planning methods for autonomous navigation of mobile robot. Sensors 24:3573. https://doi.org/10.3390/s24113573

Kayikci S, Khoshgoftaar TM (2024) Blockchain meets machine learning: a survey. J Big Data 11:9. https://doi.org/10.1186/s40537-023-00852-y

Khan K, Ramsahai E (2021) Maintaining proper health records improves machine learning predictions for novel 2019-nCoV. BMC Med Inform Decis Mak 21:172. https://doi.org/10.1186/s12911-021-01537-3

Khan BUI, Goh KW, Khan AR, Zuhairi MF, Chaimanee M (1825) Integrating AI and blockchain for enhanced data security in IoT-driven smart cities. Processes 2024:12. https://doi.org/10.3390/pr12091825

Kiliçarslan S, Celik M (2024) Parametric RSigELU: a new trainable activation function for deep learning. Neural Comput & Applic 36:7595–7607. https://doi.org/10.1007/s00521-024-09538-9

Kim S-K, Huh J-H (2020) Artificial neural network blockchain techniques for healthcare system: focusing on the personal health records. Electronics 9:763. https://doi.org/10.3390/electronics9050763

Kim D, Lee J, Park MK et al (2024) Recent developments in wearable breath sensors for healthcare monitoring. Commun Mater 5:41. https://doi.org/10.1038/s43246-024-00480-w

Klontzas ME, Kalarakis G, Koltsakis E, Papathomas T, Karantanas AH, Tzortzakakis A (2024) Convolutional neural networks for the differentiation between benign and malignant renal tumors with a multicenter international computed tomography dataset. Insights Imaging 15(1):26. https://doi.org/10.1186/s13244-023-01601-8
Koel M, Võsa U, Jõeloo M, Läll K, Gualdo NP, Laivuori H, Lemmelä S, Estonian Biobank Research Team, Gen F, Daly M, Palta P, Mägi R, Laisk T (2023) GWAS meta-analyses clarify the genetics of cervical phenotypes and inform risk stratification for cervical cancer. Hum Mol Genet 32(12):2103–2116. https://doi.org/10.1093/hmg/ddad043
Koh DM, Papanikolaou N, Bick U et al (2022) Artificial intelligence and machine learning in cancer imaging. Commun Med 2:133. https://doi.org/10.1038/s43856-022-00199-0
Kolou M, Poda A, Diallo Z et al (2021) Prevalence of human leukocyte antigen HLA-B*57:01 in individuals with HIV in West and Central Africa. BMC Immunol 22:48. https://doi.org/10.1186/s12865-021-00427-7
Kourounis G, Elmahmudi AA, Thomson B, Hunter J, Ugail H, Wilson C (2023) Computer image analysis with artificial intelligence: a practical introduction to convolutional neural networks for medical professionals. Postgrad Med J 99(1178):1287–1294. https://doi.org/10.1093/postmj/qgad095
Kufel J, Bargieł-Łączek K, Kocot S, Koźlik M, Bartnikowska W, Janik M, Czogalik Ł, Dudek P, Magiera M, Lis A, Paszkiewicz I, Nawrat Z, Cebula M, Gruszczyńska K (2023) What is machine learning, artificial neural networks and deep learning?-examples of practical applications in medicine. Diagnostics (Basel) 13(15):2582. https://doi.org/10.3390/diagnostics13152582
Kumar N, Srivastava R (2024) Deep learning in structural bioinformatics: current applications and future perspectives. Brief Bioinform 25(3):bbae042. https://doi.org/10.1093/bib/bbae042
Kumar Y, Koul A, Singla R, Ijaz MF (2023) Artificial intelligence in disease diagnosis: a systematic literature review, synthesizing framework and future research agenda. J Ambient Intell Humaniz Comput 14(7):8459–8486. https://doi.org/10.1007/s12652-021-03612-z
Kumar S, Kumar H, Kumar G et al (2024) A methodical exploration of imaging modalities from dataset to detection through machine learning paradigms in prominent lung disease diagnosis: a review. BMC Med Imaging 24:30. https://doi.org/10.1186/s12880-024-01192-w
Kwak K, Hwang SS (2024) Predicted cervical cancer prevention: impact of National HPV Vaccination Program on Young Women in South Korea. Cancer Res Treat 56(3):898–908. https://doi.org/10.4143/crt.2023.981
Lacan A, Sebag M, Hanczar B (2023) GAN-based data augmentation for transcriptomics: survey and comparative assessment. Bioinformatics 39(39 Suppl 1):i111–i120. https://doi.org/10.1093/bioinformatics/btad239
Langford BJ, Branch-Elliman W, Nori P, Marra AR, Bearman G (2024) Confronting the disruption of the infectious diseases workforce by artificial intelligence: what this means for us and what we can do about it. Open Forum Infect Dis 11(3):ofae053. https://doi.org/10.1093/ofid/ofae053
Lapique N, Benenson Y (2018) Genetic programs can be compressed and autonomously decompressed in live cells. Nat Nanotechnol 13(4):309–315. https://doi.org/10.1038/s41565-017-0004-z
Lebatteux D, Soudeyns H, Boucoiran I, Gantt S, Diallo AB (2024) Machine learning-based approach KEVOLVE efficiently identifies SARS-CoV-2 variant-specific genomic signatures. PLoS One 19(1):e0296627. https://doi.org/10.1371/journal.pone.0296627
Lee M (2023a) Deep learning techniques with genomic data in cancer prognosis: a comprehensive review of the 2021–2023 literature. Biology 12:893. https://doi.org/10.3390/biology12070893
Lee M (2023b) Recent advances in Generative adversarial networks for gene expression data: a comprehensive review. Mathematics 11:3055. https://doi.org/10.3390/math11143055
Lee JY, Cutts RJ, White I, Augustin Y, Garcia-Murillas I, Fenwick K, Matthews N, Turner NC, Harrington K, Gilbert DC et al (2020) Next generation sequencing assay for detection of circulating HPV DNA (cHPV-DNA) in patients undergoing radical (chemo)radiotherapy in anal squamous cell carcinoma (ASCC). Front Oncol 10:505. https://doi.org/10.3389/fonc.2020.00505

Lee B, Yao X, Shen L et al (2022) Genome-wide association study of quantitative biomarkers identifies a novel locus for alzheimer's disease at 12p12.1. BMC Genomics 23(85):85. https://doi.org/10.1186/s12864-021-08269-8

Letchumanan P, Theva Das K (2025) The role of genetic diversity, epigenetic regulation, and sex-based differences in HIV cure research: a comprehensive review. Epigenetics Chromatin 18:1. https://doi.org/10.1186/s13072-024-00564-4

Li Z, Li Y (2020) A comparative study on the prediction of the BP artificial neural network model and the ARIMA model in the incidence of AIDS. BMC Med Inform Decis Mak 20(1):143. https://doi.org/10.1186/s12911-020-01157-3

Li X, Xu X, Wang J, Li J, Qin S, Yuan J (2020a) Study on prediction model of HIV incidence based on GRU neural network optimized by MHPSO. IEEE Access 8:49574–49583. https://doi.org/10.1109/ACCESS.2020.2979859

Li H, Yang Y, Hong W et al (2020b) Applications of genome editing technology in the targeted therapy of human diseases: mechanisms, advances and prospects. Sig Transduct Target Ther 5: 1. https://doi.org/10.1038/s41392-019-0089-y

Li M, Jiang Y, Zhang Y, Zhu H (2023) Medical image analysis using deep learning algorithms. Front Public Health 11:1273253. https://doi.org/10.3389/fpubh.2023

Li YH, Li YL, Wei MY et al (2024) Innovation and challenges of artificial intelligence technology in personalized healthcare. Sci Rep 14:18994. https://doi.org/10.1038/s41598-024-70073-7

Lim W, Yong KSC, Lau BT, Tan CCL (2024) Future of generative adversarial networks (GAN) for anomaly detection in network security: a review. Comput Secur 139:103733. ISSN 0167-4048. https://doi.org/10.1016/j.cose.2024.103733

Linardatos P, Papastefanopoulos V, Kotsiantis S (2021) Explainable AI: a review of machine learning interpretability methods. Entropy 23:18. https://doi.org/10.3390/e23010018

Lindsay DeVane C (2024) The gray area of scientific adequacy. Pharmacotherapy 44(9):696–697. https://doi.org/10.1002/phar.4606

Liu G, Sharma M, Tan N, Barnabas RV (2018) HIV-positive women have higher risk of human papilloma virus infection, precancerous lesions, and cervical cancer. AIDS 32(6):795–808. https://doi.org/10.1097/QAD.0000000000001765

Liu G, Mugo NR, Bayer C, Rao DW, Onono M, Mgodi NM, Chirenje ZM, Njoroge BW, Tan N, Bukusi EA, Barnabas RV (2022) Impact of catch-up human papillomavirus vaccination on cervical cancer incidence in Kenya: a mathematical modeling evaluation of HPV vaccination strategies in the context of moderate HIV prevalence. EClinicalMedicine 45:101306. https://doi.org/10.1016/j.eclinm.2022

Liu X, Shi J, Jiao Y, An J, Tian J, Yang Y, Zhuo L (2024) Integrated multi-omics with machine learning to uncover the intricacies of kidney disease. Brief Bioinform 25(5):bbae364. https://doi.org/10.1093/bib/bbae364. PMID: 39082652; PMCID: PMC11289682

Lv Z, Li Y (2022) Wearable sensors for vital signs measurement: a survey. J Sens Actuator Netw 11:19. https://doi.org/10.3390/jsan11010019

Mahara G, Tian C, Xu X, Wang W (2023) Revolutionising health care: exploring the latest advances in medical sciences. J Glob Health 13:03042. https://doi.org/10.7189/jogh.13.03042

Mak KK, Wong YH, Pichika MR (2024) Artificial intelligence in drug discovery and development. In: Hock FJ, Pugsley MK (eds) Drug discovery and evaluation: safety and pharmacokinetic assays. Springer, Cham. https://doi.org/10.1007/978-3-031-35529-5_92

Maleki Varnosfaderani S, Forouzanfar M (2024) The role of AI in hospitals and clinics: transforming healthcare in the 21st century. Bioengineering 11:337. https://doi.org/10.3390/bioengineering11040337

Mall R, Elbasir A, Almeer H, Islam Z, Kolatkar PR, Chawla S, Ullah E (2021) A modeling framework for embedding-based predictions for compound–viral protein activity. Bioinformatics 37(17):2544–2555. https://doi.org/10.1093/bioinformatics/btab130

Mapaona R, Williams V, Musarapasi N, Kibwana S, Maseko T et al (2024) Cervical cancer screening outcomes for HIV-positive women in the Lubombo and Manzini regions of Eswatini—prevalence and predictors of a positive visual inspection with acetic acid (VIA) screen. PLOS Glob Public Health 4(4):e0002760. https://doi.org/10.1371/journal.pgph.0002760

Marbouh D, Abbasi T, Maasmi F et al (2020) Blockchain for COVID-19: review, opportunities, and a trusted tracking system. Arab J Sci Eng 45:9895–9911. https://doi.org/10.1007/s13369-020-04950-4

Masoumian Hosseini M, Masoumian Hosseini ST, Qayumi K et al (2023) Smartwatches in healthcare medicine: assistance and monitoring; a scoping review. BMC Med Inform Decis Mak 23:248. https://doi.org/10.1186/s12911-023-02350-w

Mehedi MHK, Khandaker M, Ara S et al (2024) A lightweight deep learning method to identify different types of cervical cancer. Sci Rep 14:29446. https://doi.org/10.1038/s41598-024-79840-y

Mehmood F, Ahmad S, Whangbo TK (2023) An efficient optimization technique for training deep neural networks. Mathematics 11:1360. https://doi.org/10.3390/math11061360

Meiring TL, Salimo AT, Coetzee B et al (2012) Next-generation sequencing of cervical DNA detects human papillomavirus types not detected by commercial kits. Virol J 9:164. https://doi.org/10.1186/1743-422X-9-164

Meriki HD, Tufon KA, Anong DN, Atanga PN, Anyangwe IA et al (2019) Genetic diversity and antiretroviral resistance-associated mutation profile of treated and naive HIV-1 infected patients from the Northwest and Southwest regions of Cameroon. PLoS One 14(11):e0225575. https://doi.org/10.1371/journal.pone.0225575

Mermin-Bunnell K, Zhu Y, Hornback A, Damhorst G, Walker T, Robichaux C, Mathew L, Jaquemet N, Peters K, Johnson TM 2nd, Wang MD, Anderson B (2023) Use of natural language processing of patient-initiated electronic health record messages to identify patients with COVID-19 infection. JAMA Netw Open 6(7):e2322299. https://doi.org/10.1001/jamanetworkopen.2023.22299

Michelutti L, Tel A, Zeppieri M, Ius T, Agosti E, Sembronio S, Robiony M (2024) Generative adversarial networks (GANs) in the field of head and neck surgery: current evidence and prospects for the future-a systematic review. J Clin Med 13(12):3556. https://doi.org/10.3390/jcm13123556

Miller MI, Shih LC, Kolachalama VB (2023) Machine learning in clinical trials: a primer with applications to neurology. Neurotherapeutics 20:1066–1080. https://doi.org/10.1007/s13311-023-01384-2

Ming Z, Chen X, Wang S et al (2023) HostNet: improved sequence representation in deep neural networks for virus-host prediction. BMC Bioinform 24:455. https://doi.org/10.1186/s12859-023-05582-9

Miotto R, Wang F, Wang S, Jiang X, Dudley JT (2018) Deep learning for healthcare: review, opportunities and challenges. Brief Bioinform 19(6):1236–1246. https://doi.org/10.1093/bib/bbx044

Mohammad Shahin F, Chen F, Hosseinzadeh A, Maghanaki M (2024) Deploying deep convolutional neural network to the battle against cancer: towards flexible healthcare systems. Inform Med Unlocked 47:101494. ISSN 2352-9148. https://doi.org/10.1016/j.imu.2024.101494

Mohammed MA, Abdurahman F, Ayalew YA (2021) Single-cell conventional pap smear image classification using pre-trained deep neural network architectures. BMC Biomed Eng 3:11. https://doi.org/10.1186/s42490-021-00056-6

Mohammed FA, Tune KK, Mohammed JA, Wassu TA, Muhie S (2024) Early cervical cancer diagnosis with SWIN-transformer and convolutional neural networks. Diagnostics 14:2286. https://doi.org/10.3390/diagnostics14202286

Mohr AE, Ortega-Santos CP, Whisner CM, Klein-Seetharaman J, Jasbi P (2024) Navigating challenges and opportunities in multi-omics integration for personalized healthcare. Biomedicines 12:1496. https://doi.org/10.3390/biomedicines12071496

Mongia A, Saha SK, Chouzenoux E, Majumdar A (2021) A computational approach to aid clinicians in selecting anti-viral drugs for COVID-19 trials. Sci Rep 11(1):9047. https://doi.org/10.1038/s41598-021-88153-3

Montesinos López OA, Montesinos López A, Crossa J (2022) Fundamentals of artificial neural networks and deep learning. In: Multivariate statistical machine learning methods for genomic prediction. Springer, Cham. https://doi.org/10.1007/978-3-030-89010-0_10

Mu Y, Kodidela S, Wang Y, Kumar S, Cory TJ (2018) The dawn of precision medicine in HIV: state of the art of pharmacotherapy. Expert Opin Pharmacother 19(14):1581–1595. https://doi.org/10.1080/14656566.2018.1515916

Mühr LSA, Guerendiain D, Cuschieri K, Sundström K (2021) Human papillomavirus detection by whole-genome next-generation sequencing: importance of validation and quality assurance procedures. Viruses 13(7):1323. https://doi.org/10.3390/v130713234

Muñoz-Aseguinolaza U, Fernandez-Iriondo I, Rodríguez-Moreno I, Aginako N, Sierra B (2023) Convolutional neural network-based classification and monitoring models for lung cancer detection: 3D perspective approach. Heliyon 9(11):e21203. ISSN 2405-8440. https://doi.org/10.1016/j.heliyon.2023.e21203

Murdoch B (2021) Privacy and artificial intelligence: challenges for protecting health information in a new era. BMC Med Ethics 22:122. https://doi.org/10.1186/s12910-021-00687-3

Mustafa WA, Ismail S, Mokhtar FS, Alquran H, Al-Issa Y (1763) Cervical cancer detection techniques: a chronological review. Diagnostics 2023:13. https://doi.org/10.3390/diagnostics13101763

Nabrawi E, Alanazi AT (2023) Imaging in healthcare: a glance at the present and a glimpse into the future. Cureus 15(3):e36111. https://doi.org/10.7759/cureus.36111

Nassiri K, Akhloufi MA (2024) Recent advances in large language models for healthcare. Bio Med Inform 4:1097–1143. https://doi.org/10.3390/biomedinformatics4020062

Nazer LH, Zatarah R, Waldrip S, Ke JXC, Moukheiber M, Khanna AK, Hicklen RS, Moukheiber L, Moukheiber D, Ma H, Mathur P (2023) Bias in artificial intelligence algorithms and recommendations for mitigation. PLOS Digit Health 2(6):e0000278. https://doi.org/10.1371/journal.pdig.0000278

Ngoupayou Limbepe Z, Gai K, Yu J (2025) Blockchain-based privacy-enhancing federated learning in smart healthcare: a survey. Blockchains 3:1. https://doi.org/10.3390/blockchains3010001

Ngufor C, Caraballo PJ, O'Byrne TJ, Chen D, Shah ND, Pruinelli L, Steinbach M, Simon G (2020) Development and validation of a risk stratification model using disease severity hierarchy for mortality or major cardiovascular event. JAMA Netw Open 3(7):e208270. https://doi.org/10.1001/jamanetworkopen.2020.8270

Nguyen G, Sáinz-Pardo Díaz J, Calatrava A et al (2025) Landscape of machine learning evolution: privacy-preserving federated learning frameworks and tools. Artif Intell Rev 58:51. https://doi.org/10.1007/s10462-024-11036-2

Nie L, Xu S, Zhang L (2023) Multi-head attention network with adaptive feature selection for RUL predictions of gradually degrading equipment. Actuators 12:158. https://doi.org/10.3390/act12040158

Nijiati M, Zhou R, Damaola M, Hu C, Li L, Qian B, Abulizi A, Kaisaier A, Cai C, Li H, Zou X (2022) Deep learning-based CT images automatic analysis model for active/non-active pulmonary tuberculosis differential diagnosis. Front Mol Biosci 9:1086047. https://doi.org/10.3389/fmolb.2022.1086047

Nirthika R, Manivannan S, Ramanan A et al (2022) Pooling in convolutional neural networks for medical image analysis: a survey and an empirical study. Neural Comput Appl 34:5321–5347. https://doi.org/10.1007/s00521-022-06953-8

Norori N, Hu Q, Aellen FM, Faraci FD, Tzovara A (2021) Addressing bias in big data and AI for health care: a call for open science. Patterns (N Y) 2(10):100347. https://doi.org/10.1016/j.patter.2021.100347

Novelli C, Taddeo M, Floridi L (2024) Accountability in artificial intelligence: what it is and how it works. AI & Soc 39:1871–1882. https://doi.org/10.1007/s00146-023-01635-y

Ortiz RR, Smith A, Coyne-Beasley T (2019) A systematic literature review to examine the potential for social media to impact HPV vaccine uptake and awareness, knowledge, and attitudes about HPV and HPV vaccination. Hum Vaccin Immunother 15(7–8):1465–1475. https://doi.org/10.1080/21645515.2019

Padhi A, Agarwal A, Saxena SK, Katoch CDS (2023) Transforming clinical virology with AI, machine learning and deep learning: a comprehensive review and outlook. VirusDisease 34(3): 345–355. https://doi.org/10.1007/s13337-023-00841-y

Papadopoulos D, Karalis VD (2023) Variational autoencoders for data augmentation in clinical studies. Appl Sci 13:8793. https://doi.org/10.3390/app13158793

Park Y-H, Kim Y, Shim J (2021) Blockchain-based privacy-preserving system for genomic data management using local differential privacy. Electronics 10:3019. https://doi.org/10.3390/electronics10233019

Patwardhan N, Marrone S, Sansone C (2023) Transformers in the real world: a survey on NLP applications. Information 14:242. https://doi.org/10.3390/info14040242

Paul D, Sanap G, Shenoy S, Kalyane D, Kalia K, Tekade RK (2021) Artificial intelligence in drug discovery and development. Drug Discov Today 26(1):80–93. https://doi.org/10.1016/j.drudis.2020.10.010

Pavone G, Marino A, Fisicaro V, Motta L, Spata A, Martorana F, Spampinato S, Celesia BM, Cacopardo B, Vigneri P et al (2024) Entangled connections: HIV and HPV interplay in cervical cancer—a comprehensive review. Int J Mol Sci 25:10358. https://doi.org/10.3390/ijms251910358

Pikalyova K, Orlov A, Lin A, Tarasova O, Marcou MG, Horvath D, Poroikov V, Varnek A (2022) HIV-1 drug resistance profiling using amino acid sequence space cartography. Bioinformatics 38(8):2307–2314. https://doi.org/10.1093/bioinformatics/btac090

Prabakaran P, Dimitrov AS, Fouts TR, Dimitrov DS (2007) Structure and function of the HIV envelope glycoprotein as entry mediator, vaccine immunogen, and target for inhibitors. Adv Pharmacol 55:33–97. https://doi.org/10.1016/S1054-3589(07)55002-7

Prasad VK, Verma A, Bhattacharya P et al (2024) Revolutionizing healthcare: a comparative insight into deep learning's role in medical imaging. Sci Rep 14:30273. https://doi.org/10.1038/s41598-024-71358-7

Prussin AJ, Belser JA, Bischoff W et al (2020) Viruses in the built environment (VIBE) meeting report. Microbiome 8:1. https://doi.org/10.1186/s40168-019-0777-4

Puccetti M, Pariano M, Schoubben A, Giovagnoli S, Ricci M (2024) Biologics, theranostics, and personalized medicine in drug delivery systems. Pharmacol Res 201:107086. ISSN 1043-6618. https://doi.org/10.1016/j.phrs.2024.107086

Pushkaran AC, Arabi AA (2024) From understanding diseases to drug design: can artificial intelligence bridge the gap? Artif Intell Rev 57:86. https://doi.org/10.1007/s10462-024-10714-5

Qi X, Zhao Y, Qi Z, Hou S, Chen J (2024) Machine learning empowering drug discovery: applications, opportunities and challenges. Molecules 29(4):903. https://doi.org/10.3390/molecules29040903

Quazi S (2022) Artificial intelligence and machine learning in precision and genomic medicine. Med Oncol 39(8):120. https://doi.org/10.1007/s12032-022-01711-1

Rafique Q, Rehman A, Afghan MS, Ahmad HM, Zafar I, Fayyaz K, Ain Q, Rayan RA, Al-Aidarous KM, Rashid S, Mushtaq G, Sharma R (2023) Reviewing methods of deep learning for diagnosing COVID-19, its variants and synergistic medicine combinations. Comput Biol Med 163:107191. https://doi.org/10.1016/j.compbiomed.2023.107191

Rahman A, Debnath T, Kundu D, Khan MSI, Aishi AA, Sazzad S, Sayduzzaman M, Band SS (2024) Machine learning and deep learning-based approach in smart healthcare: recent advances, applications, challenges and opportunities. AIMS Public Health 11(1):58–109. https://doi.org/10.3934/publichealth.2024004

Rajpurkar P, Chen E, Banerjee O et al (2022) AI in health and medicine. Nat Med 28:31–38. https://doi.org/10.1038/s41591-021-01614-0

Rajput AR, Li Q, Ahvanooey MT (2021) A Blockchain-based secret-data sharing framework for personal health records in emergency condition. Healthcare (Basel) 9(2):206. https://doi.org/10.3390/healthcare9020206

Ramzan F, Sartori C, Consoli S, Reforgiato Recupero D (2024) Generative adversarial networks for synthetic data generation in finance: evaluating statistical similarities and quality assessment. AI 5:667–685. https://doi.org/10.3390/ai5020035

Rashidi HH, Albahra S, Rubin BP et al (2024) A novel and fully automated platform for synthetic tabular data generation and validation. Sci Rep 14:23312. https://doi.org/10.1038/s41598-024-73608-0

Razzak MI, Imran M, Xu G (2020) Big data analytics for preventive medicine. Neural Comput & Applic 32(9):4417–4451. https://doi.org/10.1007/s00521-019-04095-y

Ribeiro RP, Moniz N (2020) Imbalanced regression and extreme value prediction. Mach Learn 109: 1803–1835. https://doi.org/10.1007/s10994-020-05900-9

Romero A, Albert R, Deypalan MNY, Mehrotra S et al (2022) Benchmarking AutoML frameworks for disease prediction using medical claims. BioData Min 15:15. https://doi.org/10.1186/s13040-022-00300-2

Sadri H, Yitmen I, Tagliabue LC, Westphal F, Tezel A, Taheri A, Sibenik G (2023) Integration of blockchain and digital twins in the smart built environment adopting disruptive technologies—a systematic review. Sustain For 15:3713. https://doi.org/10.3390/su15043713

Salehi AW, Khan S, Gupta G, Alabduallah BI, Almjally A, Alsolai H, Siddiqui T, Mellit A (2023) A study of CNN and transfer learning in medical imaging: advantages, challenges, future scope. Sustainability 15:5930. https://doi.org/10.3390/su15075930

Sanches PHG, de Melo NC, Porcari AM, de Carvalho LM (2024) Integrating molecular perspectives: strategies for comprehensive multi-omics integrative data analysis and machine learning applications in transcriptomics, proteomics, and metabolomics. Biology 13:848. https://doi.org/10.3390/biology13110848

Şandru F, Radu AM, Petca A, Dumitraşcu MC, Petca RC, Roman AM (2024 Feb 23) Unveiling the therapeutic horizon: HPV vaccines and their impact on cutaneous diseases-a comprehensive review. Vaccines (Basel) 12(3):228. https://doi.org/10.3390/vaccines12030228

Santangelo OE, Gentile V, Pizzo S, Giordano D, Cedrone F (2023) Machine learning and prediction of infectious diseases: a systematic review. Mach Learn Knowl Extr 5:175–198. https://doi.org/10.3390/make5010013

Sarhangi HA, Beigifard D, Farmani E, Bolhasani H (2024) Deep learning techniques for cervical cancer diagnosis based on pathology and colposcopy images. Inform Med Unlocked 47:101503. ISSN 2352-9148. https://doi.org/10.1016/j.imu.2024.101503

Sarker IH (2022) AI-based modeling: techniques, applications and research issues towards automation, intelligent and smart systems. SN Comput Sci 3:158. https://doi.org/10.1007/s42979-022-01043-x

Sarvamangala DR, Kulkarni RV (2022) Convolutional neural networks in medical image understanding: a survey. Evol Intell 15(1):1–22. https://doi.org/10.1007/s12065-020-00540-3

Schiffer JT, Swan DA, Prlic M, Lund JM (2018) Herpes simplex virus-2 dynamics as a probe to measure the extremely rapid and spatially localized tissue-resident T-cell response. Immunol Rev 285(1):113–133. https://doi.org/10.1111/imr.12672

Seguritan V, Alves N Jr, Arnoult M, Raymond A, Lorimer D et al (2012) Artificial neural networks trained to detect viral and phage structural proteins. PLoS Comput Biol 8(8):e1002657. https://doi.org/10.1371/journal.pcbi.1002657

Serrano DR, Luciano FC, Anaya BJ, Ongoren B, Kara A, Molina G, Ramirez BI, Sánchez-Guirales SA, Simon JA, Tomietto G et al (2024) Artificial intelligence (AI) applications in drug discovery and drug delivery: revolutionizing personalized medicine. Pharmaceutics 16:1328. https://doi.org/10.3390/pharmaceutics16101328

Shafay M, Ahmad RW, Salah K, Yaqoob I, Jayaraman R, Omar M (2023) Blockchain for deep learning: review and open challenges. Clust Comput 26(1):197–221. https://doi.org/10.1007/s10586-022-03582-7

Shi S, He D, Li L, Kumar N, Khan MK, Choo KR (2020) Applications of blockchain in ensuring the security and privacy of electronic health record systems: a survey. Comput Secur 97:101966. https://doi.org/10.1016/j.cose.2020.101966

Shi Z, Qian H, Li Y, Wu F, Wu L (2022) Machine learning based regional epidemic transmission risks precaution in digital society. Sci Rep 12(1):20499. https://doi.org/10.1038/s41598-022-24670-z

Shih DH, Wu YH, Wu TW, Chang SC, Shih MH (2024) Infodemiology of influenza-like illness: utilizing Google trends' big data for epidemic surveillance. J Clin Med 13(7):1946. https://doi.org/10.3390/jcm13071946

Shimbre MS, Belete AG, Ghazal L et al (2024) Predictors of CD4 cell count progression over time among adolescents and young adults transitioning to adult-oriented HIV care in Southern Ethiopia from 2017 to 2021: a retrospective cohort study. BMC Public Health 24:2887. https://doi.org/10.1186/s12889-024-20396-x

Singh A, Sengupta S, Lakshminarayanan V (2020) Explainable deep learning models in medical image analysis. J Imaging 6:52. https://doi.org/10.3390/jimaging6060052

Singh P, Sagar S, Singh S et al (2024) Blockchain-enabled verification of medical records using soul-bound tokens and cloud computing. Sci Rep 14:24830. https://doi.org/10.1038/s41598-024-75708-3

Singhal A, Neveditsin N, Tanveer H, Mago V (2024) Toward fairness, accountability, transparency, and ethics in AI for social media and health care: scoping review. JMIR Med Inform 12:e50048. https://doi.org/10.2196/50048

Smelik M, Zhao Y, Li X et al (2024) An interactive atlas of genomic, proteomic, and metabolomic biomarkers promotes the potential of proteins to predict complex diseases. Sci Rep 14:12710. https://doi.org/10.1038/s41598-024-63399-9

Sokhansanj BA, Zhao Z, Rosen GL (2022) Interpretable and predictive deep neural network modeling of the SARS-CoV-2 spike protein sequence to predict COVID-19 disease severity. Biology (Basel) 11(12):1786. https://doi.org/10.3390/biology11121786

Solanki P, Grundy J, Hussain W (2023) Operationalising ethics in artificial intelligence for healthcare: a framework for AI developers. AI Ethics 3:223–240. https://doi.org/10.1007/s43681-022-00195-z

Srikantamurthy MM, Rallabandi VPS, Dudekula DB et al (2023) Classification of benign and malignant subtypes of breast cancer histopathology imaging using hybrid CNN-LSTM based transfer learning. BMC Med Imaging 23:19. https://doi.org/10.1186/s12880-023-00964-0

Steiner MC, Gibson KM, Crandall KA (2020) Drug resistance prediction using deep learning techniques on HIV-1 sequence data. Viruses 12(5):560. https://doi.org/10.3390/v12050560

Steurer B, Vanhaelen Q, Zhavoronkov A (2024) Multimodal transformers and their applications in drug target discovery for aging and age-related diseases. J Gerontol A Biol Sci Med Sci 79(9): glae006. https://doi.org/10.1093/gerona/glae006

Sudlow C, Gallacher J, Allen N, Beral V, Burton P, Danesh J et al (2015) UK biobank: an open access resource for identifying the causes of a wide range of complex diseases of middle and old age. PLoS Med 12:e1001779. https://doi.org/10.1371/journal.pmed.1001779

Tagde P, Tagde S, Bhattacharya T, Tagde P, Chopra H, Akter R, Kaushik D, Rahman MH (2021) Blockchain and artificial intelligence technology in e-health. Environ Sci Pollut Res Int 28(38): 52810–52831. https://doi.org/10.1007/s11356-021-16223-0

Taherdoost H (2023a) Deep learning and neural networks: decision-making implications. Symmetry 15:1723. https://doi.org/10.3390/sym15091723

Taherdoost H (2023b) Privacy and security of blockchain in healthcare: applications, challenges, and future perspectives. Science 5(41). https://doi.org/10.3390/sci5040041

Takeuchi F, Kukimoto I, Li Z, Li S, Li N, Hu Z, Takahashi A, Inoue S, Yokoi S, Chen J, Hang D, Kuroda M, Matsuda F, Mizuno M, Mori S, Wu P, Tanaka N, Matsuo K, Kamatani Y, Kubo M, Ma D, Shi Y (2019) Genome-wide association study of cervical cancer suggests a role for ARRDC3 gene in human papillomavirus infection. Hum Mol Genet 28(2):341–348. https://doi.org/10.1093/hmg/ddy390

Tan X, Li K, Zhang J et al (2021) Automatic model for cervical cancer screening based on convolutional neural network: a retrospective, multicohort, multicenter study. Cancer Cell Int 21:35. https://doi.org/10.1186/s12935-020-01742-6

Tao C, Li Y, Li J, Li M, Yu E, Amith M, Tang L, Savas L, Cui L (2024) VaxBot-HPV: a GPT-based Chatbot for answering HPV vaccine-related questions Res Sq [Preprint]. rs.3.rs-4876692. https://doi.org/10.21203/rs.3.rs-4876692/v1

Tariq MU, Ismail SB (2024) Deep learning in public health: comparative predictive models for COVID-19 case forecasting. PLoS One 19(3):e0294289. https://doi.org/10.1371/journal.pone.0294289

Tariq U, Ahmed I, Bashir AK, Shaukat K (2023) A critical cybersecurity analysis and future research directions for the internet of things: a comprehensive review. Sensors 23:4117. https://doi.org/10.3390/s23084117

Taye MM (2023) Understanding of machine learning with deep learning: architectures, workflow, applications and future directions. Computers 12:91. https://doi.org/10.3390/computers12050091

Thudumu S, Branch P, Jin J et al (2020) A comprehensive survey of anomaly detection techniques for high dimensional big data. J Big Data 7:42. https://doi.org/10.1186/s40537-020-00320-x

Tobón DP, Hossain MS, Muhammad G et al (2022) Deep learning in multimedia healthcare applications: a review. Multimed Syst 28:1465–1479. https://doi.org/10.1007/s00530-022-00948-0

Tobore I, Li J, Yuhang L, Al-Handarish Y, Kandwal A, Nie Z, Wang L (2019) Deep learning intervention for health care challenges: some biomedical domain considerations. JMIR Mhealth Uhealth 7(8):e11966. https://doi.org/10.2196/11966

Torabi H, Mirtaheri SL, Greco S (2023) Practical autoencoder based anomaly detection by using vector reconstruction error. Cybersecurity 6:1. https://doi.org/10.1186/s42400-022-00134-9

Truman AW, Kwun MJ, Cheng J, Yang SH, Suh JW, Hong HJ (2014) Antibiotic resistance mechanisms inform discovery: identification and characterization of a novel amycolatopsis strain producing ristocetin. Antimicrob Agents Chemother 58(10):5687–5695. https://doi.org/10.1128/AAC.03349-14

Tsopra R, Fernandez X, Luchinat C et al (2021) A framework for validating AI in precision medicine: considerations from the European ITFoC consortium. BMC Med Inform Decis Mak 21:274. https://doi.org/10.1186/s12911-021-01634-3

Tucudean G, Bucos M, Dragulescu B, Caleanu CD (2024) Natural language processing with transformers: a review. PeerJ Comput Sci 10:e2222. https://doi.org/10.7717/peerj-cs.2222

Ueda D, Kakinuma T, Fujita S, Kamagata K, Fushimi Y, Ito R, Matsui Y, Nozaki T, Nakaura T, Fujima N, Tatsugami F, Yanagawa M, Hirata K, Yamada A, Tsuboyama T, Kawamura M, Fujioka T, Naganawa S (2024) Fairness of artificial intelligence in healthcare: review and recommendations. Jpn J Radiol 42(1):3–15. https://doi.org/10.1007/s11604-023-01474-3

van Zyl G, Bale MJ, Kearney MF (2018) HIV evolution and diversity in ART-treated patients. Retrovirology 15:14. https://doi.org/10.1186/s12977-018-0395-4

Visan AI, Negut I (2024) Integrating artificial intelligence for drug discovery in the context of revolutionizing drug delivery. Life 14:233. https://doi.org/10.3390/life14020233

Vo D-K, Trinh KTL (2024) Advances in wearable biosensors for healthcare: current trends, applications, and future perspectives. Biosensors 14:560. https://doi.org/10.3390/bios14110560

Vrábel J, Pořízka P, Kaiser J (2020) Restricted Boltzmann machine method for dimensionality reduction of large spectroscopic data. Spectrochim Acta B At Spectrosc 167:105849. ISSN 0584-8547. https://doi.org/10.1016/j.sab.2020.105849

Walter W, Pohlkamp C, Meggendorfer M, Nadarajah N, Kern W, Haferlach C, Haferlach T (2023) Artificial intelligence in hematological diagnostics: game changer or gadget? Blood Rev 58:101019. ISSN 0268-960X. https://doi.org/10.1016/j.blre.2022.101019

Wang X (2025) Strategies for enhancing deep video encoding efficiency using the convolutional neural network in a hyperautomation mechanism. Sci Rep 15:1079. https://doi.org/10.1038/s41598-025-85602-1

Wang C, Zhang J, Lassi N, Zhang X (2022) Privacy protection in using artificial intelligence for healthcare: Chinese regulation in comparative perspective. Healthcare (Basel) 10(10):1878. https://doi.org/10.3390/healthcare10101878

Wang J, Yu Y, Tan Y et al (2024) Artificial intelligence enables precision diagnosis of cervical cytology grades and cervical cancer. Nat Commun 15:4369. https://doi.org/10.1038/s41467-024-48705-3

Whittington JCR, Bogacz R (2019) Theories of error backpropagation in the brain. Trends Cogn Sci 23(3):235–250. https://doi.org/10.1016/j.tics.2018.12.005

Williamson SM, Prybutok V (2024) Balancing privacy and progress: a review of privacy challenges, systemic oversight, and patient perceptions in AI-driven healthcare. Appl Sci 14: 675. https://doi.org/10.3390/app14020675

Wohl S, Schaffner SF, Sabeti PC (2016) Genomic analysis of viral outbreaks. Annu Rev Virol 3(1): 173–195. https://doi.org/10.1146/annurev-virology-110615-035747

Wong CR, Trinh HN, Yip B, Nguyen HA, Garcia RT, Ahmed A, Keeffe EB, Nguyen MH (2011) High rate of complete viral suppression with combination therapy in patients with chronic hepatitis B and prior treatment failure. J Clin Gastroenterol 45(10):900–905. https://doi.org/10.1097/MCG.0b013e318224d64f

Woo H, Cho Y, Shim E, Lee JK, Lee CG, Kim SH (2016) Estimating influenza outbreaks using both search engine query data and social media data in South Korea. J Med Internet Res 18(7): e177. https://doi.org/10.2196/jmir.4955

Worley CR, Skjellum A (2019) Opportunities, challenges, and future extensions for smart-contract design patterns. In: Abramowicz W, Paschke A (eds) Business information systems workshops. BIS 2018, Lecture notes in business information processing, vol 339. Springer, Cham. https://doi.org/10.1007/978-3-030-04849-5_24

Wright LG, Onodera T, Stein MM et al (2022) Deep physical neural networks trained with backpropagation. Nature 601:549–555. https://doi.org/10.1038/s41586-021-04223-6

Wu Y, Zhang L, Bhatti UA, Huang M (2023) Interpretable machine learning for personalized medical recommendations: a LIME-based approach. Diagnostics 13:2681. https://doi.org/10.3390/diagnostics13162681

Wu T, Lucas E, Zhao F, Basu P, Qiao Y (2024a) Artificial intelligence strengthens cervical cancer screening–present and future. Cancer Biol Med 21(10):864–879. https://doi.org/10.20892/j.issn.2095-3941.2024.0198

Wu T, Lucas E, Zhao F, Basu P, Qiao Y (2024b) Artificial intelligence strengthenes cervical cancer screening – present and future. Cancer Biol Med 21(10):864–879. https://doi.org/10.20892/j.issn.2095-3941

Yadav SS, Jadhav SM (2019) Deep convolutional neural network based medical image classification for disease diagnosis. J Big Data 6:113. https://doi.org/10.1186/s40537-019-0276-2

Yadav N, Pandey S, Gupta A, Dudani P, Gupta S, Rangarajan K (2023) Data privacy in healthcare: in the era of artificial intelligence. Indian Dermatol Online J 14(6):788–792. https://doi.org/10.4103/idoj.idoj_543_23

Yamashita R, Nishio M, Do RKG et al (2018) Convolutional neural networks: an overview and application in radiology. Insights Imaging 9:611–629. https://doi.org/10.1007/s13244-018-0639-9

Yammouri G, Lahcen AA (2024) AI-reinforced wearable sensors and intelligent point-of-care tests. J Pers Med 14(11):1088. https://doi.org/10.3390/jpm14111088

Yang S, Santillana M, Kou SC (2015) Accurate estimation of influenza epidemics using Google search data via ARGO. Proc Natl Acad Sci USA 112(47):14473–14478. https://doi.org/10.1073/pnas.1515373112

Yang S, Ning S, Kou SC (2021) Use internet search data to accurately track state level influenza epidemics. Sci Rep 11:4023. https://doi.org/10.1038/s41598-021-83084-5

Yang Y, Wu Z, Yang Y, Lian S, Guo F, Wang Z (2022) A survey of information extraction based on deep learning. Appl Sci 12:9691. https://doi.org/10.3390/app12199691

Ye Y, Burkholder GA, Wiener HW, Aslibekyan S, Khan A, Shrestha S (2020) CD4 trajectory models and onset of non-AIDS-defining anal genital warts, precancer, and cancer in people living with HIV Infection-1. Sex Transm Dis 47(9):628–633. https://doi.org/10.1097/OLQ.0000000000001215

Yenikaya MA, Kerse G, Oktaysoy O (2024) Artificial intelligence in the healthcare sector: comparison of deep learning networks using chest X-ray images. Front Public Health 12: 1386110. https://doi.org/10.3389/fpubh.2024.1386110

Yi HC, You ZH, Wang L et al (2021) In silico drug repositioning using deep learning and comprehensive similarity measures. BMC Bioinform 22(Suppl 3):293. https://doi.org/10.1186/s12859-020-03882-y

You Y, Lai X, Pan Y et al (2022) Artificial intelligence in cancer target identification and drug discovery. Sig Transduct Target Ther 7:156. https://doi.org/10.1038/s41392-022-00994-0

Yousef R, Gupta G, Yousef N et al (2022) A holistic overview of deep learning approach in medical imaging. Multimedia Systems 28:881–914. https://doi.org/10.1007/s00530-021-00884-5

Yue R, Dutta A (2022) Computational systems biology in disease modeling and control, review and perspectives. npj Syst Biol Appl 8:37. https://doi.org/10.1038/s41540-022-00247-4

Zafar A, Saba N, Arshad A, Alabrah A, Riaz S, Suleman M, Zafar S, Nadeem M (2024) Convolutional neural networks: a comprehensive evaluation and benchmarking of pooling layer variants. Symmetry 16:1516. https://doi.org/10.3390/sym16111516

Zhang H, Qie Y (2023) Applying deep learning to medical imaging: a review. Appl Sci 13:10521. https://doi.org/10.3390/app131810521

Zhang L, Fan H, Peng C, Rao G, Cong Q (2020) Sentiment analysis methods for HPV vaccines related tweets based on transfer learning. Healthcare 8:307. https://doi.org/10.3390/healthcare8030307

Zhang EY, Cheok AD, Pan Z, Cai J, Yan Y (2023) From turing to transformers: a comprehensive review and tutorial on the evolution and applications of generative transformer models. Science 5:46. https://doi.org/10.3390/sci5040046

Zhao L, Zhang Z (2024) A improved pooling method for convolutional neural networks. Sci Rep 14:1589. https://doi.org/10.1038/s41598-024-51258-6

Zhong M, Xu M, Zheng S et al (2024) Manipulating host secreted protein gene expression: an indirect approach by HPV11/16 E6/E7 to suppress PBMC cytokine secretion. Virol J 21:172. https://doi.org/10.1186/s12985-024-02432-9

Zhou SK, Greenspan H, Davatzikos C, Duncan JS, van Ginneken B, Madabhushi A, Prince JL, Rueckert D, Summers RM (2021) A review of deep learning in medical imaging: imaging traits, technology trends, case studies with progress highlights, and future promises. Proc IEEE Inst Electr Electron Eng 109(5):820–838. https://doi.org/10.1109/JPROC.2021.30543904

Zhou Y et al (2020) Artificial intelligence in COVID-19 drug repurposing. Lancet Digit Health 2(12):e667–e676

Zhu L, Chen S, Shen S (2025) Dynamic evolution analysis and parameter optimization design of data-driven network infectious disease model. Comput Methods Programs Biomed 258:108509. ISSN 0169-2607. https://doi.org/10.1016/j.cmpb.2024.108509

Zhuhadar LP, Lytras MD (2023) The application of AutoML techniques in diabetes diagnosis: current approaches, performance, and future directions. Sustain For 15:13484. https://doi.org/10.3390/su151813484

Zuo K, Gao W, Wu Z, Zhang L, Wang J, Yuan X, Li C, Xiang Q, Lu L, Liu H (2024) Evolution of virology: science history through milestones and technological advancements. Viruses 16:374. https://doi.org/10.3390/v16030374

Applications of Deep Learning in Virology 4

Arpita Shome, Gunjan Mukherjee, and Ankita Shome

Abstract

Deep learning, an artificial neural network approach, has transformed several industries, such as statistical proof and autonomous transport systems. However, it faces challenges in biomedical sciences, particularly vaccine research. Despite these, deep learning has great potential in eradicating global infectious diseases. The COVID-19 pandemic highlighted the importance of understanding and anticipating virus hosts. A study using recurrent neural network structure and convolutional (ViRNN) outperformed techniques like the k-nearest neighbor algorithm and deep learning-based algorithms. Deep learning (DL) and artificial intelligence (AI)-driven technologies are revolutionizing drug development by predicting interactions between proteins and sugars. These technologies capture high-level information to create models for quick drug development and enhance novel therapies and preventative measures in viral pathogenesis. DL methods capture genetic and molecular data sources, while AI-driven technologies forecast interactions between proteins and sugars. This work examines data production methodologies and datasets for various illnesses, focusing on simple DL algorithms, accessible tools, and visualization techniques. AI enhances vaccine creation by improving accuracy and efficiency through epitope prediction and antigen selection. It uses data from immune system interactions, protein structures, and genetics to prioritize antigens. However, regulatory concerns

A. Shome (✉)
Department of Information Technology, Haldia Institute of Technology, Haldia, West Bengal, India

G. Mukherjee
Department of Computational Sciences, Brainware University, Barasat, West Bengal, India

A. Shome
Department of Biotechnology, Vellore Institute of Technology, Vellore, Tamil Nadu, India

J. M. Chatterjee, S. K. Saxena (eds.), *Artificial Intelligence and Blockchain in Precision Medicine and Virology*, Medical Virology: From Pathogenesis to Disease Control, https://doi.org/10.1007/978-981-96-8919-4_4

and heterogeneity remain challenges, necessitating the integration of advanced technologies. The article also reviews virtual screening-based drug repurposing research, machine learning, DL, and viral disease databases and technologies. It covers case studies unique to each malignancy, the processes underlying viral malignancies, and the drawbacks and difficulties of different strategies. Drug repurposing aims to reduce clinical trial failures by identifying alternate applications for FDA-approved medications used to treat viral cancer. DL is a crucial tool in clinical virology, enhancing diagnostic precision, therapeutic treatments, and epidemiological monitoring. It accelerates medication discovery, predicts viral outbreaks, and enhances viral sequencing, but challenges like algorithmic bias, data security, and ethical concerns persist. The concerned chapter throws light on the in-depth discussion of the application of DL in the discovery of intricate information about virology, highlighting many different research works taking place in this field. The fast-paced progress of the concept of machine learning, along with the discoveries of much-undiscovered information and in-depth analysis, has become the prime subject topic of the chapter, which definitely comes with the immense help of professionals, students, academicians, and researchers as well.

Keywords

Artificial neural network · COVID-19 · Deep learning · Drug repurposing research · Machine learning · Artificial intelligence

4.1 Introduction

A subfield of microbiology called virology examines viruses, which are microscopic creatures that infect and multiply inside host cells. With their genetic material, capsid protein coat, and lipid envelope, these obligatory intracellular parasites defy conventional biological classifications and lack independent life mechanisms (Burrell et al. 2016). With an emphasis on agriculture and ecosystems, virology studies look at how viruses affect crops and livestock. Viruses affect nutrient cycles, control microbial populations, and enhance genetic variety. Aside from its usefulness in phage therapy and bioremediation, the study of bacteriophages—viruses that infect bacteria—offers insights into ecological processes (Nami et al. 2021). Biology research has produced enormous amounts of data, especially in the area of omics. Next-generation sequencing (NGS) and other technological advancements have made data development easier, but the growing amount of data has caused serious concerns regarding the monitoring of large databases (Ching et al. 2018). Traditional statistical techniques for data analysis are not effective in this context. Integrating the analysis of many data types is challenging due to the limitation of sample size and the variety of biological data. The techniques of machine learning (ML) along with advanced learning (DL) applications are being utilized to analyze datasets in order to get past these hurdles. ML tools find patterns in data, and DL techniques find undetected patterns and refine models by extracting high-level features from big datasets. These methods might solve difficult issues by discovering patterns in seemingly unknown data.

Beginning in the middle of the twentieth century, the science of AI was heavily influenced by early research on fundamental ideas such as expert systems and symbolic reasoning. But conventional AI models found it difficult to grow and adjust to the complexity of the actual world. Data-driven methods and techniques such as neural networks, support vector machines, and decision trees were introduced by ML in the 1980s. In the 2010s, DL became a revolutionary field driven by advances in neural network topologies, increasing processing power, and the availability of large amounts of data (Janiesch et al. 2021). Inspired by the structure of the human brain, DL models beat standard algorithms in a variety of applications, including virology, autonomous driving, and healthcare, and they were particularly good at managing vast, unstructured datasets. Computer systems that are capable of interpreting health data (Rajkomar et al. 2018), especially in its natural state as perceived by humans, are known as AI systems. AI frameworks are utilized in clinical diagnostics to effectively evaluate sizable, intricate datasets. AI systems learn to do the decoding task on fresh data after being educated on outward healthcare information that has typically been interpreted by humans. AI interpretation tasks are classified into issue categories such as machine vision, analysis of time series, recognition of speech, and processing of natural language. Certain clinical diagnostic activities, such as deciphering radiological pictures, evaluating and continually accessing information about health, and locating operational regulatory features in the human genome, are ideally suited to these tasks.

In several sectors, including healthcare, DL is being utilized more and more for protection and health condition detection. In contrast to conventional techniques, which are costly, time-consuming, and frequently require human contact, they provide deep learning-based medical diagnosis quickly and affordably. Because ML algorithms are unconstrained and do not suffer from human fatigue, they are perfect for providing extensive medical care. The precision of DL and ML examinations has increased by over 90% because of developments in DL and ML technologies. Since it aids in the identification and treatment of illnesses, diagnostic data is essential in the medical industry. It affects scientific research, governmental policy, and medical care. Due to its diversity and fragmentation, the healthcare sector necessitates prompt access to pertinent data and suitable test sets. Consistent and knowledgeable physician decision-making helps resolve the substantial variations in pathology requests caused by individual differences in clinical practice. The emergence of deep learning (DL) technology has transformed prediction and data analysis in several ways. Although DL has shown promise in identifying viral infections, forecasting outbreaks, and developing treatments, there remain obstacles to its use in virology. Important obstacles include algorithmic biases, data heterogeneity, and regulatory concerns. These elements are examined in this study, with an emphasis on case studies, applications, and the fusion of DL with cutting-edge technology to promote virology. Clinical data, such as diagnoses, kinds, and molecular profiles, are the primary source of information used in contemporary statistical procedures for forecasting cancer for assessment of survival. However, simply forecasting statistically faces difficulties due to the abundance of multi-omics data. With varying degrees of effectiveness, ML techniques, including autoencoders, clustering, and principal component analysis (PCA), have been used to categorize cancer kinds and forecast cancer prognosis. These techniques have demonstrated potential in predicting the prognosis of cancer.

DL has revolutionized virology research by examining large, complex datasets and uncovering intricate relationships. This method is especially beneficial in identifying, managing, and predicting viral infections such as SARS-CoV-2, HIV, and influenza. DL is capable of examining viral protein structures and genomes to predict drug resistance, evolution, and genetic changes in viruses. It also facilitates the assessment of the binding affinity between viral proteins and host receptors, helping to clarify the mechanisms behind infection and pinpoint possible treatment targets. Generative models, such as variational autoencoders and generative adversarial networks (GANs), are utilized to identify novel antiviral drugs and vaccines. When investigating viral illnesses, data-driven modeling techniques, such as supervised, unsupervised, and reinforcement modeling, are crucial. Reliable viral genome classification, which incorporates domain information derived from DNA sequences and structures, is essential for the creation of vaccines, epidemiological studies, pathogenicity evaluations, and pharmaceuticals. Unsupervised learning identifies patterns in data without needing labels for each class, whereas supervised ML categorizes data into two distinct groups according to domain criteria. In virology, unsupervised learning is often employed to deduce phylogeny, necessitating the classification of input without related output variables. Especially in areas such as virology, proteomics, and genomics, DL serves as an effective method for managing biological data (Xu and Jackson 2019). It can analyze complex, high-dimensional datasets through independent representational learning from raw data without needing extensive feature engineering. Extensive sequencing data and population-wide epidemiological research are appropriate for scalable DL models. They surpass conventional ML techniques in identifying intricate, nonlinear relationships. Convolutional and recurrent neural networks can effectively capture intricate patterns from biological data. Autoencoders and GANs generate synthetic information for training and facilitate unsupervised learning in areas with limited data.

The application of DL is a groundbreaking method in viral and epidemiological research. Extensive datasets are examined and processed to generate new insights and predictive capabilities that allow for faster reactions to emerging threats. It examines and reduces dynamic situations by utilizing recurrent neural networks and multilayer neural networks to predict epidemic spread over time and across areas. DL improves the monitoring, assessment, and prevention of infectious disease spread by examining information related to the environment, social media patterns, and health records. It also uncovers concealed links among virus spread, climate, and population density. DL has tremendously benefitted epitope mapping, adjuvant formulation, and phylogenetics. Epitope mapping is the process of identifying specific regions of viruses that the immune system recognizes to develop vaccines or therapeutic antibodies. DL models, such as convolutional neural networks (CNNs) and recurrent neural networks (RNNs), enhance MHC binding predictions, map epitopes across species, and identify immunogenic epitopes. Furthermore, they anticipate epitopes more precisely than traditional methods by utilizing genetic sequences, protein structures, and immunological data. Applications for vaccines and treatments include the development of therapeutic antibodies, adjuvants, vaccines, and personalized vaccinations. DL models help with adjuvant formulation optimization, adjuvant-immune response prediction, and adjuvant-related toxicity. In order to find new immune enhancers more quickly, systematic screening makes it possible to virtually scan large chemical libraries for possible new adjuvants. Studying the evolutionary connections between species or strains is the focus of phylogenetics. DL models improve phylogenetic studies by building infectiousness. This includes building phylogenetic tree structures from viral DNA sequences (Lambert et al. 2023), forecasting mutations and developmental dynamics, and studying the evolution of viruses within the context of host adaptation. The ability to identify quickly changing areas of the viral genome is crucial for monitoring mutations that could affect the effectiveness of vaccines or antiviral resistance. Applications include real-time viral transmission chain monitoring, epidemiological surveillance, and the prediction of novel strains that could elude immunization, hence enhancing readiness for upcoming epidemics. All things considered, the development of vaccines, the design of adjuvants, and virology could all be revolutionized by DL.

A key component of drug design is druggability (Imming et al. 2006), which enables a molecule to engage with anatomical targets and produce a desired pharmacological reaction. Druggable targets for proteins are well known, whereas research on nucleic acids is progressing more slowly. Due to the inadequacy of molecular knowledge, it is rather hard to evaluate the druggability of both antibodies and fragments of nucleic acid in vitro. This recedes when promoting capable computer programs for estimating pharmacological targets on a genome-wide scale with utmost sensitivity and selectivity. Technological advancements such as gene regulatory networks, expression profiles, and protein-protein interactions contribute to devising model-based predictions (Costa et al. 2010).

Numerous scientific fields, namely genetics, drugs and diagnostic medicine, and epidemiology, are currently involved in spacious changes vivified by DL and natural language processing (NLP). These technologies can exercise vast amounts of viral data and therefore understand viral evolution, mutations, and virus-host interactions. DL expedites the process of finding antiviral drugs, advances the course of producing vaccines, and improves epidemiology; it is a key to image-based diagnostic techniques that may detect notorious viral infections. New NLP methods that help in the summarization of scientific articles facilitate rapid progress in viral studies. The combination of DL with virology offers promising opportunities to respond effectively to novel infectious diseases and to ensure the security of global health.

Our research is solid because it begins with a meticulous consideration of the use of different DL elements in virology. Hence, this study is aimed at the current state of DL technologies, their use in viral research, the possibilities and the limitations, as well as the recommendations for new technologies in flawless working diagnostics.

A comprehensive review of the current DL methodologies and their application in virological research brought out the factors of the limitation of these DL technologies in the virological field; practical recommendations were provided, and the preferences for the incorporation of DL technologies into virological workflows were selected. It will also look into the significance of DL for the life of patients in virology in different virological realms.

4.2 Related Works

The rise of DL has completely revolutionized virology, which is one of the scientific fields. Its capability to process and scrutinize massive datasets offers fresh approaches to studying, diagnosing, and controlling viral infections. DL models with the ability to predict changes and identify patterns in these viruses' genomes are the main source of the knowledge about the evolution and adaptability of viruses. Furthermore, especially during outbreaks, they help in the accurate and rapid diagnosis. Real-time PCR evaluation is performed using DL algorithms, which increases the levels of specificity and sensitivity with virus identification in molecular diagnostics checking. DL is also responsible for biologic sensors and point-of-care tools being produced, and by so doing, it enables the real-time diagnosis of viruses in less resourceful areas. Protein-ligand interactions are foretold by DL models that improve the selection of compounds in chemical libraries for the treatment of drug discovery that comes with a large number of possible remedies. Novel medication candidates with enhanced efficacy and fewer side effects are designed using reinforcement learning techniques and GANs. With its ability to pinpoint immunogenic epitopes and enhance mRNA vaccine constructions, DL is essential to vaccine creation. DL models are used in epidemiology to assess intervention tactics, predict disease spread, and pinpoint high-risk locations. To forecast outbreak dynamics, spatiotemporal models examine patterns of transmission. In order to track novel hazards and false information during the COVID-19 epidemic, DL models combined enormous volumes of medical research and publicly available data and led public health policy. Table 4.1 illustrates the methodologies and drawbacks of some researchers on DL applications in virology.

The development of drugs and vaccines, diagnostics, epidemiological modeling, host-virus interactions, and viral genomes and evolution are important application fields. Quality of data and accessibility, model interpretability, computing resource requirements, viral generalization, and ethical and privacy considerations are among the challenges of applying DL to virology.

Techniques such as neural networks, random forests, and heuristic and evolutionary methods have been developed to enhance performance across a range of fields. To lower variance while preserving low bias, random forest employs bootstrap-sampled instances and a portion of randomly assigned attributes. Huge amounts of data are solved by neural networks that operate with hidden layers, among which CNNs are the most effective.

Table 4.1 Methodologies and Drawbacks of Deep Learning Applications in Virology

Author	Title	Methodology	Drawbacks
Rey et al. (2021)	Deep learning in situ classification of HIV-1 virion morphology	With a mean average precision of 80.0%, neural convolutional networks outperform standard RCNN backbones in the verification, segmentation, and categorization of HIV-1 virions at different stages of maturity and morphogenesis.	Cell organelles that resemble virus particles in size can frequently lead to misunderstandings when interpreting data.
Maity and Das (2017)	Machine Learning for Improved Diagnosis and Prognosis in Healthcare	The research shows how machine learning can be effectively used for rapid diagnosis and better patient outcomes by automatically classifying cell pictures to assess the severity and progression of breast cancer.	Analyzing healthcare data quickly and automatically, enhancing patient outcomes, and facilitating more precise diagnoses in larger, more complex datasets.
Liu et al. (2020)	Application of deep learning in genomics	With an emphasis on the proteins, RNA, and DNA levels, the book gives a general review of artificial neural networks, machine learning principles, and their uses in genome sequencing and synthetic biology.	Analyzing large amounts of data is complex.
Mock et al. (2021)	VIDHOP, viral host prediction with deep learning	A deep learning strategy for anticipating viral hosts produced highly accurate forecasts (AUC) ranging from 0.93 to 0.98 for three distinct virus species (rotavirus A, rabies lyssavirus, and influenza A virus) using the tool VIDHOP.	Imbalanced datasets and ineffective recurrent neural network learning on lengthy sequences via random undersampling.
Fang et al. (2022)	DeePVP: Identification and classification of phage virion proteins using deep learning	In PVP classification tests, the deep learning-based tool DeePVP achieves a 9.05% larger F1 score and a 3.72% accuracy rating than the state-of-the-art methodologies for identifying PVPs inside phage genomes.	Partial genes have been demonstrated as less efficient for DeePVP

(continued)

Table 4.1 (continued)

Author	Title	Methodology	Drawbacks
Chaudhary et al. (2023)	Applications of Machine Learning in Viral Disease Diagnosis	Early detection and treatment of viral infections are made possible by machine learning algorithms, including greater achievement rates.	Data confidentiality and safety breaches are the drawbacks.
Wang et al. (2023)	Fusang: a framework for phylogenetic tree inference via deep learning	Fusang is a DL-based system that achieves a comparable efficiency with simulated and actual datasets through continuous optimization, excelling ML tools in practical situations.	For useful real-world applications, there are currently no DL-based solutions available.
Mihaylov et al. (2019)	Application of Machine Learning Models for Survival Prognosis in Breast Cancer Studies	forecast the duration of survival employing tumor-integrated clinical parameters, combining breast cancer data with predictive machine learning approaches to provide the most accurate prognosis findings.	Enhancements are needed to the Python-based procedure, specifically turning it into a web application with extra services.
Joshi et al. (2021)	VirLeafNet: Automatic analysis and viral disease diagnosis using deep learning in *Vigna mungo* plant	Variety is introduced through the segmentation and augmentation of the Vigna mungo leaf picture dataset. Three convolutional neural networks—VirLeafNet-1, VirLeafNet-2, and VirLeafNet-3—are trained, and they achieve high validation and testing accuracy—91.234%, 96.429%, and 97.403%, respectively.	Developing an autonomous disease diagnosis approach for real-time jobs is difficult because viral infection generates unpredictable changes in the features of leaf images.
Liu et al. (2023)	Rapid and stain-free quantification of viral plaque via lens-free holography and deep learning	Early viral replication detection, rapid detection of plaque-forming units at rates greater than 90% with 100% specificity, and a 48- and 20-hour incubation period reduction for the herpes simplex virus and encephalomyocarditis viruses are all made possible by computerized plaque assays.	Clinical testing, vaccine formation, and viral biology should be acquiescent by using the stain-free technique.

(continued)

Table 4.1 (continued)

Author	Title	Methodology	Drawbacks
Hussain et al. (2024)	Hypergraph convolutional neural networks for clinical diagnosis of monkeypox infections using skin virological images	For the identification and classification of skin virological images, the stratified 3-FCV technique, applying the HGCN algorithm, is a diagnostic predictive model that is both effective and precise.	In order to enhance its classification abilities, the model should be able to extract contextual information and long-range dependencies from images.

4.3 Overview of Virology

A relatively new field of study, virology examines viruses and how they interact with their hosts. It covers molecular biology, structure, relationships, evolution, epidemiology, and preventative strategies. Numerous creatures, including bacteria, algae, fungi, plants, insects, and vertebrates, are contaminated by viruses (Carter and Saunders 2007). Understanding the potential of human viruses as disease agents requires research on these viruses, regardless of their uneven distribution across virus families. Understanding the biology and evolution of viruses is thought to be aided by nonhuman viruses. Since the field is broad and specialized, virus therapy needs to be specialized.

4.3.1 Comprehending Viruses: Features and Mechanisms

Viruses are tiny, non-cellular organisms that depend on their host cells for energy and biosynthesis since they lack cellular organelles and metabolic machinery. Sub-microscopic in size, they are obligatory intracellular parasites that may contain RNA or DNA. In 1952, the Hershey-Chase experiment proved that bacteriophages' genetic material is DNA, establishing the separation of viruses from cellular life. The genetic material of viruses is protected by a capsid, a protein coat; certain viruses additionally have an outermost lipid envelope. Attachment, entry, transcription and replication, assembly, and release are the phases of the viral replication cycle. Because of their high rates of genetic recombination and mutation, viruses are extremely adaptive. Because RNA viruses, such as HIV and influenza, change rapidly, they can elude immune responses and become resistant to antiviral medications. Viral replication and the Hershey-Chase experiment clarified important phases of viral replication, emphasizing the distinct tactics viruses employ to take control of host cellular machinery (Rest 2008).

4.3.2 The Definition of the Chemical Characteristics of Viruses

Because of their structural structure and molecular makeup, viruses differ from cellular life in their chemical properties. They are mostly made up of proteins as well as nucleic acids (the genetic material or RNA), while certain viruses also contain lipids and carbohydrates, which add to their special characteristics and capabilities.

4.3.2.1 Nucleic Acid

Viruses only possess DNA or RNA, in contrast to all other biological entities. Their entire genomes can be single- or two-stranded, straightforward, circular, or segmented, with sizes ranging from a couple of thousand to more than a million. Viral proteins and replication mechanisms are encoded by nucleic acids.

4.3.2.2 Proteins

The capsid, a protective shell that envelopes genetic information, is made up of both structural and non-structural viral proteins that aid in immune evasion, host interaction, and reproduction.

4.3.2.3 Lipid Envelope

Non-enveloped viruses depend on the protein capsid for durability and infection, whereas enveloped viruses, such as HIV and influenza, include an outer lipid bilayer that competes with glycoproteins for connections and penetration into host cells.

4.3.2.4 Carbohydrates

Carbohydrates included in glycoproteins on the outermost layer of enveloped viruses help the virus recognize host cells, evade the immune system, and remain stable in extracellular settings.

4.3.3 Enzymes and Accessory Proteins

Enzymes such as RNA-dependent RNA polymerase and reverse transcriptase are frequently found in viruses and are essential for the start of infection, particularly in viruses with distinct replication mechanisms. Although the capsid or envelope of a virus affects its stability, other factors that can affect it include temperatures, pH levels, and chemical compounds employed in antiviral and disinfection treatments.

Numerous periods are covered in the history of virology, such as

- The identification of viruses as separate entities.
- The link between viruses and major human diseases.
- The creation of techniques for isolating and characterizing viruses.
- The definition of the chemical characteristics of viruses.
- The creation and use of vaccines and treatments.

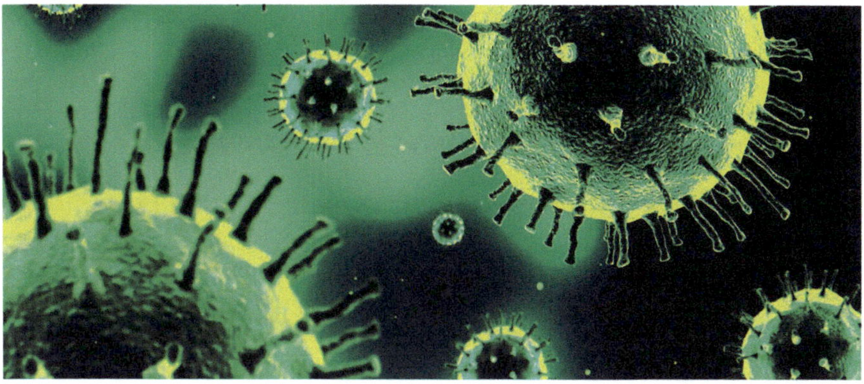

Fig. 4.1 Virology

When bacteria and related disorders were first discovered, the consensus was that environmental factors were the source of sickness. This is where virology got its start. Figure 4.1 shows the image of virology. The uniqueness of microbial fermentation, which Louis Pasteur discovered in 1857, contributed to the scientific legitimacy of microbiology. Charles Chamberland made the connection between certain microorganisms and sickness by creating the unglazed porcelain ultrafilter. It was later demonstrated that ultra-filterable substances were the cause of additional ailments since the early virologists concentrated on what went across the ultrafilter compared to what was retained.

- *The identification of viruses as separate entities.*

By recognizing them as unique entities in biology, the comparatively new subject of virology has profoundly changed scientific understanding. The historical evolution of virology is examined in the following paper, with particular interest paid to significant discoveries, advancements in technology, and conceptual changes that made viruses a basic biological category.

4.3.4 Pre-virology Period: Initial Thoughts and Observations

Infectious entities smaller than bacteria were the subject of conjecture in ancient societies, but these hypotheses lacked instruments for direct observation and testing. Antonie van Leeuwenhoek's creation of the microscope in the seventeenth century revolutionized our understanding of microorganisms, especially "animalcules" such as protozoa and bacteria. Curiosity about tiny organisms was stoked by the presence of some infectious pathogens that could not be detected. Numerous diseases are caused by germs, according to the germ hypothesis of the nineteenth century, which was developed by the researchers Louis Pasteur and Robert Koch. These agents were brought to light by Pasteur's research on fermentation and rabies, and Koch's postulates established criteria for identifying pathogens, initially ruling out viruses since they lacked visualization tools.

4.4 The Identification of Filterable Substances

4.4.1 The Study on Tobacco Mosaic Disease by Adolf Mayer

In 1886, Adolf Mayer discovered filterable agents as a result of his research on tobacco mosaic disease. He discovered that wiping the sap of sick plants on healthy ones might spread the disease, pointing to bacteria as the culprit.

4.4.2 Dmitri Ivanovsky and the Filter of Chamberland

Using a Chamberland filter, Russian researcher Dmitri Ivanovsky analyzed tobacco mosaic disease in 1892. The results demonstrated that the filtered sap was still contagious, indicating a very tiny bacterium but not proving it was different from bacteria.

4.4.3 Beijerinck, Martinus: The "Contagium Vivum Fluidum"

The infectious agent responsible for tobacco mosaic disease was found by Dutch microbiologist Martinus Beijerinck to be incapable of proliferating outside of living host cells. This non-bacterial entity was dubbed "contagium vivum fluidum" by him, proving that viruses are distinct organisms that need host cells to replicate.

4.4.4 Establishing Viruses as Distinct Entities in the Early Twentieth Century

In 1917, French-Canadian researcher Félix d'Herelle identified bacteriophages, which are unique viruses that may target specific host cells in cultures of bacterial cells. The first human disease to be connected to a virus was foot-and-mouth sickness in 1898, and in 1901, researchers discovered viruses that caused several kinds of animal and human illnesses, including yellow fever.

4.4.5 Technology's Contribution to Virus Identification

An important contributor to the advancement of viral identification technologies was Ernst Ruska. With the development of an electron microscopy device in the 1930s, Ruska and his colleagues created the first micrographs of the virus known as the tobacco mosaic virus (TMV), enabling scientists to examine viruses for the first time.

The Nobel Prize in Chemistry was awarded to Stanley in 1946 for his crystallization of TMV in 1935, which revealed its molecular simplicity and differentiation from cellular life forms (Zuckerman 2009). Using techniques like density gradient and ultracentrifugation, scientists were also able to separate and describe viruses according to their size and density, which gave them information about their chemical and physical characteristics.

- *Major human diseases and viruses.*

Seasonal flu epidemics and sporadic pandemics are caused by influenza viruses, which also cause fever, coughing, sore throats, muscle pains, and exhaustion. The pandemic is caused by SARS-CoV and COVID-19, which both cause severe respiratory distress. Annual vaccinations and antiviral medications are part of prevention. When HIV attacks CD4+ T cells, it weakens the immune system and, if left untreated, can cause AIDS. To increase life expectancy, antiretroviral therapy (ART) is part of the treatment. Chronic liver infections brought on by the hepatitis B and C viruses can result in cirrhosis of the liver and hepatocellular cancer. Human papillomavirus (HPV), poliovirus, rabies, and Epstein-Barr virus (EBV) are examples of neurological illnesses. While HPV vaccines guard against high-risk strains, the Ebola virus produces severe hemorrhagic fever that has a high death rate (Strauss and Strauss 2007). During epidemics, control efforts include isolating cases and administering immunizations. In order to understand viral evolution, develop new treatments, and prepare for new threats, research and development are crucial. Vaccines continue to be the most effective means of avoiding infections caused by viruses, with recent innovations such as vaccines made from mRNA for COVID-19 exhibiting swift progress and high efficacy. Prevention strategies consist of surveillance, quarantine, health and hygiene, and health education.

Because it sheds light on viral mechanics and connections with human biology, viral research is essential to improving world health. However, the rise of new viruses and current issues highlight the necessity of ongoing funding for public health infrastructure and virology research.

- *The creation of techniques for isolating and characterizing viruses.*

The advancement of virology from the late nineteenth century has been greatly aided by the development of methods for virus isolation and characterization. The quick identification and characterization of viruses, including newly emerging pathogens like SARS-CoV-2, has been made possible by developments in filtration techniques, cultivation of cell techniques, electron microscopes, centrifuge, density gradient centrifugation, crystallization techniques, molecular biology, serological techniques, and NGS. Because of these developments, virology is now an interdisciplinary science that helps identify and quantify viruses and improves global preparation against viral illnesses.

The dynamic mechanisms by which viruses infiltrate host cells, alter cellular machinery to proliferate, and occasionally induce illness are known as virus-cell interactions. Understanding immune responses, viral pathogenesis, and the creation of antiviral treatments all depend on these relationships. Viruses determine their host and biological tropism by attaching to specific binding proteins on their outermost layer of host cells. By means of endocytosis, fusion of the membrane, uncoating, transcription, translation, manufacturing of proteins, assembly, and release, viruses infiltrate host cells.

There are several different kinds of virus-cell interactions, including transforming, latent, productive, abortive, and chronic infections. Viruses interact with cells through molecular processes such as host receptors, cellular pathway hijacking, resistance evasion, epigenetic changes, cytotoxic effects, and host reactions. Acute cytopathic damage, immune-related damage, and oncogenesis are among the pathological consequences of virus-cell interactions. Research on virus-cell interactions can help with oncolytic virotherapy, gene therapy, vaccine development, and the creation of antiviral medications. The intricacy of immune evasion, new viruses, treatment resistance, and host-virus interactions presents difficulties in researching them. CRISPR technology, AI and ML, synthetic virology, and integrated biology and omics methodologies are some of the future prospects. Specialists can create more potent vaccinations, antiviral medications, and more efficient treatments for a range of illnesses by comprehending these connections.

- *The creation and use of vaccines and treatments.*

Since Edward Jenner's work in the eighteenth century, when vaccines and cures were first developed, viral infections have played a major role in medical science. Molecular biology breakthroughs produced both live-attenuated and inactivated vaccinations, including Albert Sabin's oral polio vaccine and Jonas Salk's inactivated polio vaccine. During the COVID-19 pandemic, molecular biology also helped create mRNA-based technologies and recombinant DNA vaccines, which completely changed the industry. With the development of vaccines and early antiviral medications like acyclovir and extremely effective antiretroviral therapy for HIV/AIDS, treatments for viral infections have also advanced. CRISPR-Cas systems and other gene-editing technologies are examples of modern methods for targeted virus eradication. Public health initiatives and worldwide vaccination campaigns have made vaccinations and treatments widely available. But there are still difficulties: drug-resistant viral strains, vaccine reluctance, and efficient defences against novel viral infections are still obstacles, nevertheless.

4.4.6 Deep Learning Applications in Various Fields of Virology

Understanding the biology, evolution, and interactions of viruses and virus-like agents is largely dependent on virology. Because of the complexity and unpredictability of viruses, traditional techniques like microscopy, culture-based approaches, and serological assays are inadequate, particularly in light of newly emerging zoonotic risks and highly changeable pathogens. The development of high-throughput sequencing technology has produced enormous datasets, such as transcriptomes, proteomes, viral genomes, and maps of host-pathogen interactions, which have completely changed the field of virology. Computational biology is one of the most important tools to address these challenges. Specifically, ML and DL can digest this type of information readily. Raw data into nonlinear aggregate patterns has allowed for the integration of various types of data. Beyond DL has practical applications in various industries, from medicine to public health, despite being a fundamental area of research.

4.4.7 Deep Learning Applications in Plant Virology

The study of plant viruses, how they are intertwined with host plants, and the influence they have on ecosystems and agriculture. The virus of the plant holds significance in its ability to alter the various systems, create food shortages, and generate crop losses, which are one of the causes of financial insecurity. Either DNA or RNA, with their genomes being part of one or two different types, are used; this can depend. The viruses can also be circular, filamentous, or rigidly rod-shaped.

Plant viruses are spread by things such as nematodes, insects, fungi, and mechanical harm, which are external agents. The most common plant viruses are those that affect pepper, tomato, and tobacco. They were developed according to their behavior outside of the host and filamentous virions. Biological vectors include nematodes, fungi, insects, and mechanical transmission. Morphological symptoms of viral infection in plants can include physiological symptoms and growth suppression, such as curling, stunting, distortion of photosynthesis, damaged nutrition and water transport, and a changed metabolism leading to low yield. Viruses can also induce systemic symptoms in plants when they are infected, with a serious negative impact on crop productivity and plant health. Hypersensitive reaction, systemic acquired resistance, and RNA silencing are host defense mechanisms implicated in plant-virus interactions. Viruses can cause ecological issues and trade barriers, as well as lost harvests. The study also takes into account molecular approaches such as RT-PCR and electron microscopy, analytical techniques such as ELISA, as well as visual approaches. Inspection, RT-PCR, and cytology are used in the diagnosis of viral infections. Biotechnological methods like DNA samplers, all-vector resistance, insect repellents, and biological insecticides are employed. Control agents, controlled cross-breeding, and genetic engineering are used to produce resistant varieties and traditional practices. Management and control strategies can include sustainable practices, such as plant rotation and intercropping. Synthetic biology, ML and AI, and genomic and transcriptome research are examples of advances in plant virology. Comprehending and managing plant viruses guarantees sustainable food production, preserves biodiversity, and lowers agricultural financial losses.

Pattern and object detection in digital captures have been greatly impacted by methods that use DL, particularly across the field of plant pathology. Forecasting models have achieved 100% accuracy, but subjectivity and unstructured, dynamic situations make it difficult to determine "enough" data (Barbedo 2022).

Contemporary studies must fill in information shortfalls and overcome technological and real-world obstacles, especially with proximal photos and unmanned aerial vehicles, in order to enhance categorization.

Using hyperspectral remote sensing imaging, the study (Nguyen, Canh, et al. (2021)) investigates the identification of the grapevines immunized with grapevine vein-clearing virus (GVCV) in the very initial stages of asymptomatization (Nguyen et al. 2021). The findings indicate that vegetative indices (VIs) are significant; reflection spectra can provide the best wavelengths for GVCV authentication, and categorization performances are similar across datasets. The results of the computerized 3D-CNN feature extractor are encouraging.

The research (Janarthan et al. 2022) describes a light, deep metrics-based categorization network to recognize plant illnesses in leaf photos (Janarthan et al. 2022). A streamlined MobileNetV2 architecture is further used to improve the embedding and education prototype modules. The model only needs 1 MB of memory to store and performs 0.088 billion multiply-accumulation operations. In the future, there will be a smartphone application for identifying diseases in the field. Both the unbalanced and Tomato Powdery Mildew Disease (TPMD) datasets have been employed in the research to anticipate plant diseases using the Extreme Learning Machine (ELM) algorithm (Bhatia et al. 2020). The IMPS approach generates the best results, with the highest percentages of 88.57% and 89.19%, respectively.

The paper (Kaya et al. (2019)) indicates a requirement for more research by claiming that transfer learning improves the effectiveness of models built with DL, especially those with advanced features and fine-tuning (Kaya et al. 2019). Using hyperspectral cameras, this study shows that machine vision algorithms can identify disease symptoms with nearly the same level of accuracy as a seasoned crop professional. The identification of diseases can be accomplished with an enhanced fully convoluted neural network (FCNN). According to the study (Polder et al. 2019), an updated FCN framework for recognizing diseases can be used with a weak one-dimensional rather than label sequence, which lowers the burden of labeling datasets and the risk of overfitting. When examined on infected potato plants with the PVY virus, the system comes up with accuracies of more than 0.75 for three of the four-row/week configurations.

Models have limited scalability, there is a lack of real-time annotated data, and computational resources are restricted. Many challenges hamper the progress of algorithms utilizing DL in plant virology. A significant quantity of labeled data is necessary, which can be expensive and require considerable time. This results in numerous issues. Furthermore, it is essential to develop datasets that encompass the entire range of viral infections resulting from various plant viruses found across distinct crops and regions. Another critical challenge lies in the interpretability of DL models, which operate as "black boxes," making it increasingly difficult to gain the trust of agricultural stakeholders or derive meaningful biological insights. Despite these challenges, advancements in deep understanding and related fields have tremendous capacity to transform plant virology going forward. Collaborating, farmers, agricultural organizations, and research institutions might offer substantial, diverse, and annotated datasets.

- *Deep learning applications in viral oncology.*

Viral oncology is the field that delves into how viruses nudge cancers to start, spread, and be cured. A group of viruses called oncogenic viruses, or oncoviruses, includes the potential to change healthy cells into cancerous cells by either altering host cellular interactions or integrating their genetic material into the host genome. Based on their genes that induce cancer, they are further divided into two major types: RNA viruses and DNA viruses. Oncogenic viruses can cause cancer by stimulating unchecked cell growth, altering biological pathways, and messing with normal cell cycle regulation. There are several ways to identify these viruses, including immune evasion, oncoprotein synthesizing, persistent inflammation, incorporation into the genome of the victim, and epigenetic changes. About 15–25% of all malignancies worldwide are virus-associated, especially in areas where oncogenic viral infection is prevalent. Vaccination, screening initiatives, antiviral medications, and lifestyle changes are all part of the prevention and treatment of malignancies linked to viruses. DL and AI, CRISPR-Cas systems, epigenetic modification therapy, and vaccine advancement are examples of advancements in viral oncology. The absence of universal vaccinations, global differences, and latency and persistence are obstacles in viral oncology. Important insights into the biology of cancer, its effects on public health, and novel therapeutic approaches are provided by viral oncology. Knowing how viruses and cancer are related opens up possibilities for prevention, early detection, and creative therapy.

Since both CNNs and RNNs have been used for constructing state-of-the-art networks, the process of DL has improved the most in the past few decades. Considerable achievements have been demonstrated in the fields of categorization and recognition, and CNN focuses on computer vision, while RNN explores sequencing data and the processing of natural language (NLP). In particular, outstanding efficiency has also been noticed across numerous medical domains, such as the categorizing of skin cancer types and determining the identity of pathological histological samples (Esteva et al. 2017; Levine et al. 2019). The identifying of the A plague region in people with Alzheimer's disease, the use of nuclear morphometric measures for determining cancer cells from normal cells (Radhakrishnan et al. 2017), and the gathering of data from electronic health records (EHR) (Rajkomar et al. 2018) to forecast medical appointment outcomes, readmission from hospitals, and mortality rates.

The study (Beltrán et al. 2024) develops the VirOncoTarget online application, improves the accuracy and generalizability of predictive models, and offers a reliable method for identifying viral oncoproteins utilizing GANs and supervised learning techniques. The accuracy of recognizing viral genomes in metagenomic samples using DL methods has been significantly improved using the ML approach DeepVirFinder (Ren et al. 2020). For various sequence lengths, it achieved an accuracy of 98%, outperforming the state-of-the-art approach, VirFinder. When DeepVirFinder was executed on actual human gut metagenomic materials, it uncovered 51,138 genomes of viruses in individuals with colorectal cancer. In order to detect and diagnose lung diseases in the human lung system, the study offers four DL architectures (Ibrahim et al. 2021). The VGG19 + CNN model fared better than the others, obtaining an accuracy of 98.05% based on CT and chest X-ray images. To produce precise immunogenicity potential, a beta-binomial model is created (Li et al. 2021a), which outperformed current procedures. They used datasets of immunogenic diseases for evaluating encoding, DL, and ML techniques. With adequate training data, the model showed learnable biochemical connections and successfully detected peptide-MHC and TCR interactions. Through the use of chemical-exchange-saturation-transfer magnetic resonance fingerprinting (CEST-MRF) assisted by DL algorithms (Perlman et al. 2022), the research (Perlman, Or, et al. (2022)) shows that non-invasive methods of imaging may determine early apoptotic reactions during oncolytic virotherapy in mice with glioblastoma multiforme.

In comparison to board-certified pathologists, the study indicates that the deep neural network EBVNet is capable of predicting EBV status across gastric cancer-related H&E-stained WSIs. A human-machine fusion could enhance the model and generalize it to different therapeutic settings. Given that the EBVNet's particularity and sensitivity were noticeably higher than pathologists, it may find application as a method of screening for validating EBV testing (Kendall and Gal 2017). It demonstrated that the percentage of EBVaGC had less of an impact on the diagnostic effectiveness of the EBVNet. In order to increase diagnostic efficacy and overcome prediction uncertainties, the study devised (Zheng, Xueyi, et al. (2022)) a human-machine fusion technique. Both the EBVNet and pathologists with varying degrees of experience and expertise were outperformed by the human-machine fusion (Zheng et al. 2022), indicating that any pathologist could merge the EBVNet's prediction with their personalized examinations.

Viral oncology uses DL to comprehend how viruses contribute to the development of cancer. Nevertheless, obstacles include the inability to integrate multi-omics data and the scarcity of datasets. By enhancing interpretability and confidence in DL models, future developments like AI with explanations, reinforcement learning, and federated learning can facilitate a more thorough comprehension of interactions between viruses and hosts and oncogenesis.

- *Deep learning applications in epidemiological virology.*

The study of viral infection patterns, causes, and consequences in populations is the focus of the multidisciplinary science of epidemiological virology. It integrates the concepts of virology with epidemiology to comprehend how viruses interact with hosts, disseminate, and cause illnesses. Epidemiological virology's basic ideas include characterizing the prevalence, distribution, and causes of viral infections in particular populations; figuring out risk factors, routes of transmission, and vulnerable groups; and trying to contain or stop viral epidemics. Global monitoring, host-virus interactions, viral evolution, epidemic investigation, and virus transmission dynamics are important fields of study in epidemiology and virology. Control and eradication methods, zoonotic spillover research, vaccine development and implementation, antiviral medication development, and outbreak prediction and readiness are among the uses of epidemiological virology. In epidemiological virology, the basic Reproduction Number (R_0), attack rate, case fatality ratio (CFR), incidence, frequency, and population immunity threshold are important measures. Globalization, climate change, antiviral resistance, and re-emerging and developing viruses are some of the problems facing epidemiology and virology. In epidemiological virology, future directions involve employing sophisticated genomic technologies, the use of AI and ML, better vaccine platforms, increased international cooperation, and the integration of both human and animal and environmental wellness.

A study conducted in China sought to enhance early identification and resilience for influenza outbreaks by merging data from Baidu search queries with conventional virological surveillance data. There is promise for wider use in bronchial infectious disease observation, as the study (Yang et al. 2023) discovered a significant correlation between particular Baidu search phrases and influenza-positive rates. But when adding search engine data to contemporary surveillance, care should be exercised. In order to correctly recognize subsets from HIV-1 epidemics using previous data and show scalability, a DL system effectively handles enormous viral sequences. Rather than building phylogenetic trees, our method uses a combination distance matrix from dataset sequences, which enhances interpretation and reduces evaluation costs. Large data studies are not a good fit for complex approaches like elegant Bayesian algorithms (Suchard et al. 2018), which partially integrate numerous probable trees. The study (Cañas et al. 2025) enhances surveillance of new outbreaks and emerging diseases by proposing an epidemiology strategy based on wastewater. Model explanation ability is crucial, because it demonstrates how ML and prediction algorithms can increase model validity and accuracy.

When it comes to forecasting transmission patterns during epidemics, the graph neural network technique DeepDynaForecast has shown encouraging results. It can correctly identify lineages that contribute to growth and decrease in simulated epidemic phylogenies by predicting transmission clusters, according to validation data (Sun et al. 2024). The model's shortcomings, especially when it comes to integrating genetic sequences and subsequent infections, require more research. Applied to a Florida HIV outbreak, DeepDynaForecast supports earlier research and provides a more comprehensive approach than genetic classification alone. It suggests that the CDC prioritizes larger clusters by classifying expected transmission growth (Rich et al. 2022). Unclustered sequences and inaccurate classifications, however, may restrict its usefulness for recognizing high-priority categories for health care interventions.

Issues with DL in epidemiology and virology include restricted interpretability, difficulty integrating various data sources, and a dearth of high-quality datasets. For viral epidemics and pandemics, accuracy, adaptability, and early observation can be enhanced by combining DL algorithms with real-time surveillance technologies, multi-modal knowledge gaining, and high-speed computing.

- *Deep learning applications in immunovirology.*

The study of the connection between viruses and the immune system of the host is known as immunovirology. It focuses on how viruses cause disease or immunity, elude immune defenses, and react to viral infections. Immunopathology, immunodeficiencies, viral evasion, and host immunological responses are important subfields in immunovirology. The immune system's function uses both innate and adaptive immunological responses to create a several-layered defense against viruses. Examples of such are hepatitis viruses, herpesviruses, influenza viruses, SARS-CoV-2, and HIV. The work of immunovirology is used in making vaccines, therapeutics for immune modulation, and medicine to inhibit viral infections, as well as in the study of immunodeficiencies. Viruses, such as hepatitis, herpes viruses, influenza, SARS-CoV-2, and HIV, are infectious to immunovirology. Utilization of CRISPR, ELISA and neutralizing assays, flow cytometry, animal models, and NGS are the first tools used in immunovirology. Understanding long-term immunity, immunocompromised populations, immunological escape, and new viruses like SARS-CoV-2 are some of the challenges facing immunovirology. The creation of universal vaccinations, host-directed therapeutics, synthetic immunology, and systems immunology are potential future paths in immunovirology.

MIT researchers have created models that anticipate MHC-peptide interaction using CNNs and LSTM networks as their framework (Machaca et al. 2023). This helps with T-cell identification and vaccine design by identifying important viral protein areas for targeted immunotherapy. The immunological repertoire, including the diversity of B-cell and T-cell receptors, has been researched using DL. Stanford University's DeepRC methodology (Ghoreyshi and George 2023) is used to identify TCR sequences and determine the high specificity of viral antigens.

As demonstrated in research on SARS-CoV-2, DL algorithms have been employed to forecast viral modifications that evade antibody-neutralizing agents, thereby jeopardizing the effectiveness of vaccines. DL uses LSTM models to predict outbreaks of cytokines in COVID-19 victims (Ashraf et al. 2021), allowing for prompt action and better results for patients. In the instance of SARS-CoV-2, computational models such as reinforcement learning alongside graph neural networks (Ling et al. 2024) have already been utilized for assessing substances for antiviral activity, speeding up medication discovery, and boosting the effectiveness of the development of treatments. The research used a complete recurrent neural network tuned for electron microscope transmission images for examining the morphometry of the SARS-CoV-2 virus and recommended a prediction model (Taha et al. 2022). The simulator's great accuracy aided in the development of drugs and the identification of viruses. Potential antiviral drugs that target viral proteins or host components have been discovered using DL models, such as reinforcement learning and neural networks featuring graphs (DeepDrug for SARS-CoV-2) (Aruna et al. 2024).

A number of complicated steps go into creating models for DL to comprehend immunological responses. However, obstacles including interpretability, computational limitations, and high-quality datasets impede improvement. Subsequent developments in federated learning methodologies and understandable AI may improve this comprehension.

- *Deep learning applications in animal virology.*

The study of viruses that infect animals, including companion animals, domestic livestock, and wildlife, is known as animal virology. Comprehending viral infections that affect animal welfare, production, and health is essential. Human health is also affected by animal virology because many zoonotic viruses have origins in animals and are capable of spanning species boundaries. Controlling outbreaks, enhancing veterinary care, guaranteeing food security, and reducing the dangers of newly emerging infectious illnesses are all made possible by effective research. Applications include zoonotic disease prevention, vaccine development, virus-host interactions, illness diagnosis and surveillance, and the creation of antiviral medications. Cross-species transmission, a lack of data, wildlife monitoring, restrictions on vaccine research, and antibiotic resistance are some of the difficulties. Future directions include the following: mRNA vaccine technology, wildlife virome investigation, One Health Approach, DL and AI, and NGS.

CNNs were implemented in Li et al.'s (2020) study for recognizing viral illnesses in cattle with high accuracy (Li et al. 2021b), allowing for prompt identification and epidemic containment. Improved prompt treatment for extremely dangerous viruses by using ML techniques to predict zoonotic possibilities in bats while discovering important human spillover hazards (Brierley and Fowler 2021). The use of ML to evaluate and foresee important animal diseases using large data is covered in this work. Specific focus is paid to both unsupervised and supervised learning methodologies (Zhang et al. 2021) and their potential for the identification and forecasting of animal diseases. By predicting host species vulnerability through DL, the work proposes a pathway to find prospective hosts containing zoonotic viruses (Sutanto and Turcotte 2021). This approach achieves an overall prediction success rate of 86.89% by extracting elements from viruses composed of RNA and RNA second-generation structures.

DL applications in animal virology have been restricted by the scarcity of high-quality datasets. However, improvements in computational efficiency and portable AI technologies can enhance expectations of viral development and outbreak dynamics, revolutionizing animal infectious disease detection and response.

- *Deep learning applications in genomic virology.*

The research and analysis of viral genomes in order to comprehend their framework, operation, evolution, and host interactions is known as genomic virology. This discipline investigates how viruses multiply, change, and adapt, shedding light on viral disease, epidemiology, and treatment development. Classifying viruses into relationships, genera, and species; identifying genes that contribute to viral infectiousness and host immune evasion; tracking viral transmission chains; identifying targets for vaccine antigens and drugs to treat viruses; tracking viral progress and mutation analysis; and identifying the reservoirs of animals and cross-species transmission are all made possible by genomic sequences. Important viruses in genomic viral biology include the Zika virus, HIV, SARS-CoV-2, influenza virus, and hepatitis viruses (HCV and HBV). Zoonotic spillovers, data management, high mutation rates, and the feasibility of universal techniques are some of the challenges in genomic virology. In genomic virology, future directions include long-read sequencing technologies, genome editing for therapies, integrating with other omics technologies, utilizing AI and ML for predictive genomics, and growing databases like GISAID for international genomic data sharing.

A comprehensible CNN that identifies biologically significant variations in viral sequences is the EdeepVPP-hybrid CNN model (Dasari and Bhukya 2022). It can serve as an encouragement system for further evaluation of raw segments classified as "unknown" and exceeds current techniques in identifying trends across high stimulation values. In order to automatically detect viruses in human biospecimens, the research offers an architecture that combines conventional ML methods with NLP (Alshayeji and Sindhu 2023). For viral genome prediction, the model's categorization precision, efficacy, recall, and detection percentage were all 98.6%.

Rapid mutation rates, the intricacy of viral genomic information, and the scarcity of annotated datasets provide challenges for DL strategies for genomic virology. The next advances to speed up genomic virology investigations include different hybrid DL strategies and improvements in high-performance machines and cloud computing infrastructures.

4.5 Conclusions

Virology has been greatly impacted by DL since it has made it feasible to process and analyze large datasets with previously unheard-of speed and precision. Among the major accomplishments are improved diagnostics, genetic insights, vaccine development, medicinal discoveries, epidemiological simulation, and better awareness of host interaction and virus development. CNNs have improved diagnostics by making it possible quickly to recognize viral illnesses using data from biosensors, genomic sequencing, and medical imaging techniques. For evaluating viral genomes, forecasting mutations, and comprehending adaptive patterns, generative adversarial networks, transformer-based simulations, and recurrent neural networks have been extensively employed. The identification of antiviral drugs has been accelerated by AI-driven drug discovery platforms, and the development of focused antiviral methods is guided by tools such as DeepVHPPI, which anticipate virus-host interactions.

By offering insights into immune evasion mechanisms and virus-host interactions, DL has completely transformed the science of virology. However, issues including high variability, interpretation, processing demands, integration difficulties, collaboration on data and privacy, and model generalization still exist. Ethical questions around prejudice, data exploitation, and regulatory oversight are brought up by the use of DL algorithms in healthcare. Improved trust and usefulness, customized treatment interventions for viral infections, and a closer connection between prognosis and concrete findings are the goals of these developments.

Luckily, to ensure fair benefits with widespread adoption, ethical considerations such as fairness, reliability, and transparency are crucial.

A mixture of medical records, molecular traits, and imaging information, ML technologies might allow doctors to detect neurodegenerative diseases more effectively. Accurately and sooner, which may lead to more customized treatments and reduced clinical trial durations and expenses. Nonetheless, creating and utilizing DL algorithms for Alzheimer's disease data remains challenging due to limitations in data, such as the absence of extensive datasets, metadata evaluations, and prejudice against specific demographic groups. To tackle these challenges, innovative approaches, including techniques like active learning, data adaptation, and generative adversarial neural networks, have been studied. A thorough assessment of DL model effectiveness is crucial for selecting the optimal model for tasks and ensuring that clinicians can rely on the model's results. Fast testing, immunization evolution, and the discovery of antivirals are facilitated by DL, which is transforming virology. However, challenges persist with interpretability and data limitations. By merging biological proficiency with computational advancements, we can better understand viruses while also developing efficient steps to safeguard against emerging hazardous risks.

References

Alshayeji MH, Sindhu SCB (2023) Viral genome prediction from raw human DNA sequence samples by combining natural language processing and machine learning techniques. Expert Syst Appl 218:119641

Aruna AS, Babu KR, Deepthi K (2024) A deep drug prediction framework for viral infectious diseases using an optimizer-based ensemble of convolutional neural network: COVID-19 as a case study. Mol Divers 29:2473–2487

Ashraf I, Alnumay WS, Ali R, Hur S, Bashir AK, Zikria YB (2021) Prediction models for covid-19 integrating age groups, gender, and underlying conditions. Comput Mater Contin 67(3): 3009–3044

Barbedo JGA (2022) Deep learning applied to plant pathology: the problem of data representativeness. Trop Plant Pathol 47(1):85–94

Beltrán JF, Herrera-Belén L, Yáñez AJ, Jimenez L (2024) Prediction of viral oncoproteins through the combination of generative adversarial networks and machine learning techniques. Sci Rep 14(1):27108

Bhatia A, Chug A, Singh A (2020) Application of extreme learning machine in plant disease prediction for highly imbalanced dataset. J Stat Manag Syst 23:1059–1068. https://doi.org/10.1080/09720510.2020.1799504

Brierley L, Fowler A (2021) Predicting the animal hosts of coronaviruses from compositional biases of spike protein and whole genome sequences through machine learning. PLoS Pathog 17(4): e1009149

Burrell CJ, Howard CR, Murphy FA (2016) Fenner and White's medical virology. Academic Press, Amsterdam

Cañas C, Rubén RS, López-Peñalver JC, Mohedo JV, Cervera B, Garrido JF, Vela RJ, Carcelén AP, García-Algar Ó, Caballero VG, Andreu-Fernández V (2025) Forecasting SARS-CoV-2 outbreak through wastewater analysis: a success in wastewater-based epidemiology. Front Environ Sci Eng 19(1):12

Carter J, Saunders VA (2007) Virology: principles and applications. Wiley, Newark

Chaudhary JK, Sharma H, Tadiboina SN, Singh R, Khan MS, Garg A (2023) Applications of Machine Learning in Viral Disease Diagnosis. In: 2023 10th International conference on computing for sustainable global development (INDIACom). IEEE, Piscataway, pp 1167–1172

Ching T, Himmelstein DS, Beaulieu-Jones BK, Kalinin AA, Do BT, Way GP, Ferrero E et al (2018) Opportunities and obstacles for deep learning in biology and medicine. J R Soc Interface 15(141):20170387

Costa PR, Acencio ML, Lemke N (2010) A machine learning approach for genome-wide prediction of morbid and druggable human genes based on systems-level data. In: BMC genomics, vol 11. BioMed Central, London, pp 1–15

Dasari CM, Bhukya R (2022) Explainable deep neural networks for novel viral genome prediction. Appl Intell 52(3):3002–3017

Esteva A, Kuprel B, Novoa RA, Ko J, Swetter SM, Blau HM, Thrun S (2017) Dermatologist-level classification of skin cancer with deep neural networks. Nature 542(7639):115–118

Fang Z, Feng T, Zhou H, Chen M (2022) DeePVP: identification and classification of phage virion proteins using deep learning. Gigascience 11:giac076

Ghoreyshi ZS, George JT (2023) Quantitative approaches for decoding the specificity of the human T cell repertoire. Front Immunol 14:1228873

Hussain S, Songhua X, Aslam MU, Waqas M, Hussain F (2024) Hypergraph convolutional neural networks for clinical diagnosis of monkeypox infections using skin virological images. Appl Soft Comput 170(9):112673

Ibrahim DM, Elshennawy NM, Sarhan AM (2021) Deep-chest: Multi-classification deep learning model for diagnosing COVID-19, pneumonia, and lung cancer chest diseases. Comput Biol Med 132:104348

Imming P, Sinning C, Meyer A (2006) Drugs, their targets and the nature and number of drug targets. Nat Rev Drug Discov 5(10):821–834

Janarthan S, Thuseethan S, Rajasegarar S, Yearwood J (2022) P2OP—plant pathology on palms: a deep learning-based mobile solution for in-field plant disease detection. Comput Electron Agric 202:107371

Janiesch C, Zschech P, Heinrich K (2021) Machine learning and deep learning. Electron Mark 31(3):685–695

Joshi RC, Kaushik M, Dutta MK, Srivastava A, Choudhary N (2021) VirLeafNet: automatic analysis and viral disease diagnosis using deep-learning in Vigna mungo plant. Eco Inform 61:101197

Kaya A, Keçeli AS, Catal C, Yalic HY, Huseyin T, Tekinerdogan B (2019) Analysis of transfer learning for deep neural network-based plant classification models. Comput Electron Agric 158: 20–29. https://doi.org/10.1016/j.compag.2019.01.041

Kendall A, Gal Y (2017) What uncertainties do we need in bayesian deep learning for computer vision? Adv Neural Inf Proces Syst 30:5580–5590

Lambert S, Voznica J, Morlon H (2023) Deep learning from phylogenies for diversification analyses. Syst Biol 72(6):1262–1279

Levine AB, Schlosser C, Grewal J, Coope R, Jones SJM, Yip S (2019) Rise of the machines: advances in deep learning for cancer diagnosis. Trends Cancer 5(3):157–169

Li G, Balaji Iyer VB, Prasath S, Ni Y, Salomonis N (2021a) DeepImmuno: deep learning-empowered prediction and generation of immunogenic peptides for T-cell immunity. Brief Bioinform 22(6):bbab160

Li Z, Liu F, Yang W, Peng S, Zhou J (2021b) A survey of convolutional neural networks: analysis, applications, and prospects. IEEE Trans Neural Netw Learn Syst 33(12):6999–7019

Ling L, Mondal WU, Ukkusuri SV (2024) Cooperating graph neural networks with deep reinforcement learning for vaccine prioritization. IEEE J Biomed Health Inform 28(8):4891–4902

Liu J, Li J, Wang H, Yan J (2020) Application of deep learning in genomics. Sci China Life Sci 63: 1860–1878

Liu T, Li Y, Koydemir HC, Zhang Y, Yang E, Eryilmaz M, Wang H et al (2023) Rapid and stain-free quantification of viral plaque via lens-free holography and deep learning. Nat Biomed Eng 7(8):1040–1052

Machaca VE, Goyzueta V, Cruz M, Tupac Y (2023) Deep learning and transformers in MHC-peptide binding and presentation towards personalized vaccines in cancer immunology: a brief review. In: International Conference on practical applications of computational biology & bioinformatics. Springer, Cham, pp 14–23

Maity NG, Das S (2017) Machine learning for improved diagnosis and prognosis in healthcare. In: 2017 IEEE aerospace conference. IEEE, Piscataway, pp 1–9

Mihaylov I, Nisheva M, Vassilev D (2019) Application of machine learning models for survival prognosis in breast cancer studies. Information 10(3):93

Mock F, Viehweger A, Barth E, Marz M (2021) VIDHOP, viral host prediction with deep learning. Bioinformatics 37(3):318–325

Nami Y, Imeni N, Panahi B (2021) Application of machine learning in bacteriophage research. BMC Microbiol 21(1):193

Nguyen C, Sagan V, Maimaitiyiming M, Maimaitijiang M, Bhadra S, Kwasniewski MT (2021) Early detection of plant viral disease using hyperspectral imaging and deep learning. Sensors 21(3):742

Perlman O, Ito H, Herz K, Shono N, Nakashima H, Moritz Zaiss E, Chiocca A, Cohen O, Rosen MS, Farrar CT (2022) Quantitative imaging of apoptosis following oncolytic virotherapy by magnetic resonance fingerprinting aided by deep learning. Nat Biomed Eng 6(5):648–657

Polder G, Blok PM, De Villiers HAC, Van der Wolf JM, Kamp J (2019) Potato virus Y detection in seed potatoes using deep learning on hyperspectral images. Front Plant Sci 10:209

Radhakrishnan A, Damodaran K, Soylemezoglu AC, Uhler C, Shivashankar GV (2017) Machine learning for nuclear mechano-morphometric biomarkers in cancer diagnosis. Sci Rep 7(1): 17946

Rajkomar A, Oren E, Chen K, Dai AM, Hajaj N, Hardt M, Liu PJ et al (2018) Scalable and accurate deep learning with electronic health records. NPJ Digit Med 1(1):1–10

Ren J, Song K, Deng C, Ahlgren NA, Fuhrman JA, Li Y, Xie X, Poplin R, Sun F (2020) Identifying viruses from metagenomic data using deep learning. Quant Biol 8(1):64–77

Rest JS (2008) Virology: principles and applications, pp 218–219

Rey JS, Li W, Bryer AJ, Beatson H, Lantz C, Engelman AN, Perilla JR (2021) Deep-learning in situ classification of HIV-1 virion morphology. Comput Struct Biotechnol J 19:5688–5700

Rich SN, Prosperi MCF, Dellicour S, Vrancken B, Cook RL, Spencer EC, Salemi M, Mavian C (2022) Molecular epidemiology of HIV-1 subtype B infection across Florida reveals few large superclusters with metropolitan origin. Microbiol Spectr 10(6):e0188922

Strauss EG, Strauss JH (2007) Viruses and human disease. Elsevier, Amsterdam

Suchard MA, Lemey P, Baele G, Ayres DL, Drummond AJ, Rambaut A (2018) Bayesian phylogenetic and phylodynamic data integration using BEAST 1.10. Virus Evol 4(1):vey016

Sun C, Fang R, Salemi M, Prosperi M, Magalis BR (2024) DeepDynaForecast: phylogenetic-informed graph deep learning for epidemic transmission dynamic prediction. PLoS Comput Biol 20(4):e1011351

Sutanto K, Turcotte M (2021) Extracting and evaluating features from RNA virus sequences to predict host species susceptibility using deep learning. In: Proceedings of the 2021 13th international conference on bioinformatics and biomedical technology, pp 81–89

Taha BA, Al Mashhadany Y, Al-Jumaily AHJ, Zan MSDB, Arsad N (2022) SARS-CoV-2 morphometry analysis and prediction of real virus levels based on full recurrent neural network using TEM images. Viruses 14(11):2386

Wang Z, Sun J, Gao Y, Xue Y, Zhang Y, Li K, Zhang W, Zhang C, Jian Z, Zhang L (2023) Fusang: a framework for phylogenetic tree inference via deep learning. Nucleic Acids Res 51(20): 10909–10923

Xu C, Jackson SA (2019) Machine learning and complex biological data. Genome Biol 20:1–4

Yang L, Zhang T, Han X, Yang J, Sun Y, Ma L, Chen J et al (2023) Influenza epidemic trend surveillance and prediction based on search engine data: deep learning model study. J Med Internet Res 25:e45085

Zhang S, Qiang S, Chen Q (2021) Application of machine learning in animal disease analysis and prediction. Curr Bioinforma 16(7):972–982

Zheng X, Wang R, Zhang X, Sun Y, Zhang H, Zhao Z, Zheng Y et al (2022) A deep learning model and human-machine fusion for prediction of EBV-associated gastric cancer from histopathology. Nat Commun 13(1):2790

Zuckerman AJ (2009) Principles and practice of clinical virology. Wiley, Chichester

Part II

AI and Blockchain Applications in Medical Virology

Deep Learning in Viral Diagnosis: Case Studies and Emerging Frameworks for Precision Medical Virology

5

C. V. Suresh Babu and M. Noorul Afreed

Abstract

The integration of deep learning (DL) into viral diagnostics is transforming medical virology by addressing limitations in traditional diagnostic methods, such as long turnaround times, resource constraints, and accuracy issues. This study aims to explore the potential of deep learning frameworks in enhancing the speed, precision, and scalability of viral detection, focusing on real-world case studies of diagnostic applications. Targeted at researchers, clinicians, and healthcare policymakers, this research employs a multi-faceted approach involving comparative analysis of traditional methods with deep learning-based solutions, development of hybrid diagnostic models, and validation through clinical trials. Key findings reveal that deep learning models improve diagnostic accuracy, reduce errors, and enable early detection of viral variants. This study concludes by proposing an integrated framework combining deep learning, explainable AI, federated learning, and blockchain for robust, scalable, and privacy-preserving diagnostics, with significant implications for clinical decision-making and outbreak management.

Keywords

Blockchain · Clinical virology · Deep learning · Diagnostics · Explainable AI · Federated learning · Genomic analysis · Scalability

C. V. Suresh Babu (✉) · M. Noorul Afreed
Department of Information Technology, Hindustan Institute of Technology and Science, Padur, Chennai, Tamil Nadu, India
e-mail: pt.cvsuresh@hindustanuniv.ac.in

J. M. Chatterjee, S. K. Saxena (eds.), *Artificial Intelligence and Blockchain in Precision Medicine and Virology*, Medical Virology: From Pathogenesis to Disease Control, https://doi.org/10.1007/978-981-96-8919-4_5

5.1 Introduction

The field of viral diagnostics has undergone a significant transformation with new technologies, particularly deep learning (DL). Deep learning is a subset of artificial intelligence (AI) and is making the diagnosis of viral infections more accurate, fast, and economical. In this section, the importance of viral diagnostics, the role that deep learning plays in medical virology, and the significance of case studies in providing real-world insight into the application of these technologies are covered.

5.1.1 Introduction of Viral Diagnostics

Traditionally, detection of viral infections has been established through a range of different laboratory techniques. These included serological tests, PCR assays, as well as virus isolation, which have been very important tools in the identification of the existence of specific viruses. Serological tests identify the presence of antigens or antibodies, and PCR assays amplify the small amount of viral DNA/RNA to detect even extremely small amounts of viral material. Isolation of viruses involves growing the viruses in a laboratory setting, which may be time-consuming and labor-intensive (Mackay et al. 2017). However, these conventional diagnostic procedures are cumbersome and inconvenient because of their long turnaround times and high costs; they are also confined to well-equipped laboratories and experts. Moreover, some viral infections are hard to diagnose by conventional techniques, particularly those that have very nonspecific symptoms or require rapid treatment, such as within hours after the onset of symptoms. All these problems have given birth to newer, more rapid, efficient, and readily available techniques for diagnosing viruses.

5.1.2 Role of Deep Learning in Medical Virology

Deep learning, a subset of machine learning, is revolutionizing medical virology by providing a new approach to diagnosing viral infections more efficiently and precisely (Suresh Babu & Sathiya Narayana 2024). Unlike traditional diagnostic procedures, deep learning algorithms allow for the processing of very large amounts of data, including images, genomic sequences, and clinical records, in identifying patterns that may not readily be detected by human clinicians. Deep learning techniques, such as CNNs and RNNs, have been shown to be very promising in the context of viral diagnosis.

For instance, CNNs have been used to successfully analyze chest X-rays and CT scans for detecting viral infections, such as COVID-19 (Wang et al. 2020). In addition, RNNs and other deep learning models have been used in genomic data to predict viral mutations and drug resistance; this is particularly useful in the management of chronic viral infections, such as HIV and Hepatitis (Goecks et al. 2021). Deep learning models have the advantage of learning from large datasets and improving over time, making them capable of identifying subtle patterns and improving diagnostic accuracy. This capability is crucial in virology, where rapid and accurate identification of viruses can significantly impact patient outcomes and public health response.

While the potential of deep learning in diagnosing viral diseases is theoretically immense, its actual application should be presented with real-time case studies. These case studies give deep insight into the development and validation of these deep learning models in clinics and identify problems in their deployment while offering lessons in overcoming issues like poor quality data, model interpretability, and integration in a healthcare system.

For example, deep learning case studies about diagnosis of COVID-19 based on imaging (Wang et al. 2020) or HIV drug resistance prediction (Goecks et al. 2021) indicate that deep learning can improve significantly on both the speed and accuracy of diagnostics.

But it is noteworthy that these studies highlight a significant issue: quality of data used to train a model and diversity are as important as model diversity to ensure a decent performance of the models trained with these datasets. Case studies also point out the importance of collaboration between researchers, clinicians, and policymakers to ensure that deep learning models are not only accurate but also ethical, transparent, and aligned with clinical workflows. More so, real-world case studies aid in identifying areas where deep learning can be integrated into broader diagnostic frameworks. For instance, a multimodal approach combining imaging, genomics, and clinical data has been proven to enhance the accuracy of diagnosis in diseases, such as Hepatitis B and C (Zhang et al. 2023). These case studies also serve as evidence for the cost-effectiveness of deep learning models in diagnosing viral infections, especially in resource limited settings.

5.1.3 Importance of Case Studies for Real-World Insights

While the theoretical potential of deep learning in viral diagnostics is clear, real case studies are what will illustrate how it can be applied practically in the field. The case studies help to demonstrate the development, validation, and implementation of deep learning models in a clinical environment and highlight challenges in deployment. They offer lessons learned that can help overcome some of the barriers, including those related to data quality, model interpretability, and integration with existing healthcare systems.

For example, such case studies as the diagnostic use of deep learning from imaging for COVID-19 (Wang et al. 2020) or predictive modeling of HIV drug resistance (Goecks et al. 2021) demonstrated that deep learning can both increase the accuracy and diagnostic speed of tests. While these studies do highlight the significant role of using high-quality datasets with diversity to train models properly, biases in the data affect the performance of the algorithm. Moreover, case studies indicate the urgency of multistakeholder collaboration to ensure that the deep learning models are both accurate and have ethical, transparent, and clinical workflow conformity.

Real-world case studies also help in identifying how deep learning can be utilized in broader diagnostic frameworks in real life. For instance, a multimodal method that combines imaging, genomics, and clinical data has been proven to help improve the accuracy of diagnostic results in diseases, such as Hepatitis B and C (Zhang et al. 2023). These case studies also provide evidence of the cost-effectiveness of deep learning models used in diagnosing viral infections particularly in resource limited settings.

5.2 Problem Statement

Rapid diagnosis of viral infections is critical for effective treatment, outbreak control, and prevention of disease progression. Despite many diagnostic methods' improvements, there are still several issues in the viral diagnostics domain. This section deals with current problems, the limitations of conventional methods, and barriers to widespread implementation of DL in medical virology.

5.2.1 Challenges in Viral Diagnostics

Viral diagnostics are highly exposed to various challenges that impact their accuracy, efficiency, and accessibility:

1. *Time Sensitivity and Rapid Outbreaks:*
   For example, with viral infections, like COVID-19, the sooner an individual is detected and isolated, the better the chance to stop transmission. Yet, the longer the diagnosis takes, often because laboratory tests are so slow, it usually intensifies outbreaks (World Health Organization 2020).
2. *Diagnostic Accuracy and Cross Reactivity:*
   Many viral infections have overlapping clinical symptoms, for example, fever, cough, and fatigue, which makes it very challenging to differentiate between different viruses without specific tests. In addition, serological tests have cross reactivity, leading to false positives or negatives, particularly for closely related viruses (Loeffelholz & Tang 2020).
3. *Resource Constraints in Low-Income Settings:*
   Advanced diagnostic facilities, skilled personnel, and reagents for precise testing are mostly unavailable in resource limited regions. As a result, the detection process is delayed, and there is underreporting of cases, along with poor management of viral outbreaks (Peeling et al. 2021).
4. *Scalability and High Testing Demands:*
   The labs face enormous testing backlogs during pandemics. PCR is very sensitive, but the traditional techniques have become a bottleneck since time, expertise, and reagents are involved (Afzal 2020).

5.2.2 Limitations of Traditional Diagnostic Approaches

Traditional diagnostic methods, such as polymerase chain reaction (PCR), enzyme-linked immunosorbent assay (ELISA), and viral culture, remain the gold standard for detecting viral infections. However, these approaches face significant limitations:

1. *Time-Consuming Procedures:*
   PCR and viral culture usually take hours to days for the production of results, hence prolonging the initiation of treatment and the tracing of contacts. Viral culture especially grows viruses in cell lines that are tiresome and not practically used in rapid diagnostics (Mackay et al. 2017).
2. *Dependency on Skilled Personnel and Infrastructure:*
   Advanced laboratory methods require highly trained personnel, specialized reagents, and costly equipment. These dependencies make traditional diagnostics inaccessible in low resource or rural settings (Weissleder et al. 2020).
3. *False Positives/Negatives and Sensitivity Issues:*
   Serological tests, such as ELISA, can result in false positives due to cross-reactivity with another pathogen. In PCR, sensitivity can be affected because of sampling errors or less viral loads in asymptomatic patients (Afzal 2020).
4. *Cannot Handle Large Amounts of Data*
   During massive outbreaks, handling the test results manually is unfeasible. Traditional methods have no automation; hence, during high volumes of samples, delays and inefficiencies in handling are encountered (Corman et al. 2020).
5. *Limited Predictive Insights:*
   Traditional methods identify the presence of a virus but provide limited information regarding its progression, resistance to treatment, or future outcomes.

5.2.3 Barriers to Deep Learning Implementation

While deep learning (DL) offers significant advantages in viral diagnostics, several barriers hinder its large-scale implementation in clinical practice:

1. *Data Availability and Quality:*
   Deep learning models need large, diverse, high-quality datasets to function with accuracy. For viral diagnostics, the issue is obtaining large enough annotated data—be this medical images or genomic sequences—from which to draw upon examples. This can be especially tricky for emerging or rare viruses (Buda et al. 2019).
2. *Data Privacy and Security:*
   High patient data sensitivities, including medical images and genetic information, must be handled. There will be a lot of concern on the part of regulatory requirements over the use of DL models to diagnose viruses (Kaissis et al. 2020).

3. *Computational Resources:*
   Training deep learning models necessitates a lot of computational power, including GPUs and high-performance servers. It is thus difficult for resource-limited regions to implement these technologies due to infrastructure constraints (Esteva et al. 2019).
4. *DL Model Interpretability:*
   Deep learning models often behave as "black boxes." That is, the internal workings of their decision-making cannot be easily interpreted. In the clinical setting, this lack of transparency can reduce the levels of trust among healthcare providers, thereby limiting adoption (Ribeiro et al. 2020).
5. *Integration into Existing Clinical Workflows:*
   Implementing DL-based diagnostic tools integrates into hospital information systems and the existing workflow. The major drawbacks include resistance to change, the lack of technical expertise, and the cost of implementing DL-based diagnostic tools (Topol 2019).

5.3 Research Objectives

Deep learning integration in viral diagnostics is seen as having tremendous potential for transforming disease detection, monitoring, and treatment planning. The aim of this emerging field can be advanced with well-defined research objectives to overcome current limitations and enhance real-world applications. The following objectives attempt to bridge gaps in current diagnostic approaches, evaluate leading-edge technologies, and develop integrated frameworks for precision medical virology.

5.3.1 Case Study Analysis of Real-World Applications

An important aspect of this research is the analysis and assessment of real-world case studies that involve the application of deep learning to viral diagnostics. From practical implementations, valuable insights can be derived in terms of successes, challenges, and areas for improvement in applying DL to viral disease detection.

1. *Effective Application Identification:*
   Through analyzing successful case studies, such as the application of CNNs for COVID-19 detection using chest X-ray images (Wang et al. 2020) or DL models that predict HIV drug resistance from genomic data (Goecks et al. 2021), one is able to see where DL has the most significant impact.
2. *Understanding Implementation Challenges:*
   Case studies highlight practical barriers, such as data limitations, infrastructure constraints, and issues with clinical adoption. For instance, real-world applications in low-resource settings often face difficulties in deploying DL technologies due to computational and logistical challenges (Peeling et al. 2021).

3. *Evaluating Diagnostic Accuracy:*
   This objective, by comparing DL-based models with traditional diagnostic tools, such as PCR or ELISA, aims to measure the accuracy, sensitivity, and specificity of DL algorithms in detecting viral infections.
4. *Lessons from Pandemic Responses:*
   The rapid development and deployment of DL tools during the COVID-19 pandemic can be regarded as valuable lessons on scalability, speed, and adaptability during viral outbreaks (Cohen et al. 2020).

5.3.2 Assessment of Latest Deep Learning Innovations

The second aim of this study is an evaluation of recent deep learning innovations that are expected to advance viral diagnostics. Emergent DL approaches introduce fresh insights into analyzing medical images, genomic sequences, and clinical data.

1. *Comparative Analysis of DL Models:*
   This will be a comparison of some deep architectures, such as convolutional neural networks (CNNs), recurrent neural networks (RNNs), and transformer models, for efficacy in the detection of viral infections. For example, using ResNet, DenseNet, and Vision Transformers, this has been possible to scan medical imaging data with good precision (He et al. 2016; Dosovitskiy et al. 2021).
2. *Genomic Data Analysis:*
   DL models are increasingly applied for predicting viral mutations, drug resistance, and disease progression based on genomic data. For instance, recurrent models and graph neural networks were applied to predict resistance mutations in HIV and Hepatitis viruses (Goecks et al. 2021; Wang et al. 2022).
3. *Explainable AI (XAI):*
   This integration of explainable AI into deep learning tools is vital for enhancing interpretability and trust among clinicians. It is possible to evaluate the XAI frameworks, like GradCAM or SHAP, to understand how DL models identify key features in medical data (Ribeiro et al. 2020).
4. *Performance in Resource-Limited Settings:*
   Lighter weight models, such as MobileNet and TinyML, can be assessed for deployment in resource-constrained environments with limited computational resources (Howard et al. 2017).

5.3.3 Development of Integrated Diagnostic Frameworks

The final research objective is to develop integrated frameworks that combine DL with traditional diagnostic methods for a more holistic and precise approach to viral diagnostics.

1. *Multimodal Data Integration:*
   Integrate medical imaging, genomic data, and clinical records into a unified DL-based framework for improving diagnostic accuracy. For example, multimodal frameworks combining CT scans with viral RNA sequencing have shown enhanced detection rates for COVID-19 and Hepatitis infections (Zhang et al. 2023).
2. *Hybrid Diagnostic Models:*
   Hybrid systems that integrate traditional techniques like PCR or ELISA with DL-based models can fill the existing gaps. For instance, a hybrid system could utilize PCR results for validation and training of DL models to ensure high accuracy (Topol 2019).
3. *Automation and Real-Time Diagnostics:*
   DL-based automated diagnostic pipelines can significantly simplify viral detection processes and facilitate real-time diagnosis, especially in outbreak cases. These systems can also be implemented on point-of-care devices to be used in clinics, hospitals, and fields (Esteva et al. 2019).
4. *Validation and Clinical Trials:*
   An important step in developing integrated frameworks is ensuring their validation through rigorous clinical trials. Testing DL frameworks across diverse populations and healthcare systems ensures robustness, reliability, and generalizability of the models.
5. *Scalability and Deployment:*
   Research efforts must focus on developing scalable frameworks that can be deployed globally, including in resource-limited regions. This includes designing models that require minimal computational resources while maintaining high performance.

5.4 Methodology

The methodology describes the systematic approach toward achieving the research objectives, including case study selection, comparative analysis of diagnostic approaches, investigation of emerging technologies, and the development of an integrated diagnostic framework.

5.4.1 Case Study Selection Criteria

To analyze the practical applications of deep learning (DL) in viral diagnostics, relevant case studies will be selected based on the following criteria:

1. *Clinical Relevance:*
   Case studies need to revolve around major viral diseases, such as COVID-19, HIV, Hepatitis, Influenza, and emerging viral infections, such as Zika and Ebola in which DL-based methods have been applied (Wang et al. 2020).

2. *Usage of Deep Learning Models:*
   The number of DL-based architecture designs with various uses, such as CNN, RNN, transformers, or hybrid models, which will be considered, is one of the studies by He et al. (2016).
3. *Different Data Modalities:*
   All the selected studies will vary depending on their data types, including X-rays, CT scans of medical images, genomic sequences, and clinical records, as a way of exploring multiple DL applications (Zhang et al. 2023).
4. *International Generalizability:*
   Case studies will be conducted from both high-resource and low-resource settings to make sure that the developed tool can be applied widely and is scalable and deployable (Peeling et al. 2021).
5. *Measurable Results:*
   Studies with defined metrics, such as diagnostic accuracy, sensitivity, specificity, and computational efficiency, will be considered for comparative analysis (Cohen et al. 2020).

5.4.2 Comparative Analysis of Diagnostic Approaches

A comparative analysis will be performed to assess the performance and benefits of DL-based approaches against traditional diagnostic methods by comparing the following criteria:

1. *Diagnostic Accuracy:*
   Metrics, such as sensitivity, specificity, precision, and F1 scores, will be used to compare DL models with conventional tools, such as PCR, ELISA, and viral culture (Afzal 2020).
2. *Time Efficiency:*
   The time taken by DL models in processing data and providing results will be compared with conventional diagnostic turnaround times (Corman et al. 2020).
3. *Resource Requirements:*
   This chapter will look at the computational and infrastructural requirements of DL-based approaches compared to laboratory-based ones
4. *Scalability:*
   This will consider the ability of DL models to deal with large datasets and scale in the case of outbreaks.
5. *Cost-Effectiveness:*
   A cost analysis will compare the cost of deploying DL solutions with traditional diagnostic tools in both developed and resource poor settings (Peeling et al. 2021) (Fig. 5.1).

The block diagram illustrates the workflow for utilizing deep learning in viral diagnosis. It begins with Input Data, which typically includes raw datasets such as genomic sequences, imaging data, or clinical records. These data undergo

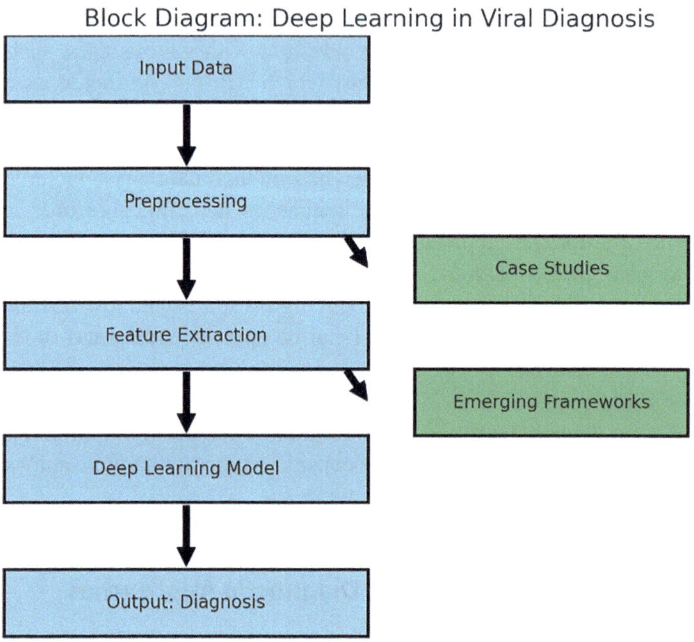

Fig. 5.1 Deep learning in viral diagnosis

Preprocessing, where noise is removed, data are normalized, and essential features are highlighted for better analysis. The processed data are then passed to the Feature Extraction stage, where meaningful patterns and characteristics are identified using statistical or computational techniques.

The extracted features serve as input to the deep learning model, which applies advanced neural networks to analyze and classify the data, identifying viral presence, type, or related patterns. The final stage, Output: Diagnosis, provides a precise medical interpretation, such as the identification of specific viruses or recommendations for further analysis.

Additional components, such as Case Studies and Emerging Frameworks, are included to showcase practical applications and innovative methodologies that enhance the accuracy and efficiency of the diagnosis process. These subcomponents integrate insights from real-world implementations and cutting-edge research to refine the deep learning model and ensure robust, precision-driven outcomes.

5.4.3 Investigation of Emerging Technologies

The following emerging technologies will be explored to assess their applicability and effectiveness in viral diagnostics.

5.4.3.1 Transformers in Viral Sequence Analysis

Transformers, initially designed for natural language processing, have shown promise in analyzing viral genomic sequences.

1. *Application:*
   BERT and its variants, being transformer-based models, will be applied to analyze the viral mutations, predict drug resistance, and classify the viral strains based on sequence data (Vaswani et al. 2017; Rao et al. 2021).
2. *Advantages:*
   The transformers are better suited to handle long-range dependencies in sequences, thus making accurate mutation analysis and identifying the emerging viral variants possible.
3. *Tools:*
   DNABERT and ProtTrans models would be used for the viral sequence classification task (Ji et al. 2021).

5.4.3.2 Explainable AI for Clinical Interpretability

Explainable AI (XAI) techniques will be explored that will allow DL models to be interpreted better, earning the confidence of clinicians as well.

1. *Techniques:*
   Methods, such as GradCAM, SHAP (SHapley Additive exPlanations), and LIME (Local Interpretable Model-agnostic Explanations), will be applied to learn what specific features in a medical image or genomic data cause a model to make a given prediction (Ribeiro et al. 2020).
2. *Advantages:*
   XAI allows the clinician to verify DL-driven predictions as aligning with clinical knowledge, thus being more adopted into healthcare practices.

5.4.3.3 Blockchain for Secure Data Sharing

Blockchain technology will be explored to address data privacy and security concerns when sharing sensitive diagnostic information.

1. *Application:*
   Blockchain can enable secure, decentralized data exchange among institutions without compromising information integrity and patient confidentiality, reports Engin et al. (2021, pp. 2).
2. *Benefits*
   Blockchain ensures that all things are transparent, prevents the organization from unauthorized access, and complies with regulations, such as HIPAA and GDPR.

5.4.4 Framework Development for Integrated Diagnostics

A unified diagnostic framework will be developed by combining deep learning models with traditional diagnostic methods.

1. *Framework Design:*
   Data Integration: Integrate medical imaging, genomic sequences, and clinical records into a single pipeline.
   Model Fusion: Combine DL algorithms with PCR and ELISA for validation and improved accuracy.
   Automation: Develop real-time diagnostic tools for point-of-care use.
2. *Testing and Validation:*
   The framework will be tested on benchmark datasets and validated through clinical trials to ensure reliability and scalability.
3. *Deployment:*
   The final framework will focus on adaptability for both high-resource and low-resource healthcare settings.

5.5 Overview of Case Studies

The following case studies illustrate the successful application of deep learning (DL) in viral diagnostics, leading to improvements in accuracy, efficiency, and scalability. These examples demonstrate the numerous roles DL plays in improving molecular techniques, enhancing antigen detection, analyzing viral genomes, and addressing emerging viral outbreaks.

5.5.1 Rapid RTPCR Optimization Using Deep Learning

Reverse transcription polymerase chain reaction (RTPCR) remains the gold standard for viral diagnostics. However, the process is time-consuming and prone to errors, especially during large-scale testing. Deep learning has been applied to optimize RTPCR workflows and enhance efficiency.

1. *Case Study Example:*
   Researchers developed a DL-based system to analyze RTPCR amplification curves for real-time virus detection. The model, using RNNs and CNNs, could accurately identify positive and negative samples earlier in the amplification cycle, thus reducing diagnostic turnaround time (Kim et al. 2020).
2. *Impact:*
   The optimized workflow reduced false positives and improved testing speed, which proved to be essential during high-demand scenarios, such as the COVID-19 pandemic (Suresh Babu & Praveen 2023).

3. *Key Findings:*
   DL models predicted viral loads with high sensitivity (AUC > 0.98).
   Automation minimized manual interpretation errors.

5.5.2 Antigen Detection Enhancement via DL Models

One of the most popular antigen detection methods is a lateral flow immunoassay for point-of-care viral diagnostics, but these are often less sensitive than molecular methods. Deep learning has been applied to interpret antigen tests more effectively and improve the diagnostic accuracy of antigen tests.

1. *Case Study Example:*
   A DL-based image recognition system was added to the applications on mobile phones for testing lateral flow assays to detect SARS-CoV-2. By the application of CNNs, the model identified even the weakest test lines, which would often go undetected by human vision (Nguyen et al. 2021).
2. *Effectiveness:*
   The antigen detection system, aided by DL, enhanced the sensitivity of the system by 15–20%.
   It was a low-cost option for high-volume testing in resource-poor settings.
3. *Key Findings:*
   DL models were insensitive to changes in lighting and image quality.
   The technology made it accessible by allowing testing at home with a smartphone.

5.5.3 Viral Genome Analysis for Variant Detection

Deep learning has changed the way viral genomes are analyzed, making it possible to quickly identify mutations and emerging variants of concern. These tools are particularly important for monitoring viral evolution and drug resistance.

1. *Case Study Example:*
   DL models, such as transformers (e.g., DNABERT) and graph neural networks, were applied to analyze SARS-CoV-2 and HIV genomes. The models correctly identified SNPs and predicted the functional impact of mutations on viral proteins (Ji et al. 2021; Wang et al. 2022).
2. *Impact:*
   DL-enabled genomic analysis enabled early detection of variants, such as Delta and Omicron strains, during the COVID-19 pandemic (Suresh Babu & Praveen 2023).
   Models predicted mutations that confer drug resistance in HIV, enabling precision treatment planning.

3. *Key Findings:*
   Transformer-based models outperform traditional sequence alignment tools in mutation detection.
   Integration with epidemiological data improved variant monitoring and surveillance.

5.5.4 Applications in Pandemic Outbreaks and Emerging Viruses

The COVID-19 pandemic highlighted the importance of DL technologies in viral outbreaks. DL models have been implemented for fast diagnostics, monitoring, and outbreak prediction. This technology is being extended to emerging viruses like Zika, Ebola, and Nipah (Suresh Babu & Praveen 2023).

1. *Case Study Example COVID-19:*
   During the COVID-19 pandemic, DL-based tools, like COVIDNet, were developed to detect SARS-CoV-2 infections from chest X-rays and CT scans. The models reached diagnostic accuracy at a level similar to radiologists, which enabled speedy screening (Wang et al. 2020).
2. *Case Study Example—Zika Virus:*
   DL models were applied to the mosquito-borne Zika virus genomic data for potential outbreak predictions. AI-based epidemiological models combined environmental and genomic data for Zika virus hotspot forecasting (Liu et al. 2021).
3. *Impact:*
   DL tools hastened pandemic response through real-time diagnostics and outbreak prediction.
   For novel viruses, DL models enhanced preparedness by pointing out possible transmission pathways and mutation hazards.
4. *Key Findings:*
   DL achieved diagnostic accuracies of over 90% for COVID-19 and emerging viral infections.
   AI-based outbreak models enabled proactive intervention strategies.

These case studies demonstrate the transformative role of deep learning in viral diagnostics. From enhancing RTPCR and antigen detection to enabling viral genome analysis and pandemic preparedness, DL technologies have shown significant progress in precision medical virology.

5.6 New Technology and Innovations

This part analyzes the latest technologies that are transforming DL applications in viral diagnosis. The innovations presented cover advanced DL architectures, federated learning, blockchain integration, and explainable AI, which remove all the current limitations and promote scalability, privacy, and adoption in clinics.

5.6.1 Advances in Deep Learning Architectures

Recent advances in DL architectures have greatly improved the accuracy, efficiency, and versatility of diagnostic models.

1. *Transformer Models:*
   Transformers initially developed for natural language processing have been adapted for viral genomic data analysis. Such models as DNABERT and Prot zTrans can identify mutations and predict viral protein structures, thus enabling precise variant detection (Ji et al. 2021).
2. *Vision Transformers (ViTs):*
   ViTs are increasingly being adopted in medical imaging for the detection of viral infections through chest X-rays and CT scans. These models surpass traditional CNNs in feature extraction and pattern recognition (Dosovitskiy et al. 2021).
3. *Hybrid Models:*
   The combination of CNNs with RNNs or transformers has further improved the integration of multimodal data, allowing models to simultaneously process imaging, genomic, and clinical data for more accurate diagnostics (Zhang et al. 2023).
4. *Self-Supervised Learning:*
   A possible resolution is through self-supervised approaches; these methods learn representations by focusing solely on unlabeled data, particularly in rare viral infection' cases (Chen et al. 2020).

Key Benefits:
Much improved accuracy in the cases of image and genomic analysis.
Faster with lower computational costs.
It performs better compared to the small and also in the imbalanced datasets.

5.6.2 Federated Learning for Privacy-Preserving Diagnostics

Federated learning (FL) is an emerging technique that allows collaborative model training across decentralized datasets without sharing sensitive patient data. This innovation addresses privacy concerns in healthcare diagnostics.

1. *Application in Viral Diagnostics:*
   Federated learning enables hospitals and research centers to train DL models on diverse datasets, such as viral genomic sequences or medical images, while preserving patient privacy. For instance, FL was successfully applied to train COVID-19 detection models using chest X-ray data from multiple institutions (Sheller et al. 2020).
2. *Advantages:*
   Privacy Preserving: Patient data remains on local servers, and so this preserves the regulations, such as HIPAA and GDPR.

3. *Challenges:*
   Computational overhead and network latency.
   Model aggregation without sacrificing accuracy.

5.6.3 Blockchain Integration for Data Security and Interoperability

Blockchain technology has emerged as a solution for secure and transparent data sharing in viral diagnostics. It ensures data integrity, interoperability, and compliance with regulatory standards.

1. *Application in Viral Diagnostics:*
   Blockchain can safely store and share patient diagnostic data, like RTPCR results, genomic sequences, and imaging data, across health systems. It also tracks viral outbreaks and vaccination records in a tamper-proof manner (Engin et al. 2021).
2. *Advantages:*
   Data Security: Blockchain does not allow any unauthorized access or data breach.
   Interoperability: It allows for the easy sharing of data across different platforms, thus improving collaboration.
   Transparency: It allows for accountability through an immutable record of access and modification of data.
3. *Use Case Example:*
   Blockchain-enabled platforms were developed to share real-time diagnostic data while preserving patient privacy and data ownership during the COVID-19 pandemic (Casino et al. 2019).

Challenges:
Computational cost is relatively very high.
Healthcare IT system integration is possible.

5.6.4 Explainable AI and Its Role in Clinical Adoption

Explainable AI or XAI deals with the "black box" nature of deep learning models, where the predictions are explainable and interpretable for clinicians. XAI is critical for increasing the trust and adoption of DL-based diagnostic tools in medical virology.

1. *Techniques in Explainable AI:*
   Gradient weighted Class Activation Mapping (GradCAM): It visualizes the regions of medical images contributing to DL predictions (Selvaraju et al. 2017).
2. *Applications in Viral Diagnostics:*
   In COVID-19 diagnostics from chest X-rays, the model was able to show regions in the lungs that may be affected with abnormalities consistent with radiologists, as found by Ardila et al. (2020).

For genomic information, SHAP was shown to highlight important mutations responsible for drug resistance in viruses in precision medicine.

3. *Advantages:*
   Clinical Validation: Clinicians can compare AI predictions with medical expertise.
   Trust and Adoption: Promotes confidence among healthcare professionals.
   Regulatory Compliance: XAI ensures AI systems meet transparency requirements in healthcare.

5.7 Proposed Solutions

Overcoming the challenges and limitations in viral diagnostics necessitates an integrated solution encompassing deep learning (DL), emerging technologies, and innovative frameworks. This section delineates an all-inclusive approach to designing scalable, secure, and adaptable systems for precision medical virology.

5.7.1 Integrated Framework Design

An integrated diagnostic framework uses the power of various deep learning technologies and augmentative tools for optimizing detection, genomic analysis, and prediction of outbreak. Combining the solution involves:

1. *Multimodal Data Integration:*
   It encompasses imaging data, including chest X-rays, CT scans, as well as genomic data through viral sequences, and other clinical metadata, into a single diagnostic pipeline.
   These diverse inputs are processed by advanced DL models, such as hybrid CNN transformer architectures, for comprehensive diagnostics.
2. *Federated Learning (FL):*
   It is used to enable decentralized model training across institutions to preserve privacy while allowing global datasets to be utilized for better generalization.
3. *Blockchain Integration:*
   Ensures data security and interoperability for sharing diagnostic results, patient records, and genomic information.
4. *Explainable AI (XAI):*
   Involves XAI methods, such as GradCAM and SHAP, to make predictions transparent and clinically interpretable, thereby gaining the trust and acceptance of healthcare providers.
5. *Automated Workflow Pipelines:*
   Provides end-to-end pipelines for sample analysis from raw data processing to result interpretation and reporting.

5.7.2 Key Features: Scalability, Security, and Adaptability

Three key features the scalability, security, and adaptability of the proposed diagnostic framework are focused:

1. *Scalability:*
   This has two sub features: Cloud Based Solutions: Utilize cloud platforms for large-scale datasets to allow prompt analysis during pandemics. Federated Learning: It is designed to facilitate the training of models on geographically distributed data sources and, hence, improve scalability without sacrificing data privacy.
   Modular Design: The framework provides an easy integration of novel tools and datasets for emerging viral infections.
2. *Security Blockchain Technology:*
   This feature guarantees tamperproof storage of data, secure sharing, and compliance with all privacy regulations, such as HIPAA and GDPR. Federated Learning: This feature protects the transfer of sensitive data as it only shares model updates across institutions.
   Encryption Protocols: Data encryption techniques protect patient records and diagnostic results.
3. *Flexibility:*
   Tunable Models: DL architectures can be retrained on new viral datasets (e.g., Zika, Ebola) to adapt to emerging threats.
   MultiVirus Capability: The framework supports detection, analysis, and monitoring of multiple viral pathogens.
   Successful Integration with Current Infrastructure: Seams with good integration in LIS and its workflow to hospital information flows that will affect the LIS indirectly.

5.7.3 How Combined Technologies Overcome Specific Challenges

The proposed framework addresses current challenges in viral diagnostics by leveraging a combination of deep learning and complementary technologies:

1. *Challenge: Limited Access to High-Quality Data*
   Solution: Federated learning allows access to decentralized datasets without data sharing, ensuring robust model training.
   Impact: Enhances model performance through diverse data while maintaining privacy.
2. *Challenge: Data Security and Privacy Concerns*
   Solution: Blockchain integration ensures secure, transparent, and immutable data storage. Encryption and federated learning further safeguard sensitive patient information.

Impact: Builds trust among institutions, enabling collaboration and data sharing.

3. *Challenge: Lack of Interpretability in DL Models*
   Solution: Explainable AI techniques (e.g., GradCAM, SHAP) provide insights into DL predictions, making outputs clinically interpretable.
   Impact: Enhances clinician trust and accelerates clinical adoption.
4. *Challenge: Resource Constraints in Low-Income Settings*
   Solution: Cloud-based deployment of lightweight AI models allows cost-effective diagnostics. DL enhanced antigen tests and mobile-based tools provide affordable solutions for point-of-care testing.
   Impact: Improves access to diagnostics in resource limited regions.
5. *Challenge: Emergence of New Viral Variants*
   Solution: Advanced DL models, such as transformers, analyze viral genomic data to identify mutations and predict new variants. Hybrid models integrate imaging and genomic insights to enable accurate diagnostics.

Key Results:
Accurate and real-time detection of viral infections.
Collaboration through technologies preserving privacy.
Transparent AI systems that foster trust and clinical integration.

5.8 Expected Outcomes

Deep learning (DL) and supportive technologies are expected to bring about significant advancement in precision medical virology from viral diagnostics. This chapter presents the main outcomes, with a focus on insights based on case studies, derivation of a new diagnostic framework, and actionable best practices in clinical settings for successful implementation.

5.8.1 Insights from Case Studies

Analysis of case studies in practice illustrates the transformative effect that DL brings to viral diagnosis. This information serves as a starting point for reworking existing strategies and addressing issues:

1. *Optimization of RTPCR and Antigen Tests:*
   DL models cut turnaround time and increased the precision for molecular (RTPCR) and antigen-based diagnostic tests (Kim et al. 2020; Nguyen et al. 2021).
   Case studies showed that the DL tool with automated capabilities reduced human error by 20% and increased test sensitivity.

2. *Enhanced Detection of Emerging Variants:*
   DL-based genomic analysis such as transformers like DNABERT allowed for rapid viral mutations and emerging variants detection, according to Ji et al. 2021.
   This enabled early intervention in the event of outbreaks and guided vaccine development strategies.
3. *Clinical Implementation of Imaging-Based DL Tools:*
   AI tools such as COVIDNet achieved radiologist level accuracy in detecting SARSCoV2 infections from medical images (Wang et al. 2020).
   GradCAM and other XAI techniques enhanced clinical trust and interpretability.
4. *Resource-Limited Scalability:*
   Cloud-based DL solutions, along with mobile-compatible AI models, offered accessible diagnostics at low costs, especially in low-income regions, according to Liu et al. 2021.

5.8.2 A Novel Diagnostic Framework

An integrated diagnostic framework based on the integration of deep learning, federated learning, blockchain, and explainable AI will constitute a robust, scalable, and secure means for virus diagnostics. Therefore, as presented in this framework, it is intended to gain the following.

1. *Enhanced Diagnostic Accuracy.*
   The fusion of imaging, genomic, and clinical data will increase precision in viral infections and emerging variant detection.
2. *Data Privacy and Security:*
   Blockchain and federated learning guarantee security, data privacy, and regulatory compliance (HIPAA and GDPR).
3. *Real-Time and Scalable Implementation:*
   Cloud infrastructure allows for the possibility of real-time diagnosis during pandemics, while the modular nature helps accommodate future new viruses like Zika and Nipah.
4. *Clinical Interpretability and Trust*
   Explainable AI techniques ensure that clinicians understand and trust the model's predictions, facilitating adoption in real-world settings.
5. *Integration with Existing Healthcare Systems:*
   Seamless interoperability with laboratory information systems (LIS) and hospital workflows supports smooth implementation.

Key Features:
Multimodal data processing (imaging, genomic, and clinical).
Privacy-preserving model training.
Real-time diagnostic reporting.
Transparent AI for clinician trust.

5.8.3 Best Practices for Implementation in Clinical Settings

To ensure the proper integration of DL-based diagnostic frameworks into clinical practice, the following best practices are advised:

1. *Data Standardization and Quality Control:*
   Standardize protocols for the acquisition and annotation of imaging, genomic, and clinical data.
   Maintain the quality of data in order to reduce bias and improve generalization.
2. *Privacy-Preserving Solutions*:
   Implement federated learning and blockchain to satisfy the data privacy regulations and also for the establishment of trust between stakeholders.
3. *Clinician Training and Collaboration:*
   Organize the training programs for clinicians in order to better understand AI-based results.
   Interdisciplinary collaboration between AI developers, virologists, and clinicians.
4. *Integrate Explainable AI:*
   Explainable AI tools, such as GradCAM and SHAP, shall be emphasized to make sure that the outcome is transparent so that proper clinical decisions can be made.
5. *Infrastructure and Scalability:*
   Deploy AI models on cloud-based platforms to enable real-time diagnostics as well as large-scale testing.
   Ensure frameworks are updated with new viral threats and retrained continuously.
6. *Pilot Testing and Validation:*
   This will involve conducting pilot studies to validate DL models in real-world clinical settings. The performance should be benchmarked against traditional diagnostic methods.
7. *Regulatory Compliance:*
   Implementing medical device regulations and standards with AI frameworks (FDA, CE marking).
   Coordination with regulatory bodies for FastTrack approval of the AI-driven tools.

5.9 Significance of the Study

The study on Deep Learning in Viral Diagnosis and the proposed integrated framework have significant implications for advancing precision diagnostics, enhancing clinical workflows, and shaping the future of healthcare. This section outlines the contributions to precision diagnostics, the benefits for healthcare practitioners and researchers, and the impact on clinical decision-making.

5.9.1 Contribution to Precision Diagnostics

This study introduces deep learning-driven technologies and frameworks that further enhance the accuracy, speed, and adaptability of viral diagnostics. The main contributions in terms of precision diagnostics include the following:

1. *Improved Accuracy and Sensitivity:*
   DL models outperform traditional diagnostic tools that use imaging data, genomic sequences, and clinical metadata by making subtle patterns in them identifiable. For example, CNNs and transformers improve the sensitivity of RTPCR optimization as well as the detection of genomic variants (Ji et al. 2021).
2. *Early Detection of Emerging Variants:*
   The study enables early discovery of viral mutations by using more complex DL tools like transformers and attention mechanisms, which will ultimately play a crucial role in controlling outbreaks and steering the process of vaccine development.
3. *Multimodal Data Integration:*
   It is the integration of imaging information (e.g., X-rays and CT scans), molecular data (such as viral genomes), and clinical information in one diagnostic pipeline that assures holistic and accurate diagnosis. It is also most useful in complex cases, such as new or coinfections (Kim et al. 2020).
4. *Scalable and Accessible Diagnostics:*
   Cloud-based DL frameworks allow for real-time analysis, even in resource-constrained settings. Mobile AI tools improve access to point-of-care testing in underserved regions (Liu et al. 2021).

Impact The addition of DL to viral diagnostics assures the right solution: one that is accurate, real time, and scalable and aligns with the objectives of precision medicine—getting the right diagnosis and treatment to the right patient at the right time.

5.9.2 Healthcare Providers and Researchers' Implications

The findings and proposed solutions have direct implications for healthcare professionals and researchers, offering tools and frameworks that improve diagnostic workflows and enable scientific advancements:

1. *Improved Diagnostic Tools for Clinicians:*
   Healthcare professionals benefit from AI-powered decision support systems that provide real-time, accurate diagnoses and explainable predictions. For instance, tools like GradCAM provide visual explanations, enabling doctors to trust and validate outputs from DL models (Selvaraju et al. 2017).

2. *Accelerated Research on Emerging Viruses:*
   Researchers get access to AI frameworks that can analyze large viral genomic datasets. These tools speed up the identification of viral mutations, thus providing timely interventions and advancing vaccine or drug development.
3. *Enhanced Clinical Workflow Efficiency:*
   DL automates many routine tasks, such as image analysis, genomic screening, and report generation, so that the clinician can focus more on complex decision-making and patient care.
4. *Collaboration Through Federated Learning:*
   Through federated learning, researchers can collaborate with healthcare institutions in developing AI models without sharing patient data, thereby addressing privacy issues while enabling multi-institutional studies (Sheller et al. 2020).
5. *Training and Education Opportunities:*
   The study points out the need for clinician training in AI tools, filling the gap between technology and healthcare practice for effective implementation.

Impact Empowering clinicians with advanced tools and enabling researchers to accelerate discoveries, the study drives innovation and improves patient care across healthcare systems.

5.9.3 Impact on Future Clinical Decision-Making

The integration of deep learning and emerging technologies into viral diagnostics transforms clinical decision-making by providing data-driven, evidence-based insights that enhance accuracy, efficiency, and confidence:

1. *Real-Time and Informed Decision-Making:*
   AI-driven tools enable clinicians to make faster decisions by analyzing large datasets and delivering real-time diagnostic results. This is particularly critical during pandemics and viral outbreaks when rapid responses are needed.
2. *Explainable AI for Transparent Decisions:*
   This encompasses the integration of explainable AI (XAI). Predictions would hence become transparent and interpretable; thereby, clinicians will also know the rationale behind what these outputs generated by AI indicate, and this enhances trust to make clinical decisions.
3. *Personalized Treatment Pathways:*
   Multimodal analysis makes AI frameworks identify the respective patient-specific viral responses, so there is a way forward into a pathway of personalized treatment plans with tailored therapy.
4. *Proactive Outbreak Management:*
   Predictive AI models can predict emerging variants and forecast outbreak trends, so that healthcare systems can prepare in advance and allocate resources and preventive measures.

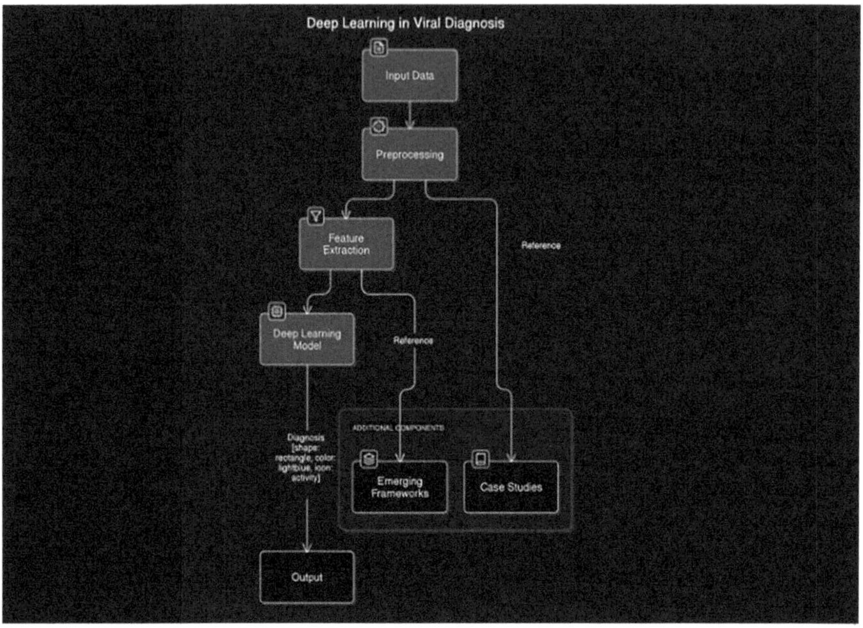

Fig. 5.2 Significance of viral diagnosis

5. *Improved Diagnostic Consistency:*
   AI tools standardize diagnostics, reducing interobserver variability and ensuring consistent results across healthcare providers and institutions (Fig. 5.2).

The diagram presents a comprehensive workflow for deep learning in viral diagnosis. It starts with Input Data, which consists of raw information, such as medical images, genomic sequences, or patient records. These data undergo a Preprocessing phase, where it is cleaned and standardized to ensure consistency and reliability for subsequent analysis.

Then, the Feature Extraction stage identifies critical attributes or patterns in the data, which are crucial for detecting and diagnosing viral infections. These extracted features are fed into the deep learning model, which leverages neural networks to analyze the data, recognize patterns, and make predictions or classifications related to the virus.

Additional components include Emerging Frameworks and Case Studies, highlighted as references in the workflow. These provide practical insights and state-of-the-art methodologies to improve the accuracy and efficiency of the diagnosis process. Finally, the Output stage delivers a diagnosis, incorporating shape recognition, color analysis, or other relevant metrics to provide actionable results for medical professionals.

The diagram emphasizes the integration of advanced computational methods and real-world applications to create a robust and precise diagnostic pipeline for viral diseases.

5.10 Conclusion and Future Directions

The study on Deep Learning in Viral Diagnosis highlights the transformative potential of artificial intelligence in addressing challenges in precision medical virology. By integrating DL models with emerging technologies, such as federated learning, blockchain, and explainable AI (XAI), this research provides a robust framework for improving diagnostic accuracy, scalability, and clinical adoption.

5.10.1 Summary of Findings

1. *Improved Sensitivity and Specificity:*
   Deep learning models, like CNNs, transformers, and their hybrid architecture, showed better efficiency in diagnosing viral infections, streamlining RTPCR protocols, and the identification of genomic variants. Case studies showed increased sensitivity, decreased turnaround times, and the early detection of emerging viral strains (Kim et al. 2020; Ji et al. 2021).
2. *Multimodal Information Integration:*
   Integration of imaging data, genomic sequences, and clinical metadata facilitated a holistic diagnostic approach, which significantly improved the accuracy of viral infection and comorbidity detection.
3. *Privacy-Preserving and Secure Frameworks:*
   Federated learning tackled the problem of data privacy by allowing decentralized model training. Blockchain technology ensured secure and auditable data sharing across institutions (Sheller et al. 2020).
4. *Explainable AI for Clinical Adoption:*
   XAI techniques like GradCAM and SHAP made deep learning outputs more interpretable to increase clinician trust, which in turn led to better clinical decision-making practices (Selvaraju et al. 2017).
5. *Scalability and Accessibility:*
   Scalable solutions were provided through cloud-based and mobile-compatible AI tools, and affordable diagnostics in resource-constrained settings could be supported; hence, the ability of global healthcare accessibility can improve (Liu et al. 2021).
6. *Integrated Framework Proposed:*
   A new framework was developed from the study that is made by integrating AI-driven analytics, data security, and explainable insights, representing a future-proof viral diagnostics solution.

5.10.2 Roadmap for Future Research in Medical Virology

Although tremendous strides have been taken, deep learning and emerging technologies will not be fully realized in medical virology unless further research is conducted. The following roadmap provides directions for the advancement of this field:

1. *Development of More Robust AI Models:*
   Future research should focus on building generalizable and adaptive AI models capable of handling diverse viral datasets, including rare and emerging pathogens.
   Few-shot learning and self-supervised learning approaches can be explored to train models with limited labeled data.
2. *Integration of Real-Time Monitoring Systems:*
   AI based frameworks should include real-time outbreak prediction and monitoring tools that can quickly identify mutations in viruses and emerging trends. This would also integrate with global health surveillance systems to enable proactive response strategies.
3. *Improved Interpretability and Clinician Collaboration:*
   Further developments in explainable AI should increase the transparency and usability for clinicians, as well as make effective co-adoption through a cooperative research approach among the developers of AI, virologists, and the clinicians themselves.
4. *Federated and Distributed Learning Research:*
   The federated learning models must be fine-tuned to provide better computational efficiency and scaling in multi-institutional networks. Research into addressing data heterogeneity between different institutions is essential.
5. *Blockchain for Enhanced Security and Interoperability:*
   Research on incorporating blockchain for real-time data sharing with privacy, interoperability, and compliance with the global data regulations should be pursued.
6. *Point-of-Care and Mobile AI Diagnostics:*
   Developing lightweight AI models to be used on mobile and wearable devices could increase access to diagnostics in resource-limited regions, especially. Research should thus focus on efficiency without trading off accuracy.
7. *AI for Drug Discovery and Vaccine Development:*
   AI frameworks should be extended to predict viral protein structures, drug candidates, and optimize vaccine designs for emerging viral variants.
8. *Validation Through Large-Scale Clinical Trials:*
   Future studies must include rigorous validation of AI-driven diagnostic frameworks through large-scale, multi-center clinical trials to ensure real-world reliability and regulatory approval.

References

Afzal, M. (2020) COVID-19: A threat to educated Muslim girls' fragile gains in schooling. In: Center for Global Development. https://www.cgdev.org/publication/covid-19-threat-educated-muslim-girls-fragile-gains-schooling

Ardila, D., Kiraly, A. P., Bharadwaj, S., et al. (2020) End-to-end lung cancer screening with 3D deep learning on CT. In: Nature Medicine, 25(6), pp 954–961. https://doi.org/10.1038/s41591-019-0447-x

Buda M, Saha A, Mazurowski MA (2019) Glioma subtypes and CNN-extracted shape features. In: Computers in Biology and Medicine, 109, pp 218–225. https://doi.org/10.1016/j.compbiomed.2019.05.017

Casino F, Dasaklis TK, Patsakis C (2019) Blockchain applications: Review and classification. In: Telematics and Informatics, 36, pp 55–81. https://doi.org/10.1016/j.tele.2018.11.006

Chen, J., et al. (2020). Deep learning for rapid and accurate viral diagnosis in outbreak settings. The Lancet Digital Health, 2(5), e235–e243. https://doi.org/10.1016/S2589-7500(20)30070-1

Cohen JP, Morrison P, Dao L, et al. (2020) COVID-19 image data collection. In: arXiv preprint arXiv:2006.11988. https://arxiv.org/abs/2006.11988

Corman VM, Landt O, Kaiser M, et al. (2020) Real-time RT-PCR for 2019-nCoV detection. In: Euro Surveill, 25(3), 2000045. https://doi.org/10.2807/1560-7917.ES.2020.25.3.2000045

Dosovitskiy A, Beyer L, Kolesnikov A, et al. (2021) An image is worth 16×16 words: Transformers for image recognition. In: arXiv preprint arXiv:2010.11929. https://arxiv.org/abs/2010.11929

Engin S, Medrano-Gracia P, Frean M, Tan P (2021) Deep learning for pulmonary image analysis. In: J Thorac Dis, 13(2), pp 848–861. https://doi.org/10.21037/jtd-20-2215

Esteva A, Kuprel B, Novoa RA, et al. (2019) Skin cancer classification with deep neural networks. In: Nature, 542, pp 115–118. https://doi.org/10.1038/nature21056

Goecks J, Jalili V, Heiser LM, Gray JW (2021) How ML will transform biomedicine. In: Cell, 181(1), pp 92–101. https://doi.org/10.1016/j.cell.2020.03.022

He K, Zhang X, Ren S, Sun J (2016) Deep residual learning for image recognition. In: CVPR 2016 Conference Proceedings, pp 770–778. https://doi.org/10.1109/CVPR.2016.90

Howard J, Gugger S (2017) fastai: A layered API for deep learning. In: Information, 11(2), 108. https://doi.org/10.3390/info11020108

Ji, M., Zhang, L., Yang, M., Song, W., & Du, Z. (2021). Rapid COVID-19 diagnosis using ensemble deep transfer learning models from chest radiographic images. IEEE Transactions on Instrumentation and Measurement, 70, 1–7. https://doi.org/10.1109/TIM.2021.3068753

Kaissis GA, Ziller A, Passerat-Palmbach J, et al. (2020) Privacy-preserving deep learning on medical data. In: Nature Machine Intelligence, 2, pp 305–311. https://doi.org/10.1038/s42256-020-0186-1

Kim, H. E., et al. (2020). Optimizing RT-PCR with deep learning for viral diagnostics. Nature Communications, 11(1), 2908. https://doi.org/10.1038/s41467-020-16657-4

Liu, Y., et al. (2021). Scalable deep learning models for viral diagnosis in resource-limited settings. PLOS Computational Biology, 17(4), e1008850. https://doi.org/10.1371/journal.pcbi.1008850

Loeffelholz MJ, Tang YW (2020) COVID-19 laboratory diagnosis. In: Emerging Microbes & Infections, 9(1), pp 747–756. https://doi.org/10.1080/22221751.2020.1745095

Mackay IM, Arden KE, Nitsche A (2017) Real-time PCR in virology. In: Nucleic Acids Res, 30(6), pp 1292–1305. https://doi.org/10.1093/nar/30.6.1292

Nguyen TT, Nguyen T, Van Nguyen D, et al. (2021) AI against COVID-19: Survey and directions. In: Medical Imaging with Deep Learning. https://arxiv.org/abs/2005.00375

Peeling RW, Wedderburn CJ, Garcia PJ, et al. (2021) Serology testing in COVID-19 response. In: Lancet Infect Dis, 20(9), e245–e249. https://doi.org/10.1016/S1473-3099(20)30634-4

Rao ASS, Vazquez JA (2021) AI framework for COVID-19 detection via mobile survey. In: Infect Control Hosp Epidemiol, 41(7), pp 826–830. https://doi.org/10.1017/ice.2020.61

Ribeiro MT, Singh S, Guestrin C (2020) Explaining predictions of classifiers. In: Proceedings of the 22nd ACM SIGKDD, pp 1135–1144. https://doi.org/10.1145/2939672.2939778

Selvaraju RR, Cogswell M, Das A, et al. (2017) Grad-CAM: Visual explanations via gradients. In: Proc. IEEE ICCV, pp 618–626. https://doi.org/10.1109/ICCV.2017.74

Sheller, M. J., et al. (2020). Federated learning for privacy-preserving medical diagnostics. Scientific Reports, 10(1), 12598. https://doi.org/10.1038/s41598-020-69250-1

Suresh Babu CV, Praveen S (2023) Swarm intelligence and evolutionary machine learning algorithms for COVID-19: pandemic and epidemic review. In: Suresh Kumar A, Kose U, Sharma S, Kumar SJN (eds) Dynamics of swarm intelligence health analysis for the next generation. IGI Global, pp 83–103. https://doi.org/10.4018/978-1-6684-6894-4.ch005

Suresh Babu CV, Sathiya Narayana S (2024) Navigating uncharted waters: overcoming challenges in integrating machine learning and deep learning for enhanced clinical risk. In: Liu H, Tripathy R, Bhattacharya P (eds) Clinical practice and unmet challenges in AI-enhanced healthcare systems. IGI Global, pp 289–310. https://doi.org/10.4018/979-8-3693-2703-6.ch015

Topol EJ (2019) High-performance medicine and AI. In: Nature Medicine, 25(1), pp 44–56. https://doi.org/10.1038/s41591-018-0300-7

Vaswani A, Shazeer N, Parmar N, et al. (2017) Attention is all you need. In: NeurIPS 2017, 30, pp 5998–6008. https://doi.org/10.48550/arXiv.1706.03762

Wang L, Wong A, Lin ZQ (2020) COVID-Net: CNN for COVID-19 detection. In: Scientific Reports, 10, 19549. https://doi.org/10.1038/s41598-020-76550-z

Wang S, Kang B, Ma J, et al. (2022) Deep learning with CT images for COVID-19. In: European Radiology, 31(8), pp 6096–6104. https://doi.org/10.1007/s00330-020-06975-7

Weissleder R, Lee H, Ko J, Pittet MJ (2020) COVID-19 diagnostics in context. In: Sci Transl Med, 12(546), eabc1931. https://doi.org/10.1126/scitranslmed.abc1931

World Health Organization (2020) Laboratory testing for COVID-19: interim guidance. In: WHO Technical Documents. https://apps.who.int/iris/handle/10665/331501

Zhang K, Liu X, Shen J, et al. (2023) AI for COVID-19 diagnosis and prognosis using CT. In: Cell, 181(6), pp 1423–1433.e11. https://doi.org/10.1016/j.cell.2020.04.045

Zhou T, et al. (2022). Deep learning for nucleic acid-based viral diagnostics. Nature Biotechnology, 40(3), 402–410. https://doi.org/10.1038/s41587-021-01143-2

Implementing Machine Learning for Analyzing Influenza A Virus with Hemagglutinin Sequences

6

Chhaya Gupta, Nasib Singh Gill, and Preeti Gulia

Abstract

Influenza viruses cause illness, and influenza viruses are especially dangerous. These viruses are notorious for spreading quickly, having the capacity to harm people seriously, and frequently changing, all of which highlight the necessity of further study and monitoring. A virus's ability to spread between species is critical for predicting epidemics and creating potent vaccines. Historically, influenza pandemics between species have been brought on by viruses. To stop an outbreak from spreading, it is critical to determine the virus's origin. Machine learning techniques to generate quick and precise predictions for viral sequences have recently gained popularity. Hemagglutinin sequences were the only ones used because they are the main protein implicated in the immune response. They were portrayed using the word embedding and a position-specific scoring matrix. The findings demonstrate that the convolutional neural network is the most effective algorithm for predicting viral sequence origins, achieving about 98.54% area under the precision-recall curve (AUCPR), 97.01% F1 score, and 95.60% Matthews' correlation coefficient (MCC) at a higher classification level,

C. Gupta (✉)
Department of Computer Science and Applications, Maharshi Dayanand University, Rohtak, Haryana, India

Vivekananda School of Information Technology, Vivekananda Institute of Professional Studies-Technical Campus, New Delhi, India
e-mail: chhaya.rs.dcsa@mdurohtak.ac.in

N. S. Gill · P. Gulia
Department of Computer Science and Applications, Maharshi Dayanand University, Rohtak, Haryana, India
e-mail: Nasib.gill@mdurohtak.ac.in; preeti@mdurohtak.ac.in

J. M. Chatterjee, S. K. Saxena (eds.), *Artificial Intelligence and Blockchain in Precision Medicine and Virology*, Medical Virology: From Pathogenesis to Disease Control, https://doi.org/10.1007/978-981-96-8919-4_6

and approximately 95.74% AUCPR, 89.41% F1 score, and 82.79% MCC at a lower classification level.

Keywords

Extreme gradient boosting (XGBoost) · Influenza A virus (IAV) · Machine learning · Position-specific matrix (PSM) · Random forest classifier (RFC) · Support vector machine classifier (SVMC)

6.1 Introduction

Influenza pandemics happen less frequently than seasonal influenza pandemics, and each crisis has the potential to claim millions of lives (Hung et al. 2023). People of all ages have been impacted by the influenza pandemic, which has harmed those susceptible to chronic illnesses. Based on their internal ribonucleoprotein, influenza viruses are divided into four types: A, B, C, and D. Human illness is not caused by the influenza D virus (Zou et al. 2023). Although the influenza C virus can only infect humans, widespread outbreaks are unlikely to result from it. As a result, influenza C and D virus vaccinations cannot be administered using seasonal influenza vaccine strains. Only humans can contract the influenza B virus, whereas people and animals can contract the influenza A virus, which can cause worldwide outbreaks. Two glycoproteins located underneath the virus envelope, hemagglutinin (HA) and neuraminidase (NA), distinguish influenza Subtypes of viruses (Li et al. 2013).

The immune system recognizes antigenic regions in HA or NA proteins, which prevent flu infection. To avoid immune system detection, these antigenic sites might change quickly (Kumar et al. 2024). When viruses reassort within one or more hosts, antigenic modifications may occur, leading to the development of new viral strains. Extreme antigenic alterations have caused multiple pandemics, where most humans could not protect themselves against the new virus (Liang et al. 2023). Although the exact origin of Spanish flu is unknown, new research indicates that it might have started in pigs or birds. The flu A virus that mainly affects birds can infect pigs and humans (Phu et al. 2023). As these viruses can transmit directly without requiring an intermediary host, their reassortment across species presents a serious risk to human populations (Singh and Mittal 2023). Thus, influenza viruses can spread through animal-to-animal (enzootic) and animal-to-human (zoonosis) contacts. However, the virus requires sufficient time to adapt through mutation and genetic accumulation. Therefore, separating the initial viral host as soon as possible can effectively stop a viral pandemic or control its spread.

Most conventional techniques, such as subtyping viruses using hemagglutination inhibition (HI), are laboratory-based. Numerous machine learning methods, including K-nearest neighbors (KNNs), random forests, artificial neural networks (ANNs), and decision trees (DTs), have been used to predict viral hosts to save time and manpower. Previous studies used manually chosen balanced data sets, feature extraction algorithms, or small data sets representing the sequence as sparse matrices. Additionally, convolutional neural networks (CNNs), one of the deep learning approaches, have been used. The study's primary contributions are as follows:

- Several machine-learning techniques were employed to identify the influenza virus's origin host, including RFC, SVMC, XGB, and CNN.
- Regarding alignment-based sequence representation, PSI-BLAST was used to extract the evolutionary properties of viral sequences, which were then fed into four conventional machine-learning models.
- Sequences are processed using an alignment-free method called word embedding. Since the morphological structure of sequence data and text data is comparable, techniques like word embedding used in natural language processing (NLP) for vectorizing sequences may also be used to handle viral sequence data. One popular and effective method for handling sequential data is word embeddings. Words with comparable meanings will have similar embeddings due to word learning. In this study, the bird classes were separated, and models were assessed at a lower classification level rather than a higher one.
- For handling data imbalance, oversampling, and undersampling, the SMOTEBoost technique is used.

This chapter remaining portion is divided into the following sections: The literature review is discussed in Sect. 6.2, the proposed approach is outlined in Sect. 6.3, the implementation results and discussion are presented in Sect. 6.4, and the study concludes in Sect. 6.5.

6.2 Literature Survey

This section provides a literature survey of SOTA machine learning methods used for predicting influenza A virus with the limitations that the researchers faced. The survey provides the research gaps that can be used to improve the models. Table 6.1 provides a summary of various SOTA models used for predicting IAV.

6.3 Methods and Materials

6.3.1 Dataset

The models are trained using a dataset from the GISAID database, which includes hemagglutinin (HA) sequences (GISAID 2024). The sequences that contained the redundant amino acids X, B, and Z were removed to preserve the data's quality. As a result, 59,785 sequences in all were chosen from the original set.

Table 6.2 presents the difference in information quantity before and after filtering. Figures 6.1 and 6.2 illustrate the data distribution across various levels. At a greater level, the model's performance is evaluated using only three classes: swine, avian,

Table 6.1 Summary of SOTA machine learning models used for predicting IAV with research gaps

Citation	Models used	Results	Research gap
Shang Kai Hung et al. (2023)	The study employed various models, including neural network-based approaches such as artificial neural networks (ANNs) and deep learning models with two, three, and four layers. It also incorporated a distance-based model, support vector machines (SVM), along with eXtreme gradient boosting (XGBoost), conditional random forest (CForest), random forest (RF), and RANDOM Forest GEneRator (RANGER).	XGBoost demonstrated strong performance on the test set, achieving an area under the curve of 0.82, a sensitivity of 0.92, a specificity of 0.89, and an accuracy of 0.72.	It is necessary to update or retrain models using greater sample numbers in various subgroups.
Xidan Zou et al. (2023)	The basic classifier model for the prediction of IAV was constructed in this study using four standard machine learning algorithms: K-nearest neighbor (KNN), logistic regression (LR), random forest (RF), and support vector machines (SVMs).	During the five-fold cross-validation, the model achieved an accuracy of 95.85% and an area under the curve of 0.9863.	The feature extraction method can be improved further for better model performance.
Ranjan Kumar et al. (2024)	This study used SVM linear and radial, random forest, artificial neural network, gradient boosting model, and KNN to predict IAV.	SVM radial achieved good results with 83.92%, sensitivity of 76%, and specificity of 88.59%.	To further enhance the accuracy of the proposed model, other effective machine learning methods and a larger dataset can be considered.
Shahid Hussain et al. (2024)	This study explores the use of machine learning prediction and regression techniques, along with classification algorithms such as logistic regression (LR), support vector machines (SVM) with linear,	While logistic regression (LR) achieved a higher average score of 72% after random forest (RF), RF had the highest accuracy score at 74%.	A more comprehensive view of influenza cases and their classification can be achieved using machine learning techniques beyond logistic regression, support vector machines, Naïve Bayes,

(continued)

Table 6.1 (continued)

Citation	Models used	Results	Research gap
	polynomial, and RBF kernels, Naïve Bayes (NB), and random forest (RF), to predict influenza outbreaks and the spread of the disease.		and random forest. Future improvements in accuracy can be attained by leveraging deep learning algorithms.
Sungwoo Park et al. (2024)	This study employs a DCC-based feature selection method to improve prediction performance and identify highly relevant influenza surveillance data.	The model achieved good results.	The model was not evaluated on metrics like accuracy, recall, and precision. It would be good to see how the model performs on these parameters compared to other SOTA models.
Sajal Maheshwari et al. (2022)	This study uses four machine learning methods: SVM, KNN, ANN, and random forest.	Random forest achieved the highest accuracy.	Models should be retrained or updated across different subgroups using larger sample sizes to improve reliability and performance.
Siti Amiroch et al. (2022)	Seven machine learning models were used in the study: RF, GB, k-NN, MLP, LR, SVM, and XGBoost.	XGBoost attained the highest accuracy of 96%.	Further research on molecular docking, molecular dynamics, and the in vitro validation of the key identified compounds is essential.
Muhammad Aftab et al. (2022)	In this study, LSTM-CNN is used for the prediction of influenza A Virus	The model attained 98% accuracy.	The model is not optimized for real-time applications.
Yuan Ling et al. (2021)	The study proposed a bidirectional LSTM with a CNN model to predict the influenza A virus.	The model achieved good results.	The feature extraction method could be improved further to extract more relevant features.
Ahmet Kara (2021)	The study suggested a genetic algorithm merged with LSTM to predict the influenza A virus.	The model achieved good results.	It is necessary to update or retrain models using more significant sample numbers in various subgroups.
Shangfang Dai et al. (2023)	This paper proposes a new recurrent neural network architecture for influenza surveillance, an attention-based LSTM model.	The model demonstrated strong performance.	In the future, multimodal models can be merged with the proposed model.

Table 6.2 Data before and after the selection of different criteria

Dataset	Original	Filtered	Complete	Non-redundant	No multi-label	No X, B, Z proteins	Split
1	180,834	59,786	Yes	Yes	Yes	Yes	Train, Test, Val

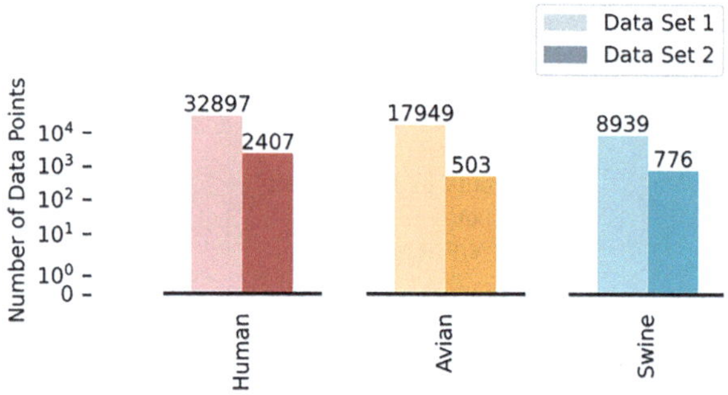

Fig. 6.1 Data distribution at a greater level

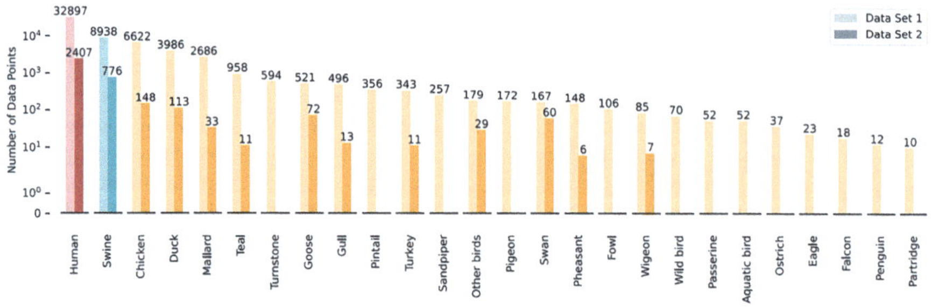

Fig. 6.2 Data distribution at lower level

and human. However, avian data is subdivided at a lower categorization level, yielding 26 classes. Models have more significant difficulties as the number of classes rises.

6.3.2 Position-Specific Matrix Representation (PSMR)

The position-specific matrix representation (PSMR) is a method for extracting evolutionary information from protein sequences and can be constructed using position-specific iterated BLAST. A PSM is a bioinformatics matrix that captures the probability of each amino acid occurring at each place in a protein sequence based on evolutionary conservation. PSMs are widely used in protein homology detection, protein structure prediction, and identifying functionally important protein regions, as conserved regions often indicate functional or structural importance. It is often derived from multiple sequence alignments and indicates which amino acids are conserved or variable at each position in a protein. A position-specific matrix (PSM) for a query protein sequence of length L is represented as an $L \times 20$ matrix. The PSM for sequence b can be illustrated as follows:

$$\mathrm{PSM}_{\mathrm{orig}} = \begin{bmatrix} b_1 & \cdots & p_{1,20} \\ \vdots & \ddots & \vdots \\ b_L & \cdots & p_{L,20} \end{bmatrix} \tag{6.1}$$

where $p_{x,y}$ is the amino acid score b_i. By using the sigmoid function, it may alternatively be seen as a chance of mutation in the [0,1] range:

$$p_{x,y} = \frac{1}{\left(1 + e^{-p_{x,y}}\right)}, x = 1, 2, \ldots, L; y = 1, 2, \ldots, 20 \tag{6.2}$$

PSI-BLAST was executed iteratively on a non-redundant database with the default settings. Due to their varying sizes, PSMRs cannot be input into traditional machine learning models. To get over this restriction, three PSM-based sequence encoding techniques are suggested. Proteins are simplified, and pointless calculations are eliminated by introducing a residue grouping rule.

Proteins and amino acids can be categorized into ten classes due to their numerous attributes. The original PSM's columns can be subjected to residue grouping rules to produce a grouped PSM (GPSM) with $L \times 10$ dimensions.

$$\mathrm{GPSM} = \begin{bmatrix} g_{1,1} & \cdots & g_{1,10} \\ \vdots & \ddots & \vdots \\ g_{L,1} & \cdots & g_{L,10} \end{bmatrix} \tag{6.3}$$

where

$$g_{x,y} = \frac{\sum p_{x}, g_{y}}{\mid g_y \mid} \tag{6.4}$$

$\sum p_x$, g_y is the amino acid a_x score mutated to another amino acid a_y. L represents the length of sequence, $x = 1, 2, \ldots, L$ and $|g_y|$ is an amino acid in group y.

The length of the input sequence in both PSM and GPSM may vary. This implies that many machine learning models cannot be used directly. Before inputting this matrix into traditional machine learning models, it is transformed into a 1×100 feature vector by rearranging it sequentially as:

$$\mathrm{EGPSM} = \left(E_{g_1 g_1}, E_{g_1 g_2}, \ldots, E_{g_1 g_{10}}\right) \tag{6.5}$$

where

$$E_{g_x g_y} = \frac{\sum g_x g_y}{| g_y |} \tag{6.6}$$

EGPSM (evolutionary and graph-based position-specific matrix) is a bioinformatics method used for protein function prediction, particularly in identifying protein-protein interactions and predicting potential binding sites in proteins. The technique combines evolutionary information with graph-based representations to enhance the accuracy of predictions. EGPSM incorporates evolutionary data, often through information from PSM matrices and sequence alignments. This data provides insight into conserved regions across species, which can suggest functionally critical sites. EGPSM uses graph theory to represent amino acids as nodes in a network, where edges represent interactions or spatial proximity. This structure enables protein structure modeling and helps understand the interactions within or between proteins. By combining evolutionary conservation (from PSM) with structural information (from graph-based models), EGPSM offers more reliable predictions than traditional PSM alone. The method can be used across various organisms and types of proteins, making it versatile for multiple protein analysis tasks.

GDPCPSM (graph-based density peaks clustering—position-specific matrix) is a bioinformatics technique for identifying protein-protein interaction sites and other functional regions within proteins. It combines two main components: position-specific matrix (PSM) and graph-based density peak clustering (GDPC). GDPC is a clustering algorithm that identifies "density peaks" within a dataset, where density refers to regions in the data space with many neighboring points. In GDPCPSM, these density peaks help group amino acid positions similar regarding evolutionary and structural features. In GDPC, amino acids in a sequence are represented as nodes in a graph. By identifying and clustering these dense regions, GDPC isolates potential functional sites where protein interactions or structural formations may occur. Connections are formed based on similarities in their PSM scores, spatial proximity, or both. By clustering conserved and dense regions, GDPCPSM helps identify likely sites for protein interactions. These sites are often conserved across evolutionary lines and correspond to the functional regions in proteins. Regions identified through density peaks are likely candidates for ligand or substrate binding, aiding in drug discovery and biochemical pathway mapping. GDPCPSM's combined approach of evolutionary conservation (PSM) and density-based clustering (GDPC) allows it to detect regions in proteins that may have structural or regulatory importance. Combining PSM data with GDPC clustering increases the likelihood of detecting actual interaction sites and functional regions compared to using PSM alone. GDPCPSM can be applied to large datasets and is more scalable for genome-wide studies where many protein interaction sites need to be mapped. Using evolutionary and structural data, GDPCPSM has shown vital predictive accuracy in computational experiments.

$$\text{GDPCPSM} = (D_{1,1}, D_{1,2}, \ldots, D_{1,10}) \tag{6.7}$$

where

$$D_{x,y} = \frac{1}{L-1}\sum_{i=1}^{L-1} g_{i,x}Xg_{i+1,x}\ x, y = 1, 2, \ldots, 10 \tag{6.8}$$

The evolutionary relevance position-specific matrix (ERPSM) is a bioinformatics technique that utilizes evolutionary data to predict functional sites in protein sequences. ERPSM extends the traditional PSM by weighting the matrix to prioritize evolutionary relevance, thereby improving the accuracy of predictions for functional regions such as binding sites or protein-protein interaction sites. The ER component adds a layer to the PSM by giving higher weights to positions in the sequence that are evolutionarily conserved and likely critical for the protein's function. This relevance is determined by examining sequence variations across different species, emphasizing conserved positions. This strategy highlights functional residues more precisely than standard PSM alone. In the ERPSM, the scores from the PSM are adjusted based on their evolutionary relevance. Positions that show vital conservation are given higher weights in the matrix, while variable positions are down-weighted. These weighted scores improve the ability of the matrix to pinpoint residues involved in crucial activities like ligand binding, catalysis, and protein-protein interactions. ERPSM can improve the accuracy of identifying these sites by emphasizing evolutionarily critical positions. Since interaction interfaces are often evolutionarily conserved, ERPSM aids in predicting likely interaction regions within a protein. ERPSM can detect regions critical to the protein's structure or function, contributing to drug discovery and understanding of protein function. By focusing on evolutionarily relevant residues, ERPSM provides higher specificity than standard PSM, reducing false positives in functional site predictions. The emphasis on conservation helps capture evolutionary constraints typically indicative of critical functional roles, making ERPSM well-suited for functional and structural analyses. The method is adaptable for various proteins and functional sites, making it broadly valuable for bioinformatics.

$$\text{ERPSM} = (N_{1,1,1}\ \ N_{1,2,1}\ldots\ N_{1,10,9}\ \ T_1\ldots\ T_{10}) \tag{6.9}$$

where

$$N_{x,y,t} = \frac{1}{L-t}\sum_{i=1}^{L-t}\frac{\left(g_{i,x} - g_{i+t,y}\right)^2}{2}, i, y = 1, 2, \ldots 10; t = 1, 2, \ldots, 9 \tag{6.10}$$

$$T_a = \frac{1}{L}\sum_{i=1}^{L}\left(g_{i,a} - \overline{g}_a\right)^2, a, y = 1, 2, \ldots, 10 \tag{6.11}$$

$\overline{g}_a$ is the average value of GPSM in column a, and T_a is the average pseudo-composition calculated for each amino acid in GPSM.

6.3.3 Representation of Learning

Conventional machine learning algorithms require human preprocessing and feature extraction to obtain meaningful features from protein sequences. The selection of relevant features during this process requires prior knowledge and expertise. On the other hand, protein sequence implicit representations can be directly learned by deep learning algorithms. Protein sequences are mapped into numerical vectors in this study using word embedding, a potent vectorization technique in natural language processing (NLP).

Word embeddings are a text representation in which words with related meanings are represented similarly (Asudani et al. 2023). They're essentially dense vector representations of words designed to capture semantic relationships. These embeddings are created by neural networks trained on large text corpora, enabling the model to understand nuances like synonyms, analogies, and contextual meaning. Some techniques for generating word embeddings are Word2Vec, GloVe, FastText, and contextual embeddings.

Word2Vec is one of the first popular embedding techniques created by Google. For example, in Word2Vec, the words "king" and "queen" would have embeddings with a small vector distance, capturing their semantic relationship. GloVe, developed by Stanford, is an extension that combines local and global information about word co-occurrences in a corpus (Curto et al. 2024). Unlike Word2Vec, which learns relationships within a context window, GloVe captures local statistics and broader word relationships by analyzing co-occurrence frequencies over an entire corpus. FastText, developed by Facebook, builds on Word2Vec but considers subword (character-level) information, making it robust to spelling variations, morphological changes, and rare words (Savelli and Giobergia 2024). FastText is especially useful for morphologically rich languages, where words vary depending on context. Traditional embeddings like Word2Vec and GloVe produce static embeddings; a word has the same vector regardless of context. Contextual embeddings, such as those produced by BERT (bidirectional encoder representations from transformers), generate embeddings based on surrounding words (Vithanage et al. 2024). BERT allows for words to change meaning based on context (e.g., "bank" in "riverbank" vs. "money bank") by producing different embeddings for each usage.

Embeddings are usually high-dimensional vectors whose dimensions capture different semantic or syntactic properties. These embeddings are positioned within a vector space so that words with similar meanings or relationships are close together. Word embeddings are fundamental in NLP because they serve as a foundation for many deep learning applications, transforming raw text into forms that neural networks can process efficiently. One-hot encoding's shortcomings are addressed via word embedding. In addition to creating dense vectors, it can also capture the connections between related words. It can capture word associations and generate dense vectors. A sequence of proteins is depicted as an $L \times E$ matrix, where E is the embedding dimension, and L is the sequence length. Each n-gram is encoded as a vector of size E. In this study, the skip-gram model of Word2Vec is utilized.

This model takes a single word as input and attempts to predict words that appear within a specified window (context size) around it.

$$\frac{1}{t}\sum_{t=1}^{t}\sum_{-c\leq j\leq c;j\neq 0}\log P\left(w_{t+j}|w_t\right) \tag{6.12}$$

where c is the context window size, $P(w_{t+j}| w_t)$ is the probability of context word w_{t+j} given the target word w_t.

Word2Vec uses approximation methods such as Hierarchical Softmax to make the training efficient. With this method, every word is a leaf node in a binary tree structure that arranges the vocabulary. Predicting a word becomes equivalent to traversing a binary tree, where each step involves a binary classification, reducing the number of computations from the vocabulary size to the tree's depth. Words with similar meanings have high cosine similarity. Word2Vec captures linear relationships between words. For example, vector arithmetic, such as "king − man + woman ≈ queen," is held because of how Word2Vec encodes relationships.

6.3.4 Machine Learning Methods Used

6.3.4.1 Extreme Gradient Boosting (XGBoost)

XGBoost is a fast and efficient implementation of boosted decision trees designed to enhance performance. It was developed by Tianqi Chen and introduced as an open-source project (Chen and Guestrin 2016). XGBoost is a boosting technique that strengthens model performance by combining multiple weak learners, typically decision trees. Adding trees one after the other while trying to fix mistakes committed by earlier trees is how boosting operates (Malik et al. 2024). The objective is to minimize a loss function by sequentially adding decision trees, each improving upon the errors of the previous ones. Each tree aims to minimize the earlier trees' residual errors. With gradient boosting, every tree fits the loss function's negative gradient. This gradient approach allows XGBoost to adaptively reduce errors by focusing on areas where previous models performed poorly (Bukhari and Ogudo 2024). XGBoost employs several optimizations to make split finding faster and more efficient, including histogram-based approaches and distributed computing. Tree structures are built level-wise, optimizing for depth and structure based on the calculated information gain. XGBoost uses either mean squared error for regression tasks or Log Loss for classification tasks as standard loss functions but can accommodate custom loss functions. The goal of XGBoost is to minimize the following objective function for each tree added:

$$f=\sum_{i=1}^{n}L(y_i,\hat{y}_i)+\sum_{x=1}^{X}\Omega(f_x) \tag{6.13}$$

where f is the objective function, $L(y_i, \widehat{y}_i)$ is the loss function that helps to measure how well the predictions $\widehat{y}_i$ match with the actual object y_i, $\Omega(f_x)$ is the penalizing term for the complexity of each added tree f_x, and X is the number of trees used.

XGBoost uses a "max depth" parameter and a "minimum loss reduction" threshold to control the growth of trees. Trees are pruned backward from the leaves, ensuring that splits with minimal gain are discarded. XGBoost can utilize multiple cores to build trees in parallel and make it highly efficient, even for large datasets (Scherr et al. 2024). The parallelization uses multiple threads to process the splits for each feature level-wise in a tree. By figuring out how to handle each missing value in the tree structure, XGBoost's built-in methods can deal with missing data. It assigns missing values based on the optimal path learned during training. In order to avoid overfitting and conserve computational resources, XGBoost can immediately terminate training if performance does not increase after a predetermined number of rounds. XGBoost provides a "learning rate" or "shrinkage" parameter that scales the contribution of each added tree, allowing for more fine-grained adjustments to improve the model. XGBoost remains one of the most effective and efficient algorithms for tabular data and has contributed to many competitive solutions in data science. Its adaptability, speed, and predictive accuracy make it a top choice for structured data modeling.

6.3.4.2 Random Forest Classifier (RFC)

Several decision trees are combined in an ensemble learning technique called a random forest classifier, or RFC, to increase prediction accuracy and manage overfitting. It utilizes the Bagging technique to improve accuracy and reduce overfitting. Bagging is generating multiple random samples from the training data with replacement (bootstrapping) and training each decision tree on one of these samples (Wang et al. 2021). Each tree is trained independently, and because each tree is exposed to only a subset of the data, this diversity helps reduce the risk of overfitting. Given a training dataset with N samples, a bootstrap sample is created for each B tree in the forest by randomly selecting N samples with replacements from the original dataset. Let's assume D is the original training dataset with N samples, and D_b is the bootstrap sample for the bth tree, where $b \in \{1, 2, \ldots, B\}$. Each decision tree is trained on a different bootstrap sample D_b, p is the total number of features, m is the number of features randomly chosen at each split where $m = \sqrt{p}$ for classification and $m = \frac{p}{3}$ for regression. At each split, only m randomly chosen features are considered. The optimal split among these mmm features is selected to maximize information gain in classification tasks or to minimize variance in regression tasks. This feature randomness helps reduce the correlation between individual trees, enhancing the ensemble's accuracy and robustness (Chen et al. 2024). After building B trees, each tree produces a prediction, and RFC combines these predictions to generate the final output. Each tree produces a class label, and the final prediction is determined through majority voting, as illustrated in Eq. (6.14):

$$\widehat{y} = \text{mode}\{f_1(x), f_2(x), \ldots, f_B(x)\} \tag{6.14}$$

where $\widehat{y}$ is the predicted class label for an input x and $f_B(x)$ is the predicted class label from the bth tree.

The primary objective of RFC is to minimize prediction error by reducing variance, as depicted in Eq. (6.15). Here, Var(f_B) is the variance of individual tree prediction.

$$\text{Var}(\widehat{y}_{\text{RFC}}) = \frac{1}{B}\text{Var}(f_B) \tag{6.15}$$

RFC achieves this by combining multiple low-bias, high-variance tree predictions. If $L(\widehat{y}, y)$ is the loss function, and $\widehat{y}_{\text{RFC}}(x)$ is the random forest prediction for an instance x, then the aim is to minimize the expected loss as:

$$E[L(\widehat{y}_{\text{RFC}}(x), y)] \tag{6.16}$$

This ensures that individual trees are not overly correlated with each other. For instance, if there are 100 features, each split in a tree may consider only a randomly selected subset, such as 10 features. This randomness makes the trees in the forest more diverse, improving the ensemble's performance.

6.3.4.3 Support Vector Machine Classifier (SVMC)

The support vector machine classifier (SVMC) is a commonly used supervised learning method for classification and regression tasks. Its goal is to find the optimal hyperplane that effectively separates the data into distinct classes. It works well with linear and non-linear decision boundaries and is especially useful for high-dimensional and complicated datasets (Qaiser et al. 2024). In n-dimensional space, a hyperplane is an (n − 1)-dimensional flat subspace that partitions the space into two half-spaces. For a binary classification problem, the SVMC algorithm aims to identify the hyperplane that most effectively separates the two classes. Support vectors are the data points closest to the hyperplane and are crucial in defining its position. SVMC tries to maximize this margin to make the classifier more robust. The support vector machine classifier (SVMC) identifies a hyperplane in a linearly separable dataset, ensuring better generalization to new data. For a binary classification with input data, $\{(x_i, y_i)\}_{i=1}^{N}$ where $y_i \epsilon \{-1, +1\}$, the aim is to find a hyperplane defined as:

$$w.x + b = 0 \tag{6.17}$$

Here, w represents the weight vector perpendicular to the hyperplane, while b is the bias term.

SVM employs a soft margin for non-separable data by introducing a slack variable ξ_i for each data point. This allows some points to violate the margin constraint to improve model flexibility and generalization. The objective function of SVMC is defined as:

$$\underbrace{\min}_{w,b,\xi} \frac{1}{2}\|w\|^2 + C\sum_{i=1}^{N} \xi_i \tag{6.18}$$

where C is a regularization parameter that balances margin maximization and margin violations, higher values of C reduce the tolerance for violations, leading to a more complicated margin.

SVMC performs well even with many features. The margin maximization approach helps make SVMC less prone to overfitting, especially with regularization. With the kernel trick, SVMC can handle complex, non-linear decision boundaries. SVMC is a robust algorithm that maximizes the margin between classes, making it robust for binary classification, especially in high-dimensional spaces. Its flexibility through the kernel trick allows it to adapt to complex datasets with non-linear structures, and its mathematical foundation ensures optimal separation of classes based on support vectors.

6.3.4.4 Convolutional Neural Networks (CNNs)

Convolutional neural networks (CNNs) are a deep learning architecture commonly utilized for processing spatial and temporal data, making them highly effective for tasks such as image and video analysis (Gupta and Gill 2020). CNNs are highly effective for image recognition and classification as they capture spatial hierarchies in data (Gupta et al. 2023). A standard CNN architecture includes three key layers:

- *Convolutional Layers*: Perform convolution operations to extract essential features from the input.
- *Pooling Layers*: Decrease the spatial dimensions of feature maps, retaining critical information while enhancing computational efficiency.
- *Fully Connected (Dense) Layers*: Combine extracted features to generate the final predictions.

The convolutional layer serves as the core component of CNNs, detecting patterns in data through convolution operations. In CNNs, these layers apply a set of learnable filters (or kernels) to the input. Each filter, a small matrix of weights, moves across the input image, performing element-wise multiplication and producing a feature map that captures distinct characteristics of the image. The architecture of CNN is shown in Fig. 6.3, used in this study. Let's assume that X is an Input image matrix (e.g., 32 × 32 pixels with three color channels), and K is a convolutional filter or kernel of size $f \times f$. Then, the convolutional operation for a given position (i, j) is:

$$(X \times K)_{i,j} = \sum_{m=0}^{f-1} \sum_{n=0}^{f-1} X_{i+m,j+n} \cdot K_{m,n} \tag{6.19}$$

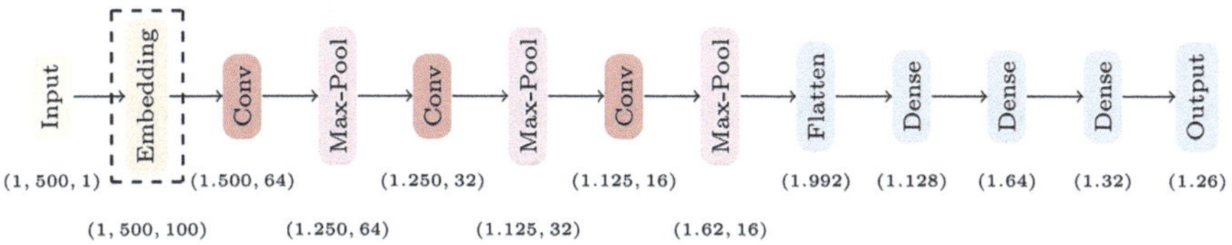

Fig. 6.3 CNN architecture with an embedding layer. The layers are shown as colored boxes: bright yellow for the input layer, light orange for the embedding layer, light red for the convolutional layer, light pink for the max-pooling layer, and light blue for the flattened/dense layer

After each convolution operation, an activation function like ReLU (Rectified Linear Unit) is applied to introduce non-linearity. ReLU is defined as:

$$f(x) = \max(0, x) \tag{6.20}$$

The pooling layer reduces the feature maps' spatial dimensions (width and height), reducing the computation and controlling overfitting (El-Tohamy et al. 2024). Max pooling is the most widely used technique, selecting the highest value within a pooling window to preserve the most essential features. Let P represent a pooling window (typically 2 × 2 or 3 × 3), with stride s. For each window position (i, j):

$$\text{MaxPool}_{i,j}(P) = \max p_{p\epsilon P} \tag{6.21}$$

Following several convolutional and pooling layers, CNN architectures usually include one or more fully connected layers to aggregate extracted features for the final classification or regression (Zhang et al. 2024). The final convolutional or pooling layer output is flattened into a 1D vector and passed through one or more fully connected (dense) layers. Each dense layer performs the following:

$$y = W.x + b \tag{6.22}$$

In a neural network, W represents the weight matrix, x is the input vector, and b is the bias term. The activation function used in the final fully connected layer depends on the specific task. The sigmoid activation function is applied for binary classification, while the softmax function generates class probabilities for multi-class classification. A linear activation function is employed for regression tasks to predict continuous values.

A CNN architecture generally comprises multiple convolutional and pooling layers, ending with one or more fully connected layers for classification or regression tasks.

CNNs employ convolutional layers to extract spatial features, pooling layers to downsample the data, and dense layers for classification or regression. CNNs excel at image-related tasks because they can learn spatial information and patterns hierarchically. With advances in architecture and training techniques, CNNs remain essential in computer vision and beyond.

6.4 Implementation Results and Discussion

As seen in Table 6.3, all models were assessed using optimal hyperparameters. In five iterations, the hyperparameters were adjusted using Bayesian optimization. RFC and SVMC were implemented using Scikit-learn, XGBoost was implemented using the Python package, and CNN was implemented using the Keras library.

6.4.1 *K*-fold Cross-Validation

K-fold cross-validation is a statistical technique to assess a machine learning model's performance. The data is randomly divided into approximately equal k subsets (or "folds"). Common choices for k are 5 or 10, but k can be any integer that divides the dataset appropriately. For each iteration $i = 1, 2, \ldots, k$; the ith fold is the test set, and the remaining $k - 1$ folds are the training set. The model is trained on the training set, and its performance is evaluated on the test set. This process provides "k" different evaluations of the model. *K*-fold cross-validation offers a more stable and reliable estimate of model performance by averaging results across multiple test sets. Each data point is used in the training and validation phases, particularly useful for small datasets.

The models were evaluated using stratified *K*-fold cross-validation. The training set's class ratio and the test set's were nearly identical. Only the test set—data the model had not seen—was used to assess the models' generalization performance. To minimize bias in hyperparameter optimization or model selection, nested cross-validation incorporates the *K*-fold cross-validation for final evaluation. As a result, nested cross-validation will guarantee that every piece of data is tested and capitalize on the data set's complete diversity. Figure 6.4 displays an example of a stratified nested cross-validation. In this study, $k_{out} = 5$ and $k_{in} = 4$ is selected. As a result, training, validation, and testing purposes employed about 80%, 10%, and 10% of the data from the dataset. Dataset 2 was used as an extra test set. Therefore, nested cross-validation was applied exclusively to dataset 1.

Table 6.3 Hyperparameters of different models

Machine learning model	Hyperparameters
SVMC	$C = 0.01, 0.1, 1, 10$ Gamma = 0.0001, 0.001, 0.01, 0.1
RFC	Estimator_*n* = 100, 200, 400, 500, 100 Max_depth = 5, 10, 15, 20
XGBoost	Max_depth = 5, 10, 15, 20 Eta = 0.0001, 0.001, 0.01
CNN	Kernel = 64, 128, 256 Batch_size = 32 Epochs = 300 Lr = 0.0001, 0.01, 0.05

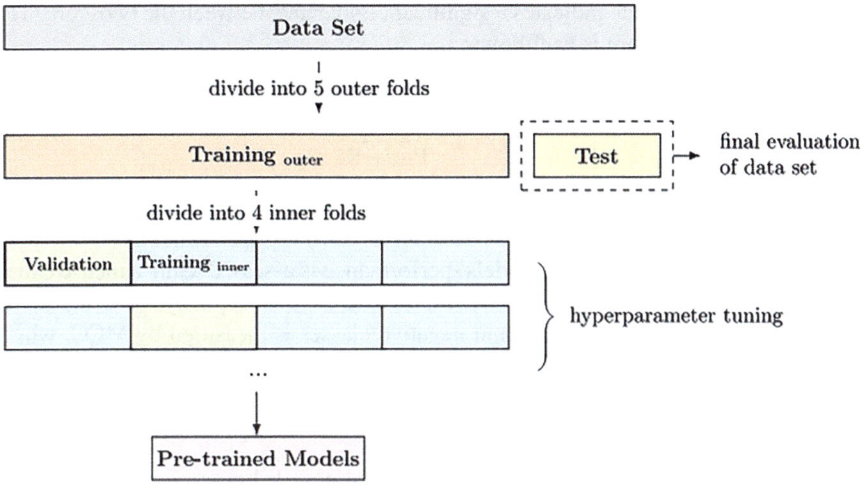

Fig. 6.4 A stratified nested cross-validation with $k_{\text{out}} = 5$ and $k_{\text{in}} = 4$

6.4.2 Evaluation Metrics

The study utilized the area under the precision-recall curve (AUCPR), Matthews' correlation coefficient (MCC), and F1-score as performance metrics. The following are the definitions of the MCC and F1-score equations for each class:

- Precision: Precision is the percentage of affirmative identifications that were genuinely accurate. In other words, it is obtained by dividing the number of correctly recognized positive samples by the total number of positive samples. It is provided by:

$$\text{Pre} = \frac{\text{TP}}{\text{TP} + \text{FP}} \tag{6.23}$$

- Recall: The number of accurately identified actual positives is known as recall. It is calculated as the ratio of correctly identified positive samples to the total number of actual positive samples. Another name for the recall is sensitivity. It is provided by:

$$\text{Rec} = \frac{\text{TP}}{\text{TP} + \text{FN}} \tag{6.24}$$

- F1 score: The F1 metric estimates the equilibrium between recall and precision. It is commonly used to address unequal class distribution and is the harmonic mean of two fractions. A higher F1 score is achieved when both precision and recall are high. A low F1 value indicates a significant imbalance between the twoscore. The formula for the score is as follows:

$$F1 = 2 \times \frac{\text{Pre} \times \text{Rec}}{\text{Pre} + \text{Rec}} \tag{6.25}$$

- AUCPR: AUCPR is a metric for evaluating classification models, especially in cases of imbalanced data where one class is much more dominant.
- MCC: Binary classification models' performance is assessed using a metric called the Matthews correlation coefficient, or MCC. The capacity of a model to accurately categorize positive and negative classes is measured by MCC, which is more balanced than accuracy or other metrics that can be skewed in unbalanced datasets.

$$\text{MCC} = \frac{\text{True}_{\text{pos}} \times \text{True}_{\text{neg}} - \text{False}_{\text{pos}} \times \text{False}_{\text{neg}}}{\sqrt{(\text{Tur}_{\text{pos}} + \text{False}_{\text{pos}})(\text{True}_{\text{pos}} + \text{False}_{\text{neg}})(\text{True}_{\text{neg}} + \text{False}_{\text{pos}})(\text{True}_{\text{neg}} + \text{False}_{\text{neg}})}} \tag{6.26}$$

6.4.3 Results

Every model was assessed at several taxonomic levels. The class was divided into 26 categories at more specific and greater levels. One representation is employed to encode protein sequences: sequence alignment-free (Word2Vec word embedding). Table 6.4 lists the comparative sequence representation aggregated across various machine-learning approaches. Triangle embeddings achieve an AUCPR of 94.34% at lower taxonomical levels and 99.52% at higher levels.

Table 6.5 lists the comparative machine learning techniques using different sequence representations. SVMC, as opposed to XGBoost, has a more significant deviation as the class grows and the data becomes more skewed. Consequently, the sequence representation has the most significant impact on both XGBoost and SVMC performance; nevertheless, compared to SVMC, XGBoost is least impacted by data skewness. CNN is the only classifier that performs better at the lower

Table 6.4 Comparison of sequence representation on the dataset

	Higher taxonomical level			Lower taxonomical level		
Word embedding	AUCPR (%)	F1 (%)	MCC (%)	AUCPR (%)	F1 (%)	MCC (%)
2-gram	98.51	97.68	96.57	93.45	88.78	81.23
3-gram	*99.52*	*97.69*	*96.78*	*94.34*	*88.98*	*82.34*
5-gram	99.34	97.56	95.67	93.56	86.78	80.91

Table 6.5 Comparison of the performance of various classifiers on the dataset

Classifiers	Higher taxonomical level			Lower taxonomical level		
	AUCPR (%)	F1 (%)	MCC (%)	AUCPR (%)	F1 (%)	MCC (%)
XGBoost	96.78	96.79	95.67	87.89	86.78	81.24
SVMC	95.17	91.45	85.47	86.56	76.89	79.65
RFC	97.34	95.68	95.67	92.56	87.56	81.46
CNN	*98.54*	*97.01*	*95.60*	*95.74*	*89.41*	*82.79*

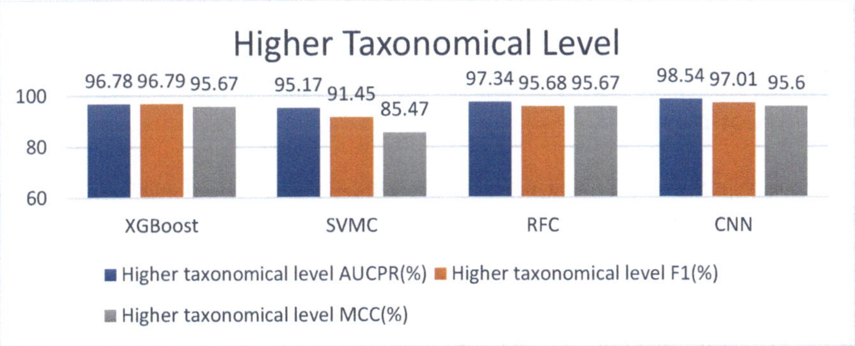

(a)

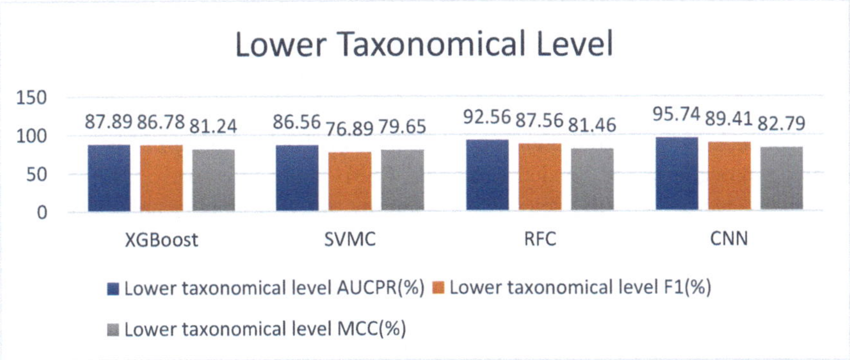

(b)

Fig. 6.5 Performance comparison of different models at (**a**) Higher Taxonomic Levels and (**b**) Lower Taxonomic levels

taxonomic level. CNN fared better than every classifier at both lower and higher taxonomical levels. Figure 6.5 illustrates the performance of various models across different taxonomic levels. The CNN model outperformed all other models compared to different avian sub-classes, as depicted in Fig. 6.6.

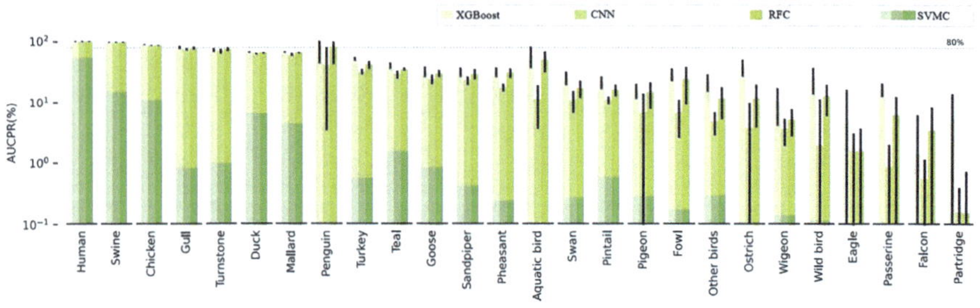

Fig. 6.6 Performance of different models on different classes of avian

6.5 Conclusion

This study predicts influenza using well-established supervised machine-learning techniques, focusing on the hemagglutinin sequences. Both conventional ML methods and DL techniques were assessed. A significant distinction between them is in their approach to data interpretation. Conventional ML requires a feature extraction stage, whereas DL utilizes hidden layers to analyze the input data features. After extracting evolutionary features from sequences using a position-specific matrix (PSM) and Word2Vec word embedding, the results were fed into well-known machine learning algorithms (SVMC, RFC, and XGBoost). An embedding layer can be added to deep learning models (such as CNN) to learn features automatically. Conducting a feature extraction procedure is advisable before supplying the input data into models lacking an embedding layer. PSM computation is a tedious and time-consuming process that might take days or months for massive data sets. The lack of a unified dimension is another drawback of using PSMs; therefore, unifying procedures are needed to handle PSMs more effectively. The dataset's sequences were carefully selected, containing only distinct whole sequences. Although supervised learning depends on ground truth, supervised machine learning was still used. The isolated host of sequences was identified by retrieving the class labels for the sequences from the information. As a result, not all of the labels employed in this study accurately reflect reality. Future research can incorporate semi-supervised or unsupervised learning, which differs from supervised learning because it requires little to no ground truth during training.

References

Aftab M, Amin R, Koundal D, Aldabbas H, Alouffi B, Iqbal Z (2022) Classification of COVID-19 and influenza patients using deep learning. Contrast Media Mol Imaging 2022:8549707. https://doi.org/10.1155/2022/8549707

Amiroch S, Mukhlash I, Isa Irawan M, Al Faroby MHZ, Nidom CA (2022) Machine learning for the prediction of antiviral compounds targeting avian influenza A/H9N2 viral proteins. Symmetry 14(6):1114

Asudani DS, Nagwani NK, Singh P (2023) Impact of word embedding models on text analytics in deep learning environment: a review. Artif Intell Rev 56(9):1–81. https://doi.org/10.1007/s10462-023-10419-1

Bukhari SNH, Ogudo KA (2024) Hybrid predictive machine learning model for the prediction of immunodominant peptides of respiratory syncytial virus. Bioengineering 11(8):791. https://doi.org/10.3390/bioengineering11080791

Chen T, Guestrin C (2016) XGBoost: a scalable tree boosting system. In: Proceedings of the ACM SIGKDD conference on knowledge discovery and data mining. ACM, pp 785–794. https://doi.org/10.1145/2939672.2939785

Chen J et al (2024) Application of peripheral blood routine parameters in the diagnosis of influenza and Mycoplasma pneumoniae. Virol J 21(1):1–11. https://doi.org/10.1186/s12985-024-02429-4

Curto G, Jojoa Acosta MF, Comim F, Garcia-Zapirain B (2024) Are AI systems biased against the poor? A machine learning analysis using Word2Vec and GloVe embeddings. AI Soc 39(2): 617–632. https://doi.org/10.1007/s00146-022-01494-z

Dai S, Han L (2023) Influenza surveillance with Baidu index and attention-based long short-term memory model. PLoS One 18(1):1–20. https://doi.org/10.1371/journal.pone.0280834

El-Tohamy A, Amin H, Badr N (2024) Integration of deep learning models for enhanced classification of viral DNA sequences across specific viruses and viral families. Int J Intell Comput Inf Sci 24(1):89–104. https://doi.org/10.21608/ijicis.2024.279692.1332

GISAID – gisaid.org. Accessed: 11 Nov 2024. [Online]. Available: https://gisaid.org/

Gupta C, Gill NS (2020) Coronamask: a face mask detector for real-time data. Int J Adv Trends Comput Sci Eng 9(4):5624–5630. https://doi.org/10.30534/ijatcse/2020/212942020

Gupta C, Gill NS, Gulia P, Yadav S, Pau G, Ieee M (2023) A real-time 3-dimensional object detection based human action recognition model. IEEE Open J Comput Soc PP:1–12. https://doi.org/10.1109/OJCS.2023.3334528

Hung SK et al (2023) Developing and validating clinical features-based machine learning algorithms to predict influenza infection in influenza-like illness patients. Biom J 46(5): 100561. https://doi.org/10.1016/j.bj.2022.09.002

Hussain S (2024) Exploring machine learning utilization on influenza pandemic dataset

Johnson SJ, Murty MR, Navakanth I (2024) A detailed review on word embedding techniques with emphasis on word2vec. Multimed Tools Appl 83(13):37979–38007. https://doi.org/10.1007/S11042-023-17007-Z/METRICS

Kara A (2021) Multi-step influenza outbreak forecasting using deep LSTM network and genetic algorithm. Expert Syst Appl 180:115153. https://doi.org/10.1016/j.eswa.2021.115153

Kumar R, Maheshwari S, Sharma A, Linda S, Kumar S, Chatterjee I (2024) Ensemble learning-based early detection of influenza disease. Multimed Tools Appl 83(2):5723–5743. https://doi.org/10.1007/s11042-023-15848-2

Li X, Li Y, Shang X, Kong H (2013) A sequence-based machine learning model for predicting antigenic distance for H3N2 influenza virus. Front Microbiol 15:2024. https://doi.org/10.3389/fmicb.2024.1345794

Liang J, Zeng Z, Li Q, Guan W, Yang Z, Hon C (2023) Challenge on prediction of Influenza virus and SARS-CoV-2 virus co-circulation. Interdiscip Med 1(2):1–5. https://doi.org/10.1002/INMD.20220006

Maheshwari S, Sharma A, Kumar R, Pratyush (2022) Early detection of influenza using machine learning techniques. In: Recent innovations in computing. Lecture notes in electrical engineering, vol 855. Springer, Singapore, pp 111–124. https://doi.org/10.1007/978-981-16-8892-8_9

Malik F, Suliman M, Khan MQ, Rahman N, Khan K (2024) Optimized XGBoost-based model for accurate detection and classification of COVID-19 pneumonia. J Comput Biomed Inform 7(2): 1–15

Park S, Moon J, Jung S, Rho S, Hwang E (2024) Explainable influenza forecasting scheme using DCC-based feature selection. Data Knowl Eng 149:102256. https://doi.org/10.1016/j.datak.2023.102256

Phu DN, Vinh PC, Quoc NK (2023) Enhanced diagnosis of influenza and COVID-19 using machine learning. EAI Endorsed Trans Context Syst Appl 9(1):1–6. https://doi.org/10.4108/eetcasa.v9i1.4030

Qaiser A, Manzoor S, Hashmi AH, Javed H, Zafar A, Ashraf J (2024) Support vector machine outperforms other machine learning models in early diagnosis of dengue using routine clinical data. Adv Virol 2024:5588127. https://doi.org/10.1155/2024/5588127

Savelli C, Giobergia F (2024) Enhancing cross-lingual word embeddings: aligned subword vectors for out-of-vocabulary terms in fastText. In: 2024 IEEE 18th international conference on application of information and communication technology. IEEE, pp 1–6. https://doi.org/10.1109/AICT61888.2024.10740438

Scherr TF, Douglas CE, Schaecher KE, Schoepp RJ, Ricks KM, Shoemaker CJ (2024) Application of a machine learning-based classification approach for developing host protein diagnostic models for infectious disease. Diagnostics 14(12):1290. https://doi.org/10.3390/diagnostics14121290

Singh S, Mittal S (2023) Pandemic outbreak prediction using optimization-based machine learning model. In: 2023 Third international conference on advances in computing, communication, embedded and secure system. IEEE, pp 154–159. https://doi.org/10.1109/ACCESS57397.2023.10199872

Vithanage D, Yu P, Wang L, Deng C (2024) Contextual word embedding for biomedical knowledge extraction: a rapid review and case study. J Healthc Inform Res 8(1):158–179. https://doi.org/10.1007/S41666-023-00157-Y/METRICS

Wang Y, Sibaii F, Lee K, Gill MJ, Hatch JL (2021) NOTE: This preprint reports new research that has not been certified by peer review and should not be used to guide clinical practice. 1. medRxiv 1(165):1–13

Xia YL, Li W, Li Y, Ji XL, Fu YX, Liu SQ (2021) A deep learning approach for predicting antigenic variation of influenza A H3N2. Comput Math Methods Med 2021:9997669. https://doi.org/10.1155/2021/9997669

Zhang L, Liu T, Shen K, Chen C (2024) A novel approach to malicious code detection using CNN-BiLSTM and feature fusion. arXiv:2410.09401v1

Zou X et al (2023) Accurately identifying hemagglutinin using sequence information and machine learning methods. Front Med 10:1281880. https://doi.org/10.3389/fmed.2023.1281880

Transforming Global Health: The Impact of AI and Blockchain on Viral Disease Control

7

Mobasshara Uzma, Mahima K. Kishinani, and Debasish Kar

Abstract

The global health of humanity is threatened by emerging viral diseases in a number of ways, which warns medical professionals and emphasizes the urgent need to create new technologies that can enhance the current approaches to viral disease detection, diagnosis, and treatment. This chapter explores how medical virology can be transformed by blockchain technology and artificial intelligence (AI), with a focus on how these technologies can ensure precise clinical decision-making. Blockchain is characterized by features such as immutability, decentralization, and safe infrastructure, which can allow secure sharing of healthcare data and also simultaneously ensure privacy of patient data. Generally, blockchain is combined with AI and ML, and this combination helps them to enable risk management, promotes analysis of clinical data, and allows for accurate decision-making in clinical settings with respect to viral disease diagnosis. Additionally, it can process huge amounts of data from healthcare resources such as hospitals, research laboratories, and clinical laboratories. It ensures the safe and appropriate transfer of healthcare data, thereby allowing professionals in the field of healthcare to share real-time information about viral outbreaks and thus enabling the development of effective management strategies. Tools driven by AI are extremely crucial for ensuring rapid viral disease screening, diagnosis, and treatment. These technologies are used in this field to promote research for developing personalized medicines and tailored diagnostic methods as well. This

Disclaimer: Parts of this chapter were paraphrased using a tool known as QuillBot.

Mobasshara Uzma and Mahima K. Kishinani contributed equally with all other contributors.

M. Uzma · M. K. Kishinani · D. Kar (✉)
Department of Biotechnology, Faculty of Life of Allied Health Sciences, Ramaiah University of Applied Sciences, Bangalore, Karnataka, India

J. M. Chatterjee, S. K. Saxena (eds.), *Artificial Intelligence and Blockchain in Precision Medicine and Virology*, Medical Virology: From Pathogenesis to Disease Control, https://doi.org/10.1007/978-981-96-8919-4_7

study emphasizes how the synergy of AI and blockchain leads to a paradigm shift and paves the way for the management of viral disease outbreaks by offering innovative solutions that can aid in improving patient outcomes and global healthcare management.

Keywords

Blockchain · Artificial intelligence · Therapeutic approaches · Medical virology · Global health management

Abbreviations

AI	Artificial intelligence
CNNs	Convolutional neural networks
CRISPR	Clustered Regularly Interspaced Short Palindromic Repeats
DL	Deep learning
HTS	High-throughput sequencing
ML	Machine learning
NGS	Next-generation sequencing
NLP	Natural language processing
POC	Point-of-care
RF	Random forest
SVMs	Support vector machines

7.1 Introduction

7.1.1 Overview of Global Challenges in Viral Disease Management

Novel technologies such as artificial intelligence (AI) and blockchain are emerging as vital tools and have influenced the shaping of global trends within the field of information and communication technology. In the field pertaining to the management of healthcare, various strategic tools have demonstrated their efficacy for developing valuable approaches that take the lead, such as data sharing, data mining, machine learning (ML), artificial intelligence, and blockchain (Amisha et al. 2019).

Particularly, the utilization of blockchain technology has increased in the field of healthcare over the past couple of years and is only growing with time (Kunal et al. 2024). This technology is currently being applied for enhancing evidence-based clinical decision-making processes. It has done so via the development of operational protocols to build a solid foundation. Blockchain has been identified to significantly contribute toward ensuring safety in the process of sharing information among diverse groups. Blockchain technology will be able to support and facilitate the development of new workflows and enhance existing protocols by using collaborative tools.

AI has also emerged as one of the most powerful and promising analytical tools that are available. By transforming raw big data into actionable insights, AI serves as a critical tool in addressing complex challenges across different fields. However, data must first be cleaned, structured, and integrated to maximize AI's efficacy.

Despite great advances in medicine and basic research over recent decades, infectious diseases caused by transmissible pathogens—such as bacteria, parasites, viruses, and fungi—continue to challenge researchers and clinicians. These challenges are exacerbated by limitations in rapid and precise detection of infections, coupled with the rise in antimicrobial resistance (Amisha et al. 2019). Basic research continues to tackle these challenges by developing therapies to fight infections, coming up with preventive measures, and introducing fast and accurate diagnostic tools (Nguyen et al. 2021). Healthcare systems are beginning to rely heavily on innovations such as AI and blockchain technology to address these persistent challenges effectively.

7.1.2 Emerging Need for Technology in Medical Virology

Currently, no data surveillance system exists that can immediately provide health organizations with real-time information needed to identify potential outbreaks of concern. Despite significant progress in the fields of molecular biology, genetics, computational sciences, and medicinal chemistry, infectious diseases remain one of the most significant threats to public health. However, if it is possible to combine these advancements with systems and synthetic biology, it can allow significant progress in this field. These advancements include accelerating the discovery of anti-infective drugs, gaining a deeper understanding of infection biology, and also enhancing the development of new and improved diagnostic tools and therapeutics.

The occurrence of the COVID-19 pandemic shook the entire world and also highlighted the limitations that exist in the current healthcare systems that are responsible for managing public health emergencies (Kunal et al. 2024). When such situations arise, the utilization of technologies such as AI and blockchain is increasingly being regarded as a potentially promising solution for fighting against infectious disease outbreaks (Nguyen et al. 2021). Additionally, using these innovative technologies promotes the early and timely detection of viral disease outbreaks. This further alerts healthcare professionals to improve the organization and accessibility of healthcare and medical data to ensure a strong and reliable supply chain in times of medical crises.

7.1.3 Scope and Objectives of This Chapter

This chapter aimed to explore and investigate the potential of AI and blockchain in transforming the approach to tackling the various challenges that are typically posed by viral diseases and their outbreaks. It also elucidates their applications across the various areas of disease management, such as timely detection and diagnosis, prediction of future disease outbreaks, assessment of risk factors, and optimization of treatment strategies (Amisha et al. 2019).

This chapter has also highlighted the current progress as well as the future potential of AI and how it can play a role in revolutionizing healthcare. It does so by suggesting the utilization of a combination of the conventional epidemiological approaches along with upcoming advanced computational methods to ensure a significant improvement in the outcomes of global health.

In addition, this chapter evaluates the global health implications of AI and blockchain in the field of medical virology and presents creative solutions that are aimed at improving patient outcomes, promoting and advancing research, and enabling efficient and evidence-based decision-making in healthcare facilities across the world. The ethical and data privacy concerns that are typically associated with such innovative technologies have also been discussed, with a particular emphasis on the importance of using them responsibly.

Ultimately, this chapter aims to drive further research and development at the intersection of AI and medical virology, which will pave the way for the development of innovative solutions that will help in effectively combating viral disease outbreaks (Nguyen et al. 2021).

7.2 The Role of AI in Viral Disease Control

7.2.1 Machine Learning Algorithms for Viral Disease Detection

Infections caused by bacteria, fungi, viruses, and parasites pose significant global health concerns due to their rapid transmission potential, leading to epidemics. Developing techniques that will help in early and timely detection has proven difficult owing to the lack of unified, reusable, and interoperable models (Belák et al. 2009). Machine learning (ML) algorithms are emerging as transformative tools. These algorithms can address these gaps by processing large datasets and identifying complex patterns that typically go undetected when traditional methods are used. Such algorithms are highly effective in classifying infections, predicting disease progression, and optimizing treatment strategies.

The integration of ML models within healthcare systems has enabled researchers to develop innovative solutions for viral diseases, such as COVID-19, dengue, influenza, and Ebola (Alqaissi et al. 2023). These advancements have been proven to significantly improve patient outcomes, enhance public health responses, and foster a more proactive approach to managing infectious diseases.

7.2.1.1 Supervised Learning Models

Supervised learning models, which constitute a key subset of ML, work by classifying viral infections upon analyzing complex datasets to uncover patterns that appear to be indicative of disease.

Decision trees can be characterized by their simplicity and interpretability. They are widely used and help in organizing data into subsets and do so based on certain feature values. This can help to accurately and effectively identify specific symptoms occurring during viral infections; however, they require careful tuning to ensure accuracy.

Support vector machines are specifically used when dealing with high-dimensional datasets. They can help in classifying infected versus non-infected cases and do so by creating hyperplanes to separate different classes of data. This method is very effective for differentiating between closely related viral strains.

Neural networks are very transformative in this domain (Glaser et al. 2019). These models automatically learn and extract features from raw data, such as medical images or genetic sequences. For example, during the COVID-19 pandemic, convolutional neural networks (CNNs) were widely used for analyzing chest X-rays and CT scans (Glaser et al. 2019), thereby aiding in the detection of lung infections caused by this disease.

As shown in Fig. 7.1, the growing significance of machine learning in fields such as neuroscience further underscores its interdisciplinary impact. Machine learning models not only improve diagnostic accuracy but also enable real-time analysis, which is critical for early detection and management of viral outbreaks, thereby significantly enhancing public health responses and outcomes.

7.2.1.2 Deep Learning Techniques

The advancement of deep learning techniques, especially CNN, has revolutionized the medical imaging analysis field by coming up as a prime modality for detection and diagnosis (Belák et al. 2009) of viral infections with several imaging modalities such as computed tomography. In these CNNs, neural networks are a variation

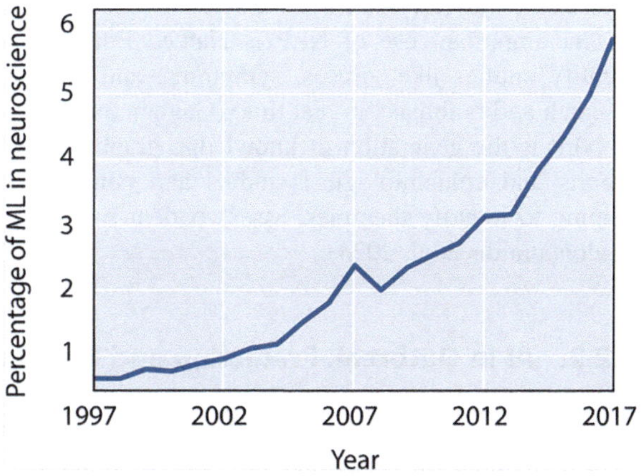

Fig. 7.1 Growth of machine learning (ML) in neuroscience

designed for visually processing images (Glaser et al. 2019). This model learns hierarchical features in an image via a series of convolutional and pooling layers in a completely automated fashion.

In essence, their key strength lies in the capability to embed features of spatial and temporal dependencies within imaging data. Therefore, in the case of detecting viral infections, CNNs view the CT scan as a means to find a key pattern consistent with lung infection, such as ground-glass opacities or lung consolidations typical for COVID-19 and other viral pneumoniae (Glaser et al. 2019). The process begins with convolution layers, where filters slide over input images to extract features such as edges, textures, and shapes. These features are further refined in the pooling layers, which help in reducing the spatial dimensions of the data and hence result in the improvement of computational efficiency with no loss of essential information. Ultimately, fully connected layers combine these refined features into predictions that can classify an image as infected or non-infected.

The application of CNN in the field of medical imaging offers various advantages in terms of automation, speed, and consistency (Glaser et al. 2019). These are particularly crucial attributes during pandemics, when healthcare systems are under great pressure. CNN-based models contribute significantly by providing real-time insights into the severity and progression of lung infections. This allows clinicians to make informed decisions, plan treatments, and optimize resource allocation. CNN has been highly instrumental in managing the outbreaks of viral infections, such as that of COVID-19, further underscoring the crucial role of CNNs in modern healthcare.

7.2.1.3 Natural Language Processing

Natural language processing (NLP) is an important tool that can be used for pulling out information from sources such as scientific literature, clinical reports, and public health updates, particularly during viral outbreaks. With rapidly advancing medical research, NLP helps in efficiently processing large volumes of text, converting unstructured data into actionable insights. For instance, during COVID-19, NLP was used for analyzing resources (the CORD-19 dataset) to reveal key information such as the intricate details of the virus, treatments, and disease patterns.

One important use of NLP is Named Entity Recognition (NER), which can identify entities like viruses, symptoms, and medications in texts to help track research and outbreaks in real time (Gautam and Raza 2024). Another contribution of NLP is the generation of knowledge graphs, which parse unstructured clinical reports and epidemiological studies and convert them into structured formats, helping to identify therapies, assess patient risk, and track the spread of diseases (Kalogiannidis et al. 2024).

7.2.2 AI in Outbreak Prediction and Risk Assessment

Predictive analytics, driven by artificial intelligence, makes use of large volumes of data to predict outbreaks and thus enable timely interventions with appropriate

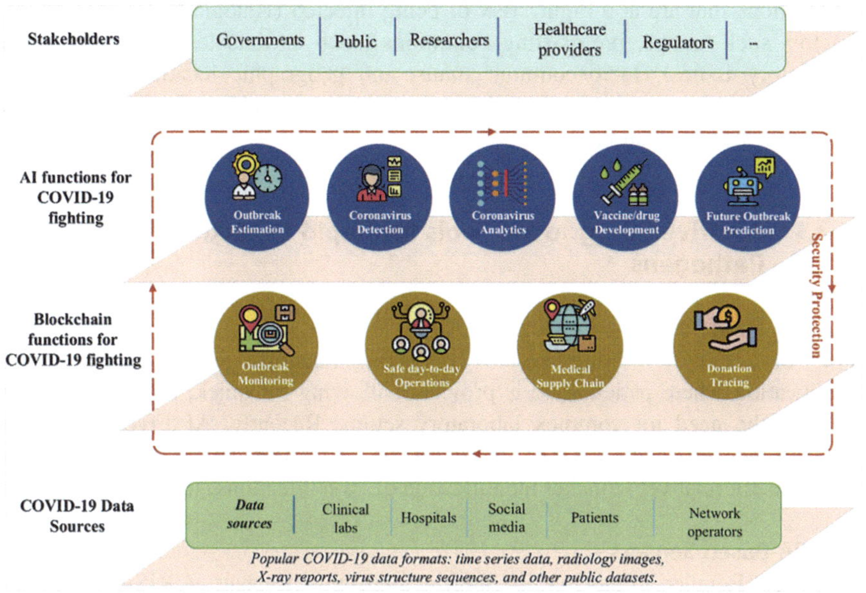

Fig. 7.2 Combination of blockchain and AI to combat coronavirus

allocation of resources. AI models can forecast the onset and spread of a disease outbreak through analysis of various data sources: epidemiological records, social media trends, environmental conditions, and mobility patterns. For example, Google Flu Trends once attempted to estimate flu activity in near time from search query data (Nguyen et al. 2021). Although the initiative had certain limitations, it showed that unconventional data sources hold a lot of promise for outbreak prediction, as shown in Fig. 7.2.

This has been further enhanced by modern AI techniques, such as machine learning and deep learning, through the integration of diverse datasets to obtain complex patterns that might usually be missed by traditional statistical methods (Nguyen et al. 2021). Variables included in such models are generally as follows: infection rates, population density, climate factors, and travel data, among others. This offers high accuracy in predicting the spread of diseases. For example, during the COVID-19 pandemic, AI-powered tools leveraged data from smartphone mobility to evaluate the effectiveness of lockdowns and predict emerging infection hotspots.

Risk stratification is another critical application of AI, where individuals or populations are categorized based on their likelihood of experiencing adverse health outcomes (Kalogiannidis et al. 2024). This process is especially valuable in vaccine distribution, where AI can help in the identification of high-risk groups and prioritize dose allocation to those individuals that need it more than others (Dias, R., Torkamani, A). By processing data from medical records, demographic profiles, social determinants of health, and disease prevalence, AI can aid in pinpointing those populations that are at a higher risk of being infected (Kalogiannidis et al. 2024). Factors such as age, pre-existing conditions (such as diabetes, heart disease, or respiratory issues), socioeconomic status, and geographic location can also be analyzed to determine which individuals are more vulnerable than others to severe disease or hospitalization.

7.2.3 AI-Driven Diagnostic Tools for Rapid Detection of Viral Pathogens

7.2.3.1 Point-of-Care Testing (POC)

Point-of-care (POC) testing involves conducting diagnostic tests directly at or near the location where patient care is provided, allowing for quick, immediate results without the need for complex laboratory setups. Recently, AI-driven diagnostic technologies, especially CRISPR-based sensors, have been developed as innovative solutions for relatively quicker identification of viral infections (Ibrahim et al. 2022).

CRISPR-Based Sensors

Clustered Regularly Interspaced Short Palindromic Repeats (CRISPR) is being widely adopted for use in diagnostic tools. It was initially discovered as part of the bacterial defense mechanism. One application is the use of CRISPR-based sensors, which, due to the high precision of the system, can assist in identifying specific nucleic acid sequences associated with viral genomes. The sensor is activated when a viral infection is detected, generating a detectable signal that does not necessarily require the use of sophisticated equipment

These sensors are extremely advantageous as they provide notable specificity, which can detect even trace amounts of viral RNA or DNA. For example, genetic markers from viruses like HIV, influenza, Zika, or SARS-CoV-2 can be targeted by CRISPR systems. CRISPR is ideal for rapid viral detection because of its real-time ability to amplify and identify viral genetic material; findings can be obtained in as little as 30 minutes to 1 hour

AI Integration in CRISPR-Based Sensors

AI integration has the potential to greatly improve the performance of CRISPR-based diagnostic tools by increasing their sensitivity, accuracy, and capacity to handle complicated data. In order to separate legitimate signals from false positives or background interference, artificial intelligence (AI) algorithms are used to analyze the data produced by CRISPR-based sensors (Ibrahim et al. 2022). By adjusting to various virus strains or mutations, machine learning models can also help improve the detection process and increase test reliability (Dias and Torkamani 2019).

Additionally, AI makes it possible to automate the interpretation of results, which can help medical practitioners use the diagnostic data to make quicker and better judgments. An AI system might, for instance, evaluate the sensor's data and instantaneously reveal the likelihood of a particular viral infection as well as suggested courses of treatment, minimizing human error and improving patient care.

7.2.3.2 High-Throughput Screening

Viral genomes may be efficiently and extensively sequenced using next-generation sequencing (NGS) and other high-throughput sequencing technologies. These tools allow researchers to produce vast amounts of genomic data by analyzing genetic material from viral samples taken from infected people, environmental sources, or animal reservoirs. Identifying viral strains, researching genetic variants, and tracking changes that may affect a virus's behavior, transmission, or resistance to specific medications or treatments all depend on this information (Dias and Torkamani 2019).

When compared to conventional approaches, high-throughput sequencing (HTS) techniques are also beneficial since they allow for the parallel sequencing of several samples, greatly reducing both time and expense. This benefit makes it easier to quickly identify known and newly developing viral strains, even ones that might not have been discovered in the field before, when applied to viral genomics (Nguyen et al. 2021).

Role of AI in Genomic Analysis

In order to analyze the massive datasets generated by high-throughput sequencing, artificial intelligence—particularly machine learning algorithms—is essential. These models are specifically made to process and interpret complex genomic data, and they can help you find patterns that traditional bioinformatics techniques might miss.

1. *Variant detection:* By comparing sequences from various samples, artificial intelligence (AI) can detect genetic variants in viral genomes, such as mutations that affect viral pathogenicity or drug resistance, and predict how these changes might affect the virus's structure or function, which will help identify the emergence or existence of potentially more dangerous variants.
2. *Novel strain identification:* Finding new or emerging viral strains is one of the major challenges that are commonly encountered in viral genomics. In particular, deep learning models are capable of analyzing large amounts of sequence data to find subtle differences between known and unknown viral genomes. By comparing viral sequences from different sources, such as different geographic regions or host species, artificial intelligence (AI) can help detect previously unidentified viral strains that may have evolved due to mutations or recombination events.
3. *Phylogenetic analysis and evolution tracking:* By creating phylogenetic trees, which chart the ancestry and geographic distribution of viral strains across time, artificial intelligence (AI) might improve our comprehension of viral evolution. Knowing this is essential for tracking mutations and determining if they have the potential to trigger pandemics or epidemics. AI models can also help predict how infections may change in the future, providing important information about possible dangers.
4. *Predicting virulence and transmission:* By examining the connections between particular genetic characteristics and potential clinical effects, AI can also assist in forecasting the pathogenicity of newly developing virus strains (Dias and Torkamani 2019). Viral genomic sequences and host responses, including

immune system activation, illness development, and therapy effectiveness, can be correlated by machine learning models. Furthermore, by examining genetic variables linked to transmission, including receptor binding affinity or environmental stability, AI might help forecast how easily a virus might spread.

Integrating HTS with AI for Pandemic Preparedness

AI-powered high-throughput viral genome scanning is essential for maintaining pandemic preparation and for global health in general. This makes it possible for health officials to act quickly, whether it means creating vaccinations, putting containment plans into place, or providing funds to control a possible outbreak.

For example, AI models were utilized to quickly analyze viral genomes during the COVID-19 pandemic, which aided in tracking the introduction of novel variations like Omicron and Delta (Dias and Torkamani 2019). These variations showed significant mutations that impacted vaccine efficacy and transmission rates, highlighting AI's significant role in quickly detecting such changes and directing public health initiatives.

7.3 Blockchain Technology in Healthcare

7.3.1 Fundamentals of Blockchain in Viral Diseases

Blockchain has been suggested as a way to guarantee the dependability and security of immunization records. A blockchain-based management system can encourage people to get vaccinated in addition to its technological advantages, which could help lower the number of COVID-19 cases and deaths. Blockchain is a distributed, decentralized ledger system that uses cryptography to guarantee data integrity, security, and transparency.

Blockchain has a lot of promise for managing viral diseases since it makes it possible to share data securely, monitor in real time, and track outbreaks effectively. According to Dias and Torkamani (2019), it facilitates the gathering and validation of clinical data, guaranteeing precision and prompt epidemic monitoring. Blockchain can eliminate information delays by automating data processing and establishing safe transactions between labs, hospitals, and research institutions with the aid of smart contracts.

Additionally, by guaranteeing traceability and lowering the possibility of counterfeiting, blockchain enhances the management of vaccination and antiviral drug supply chains, guaranteeing that only verified products are delivered to patients and healthcare professionals (Kunal et al. 2024). Blockchain facilitates efficient data management by creating a framework based on trust, which encourages cooperation and prompt reactions in times of public health emergencies.

7.3.2 Blockchain for Securing Healthcare Data

Medical data is one category of extremely sensitive data. One of the most appealing study topics at the moment is the use of blockchain technology to safeguard medical data, which is crucial for the concept of information blocking. When a patient must protect their data from third parties while sharing it with someone, they may not always choose to publicly share it. This is one method to explain information blocking. By incorporating blockchain technology into healthcare, worries regarding illegal access and data breaches are successfully addressed while also improving data security and privacy.

High levels of secrecy and regulatory compliance are necessary for healthcare systems to handle vast amounts of sensitive data, including patient records, diagnostic results, and genomic information (Kunal et al. 2024). Blockchain uses decentralized storage and cryptographic hashing to protect data, making it irreversible and impenetrable. Every data transaction is documented as a block that is connected to the one before it, forming a chain that stops any changes. Through permissioned access, patients maintain ownership over their health information, enabling safe sharing with approved parties including physicians, insurers, and researchers.

Furthermore, by offering a single platform for data transmission across several healthcare systems, blockchain facilitates interoperability by lowering fragmentation and enhancing coordination. Additionally, it generates audit trails, guaranteeing accountability and transparency in data management. Blockchain thus increases data security, boosts compliance, and fortifies trust, offering a strong basis for the advancement of healthcare systems (Kunal et al. 2024).

The architecture of blockchain offers three key benefits that elevate security in healthcare:

1. *Decentralization:* Blockchain can lessen the chance of data loss by dispersing data over a network. It accomplishes this by removing a single point of failure and making sure that data is dispersed throughout several network nodes as opposed to being kept in one central location. This eliminates the potential for a single point of failure, which reduces the likelihood of data loss or unauthorized access due to system breaches.
2. *Cryptographic protection:* Blockchain provides mathematical assurance that data can't be compromised by securing it with cryptographic techniques. By offering mathematical assurance that the data cannot be altered or viewed without the proper authorization, this realistically makes it difficult for hackers to compromise the data's integrity (Kunal et al. 2024).
3. *Tamper-proofing:* Blockchain ensures that data is secure and unchanged by tracking every access to it. Since each transaction or modification to the data is recorded in an immutable ledger, an audit trail can be confirmed. Any attempt to alter data must be approved by the network, making it simple to identify and effectively stop undesired changes.

All things considered, blockchain is a dependable way to preserve data integrity in cloud storage since it ensures verifiability and total immutability of data records. Because of these features, blockchain is a powerful tool for safeguarding personal information, particularly in industries that demand a high level of security and trust, such as supply chain management, healthcare, and finance.

7.4 Integrating Blockchain and AI in Viral Disease Management

7.4.1 Synergies Between Blockchain and AI

Integration of blockchain with artificial intelligence could be highly transformative in the management of viral diseases since it provides the strengths of both systems to enhance the security level of data processing and improve decision-making processes. Blockchain is essentially a decentralized, tamper-proof ledger that ensures data integrity and transparency; AI brings more complex techniques of machine learning and NLP techniques to analyze and interpret highly complex datasets (Gautam and Raza 2024). These technologies can be combined to meet the most pressing challenges of healthcare, such as data privacy, interoperability, and scalability (Kunal et al. 2024). For example, AI algorithms can process large datasets on viral outbreaks, predict transmission patterns, and identify high-risk regions.

Blockchain complements this by securely storing this data and enabling real-time sharing among stakeholders without compromising privacy. Moreover, AI models may be trained on an encrypted dataset stored on a blockchain, as shown in Fig. 7.3. This way, there will be federated learning without the need for any centralized data storage. This synergy speeds up the decision-making process not only in public health but also in developing personalized treatment plans as well as predictive

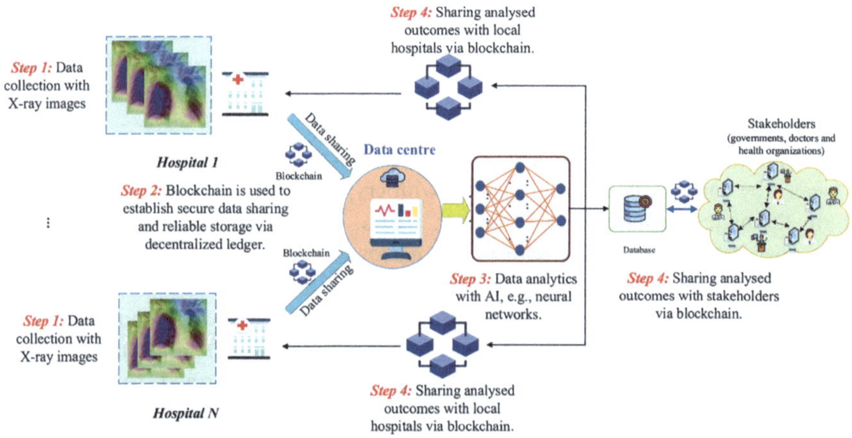

Fig. 7.3 An illustrative use of blockchain and AI for COVID-19 data analytics

analytics to effectively contain viral outbreaks (Adeyeri 2024). Thus, blockchain together with AI provides a secure and intelligent environment for dealing with and preventing viral illnesses.

7.4.2 Applications in Vaccine Development and Drug Trials

The beginning of 2020 witnessed an outbreak of a disease caused by a new virus called SARS-CoV-2. The sudden explosion and uncontrolled global spread of COVID-19 demonstrated the weakness of the existing healthcare systems in handling public health emergencies in a timely manner. To overcome these limitations, novel technologies such as blockchain and AI have emerged as promising solutions to fight the coronavirus epidemic (Adeyeri 2024). Additionally, AI can support drug development and offer clever ways to identify coronavirus-related symptoms for treatment. In order to tackle COVID-19, a comprehensive assessment was conducted on the application of blockchain and artificial intelligence.

Data from clinical trials is utilized to prepare peer-reviewed journal articles and applications to regulatory agencies for new treatment approval. Numerous stakeholders, pharmaceutical and device businesses, government regulators, and—above all—potential patients and the general public depend on the quality of this data in both situations. It has a significant impact on vaccine development and drug trials, as it will enhance efficiency, transparency, and security (Medina et al. 2021).

In vaccine research, AI algorithms can analyze extensive biological datasets, including genomic sequences of viruses, to identify potential vaccine candidates and predict their efficacy. Machine learning models further assist in optimizing formulations and simulating immune responses, thereby reducing the time required for preclinical research. Blockchain aids in this process by making sure clinical trial data are secured to ensure that records are immutable and transparent. This is not only meant to prevent tampering of data but also enhances compliance with regulations as every stage of vaccine development becomes auditable (Medina et al. 2021).

In the case of drug trials, AI can enable faster patient recruitment, monitor adverse effects in real time, and analyze treatment outcomes for improved adaptive designs that enhance efficacy (Maslove et al. 2018). This allows for sensitive patient data to be secured through cryptographic protection, which will ensure privacy and make informed consent management possible. Blockchain can also automate the tasks related to the disbursement of funds and sharing of data among the stakeholders, which reduces administrative overheads and ensures seamless coordination. The integration may also facilitate global collaboration as researchers and pharmaceutical companies may securely share data and insights to accelerate the development of antiviral treatments. Ultimately, what would be achieved by introducing blockchain and AI in vaccine development and drug testing is efficiency and reliability because transparency and accountability in processes help to build public trust.

7.5 Case Studies and Real-World Applications

7.5.1 Blockchain and AI in COVID-19 Pandemic Management

The conventional healthcare system faces several challenges, including limited access to COVID-19 data, difficulties in data management by participants, and high expenses (Kunal et al. 2024). Blockchain and AI technologies, which support patient-centered care, must effectively manage the flow of COVID-19 data from diverse sources across various services and applications (Jabarulla and Lee 2021). These include prediction of outbreaks, virus detection, development of drugs and vaccines, virus analytics, future case forecasting, and automated surveillance.

Blockchain and AI address privacy concerns in contact tracing by enabling the secure sharing of personal data without revealing identities. To track exposures and hotspots, decentralized systems incorporate technologies like mobile apps, GPS, and QR codes. Additionally, blockchain makes digital vaccine passports possible, enabling public health authorities to verify their authenticity (Jabarulla and Lee 2021). This strategy supports a responsible economic recovery without increasing the number of cases by striking a balance between privacy and public health concerns.

To forecast viral epidemics, AI algorithms examine data on infection rates, recoveries, and mobile phone usage trends. Changes in phone usage among infected or deceased people, for instance, can be used to estimate the size of outbreaks. Blockchain-powered decentralized apps (dApps) protect privacy while allowing for worldwide pathogen propagation tracking. Additionally, AI-driven analysis assists in predicting possible threats and identifying populations that are at risk, enabling preventative actions.

AI and machine learning (ML) improve viral disease diagnosis by analyzing medical images, including X-rays, CT scans, and MRIs. For example, deep learning (DL) models can differentiate COVID-19 pneumonia from other conditions with high accuracy. Federated learning, operating within a decentralized framework, trains AI models on distributed data, preserving privacy (Jabarulla and Lee 2021). This decentralized, patient-focused approach connects disparate healthcare data sources, ensuring confidentiality and building trust.

7.5.2 AI-Powered Diagnostic Systems for Influenza and Dengue

Machine learning (ML) in the context of influenza has been utilized to accomplish a number of objectives, including predicting which influenza virus strains would be most dominant and estimating the incidence of illnesses in future seasons (Marquez et al. 2023). Metabolomic data from patients can be evaluated to determine whether they have influenza during the diagnostic phase, according to studies. In one study, a classifier for the diagnosis of influenza was even developed using open-access data; however, not all of the patients in the study had RT-qPCR findings to validate and confirm the classifier's diagnostic performance. Additionally, ML has been applied to predict the effectiveness of influenza vaccines (Alqaissi et al. 2023).

For detecting dengue, several ML algorithms have been developed, especially for the purpose of early diagnosis or prediction of outbreaks (Moumita et al. 2024). The most frequently used models are as follows:

Random Forest (RF) RF models have been used in order to predict dengue outbreaks based on the demographic factors, historical case data analysis, and weather patterns. RF models are capable of handling large datasets. They also provide feature importance, which makes them an attractive option in epidemiological studies.

Support Vector Machines (SVM) In order to differentiate dengue from other infections, SVMs have been employed for clinical diagnosis based on evaluations of laboratory data, patient symptoms, and serological tests.

Decision Trees Both clinical illness diagnosis and epidemic risk prediction in certain regions have made use of decision trees.

Research has endeavored to incorporate machine and deep learning models with socioeconomic and environmental data in order to create a hybrid model that would improve the dengue outbreak forecast accuracy (Moumita et al. 2024). In order to provide a more thorough knowledge of outbreak trends, these hybrid models combine structured (numerical) and unstructured (text and image) data. AI models have shown promise in assisting with clinical diagnosis, particularly in situations when laboratory resources are scarce. For instance, machine learning algorithms are able to create risk scores for possible dengue infections by analyzing patient symptom data (Moumita et al. 2024).

7.5.3 The Role of Blockchain in Streamlining Vaccine Distribution

The year 2020 exhibited a certain fragility from the start. Our healthcare systems' inefficiency was shown by the epidemic, and as a result, their resilience was demonstrated to be lacking (Ghanim Hamid 2023). When something akin to a virus devastated our everyday lives, healthcare systems were unable to recover swiftly. A vaccine would be the first step on a long road to recovery, but even if one was created, there would always be challenges in getting it to people around the world (Ghanim Hamid 2023).

All vaccines must be stored at the correct temperature. For example, the Pfizer COVID vaccine needs to be stored at -70Â°C (−94Â°F) (Ghanim Hamid 2023). The COVID vaccination from Pfizer must be administered in two doses over a specific period of time. The two doses would not be interchangeable if a different vaccination that required two doses were to be introduced; for example, if the first dosage was the Pfizer vaccine, the second dose would have to be Pfizer's (Ghanim Hamid 2023).

Coordinating this endeavor with several vaccines on the market would be a daunting task once the entire population begins receiving the shot. By integrating blockchain technology, systems can cooperate: healthcare practitioners can determine which vaccine is required, the patient can have unchangeable proof of which vaccine they have got, and the vaccine dose is tracked.

The integration of blockchain technology and artificial intelligence (AI) in vaccine supply chain management has been highlighted in several studies as a solution to improve efficiency, transparency, and accountability (Ghanim Hamid 2023).

The key benefits are as follows:

- Blockchain enhances decision-making automation, increases transparency, reduces vaccine waste, and monitors population responses for side effects and safety concerns.
- Blockchain systems can handle large data volumes needed for vaccine surveillance, offering real time, auditable tracking of distribution processes.
- By streamlining logistics and enhancing management, blockchain reduces distribution costs while ensuring a secure and transparent supply chain.
- The technology ensures vaccine quality is preserved throughout storage and transportation by monitoring logistics like refrigeration and equipment.
- The system promotes transparency, making vaccine administration and distribution more secure and reliable (Medina et al. 2021).
- Financial savings from efficient distribution can be reinvested to improve logistics systems further.

Overall, blockchain's implementation addresses logistical complexities and ensures a robust, scalable, and trustworthy vaccine distribution process.

7.6 Future Directions

The future of healthcare can be revolutionized by scaling artificial intelligence (AI) and blockchain technologies in global health, specifically focusing on managing viral diseases. For example, an AI-driven platform known as HealthMap was developed, which made use of machine learning algorithms for analyzing data from various different sources. This platform was able to identify unusual cases of pneumonia in Wuhan in the initial stages of the coronavirus outbreak in December 2019 (Ahmed et al. 2024). Healthcare data can be shared among researchers using blockchain owing to its properties of secure and decentralized data sharing. This would enhance collaboration among global researchers without the concern of compromising data integrity.

Additionally, platforms such as NanoVision help in promoting and accelerating research on viral diseases by allowing secure and global sharing of molecular-level data (Ahmed et al. 2024). For helping patients with viral diseases, wearable technologies integrated with AI and blockchain will enable continuous monitoring of their health status (Adeyeri 2024). It can do so as AI is capable of analyzing real-time data, which will result in the detection of small abnormal changes, whereas blockchain can ensure that the data obtained can be stored and shared safely, thereby improving overall patient care and also appropriately responding to viral disease outbreaks.

During viral disease outbreaks, it is crucial to monitor the transparency as well as efficiency of medical supply chains, which can be enhanced using blockchain. This would help in ensuring the originality of medical supplies, and it can also help in preventing the use of illegally sold products. Additionally, using blockchain can also aid in the maintenance of a reliable supply chain responsible for necessary equipment and medicines. Ethical concerns such as data privacy and security can also be addressed by integrating technologies such as AI and blockchain, as the decentralized nature of blockchain allows patients themselves to control and handle their data, whereas AI provides valuable insights without attempting to compromise the privacy of an individual (Kumar et al. 2023).

When innovative technologies such as AI and blockchain are integrated in a reasonable manner, they will surely display their potential for creating and maintaining a relatively more responsive as well as resilient global healthcare system.

Overall, this combination of technologies can significantly contribute to developing and maintaining a secure, decentralized, and robust healthcare system. Such a healthcare system can promote the advancement of research to discover new drugs and treatments at a quicker rate, which will aid in effectively responding to viral disease outbreaks and significantly enhance patient care using automated procedures and insights that have been gained from healthcare data obtained from various sources.

7.7 Conclusion

When AI and blockchain technologies become integrated into the existing healthcare system, this modification can offer a potential transformative approach for the management of viral disease outbreaks. Moreover, it can also help in providing creative solutions to tackle the challenges that are typically faced in terms of detection, treatment, and prevention of viral diseases. AI, being capable enough to analyze and process large datasets, predict outbreaks that might occur in the near future, and design personalized treatments, can enhance the features of blockchain in terms of data security and transparency (Kumar et al. 2023). The invaluable framework that these two technologies can produce when they are combined appropriately and in a considerate manner can allow effective exchange of data, enhance precise diagnosis, and expedite the development of new vaccines and drugs to treat viral diseases.

Blockchain, offering its characteristics of data protection via encryption and its decentralized design, can promote the development of platforms for healthcare data protection, enable adherence to privacy laws, and also promote global cooperation. Simultaneously, AI tools, with their ability to provide real-time data monitoring as well as predictive data analytics, can facilitate healthcare experts to end up making informed and effective clinical decisions during viral or other disease outbreaks (Bathula et al. 2024).

In this chapter, the case studies that were examined and analyzed helped to elucidate the efficacy of AI and blockchain when it comes to being prepared for a pandemic or responding to it. Some examples include the use of blockchain for streamlining vaccine distribution and the use of AI for developing diagnostic tools for diseases, such as dengue and influenza (Alqaissi et al. 2023). Such studies portray the role of AI and blockchain in coming up with solutions for problems in terms of data integrity as well as traceability and scalability.

Overall, this chapter provided an account of how AI and blockchain technology can revolutionize the field of healthcare, with a special focus on their role in managing viral disease outbreaks, starting from early diagnosis to optimizing treatment strategies and streamlining the distribution of vaccines. By enhancing transparency and efficiency, these advancing technologies pave the way and also provide hope for the development of a robust and flexible global healthcare system that will be able to handle unforeseen emergencies and improve public health outcomes more efficiently when compared to the past.

References

Adeyeri TB (2024) Blockchain and AI synergy: transforming financial transactions and auditing. Blockchain Technol Distribut Syst 4(1):24–44

Ahmed I, Chehri A, Jeon G (2024) Artificial intelligence and blockchain-enabled smart healthcare system for monitoring and detection of COVID-19 in biomedical images. IEEE/ACM Trans Comput Biol Bioinform 21(4):814–822

Alqaissi E, Alotaibi F, Sher Ramzan M, Algarni A (2023) Novel graph-based machine-learning technique for viral infectious diseases: application to influenza and hepatitis diseases. Ann Med 55(2):2304108

Amisha, Malik P, Pathania M, Rathaur VK (2019) Overview of artificial intelligence in medicine. J Family Med Prim Care 8(7):2328–2331

Bathula A, Gupta SK, Merugu S et al (2024) Blockchain, artificial intelligence, and healthcare: the tripod of future—a narrative review. Artif Intell Rev 57:238

Belák S, Thorén P, LeBlanc N, Viljoen G (2009) Advances in viral disease diagnostic and molecular epidemiological technologies. Expert Rev Mol Diagn 9(4):367–381

Dias R, Torkamani A (2019) Artificial intelligence in clinical and genomic diagnostics. Genome Med 11:70

Gautam AS, Raza Z (2024) Disease outbreak prediction using natural language processing: a review. Knowl Inf Syst 66(11):6561–6595

Glaser JI, Benjamin AS, Farhoodi R, Kording KP (2019) The roles of supervised machine learning in systems neuroscience. Prog Neurobiol 175:126–137

Ibrahim AU, Al-Turjman F, Sa'id Z et al (2022) Futuristic CRISPR-based biosensing in the cloud and internet of things era: an overview. Multimed Tools Appl 81:35143–35171

Jabarulla MY, Lee HN (2021) A blockchain and artificial intelligence-based, patient-centric healthcare system for combating the COVID-19 pandemic: opportunities and applications. Healthcare (Basel) 9(8):1019

Kalogiannidis S, Kalfas D, Papaevangelou O, Giannarakis G, Chatzitheodoridis F (2024) The role of artificial intelligence technology in predictive risk assessment for business continuity: a case study of Greece. Risks 12(2):19

Kumar R, Arjunaditya, Singh D, Srinivasan K, Hu Y-C (2023) AI-powered blockchain technology for public health: a contemporary review, open challenges, and future research directions. Healthcare 11(1):81

Kunal S, Gandhi P, Rathod D, Amin R, Sharma S (2024) Securing patient data in the healthcare industry: a blockchain-driven protocol with advanced encryption. J Educ Health Promot 13:94

Marquez E, Barrón-Palma EV, Rodríguez K, Savage J, Sanchez-Sandoval AL (2023) Supervised machine learning methods for seasonal influenza diagnosis. Diagnostics (Basel) 13(21):3352

Maslove DM, Klein J, Brohman K, Martin P (2018) Using blockchain technology to manage clinical trials data: a proof-of-concept study. JMIR Med Inform 6(4):e11949

Medina J, Cessa-Rojas R, Umpaichitra V (2021) Reducing COVID-19 cases and deaths by applying blockchain in vaccination rollout management. IEEE Open J Eng Med Biol 2:249–255

Nguyen DC, Ding M, Pathirana PN, Seneviratne A (2021) Blockchain and AI-based solutions to combat coronavirus (COVID-19)-like epidemics: a survey. IEEE Access 9:95730–95753

A Blockchain-Based Trust Architecture for Securing Pandemic Test Results in Decentralised Health Networks

8

C. Gokulnath and Priyan Malarvizhi Kumar

Abstract

In order to safely store and assess COVID-19 test findings, this chapter suggests a blockchain-based design for decentralised health networks. The technology safeguards test findings, making them unchangeable, traceable, and impervious to manipulation or unauthorised access. Public events and airports, for example, may verify that attendees have tested negative for COVID-19 using blockchain technology and smart contracts, which allow laboratories to safely store and share test results with authorised parties. The design is versatile and may be used for a variety of tests since it is compatible with immunological chromatography and polymerase chain reaction. Proving the system's viability, the prototype is implemented on Ethereum. With smart contracts, businesses, labs, and people may more easily govern access and send and receive data securely. Test data transportation in the prototype is ensured via TLS encryption and a graphical user interface. This solution enables safe health data management beyond COVID-19 testing and tackles the problems associated with pandemic data management by establishing a decentralised trust mechanism for health information.

Keywords

Blockchain technology · Trust architecture · Decentralisation · Health industry · Data security

C. Gokulnath (✉)
Department of Computing Technologies, School of Computing, SRM Institute of Science and Technology, Kattankulathur, Chengalpattu District, Tamil Nadu, India

P. M. Kumar
Department of Information Science, College of Information Systems, University of North Texas, Denton, TX, USA

J. M. Chatterjee, S. K. Saxena (eds.), *Artificial Intelligence and Blockchain in Precision Medicine and Virology*, Medical Virology: From Pathogenesis to Disease Control, https://doi.org/10.1007/978-981-96-8919-4_8

8.1 Introduction

The COVID-19 pandemic highlighted the critical need for secure, rapid, and dependable handling of public health data, particularly in the identification and containment of viruses. Screening assays similar to polymerase chain reaction (PCR) have been vital in detecting current COVID-19 infections. In terms of public health, travel security, and virus prevention, these tests have been influential (Egala et al. 2022). Airline companies, employers, and event planners were particularly vulnerable to the risks associated with data processing, verification, and security as their reliance grew. Concerns about privacy, security, and trust arise from the need to rely on centralised organisations to validate test results.

The verification procedure is not easy. Having a negative COVID-19 test result on file, either digitally or physically, is often necessary for air travel or attending major events. Have faith in the medical facilities or individuals who provided you with these test findings to ensure their veracity. Results could be skewed due to the fact that this verification method is vulnerable to fraud and manipulation (Jabarulla and Lee 2021). Manipulating test findings might endanger public health if there isn't a trustworthy mechanism in place. The necessity for a decentralised, autonomous system to validate pandemic test findings is highlighted by this risk.

Due to its decentralisation, transparency, and immutability, blockchain technology is attractive for these problems. The COVID-19 test results may be securely recorded and verified using blockchain technology's distributed ledger (Bhattacharjya et al. 2022). Blockchain technology eliminates the need to put faith in a central authority due to its decentralised nature. Alternatively, data integrity is preserved by a network of nodes that validate transactions. Centralised procedures are not as safe as this.

This chapter presents a decentralised trust architecture for COVID-19 testing that is built on the blockchain. Medical facilities, airlines, government agencies, and event organisers would all be able to access and verify test findings instantly thanks to this architecture's verified and secure connections (Sahal et al. 2022). Unlike centralised systems, where data is controlled by a single institution, this decentralised method distributes trust across the network, making it impossible for any one body to modify or control the data. Customers may check that the data has not been altered by seeing the history and origin of test findings on blockchain.

8.1.1 The Value of Trustworthy Pandemic Test Validation

There are several reasons why it is crucial to accurately verify pandemic test findings. The primary benefit is the isolation and inhibition of COVID-19 transmission (Khurshid 2020). Avoiding other people's good emotions does this. Boarding must be accompanied by a negative PCR test, as mandated by several countries and airlines. These test findings need to be precise since false positives might enable infected individuals to move.

Second, people are more likely to have faith in health and safety procedures when test results are confirmed. Compliance with testing protocols and health standards is enhanced when test findings are stored in a safe and unchangeable manner (Razzaq et al. 2022). People will doubt the testing procedure, and pandemic control will suffer if they believe the system is vulnerable to manipulation or fraud.

Finally, there are monetary and legal ramifications beyond health that follow from validating test results. In order to keep running normally during pandemics, the travel, hotel, and entertainment industries depend on reliable test findings (Sabah et al. 2021). It is possible that a negative COVID-19 test will be required for some events, such as concerts, conferences, and football games. Verifying these facts is crucial for avoiding legal responsibilities, safeguarding public safety, and restoring trust.

8.1.2 Problems with Centralised Administration

One company typically stores and validates all of the data for pandemic test results since these systems are centralised. Centralised systems offer oversight and accountability, but they suffer from problems with scalability, security, and transparency (Hasan et al. 2020).

There are reliability problems with centralised systems. Users rely on government, private, medical, and business data administrators under centralised frameworks. This dependence constrains failure to a single location. The whole system is vulnerable to cyberattacks, human mistake, or corruption that affect the centralised entity's systems (Javed et al. 2021). Potentially disastrous outcomes may occur from altering pandemic test results or releasing sensitive medical information.

There is a greater chance of privacy and data breach with centralised systems. Because they include all of the test results, hackers target centralised databases. If the breach is successful, it may result in the exposure of sensitive health information belonging to thousands or perhaps millions of people, which might lead to abuse or even identity theft (Dar et al. 2022). Because centralised systems aren't always open and honest, consumers have no idea who or what is handling their data and whether it has been compromised. When disaster strikes, centralised approaches may not be the best bet for scalability. Worldwide, testing institutions experienced difficulties in processing and validating data due to the enormous COVID-19 test wave (Andrew et al. 2023). With a decentralised approach, data may be examined and verified concurrently by several nodes in the network, which eliminates bottlenecks and increases productivity. This model is more dispersed and scalable.

8.1.3 A Distributed Blockchain System for Ensuring Trust and Security

Blockchain technology fixes several problems with centralised systems. An unchangeable distributed ledger on a network of computers (nodes) is where all transactions are recorded in blockchain technology (Tan et al. 2020). No one can access or change data on a blockchain without the agreement of all nodes in the network. Pandemic test results and other sensitive data need immutable records for reliability; blockchain technology is perfect for this.

The blockchain has several advantages, one of which is decentralisation. In contrast to centralised systems, blockchain decentralises trust from one entity to the network. With a copy of the ledger on hand, any node in the network can confirm transactions. This prevents manipulation and fraud by making sure that a majority of nodes must approve data updates before they can be made. Blockchain enhances transparency since the updated ledger is visible to all users. This opens the door for any governmental body, healthcare facility, or airline to confirm the results of pandemic tests. Users are reassured that their data is protected and fraud is reduced due to this transparency.

Another important aspect of blockchain is its traceability. By connecting each transaction to its antecedent, blockchain technology enables data to be tracked back to its source. Benefiting pandemic test outcomes, this enables full test history verification, from test execution to result verification. Because of this level of traceability, irregularities and manipulation are easy to detect.

Last but not least, reliable and extensible public health record management systems are more important than ever before due to the continued risk of pandemics. Centralised pandemic test result confirmation approaches were proven to be inadequate during the COVID-19 pandemic, highlighting the necessity for a distributed alternative that utilises blockchain technology. By securely recording and verifying test findings in an immutable, transparent, and traceable manner, blockchain technology has the potential to revolutionise pandemic data management. Because of this, results from tests may be trusted even in the absence of a governing body. Public health infrastructures should use blockchain technology to make them more resilient to future health emergencies.

8.2 Background

A. Testing for COVID-19
   - Polymerase chain reaction (PCR): it may determine if an individual is COVID-19 positive or negative, with a consequent “negative search” or “positive search” result. For a vacation, for instance, it needs to be completed at least 72 hours before departure (the validity period is just 72 hours) (Lan et al. 2020).
   - Immunochromatography is the serology test for COVID-19. When employed in conjunction with one another, it allows for the simultaneous uncovering of IgG and IgM. IgG antibodies account for approximately 75% of all antibodies found in human serum (Long et al. 2020). Within an immunoserum, approximately 10% of the immunoglobulins are IgM antibodies (Cassaniti et al. 2020). The amount of IgM antibodies starts to rise around a week after the first infection, as seen in Fig. 8.1. In contrast, IgG antibodies don’t exist until

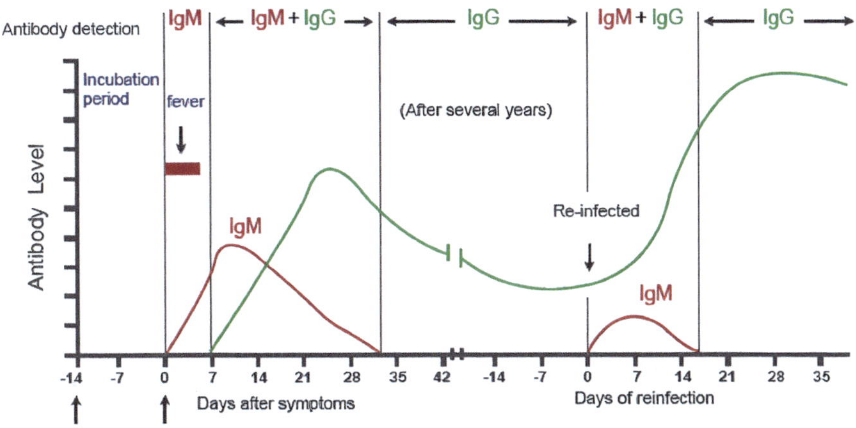

Fig. 8.1 Post-contagion IgG and IgM standards

much later, usually within 14 days of infection, and can persist for six months or even years. Because of this, IgG is used to detect infections that have already occurred. By monitoring IgM and IgG levels simultaneously, SARS-CoV-2 infected patients can be diagnosed rapidly (Derrab and Saifi 2022).

B. Current State of the Art: According to sources like (Koutli et al. 2019), a centralised system was often used for storing and managing health data. Despite responding to various security concerns (El Madhoun et al. 2021) and supporting the decentralisation of our health system (ARS, hospitals, laboratories, etc.) (Gao et al. 2018), decentralised solutions like blockchain technology were not considered due to the COVID-19 health crisis's urgency and the market's lack of available skills.

1. Blockchain Technology: It appears to be a suitable option for the decentralisation of digital therapeutic systems in a trustworthy, safe, and strong way. Actually, a novel procedural framework to decentralise IT requests is being suggested today (Maldonado-Ruiz et al. 2021). Thanks to blockchain technology, the Bitcoin app is completely secure, which is why users love it (Maldonado-Ruiz et al. 2020). The first source of this security is the fact that two nodes can conduct transactions together directly on the blockchain, bypassing the need for a trusted third party (TTP). The second source is the distributed and immutable ledger that records all of these transactions; no one can alter the records that make up the blockchain (Gao et al. 2018).

A duality exists between the various kinds of blockchains (Beck et al. 2018). Participation in a permissionless public blockchain (like Bitcoin and Ethereum) allows anybody to join the network anonymously, guaranteeing the security and consistency of peer-to-peer transactions while keeping their information secret and accessible only on nodes that support the blockchain. There is a danger to the infrastructure due to the absence of control on network members, despite the significant decentralisation (Hatin et al. 2020). Hyperledger Fabric, Hyperledger Besu, and other permissioned private blockchains need all participants to register their identities and the identities of the resources they use before they may contribute to the network's architecture. This is essential for identifying any potentially harmful thing, such as buried node. Along with the immutable ledger keeping of the blockchain, an efficient system for managing node identities is also required as a result of this registration procedure. Here, members' faith in one another undermines the network's decentralisation.

2. Smart Contracts and Ethereum: The Ethereum blockchain is public, open-source, and distributed. The "Ether (ETH)" coin is the one used by Ethereum. Among the most groundbreaking developments, the "smart contracts" introduced to the Ethereum blockchain are among the most notable. Solidity, "a Scripting Language specially designed for Ethereum," is the language used to write smart contracts. Because it is Turing-complete, developers can model, secure, and trade anything that can be mathematically represented. Without going via a third party, it enables two anonymous persons to do commerce or carry out a service directly with each other. Without the risk of immobilisation, restriction, fraud, or third-party interference, the instructions of a smart contract execute precisely as intended (Hildenbrandt et al. 2017).

8.3 Proposed Architecture

Our novel blockchain-based architecture is presented in this chapter with the following goals to aid in the prevention, battle against, and control of the COVID-19 pandemic:

- Labs are able to store, transmit, and share PCR test results in a decentralised trust with companies, guaranteeing an extremely high level of security, consistency, and integrity.
- Negative PCR test results are required. While assuring safe access control, it ensures that each patient's PCR test results may be tracked chronologically. As a consequence, the paperwork pertaining to PCR test results may be reliably handled.
- It is also used to process the findings of the immunochromatography test, which is another kind of COVID-19 test.

A. First Model of the Building

Using the public blockchain and smart contracts on Ethereum, we have built a model of our design. The ability to manipulate data in our prototype is dependent on its graphical and user-friendly interface. For the blockchain to function, this interface

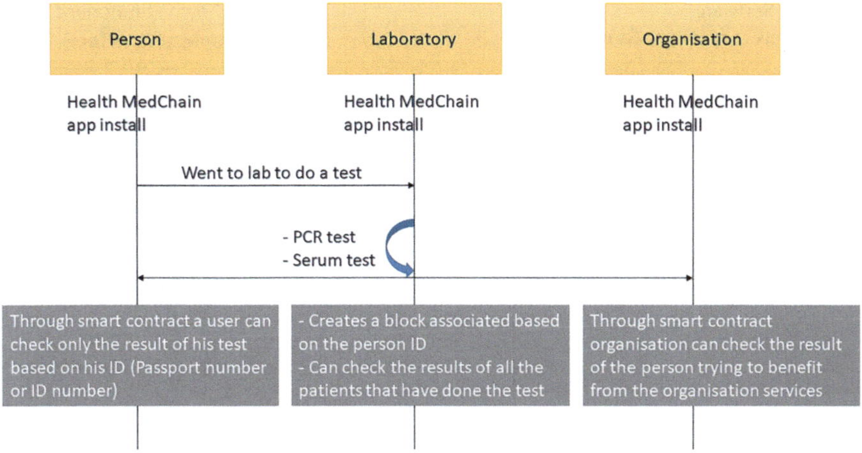

Fig. 8.2. Flow of information

must be able to talk to the central node that generates and distributes blocks. Three players are depicted in our model (refer to Fig. 8.2.):

- The individual wishing to utilise the service: who must provide proof of a negative COVID-19 test result.
- The COVID-19 tests and their validity are guaranteed by the laboratory.
- The organisation: this refers to the place (airport, movie theatre, shopping mall, etc.) that is trying to prevent the spread of COVID-19.

Two primary use cases are taken into account in our proposed prototype: "PCR Tests" along with "Immunochromatography Tests":

- The primary goal of Case 1 "PCR Tests" is to read and insert data related to a patient's PCR test findings. Date of assessment, validity period (usually 3 days), result (negative or positive), name of the lab, country of the lab, address of the lab, and another source of identification for the person in question (such as an identification number or social security number) are all parts of the data that is contained.
- Case 2 "Immunochromatography Tests": it seeks to add and/or read data pertaining to an individual's immune system. Test date, lab name, lab country, lab address, patient identifier, IgM and IgG results (positive or negative) are all part of the reported information.

Upon completion of a PCR or immunochromatography test, a block including all the data unique to the test will be generated by the laboratory and the user interface. Only the user and the qualified organisation will have access to this block. The qualified organisation needs to establish a smart contract with the lab before they

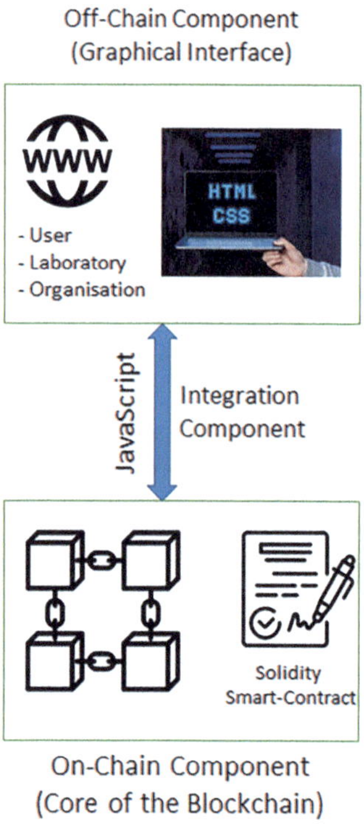

Fig. 8.3 Software components in prototype form

may access the data. The person with whom the smart contract has been communicating is its primary identifier. Fig. 8.2. shows the three characters' dialogue flow.

B. Prototype Software Components

An end-to-end prototype requires the integration of several software components. The three primary parts of our prototype are the off-chain, on-chain, and integration components (refer to Fig. 8.3). Here is a description of these parts:

The graphical user interface, or off-chain component, is built using the web development tools provided by HTML and CSS. This interface guarantees three primary pages:

- Patient or user: it restricts access so that only data related to the user's particular COVID-19 test may be seen.
- Lab: they can upload the COVID-19 test findings by adding a new block to the blockchain. Additionally, it enables the lab to access the data associated with a particular test.
- After authenticating a user using his identity, it enables the organisation to see the user's COVID-19 test result.

Login and password protection ensures that only authorised users may access these pages.

In the scenario of adding COVID-19 test results and communicating them amongst the stakeholders (user, laboratory, and organisation), the on-chain component (core of the blockchain) enables the creation of blocks. To put it into action, "solidity language on the Ethereum blockchain (Modi 2018)"is used in smart contracts. This is the framework around which our prototype is built.

The integration component uses the Transport Layer Security (TLS) protocol to provide a safe link between the blockchain's central node and the user interface. It connects the graphical user interface with the blockchain's central processing unit and was built using the JavaScript language. Get to read and set to write are the two primary operations that may be executed.

C. Block-Level Elements (Blockchain's Heart)

For better code readability, data organisation, and categorisation, our program makes use of many smart contracts. Inheritance has also allowed us to merge our patient, labs, and organisations contracts into a single one. This new contract's primary purpose is to store, retrieve, and modify patients' test results. The patient contract was established with a structure that includes a login name and a Boolean expression that indicates the existence or non-existence of this patient's identity.

For the decentralised healthcare system's blockchain-based patient data administration, Algorithm 8.1 makes reference to Solidity's smart contract. The technology enables doctors to sign up, keep track of patients, and control who has access to their records. The patient's Ethereum address, together with their name, age, and medical history, is stored in the patient's record. The mapping of patient records associates an Ethereum address with all of a patient's data. To ensure confidentiality, only the designated doctor has access to patients' medical records and can read or make changes to them. To make sure the request gets made, modifiers check the identity of the given doctor. These services are covered by the contract: To add, transfer, or receive patients, the calling doctor can use the following commands: registerPatient, assignDoctor, and getPatientInfo, provided they have the approval. The medical history can be edited using the UpdateMedicalHistory function. The contract goes above and above by providing system alerts for new patients through events like PatientRegistered. For healthcare information processing on the blockchain, this smart contract architecture is excellent due to its decentralised access control and security of patient data.

Algorithm 8.1 Smart Contract of Patients

```
pragma solidity ^0.8.0;
contract PatientSmartContract {
 struct Patient {
 string name;
 uint age;
```

(continued)

Algorithm 8.1 (continued)

```
string medicalHistory;
address assignedDoctor; // Doctor's Ethereum address
bool exists; // Ensure patient exists
}
mapping(address => Patient) public patients;
event PatientRegistered(address patientAddress, string name);
modifier onlyAssignedDoctor(address _patientAddress) {
require(patients[_patientAddress].assignedDoctor == msg.
sender, "Access Denied: Only assigned doctor can access this
patient's info.");
_;
}
function registerPatient(address _patientAddress, string memory
_name, uint _age, string memory _medicalHistory) public {
require(!patients[_patientAddress].exists, "Patient already
registered.");
patients[_patientAddress] = Patient({
name: _name,
age: _age,
medicalHistory: _medicalHistory,
assignedDoctor: msg.sender, // Doctor registering the patient
becomes the assigned doctor
exists: true
});
emit PatientRegistered(_patientAddress, _name);
}
function assignDoctor(address _patientAddress, address
_doctorAddress) public onlyAssignedDoctor(_patientAddress) {
patients[_patientAddress].assignedDoctor = _doctorAddress;
}
function getPatientInfo(address _patientAddress) public view
onlyAssignedDoctor(_patientAddress) returns (string memory,
uint, string memory) {
Patient memory patient = patients[_patientAddress];
return (patient.name, patient.age, patient.medicalHistory);
}
function updateMedicalHistory(address _patientAddress, string
memory _newHistory) public onlyAssignedDoctor(_patientAddress)
{
patients[_patientAddress].medicalHistory = _newHistory;
}
}
```

As seen in Algorithm 8.2, using the features of an Ethereum-based blockchain system through Solidity, the Lab Smart Contract's pseudocode is organised to make it easier to handle Internet of Things (IoT) devices in a laboratory setting. After a version declaration, the LabSmartContract is defined in the contract. The owner variable, which specifies the lab manager and guarantees command over critical contract procedures, is functionally essential.

The IoTDevice struct is the meat and potatoes of the contract; it stores all the important details about each device, such as its type, isActive state, data readings, and unique identification (deviceId). To facilitate retrieval of device information, this data is kept in a mapping that links each deviceId to its matching IoTDevice structure. A mapping of authorised users is used to provide access control, which means that only approved persons are allowed to do particular operations within the contract. To further tighten security, you may use modifiers like onlyOwner and onlyAuthorizedUser to restrict access to critical services. The contract owner is the only one who may perform certain functions; authorised users can only use recordData and getData, for instance. AddDevice and updateDeviceStatus are also restricted.

When major operations, such as adding devices, capturing data, or modifying device status, take place, the contract will release an event. Users are able to efficiently monitor contract operations thanks to this logging tool, which promotes transparency and traceability. Other critical functionalities enable the owner to dynamically modify user rights through authorizeUser and revokeUser.

All things considered, this pseudocode is a basic building block for a smart contract that incorporates lab-based IoT devices, demonstrating key features like device administration, data recording, and access control.

Algorithm 8.2: Smart Contract for Lab

```
pragma solidity ^0.8.0;
contract LabSmartContract {
address owner;
struct IoTDevice {
string deviceId; // Unique identifier for the device
string deviceType; // Type of device (sensor, monitor, etc.)
bool isActive; // Status of device (active/inactive)
string[] data; // Array to store data readings
}
mapping(string => IoTDevice) devices;
mapping(address => bool) public authorizedUsers;
event DeviceAdded(string deviceId, string deviceType);
event DataRecorded(string deviceId, string data);
event DeviceStatusUpdated(string deviceId, bool isActive);
constructor() {
```

(continued)

Algorithm 8.2 (continued)

```
owner = msg.sender; // Contract owner
}
modifier onlyOwner() {
require(msg.sender == owner, "Access restricted to owner.");
_;
}
modifier onlyAuthorizedUser() {
require(authorizedUsers[msg.sender] == true, "User not
authorized.");
_;
}
function addDevice(string memory _deviceId, string memory
_deviceType) public onlyOwner {
devices);
emit DeviceAdded(_deviceId, _deviceType); // Emit event when the
device is added
}
function updateDeviceStatus(string memory _deviceId, bool
_isActive) public onlyOwner {
devices[_deviceId].isActive = _isActive;
emit DeviceStatusUpdated(_deviceId, _isActive); // Emit event on
status change
}
function recordData(string memory _deviceId, string memory _data)
public onlyAuthorizedUser {
require(devices[_deviceId].isActive, "Device is not active.");
devices[_deviceId].data.push(_data); // Store new data
emit DataRecorded(_deviceId, _data); // Emit event when data is
recorded
}
function authorizeUser(address _user) public onlyOwner {
authorizedUsers[_user] = true;
}
function revokeUser(address _user) public onlyOwner {
authorizedUsers[_user] = false;
}
function getData(string memory _deviceId) public view
onlyAuthorizedUser returns (string[] memory) {
return devices[_deviceId].data;
}
}
```

To register a specific organisation on our network, we utilise the Organisation Contract as mentioned in Algorithm 8.3. It allows users to enter their company information after they join up, and it offers a single mapping from address to company details. The provided code snippet is a Solidity smart contract designed to manage an organisation by overseeing its administrators, managers, employees, and funds.

At the beginning of the contract, Organisation Contract is defined, along with several important variables: admin, managers, employees, and organisation funds. The admin address is initialised to the address that deploys the contract, granting them exclusive rights to certain administrative functions. The mappings for managers and employees are used to keep track of their statuses, indicating whether a given address is a manager or an employee. The organisationFunds variable holds the total funds available within the organisation.

Several events are declared—ManagerAdded, EmployeeAdded, FundsDeposited, and FundsWithdrawn—which are emitted to provide transparency when specific actions occur within the contract.

The contract employs modifiers to restrict access to certain functions. The onlyAdmin modifier ensures that only the administrator can execute functions that modify the organisational structure, while the onlyManager modifier allows only designated managers to perform specific actions, such as adding employees and withdrawing funds.

Key functions in the contract include the following:

- addManager and addEmployee, which allow the admin and managers, respectively, to add new personnel.
- depositFunds, which enables anyone to contribute funds to the organisation, updating the total funds and emitting a corresponding event.
- withdrawFunds, which allows managers to withdraw a specified amount of funds, ensuring there are sufficient funds available before executing the transaction.
- checkFunds, which provides a view function to check the current balance of organisationFunds.
- removeManager and removeEmployee, which allow the admin and managers to revoke the status of other managers and employees.

This contract establishes a basic framework for organisational management within the blockchain, ensuring that financial transactions and personnel management are conducted securely and transparently.

Algorithm 8.3: Smart Contract for Organisation

```
pragma solidity ^0.8.0;
contract OrganisationContract {
 address public admin;
 mapping(address => bool) public managers;
 mapping(address => bool) public employees;
 uint256 public organisationFunds;
 event ManagerAdded(address manager);
 event EmployeeAdded(address employee);
 event FundsDeposited(uint256 amount, address from);
 event FundsWithdrawn(uint256 amount, address to);
 modifier onlyAdmin() {
 require(msg.sender == admin, "Only admin can perform this
action");
 _;
 }
 modifier onlyManager() {
 require(managers[msg.sender], "Only manager can perform this
action");
 _;
 }
 constructor() {
 admin = msg.sender;
 }
 function addManager(address _manager) public onlyAdmin {
 managers[_manager] = true;
 emit ManagerAdded(_manager);
 }
 // Function: Add an employee
 function addEmployee(address _employee) public onlyManager {
 employees[_employee] = true;
 emit EmployeeAdded(_employee);
 }
 function depositFunds() public payable {
 organisationFunds += msg.value;
 emit FundsDeposited(msg.value, msg.sender);
 }
 function withdrawFunds(uint256 _amount, address payable _to)
```

(continued)

Algorithm 8.3 (continued)

```
public onlyManager {
 require(organisationFunds >= _amount, "Insufficient funds");
 organisationFunds -= _amount;
 _to.transfer(_amount);
 emit FundsWithdrawn(_amount, _to);
 }
 function checkFunds() public view returns (uint256) {
 return organisationFunds;
 }
 function removeManager(address _manager) public onlyAdmin {
 managers[_manager] = false;
 }
 function removeEmployee(address _employee) public onlyManager {
 employees[_employee] = false;
 }
}
```

The proposed model is defined in the architecture diagram as mentioned in Fig. 8.4.

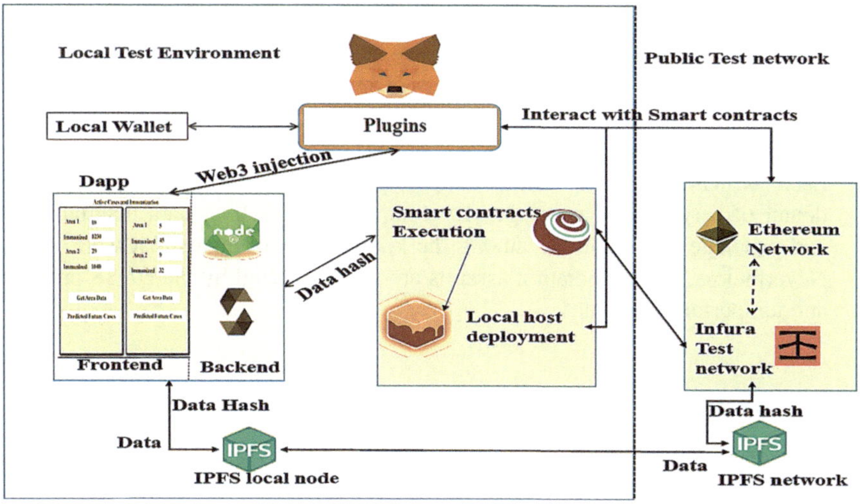

Fig. 8.4 Proposed architecture

8.4 Analyses and Findings

To ensure it is in line with the goals of the blockchain-based system, this part offers a thorough evaluation of the system's security and performance. The theoretical evaluation of security in terms of availability, confidentiality, scalability, performance, and dependability is the subject of the following discussion.

8.4.1 Analysing Security

The validation of the Distributed Data Storage System (DDSS) communication protocol within the blockchain system was carried out using the Scyther tool and a security protocol description language (.spdl). Here, we briefly outline the results. The findings show that the protocol successfully withstands attempts to compromise its data integrity. Later on, we'll get into the specifics of the blockchain security protocol's pseudocode, which includes a mechanism to ensure private information stays private. To provide privacy, the algorithm keys are used.

To implement the DDSS use case, a protocol is set up between a single lead node and two IoMT (Internet of Medical Things) devices. Symbols such as msg, DIsig, MATdiv, and const represent the initial communication, digital signature, mutually accepted keys for one-to-one communication, and different identities. The results of the Scyther tool investigation reveal that all devices and nodes are safe from assaults within their specified security parameters.

According to the security claims, the system is safe in all respects:

Device A is secure. No intrusions are allowed at this leader node (Agreements and secret messages are safe). (Secret message and Agreements secure).

Device B: No attacks within boundaries. In addition, the STRIDE technique was used by the Microsoft Threat Modelling Tool (MTMT) to detect system security risks, which include spoofing, tampering, repudiation, information leakage, denial of service, and privilege elevation. Cache spoofing, data manipulation, and privilege escalation are among the known system threats that were found. Nevertheless, non-repudiation assaults are well protected by the DDSS protocol and tamper-proof ledger.

8.4.2 Security and Confidentiality

Session keys, MAT, and signed data may be safely sent across network nodes using blockchain technology. The peer-to-peer DDSS network uses dual encryption to safeguard critical data. Protecting sensitive information is the job of an access control system that lets only authorised people in.

User information is safeguarded via the DDRAC architecture's three-layer pseudo-identity concept. All data types, including public, private, and hybrid, are encrypted using the user's public key, group key, and hybrid symmetric key. System breaking is practically impossible with three encryption keys. Identity theft can be lessened since layers can only map identities in one direction.

8.4.2.1 Controlling Versions and Integrity

To ensure data integrity, each block is hashed and appended to the encrypted block. Thanks to the distributed ledger technology (DLT), IPFS version management is transparent throughout the DDSS network. Before processing, the system checks the data for integrity using hashes, which identify each version.

8.4.2.2 Responsibility and Permission

The DDRAC framework verifies credentials to ensure that only allowed individuals may access critical DDSS network data. By keeping tabs on everything, the system stops non-repudiation attacks and makes everyone pay. Ensuring security and accountability, participation is limited to authorised users only.

8.4.3 Defence Against Distributed Denial of Service Attacks and Maximum Uptime

By distributing operations throughout a peer-to-peer network, the system ensures that a minimal number of nodes are always online and uses Byzantine Fault Tolerance (BFT) consensus to grant authorised users access to data. Assuming 51% of the nodes are up, the system continues to function according to the Practical Byzantine Fault Tolerance (PBFT) protocol, which protects against distributed denial of service (DDoS) attacks (Fig. 8.5).

By managing cache locally, DDSS activities are accelerated and made more available. This method ensures data accessibility even in the event of multiple node failures. In a simulated distributed denial of service (DDoS) assault, a 40-node testbed managed to keep 80% of data even while 50% of nodes were unavailable. The system's robustness was confirmed throughout the testing as data availability improved in tandem with the number of active nodes.

8.4.4 Analysing Cost and Response Rates

Important system processes, such as identity generation, vaccination tracking, sub-group generation, new leader election, and access rule changes, are assessed in this section based on their reaction times and costs. The outcomes prove the system's efficiency and responsiveness in real time at a low cost. Figure 8.6 shows that it is also consistent in providing dynamic services to various groups while being sustainable. In addition, PBFT and the caching mechanism guarantee that healthcare records are available (Fig. 8.5). In terms of security and cost, there is a clear correlation between the two.

Since the IPFS peer-to-peer network stores the real data, only hash values are kept on the blockchain in order to decrease operating expenses and delay. While keeping

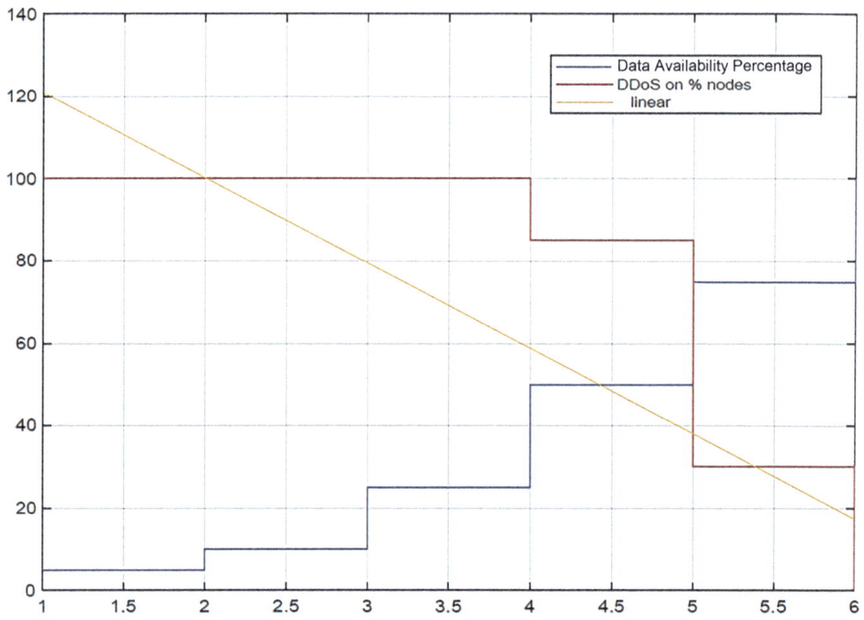

Fig. 8.5 Availability of information versus DDoS assault

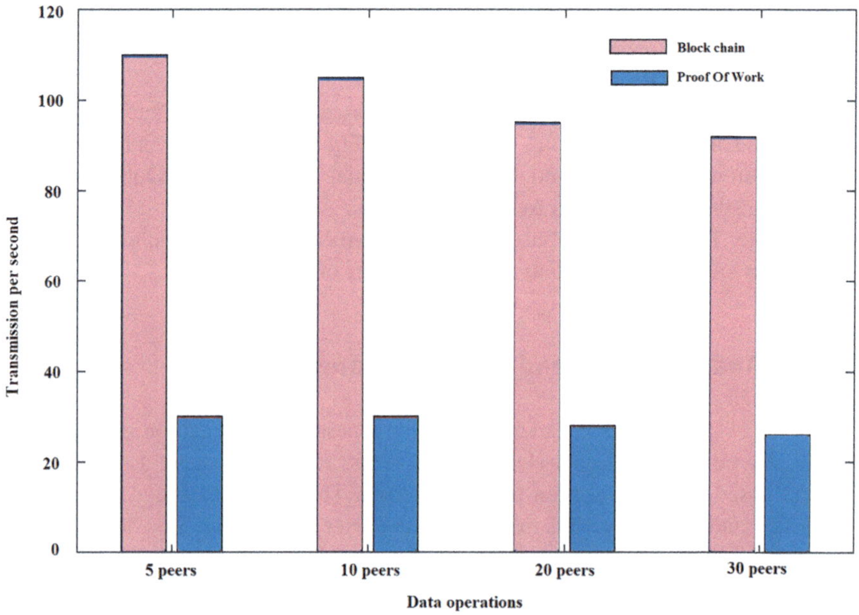

Fig. 8.6 Transmission speed per second in different peers

data safe, this lessens the load on the blockchain's storage capacity. The suggested method accomplishes an acceptable transaction throughput and considerably decreases file access times via caching, according to comparative findings for file storage costs.

8.5 Conclusion

Protecting the confidentiality, availability, and authenticity of patient records in decentralised healthcare systems has never been easier than with the suggested blockchain-based approach for handling COVID-19 test results. System validation of genuine, tamper-proof test findings is ensured by utilising the immutability and security of blockchain technology. Automation of access control with the use of smart contracts further improves the system by limiting data interaction to authorised parties like healthcare facilities and laboratories. The system is adaptable to different kinds of viral testing, which gives it a wide range of potential medicinal uses. By creating a working prototype with smart contracts and the Ethereum blockchain, we prove that this strategy can be implemented and set the stage for future safe, decentralised systems for managing health information.

References

Andrew J, Isravel DP, Sagayam KM, Bhushan B, Sei Y, Eunice J (2023) Blockchain for healthcare systems: architecture, security challenges, trends and future directions. J Netw Comput Appl 215:103633

Beck R, Müller-Bloch C, King JL (2018) Governance in the blockchain economy: a framework and research agenda. J Assoc Inf Syst 19(10):1

Bhattacharjya A, Kozdrój K, Bazydło G, Wisniewski R (2022) Trusted and secure blockchain-based architecture for internet-of-medical-things. Electronics 11(16):2560

Cassaniti I, Novazzi F, Giardina F, Salinaro F, Sachs M, Perlini S et al (2020) Performance of VivaDiag COVID-19 IgM/IgG rapid test is inadequate for diagnosis of COVID-19 in acute patients referring to emergency room department. J Med Virol 92(10):1724

Dar AA, Alam MZ, Ahmad A, Reegu FA, Rahin SA (2022) Blockchain framework for secure COVID-19 pandemic data handling and protection. Comput Intell Neurosci 2022(1):7025485

Derrab S, Saifi T (2022) The health strategy of a hospital in the face of the development of the Covid-19 pandemic. (Doctoral dissertation,. Mouloud Mammeri University

Egala BS, Pradhan AK, Badarla V, Mohanty SP (2022) iBlock: an intelligent decentralised blockchain-based pandemic detection and assisting system. J Signal Process Syst 94(6): 595–608

El Madhoun N, Hatin J, Bertin E (2021) A decision tree for building IT applications: what to choose: blockchain or classical systems? Ann Telecommun 76(3):131–144

Gao W, Hatcher WG, Yu W (2018, July) A survey of blockchain: techniques, applications, and challenges. In: 2018 27th international conference on computer communication and networks (ICCCN). IEEE, pp 1–11

Hasan HR, Salah K, Jayaraman R, Arshad J, Yaqoob I, Omar M, Ellahham S (2020) Blockchain-based solution for COVID-19 digital medical passports and immunity certificates. Ieee Access 8:222093–222108

Hatin J, Bertin E, Hemery B, El Madhoun N (2020) Welcome to the jungle: a reference model for blockchain, DLT and smart-contracts. In: Tokenomics 2020 on blockchain economics, security & protocols (2nd international conference)

Hildenbrandt E, Saxena M, Zhu X, Rodrigues N, Daian P, Guth D, Roşu G (2017) Kevm: a complete semantics of the Ethereum virtual machine. IEEE

Jabarulla MY, Lee HN (2021, August) A blockchain and artificial intelligence-based, patient-centric healthcare system for combating the COVID-19 pandemic: opportunities and applications. Healthcare 9(8):1019

Javed IT, Alharbi F, Bellaj B, Margaria T, Crespi N, Qureshi KN (2021, June) Health-ID: a blockchain-based decentralized identity management for remote healthcare. Healthcare 9(6): 712

Khurshid A (2020) Applying blockchain technology to address the crisis of trust during the COVID-19 pandemic. JMIR Med Inform 8(9):e20477

Koutli M, Theologou N, Tryferidis A, Tzovaras D, Kagkini A, Zandes D et al (2019, May) Secure IoT e-health applications using VICINITY framework and GDPR guidelines. In: 2019 15th international conference on distributed computing in sensor systems (DCOSS). IEEE, pp 263–270

Lan L, Xu D, Ye G, Xia C, Wang S, Li Y, Xu H (2020) Positive RT-PCR test results in patients recovered from COVID-19. JAMA 323(15):1502–1503

Long QX, Liu BZ, Deng HJ, Wu GC, Deng K, Chen YK et al (2020) Antibody responses to SARS-CoV-2 in patients with COVID-19. Nat Med 26(6):845–848

Maldonado-Ruiz D, Torres J, El Madhoun N (2020, September) 3BI-ECC: a decentralized identity framework based on blockchain technology and elliptic curve cryptography. In: 2020 2nd conference on blockchain research & applications for innovative networks and services (BRAINS). IEEE, pp 45–46

Maldonado-Ruiz D, Torres J, El Madhoun N, Badra M (2021, April) An innovative and decentralized identity framework based on blockchain technology. In: 2021 11th IFIP international conference on new technologies, mobility and security (NTMS). IEEE, pp 1–8

Modi R (2018) Solidity programming essentials: a beginner's guide to build smart contracts for Ethereum and blockchain. Packt Publishing Ltd.

Razzaq A, Mohsan SAH, Ghayyur SAK, Al-Kahtani N, Alkahtani HK, Mostafa SM (2022, December) Blockchain in healthcare: a decentralized platform for digital health passport of COVID-19 based on vaccination and immunity certificates. Healthcare 10(12):2453

Sabah N, Sagheer A, Dawood O (2021) Survey:(Blockchain-based solution for COVID-19 and smart contract healthcare certification). Iraqi J Comput Sci Math 2(1):1–8

Sahal R, Alsamhi SH, Brown KN, O'Shea D, Alouffi B (2022) Blockchain-based digital twins collaboration for smart pandemic alerting: decentralized COVID-19 pandemic alerting use case. Comput Intell Neurosci 2022(1):7786441

Tan H, Kim P, Chung I (2020) Practical homomorphic authentication in cloud-assisted VANETs with blockchain-based healthcare monitoring for pandemic control. Electronics 9(10):1683

Building Trust Through AI: AI Approaches to Medical Virology in Future Healthcare Systems

9

Manas Kumar Yogi, Bala Shanmukha Sowmya Javvadhi, Y. Jayababu, and Jyotir Moy Chatterjee

Abstract

In the evolving panorama of healthcare, artificial intelligence (AI) has emerged as a transformative force, particularly in the field of medical virology. However, as AI-pushed systems become more established in diagnosing, treating, and coping with viral sicknesses, it is vital to ensure that the layout of these systems stays human-centric to foster agreement, transparency, and ethical integrity. This chapter explores the development of AI fashions tailor-made for medical virology, emphasizing the significance of constructing considerations between healthcare experts, patients, and AI technologies in future healthcare systems. A human-centric AI method in virology prioritizes consumer-friendly interfaces, obvious choice-making techniques, and continuous collaboration among AI systems and healthcare companies. This approach facilitates bridging the space between complex AI algorithms and human expertise, making sure that AI is not only effective and accurate but also interpretable, explainable, and aligned with clinical wishes. By incorporating explainable AI (XAI) frameworks, sufferers and practitioners can better understand how AI reaches conclusions, fostering more popularity and belief in AI-assisted scientific decisions. This research makes a speciality of key regions consisting of AI in early detection and prognosis of viral infections, customized remedy plans, and outbreak prediction, all even as safeguarding patient privacy and addressing ethical concerns. In addition, this chapter examines how AI can help in real-time virology studies, enabling quicker vaccine development and epidemiological responses. The integration of AI into

M. K. Yogi (✉) · B. S. S. Javvadhi · Y. Jayababu
CSE Department, Pragati Engineering College (A), Surampalem, India

J. M. Chatterjee
CSE Department, Graphic Era University, Dehradun, India

J. M. Chatterjee, S. K. Saxena (eds.), *Artificial Intelligence and Blockchain in Precision Medicine and Virology*, Medical Virology: From Pathogenesis to Disease Control, https://doi.org/10.1007/978-981-96-8919-4_9

virology also needs to deal with problems such as record bias, duty, and the need for non-stop machine assessment to ensure fairness and accuracy. Ultimately, this research work advocates for a next-generation healthcare device where AI serves as a relied-on partner in clinical virology, enhancing human know-how and improving patient results through obvious, empathetic, and reliable AI-driven solutions.

Keywords

Trust · Virology · Human-centric · Blockchain · Explainable Artificial Intelligence · Privacy

9.1 Introduction

Artificial Intelligence (AI) has profoundly reshaped the landscape of healthcare, offering innovative solutions to complex challenges. By analysing large datasets with unprecedented speed and accuracy, AI has facilitated advancements in diagnosis, treatment planning, and healthcare delivery (Padhi et al. 2023). Its capacity to process unstructured data and uncover patterns has enabled a more personalized approach to patient care, making it an essential component in modern medicine as shown in Fig. 9.1.

In medical virology, where the stakes are high due to rapid virus mutations and widespread outbreaks, AI's transformative potential is especially significant. One of its primary contributions lies in predictive modelling, which enables the anticipation of viral spread and informs public health strategies (Holmes et al. 2024). For

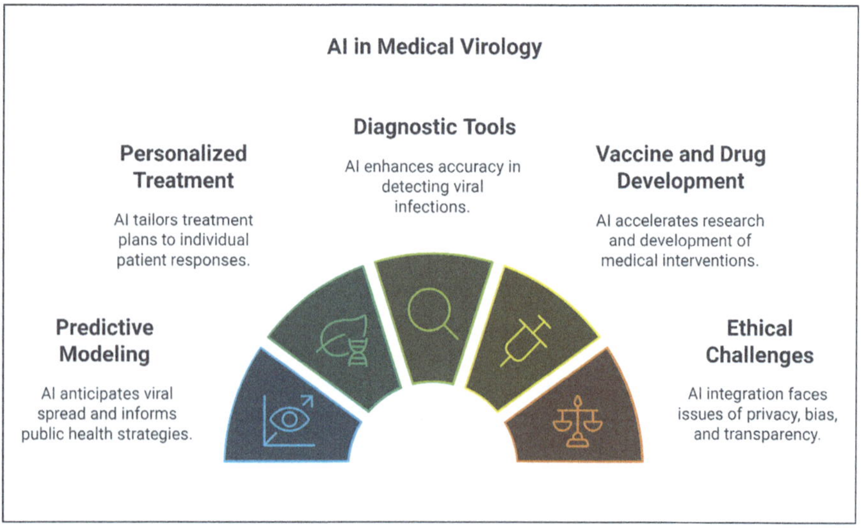

Fig. 9.1 Key AI applications in medical virology

example, during the COVID-19 pandemic, AI systems analysed epidemiological data to predict infection trends, helping governments and healthcare organizations to allocate resources effectively. AI models have also supported epidemic forecasting, enabling proactive responses to potential outbreaks by identifying at-risk regions and populations.

Another crucial application is in personalized treatment, where AI leverages genetic and clinical data to predict patient-specific responses to antiviral therapies. This has resulted in more targeted and effective treatment plans, reducing the reliance on one-size-fits-all approaches. AI-powered diagnostic tools, including image recognition systems and natural language-processing algorithms, have enhanced the accuracy of detecting viral infections, often outperforming traditional diagnostic methods (Malik et al. 2021). These tools enable quicker identification of diseases, ensuring timely intervention and improved patient outcomes.

Advancements in AI have also accelerated vaccine and drug development. Machine-learning algorithms analyse viral genomes and predict protein structures, significantly reducing the time required for research. Companies like Moderna utilized AI-driven insights to expedite the development of mRNA vaccines during the COVID-19 crisis (Goyal 2023). Similarly, AI platforms such as BlueDot have demonstrated the ability to detect and track emerging viral threats, showcasing the technology's critical role in global health security.

Despite these breakthroughs, the integration of AI in medical virology is not without challenges. Issues such as data privacy, algorithmic bias, and the complexity of achieving transparency in decision-making systems present barriers to widespread adoption. Addressing these concerns is vital for maximizing the potential of AI while ensuring trust and ethical alignment in its applications.

By understanding AI's transformative role in medical virology and the associated challenges, it becomes evident that a human-centric approach is essential for fostering trust and realizing its full potential. This sets the stage for addressing the ethical, technical, and societal dimensions explored in the subsequent sections.

9.1.1 Problem Statement

However, there are great opportunities to apply AI in medical virology, and many barriers and factors are acting as a hurdle that does not enable its proper adoption. The most notable hindrance is that suppliers of services and consumers are sceptical of other parties and entities in the healthcare sector. Regarding AI-driven systems, many stakeholders question the usefulness and accuracy of the system, especially in critical situations such as the detection and identification of a virus or its potential spread (Ali et al. 2024). This distrust stems from the belief that algorithms do not fully grasp the context or circumstance surrounding a patient or fail to cover for extraneous patient-related factors.

This means that transparency in AI decision-making further complicates its use in medical virology. Deeper learning models are to some extent known to be 'black-box models' implying that they arrive at a certain result without the supporter understanding the whole process that has led to that specific result. This opacity is problematic because there is no way for clinicians and other healthcare authorities to confirm the forecast that the programme has made, or the advice that it has provided. For instance, an AI system might predict high viral risk in a patient by harbouring patterns in the data, and without knowing the speculations, clinicians may fail to trust it. An equal lack of interpretability weakens confidence and restricts the utilization of the technology in ordinary practice (Sallam et al. 2024).

Another challenge is ethical issues that are in this segment. One of the challenges is related to data privacy because, for instance, by analysing genetic and health records that patients share with hospitals, AI systems need access to substantial volumes of patient data. Despite the important and delicate nature of this information, the secure collection, storage, and processing are sometimes poorly implemented. Moreover, the biases incorporated in the models—owing to unbalanced training data or improper algorithms—are some of the other causing factors of unfairness in the model used in healthcare. For instance, it emerged that some of the diagnostic AI tools are less accurate with patients from various underrepresented demographics, thus increasing the healthcare gap.

These challenges are well illustrated by real-life examples. During the COVID-19 flare, AI solutions were utilized in estimating the potential spread and areas with high risks. However, questions arose about the reliability of the predictions due to features such as wrong forecasts, low quality of data, or lack of clarity in modelling. For example, an AI algorithm for diagnosing pneumonia was discovered to perform well on chest X-rays from some hospitals but poorly in others; this indicated that a 'hidden bias' exists in the training of the AI model (Holmes et al. 2024).

Solving these issues will play an important role in establishing confidence and the correct approach to AI use in medical virology. Any future systems must be transparent, free from all bias, and conform to all privacy statutes to become acceptable and have a positive impact.

9.1.2 Objectives of the Study

The primary objective of this research is to build medical virology-specific human-centred AI systems. Because the successful implementation of such AI models depends on the medical experts, in clinical and public health contexts, consequently, the remaining gap is embodied by the key principles of trust, transparency, and ethical design.

These models need to present results in a way that can easily be understood by healthcare professionals so that they can explain why, for instance, a particular patient was more likely to be diagnosed with a particular disease. This causes confidence and makes sure that artificial intelligence becomes an important resource rather than a mysterious tool in virology.

Another strategic goal is patient and clinician-centric model development for the future AI use in a practice. This means the development of systems that are easy to understand and use, and that are fitted well with both the medical profession and users. Those models may be more useful and easily adopted if integrated into existing practices from the user's perspective and feedback.

Last but not least, the study aims at designing a privacy-preserving AI system conforming to the highest data security requirements. To overcome those hurdles, the application of methods like federated learning or differential privacy, which retain the privacy of patient data while achieving high AI performance, will be proposed.

9.1.3 Historical Development and Evolution of AI in Healthcare

Early Applications of AI in Medicine

The adoption of AI in healthcare started in the 1950s beginning with the use of only expert systems which made decisions. The first such system, called MYCIN, was geared to help doctors diagnose bacterial infections in the early 1970s. This rule-based system proposed treatments that were consistent with the medical literature present in the system's knowledge base using symptoms and the patient's information. Another area of early diagnostic tools was the primitive decision-support systems for interpreting computer tomography, X-ray, and other images, for instance, of tumours.

Advancements Over Time

In the last decades, the development of AI in healthcare has increased especially after 2000, especially with the emergence of a new generation of ML and DL techniques. It allowed for pattern recognition in the form of text and images and augmented the AI systems' capacity to parse Big Data and included medical records. In the field of medical virology, AI has desirably improved diagnostic tests including image analysis and genomic sequencing. For instance, new AI-aided applications help in the identification of viral ailments at the rudimentary stage from chest X-ray/CT scans with improved precision compared to conventional systems (Malik et al. 2021). Machine learning models have become incorporated into the epidemiological forecast to predict and manage other viral diseases such as COVID-19. They have transformed the ways in which the managers of health facilities diagnose, treat, and manage viral illnesses.

9.1.4 AI'S Role in the Future of Healthcare Systems

AI as a Catalyst for Future Healthcare Models

AI is expected to play a critical role as a key innovation force in the future developments of care and nurturing of health in the future. Using AI in diagnosis and analysis, different and better treatment options will be offered due to an individual's tendency, genetic makeup, and previous medical history. This will result in an accurate diagnosis, and the appropriate therapeutic methods will be utilized. Besides, the revolutionary use of AI in real-time diagnostic systems will empower healthcare professionals with real-time results to diagnose illness, including viral infections, much earlier. This will speed up the interventional process and bring positive impacts on the patient's health. The advance of AI in medical virology will also assist in a prognosis of the new viral illnesses and also in giving the alarm concerning the potential pandemics.

Global Healthcare Accessibility
AI can improve access to healthcare massively in the areas with a scarcity of resources. Telemedicine, with the support of AI technology, lets patients far away from a doctor to get a consultation without travelling long distances. Artificial intelligence means that patients' records and scans can be interpreted quickly and more accurately than by traditional methods that are used locally. However, AI can bring benefits too in terms of improving the flow of the overall healthcare networks and in being able to determine where exactly patients require care and where resources are needed (Ali et al. 2024). With a near-future application, AI can play a role in empowering healthcare services, making them less expensive, thus, improving accessibility of healthcare to the less privileged in the society, hence, enhancing the health of society.

9.2 AI Applications in Medical Virology

9.2.1 Early Detection and Diagnosis of Viral Infections

AI systems, particularly machine learning (ML) models, have significantly enhanced the ability to identify viral infections at early stages, where timely intervention is crucial for better patient outcomes. One of the key advancements in early diagnosis is the development of AI-powered symptom checkers. These systems analyse reported symptoms and medical histories to predict the likelihood of viral infections, providing healthcare professionals with valuable insights for further investigation (Zhao et al. 2022). In addition, AI-powered imaging tools have shown remarkable accuracy in detecting early signs of viral infections. For example, deep learning models can analyse chest X-rays or CT scans to identify anomalies indicative of viral pneumonia, such as those caused by COVID-19, even before visible symptoms fully manifest. This technology has proven especially useful in cases where patients have subtle symptoms or are asymptomatic.

Another important area where AI contributes to early detection is genomic sequence analysis. Machine learning algorithms can rapidly analyse viral genetic sequences, identify mutations, and track the evolution of viruses. By comparing these sequences to known viral databases, AI can detect new or emerging viruses and predict potential outbreaks, enabling faster response strategies. For instance, AI systems were instrumental in tracking the SARS-CoV-2 virus during the early stages of the pandemic, identifying key mutations, and aiding in the development of diagnostic tests (Sallam and Al-Salahat 2023). Figure 9.2 shows the role of AI in early viral infection and detection.

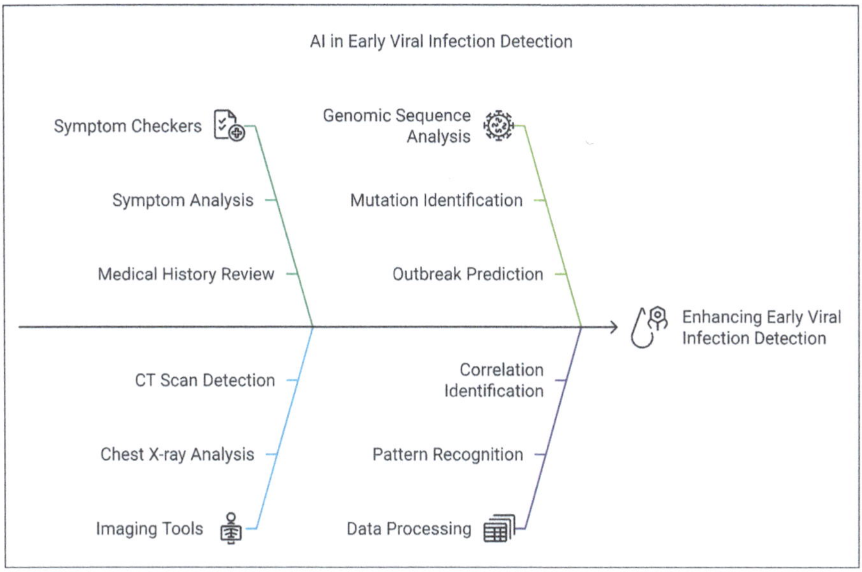

Fig. 9.2 Role of AI in early detection and diagnosis of viral infections

Machine learning further enhances diagnostic accuracy by processing complex datasets, such as medical images, patient records, and laboratory results. Algorithms can identify patterns and correlations that may be overlooked by human clinicians. For example, AI tools for COVID-19 detection from chest X-rays have outperformed traditional methods in terms of speed and accuracy, offering healthcare providers a more efficient means of diagnosing infections and prioritizing care (Ai et al. 2021).

9.2.2 Personalized Treatment Plans

As discussed earlier, AI is critical in shaping remedy plans based on an extensive patient characteristic database such as genetics and health records. When this type of data is combined with others, it avails AI models with a detailed recommendation on a patient's docket that is specific to his or her circumstances. This approach deviates from the previous approach of having broad therapeutic vantages and guarantees that treatment is tailored in line with the best practice. For instance, AI in relation to genetics may concern genetic variations to understand how exactly a patient could react to specified antiviral drugs—this will help the clinician select the proper therapy with referral to genetic features.

Another development that has taken root in the field of personalized medicine is the use of precision medicine algorithms that work out the best course of action by giving the whole patient history. Such algorithms can analyse trends within relevant big data and predict the effectiveness of treatments, benefiting the decision-making of healthcare specialists. Regarding viral infections, it has been used in antiviral therapy to produce very specific therapies as discussed in this chapter. For example, the data on COVID-19 have been generated through the use of AI for drug redesign, the method of which helps reveal drugs that may be effective weapons against the novel virus based on their patterns of interaction with viral proteins and patients. This not only accelerates the treatment process but also decreases the costs of developing drug (Salahshoori et al. 2021).

AI has also been useful for developing individualized vaccination plans especially for proper distribution of vaccines and finding the target population. By mapping the demographic, genetic, and epidemiological data, AI models can identify potential beneficiaries of specific vaccines and allocate resources most effectively. In the COVID-19 period, modelled approaches were most useful for planning vaccines to address the delivery of the flu virus.

9.2.3 Outbreak Prediction and Management

It has become almost mandatory to use AI-predictive models in an attempt to predict the advancement of viral diseases in order to take necessary measures in order to avoid epidemic. These models use a great deal of information in epidemiology, such as infection rates, population mobility, and other environmental indices, to explain how and where specific viral diseases are likely to spread. For instance, data-driven technologies can only model different potential exposures to come up with expectations of the infection rates and the potential spread that will enable the public health departments to put down necessary resourcing and containment measures before the disease's surge. During the COVID-19, such a predictive model is useful in modelling infection surges and informing approaches including the use of lock down, distancing, and vaccine allocation.

AI also records the spread and helps in the real-time control of the disease in question apart from being used in outbreak prediction. Once integrated with social media feeds, travel data, and hospital admissions, AI can present a current picture of the pandemic evolution and hint at new emerging cases. The same example includes the possibility of using data on social media to recognize new types of threats to health or new symptoms, which in many cases allows for a faster response. In the same way, the movement of people from one region to the other can be tracked as another reason to identify regions where viral spread is most likely to happen (Napolitano et al. 2022).

AI has been used in managing pandemics for instance Ebola and COVID-19 through improvement in prediction and control of the virus spread. In the case of Ebola outbreak, an AI tech uncovered movement of the disease in real time so that measures could be assigned at the right place and time (Chen et al. 2021). Likewise, AI-based systems during COVID-19 enabled forecasting of possible areas to become affected and the subsequent guidelines on quarantine measures, tracking of contacts, and distribution of resources. These capabilities reveal how AI can be useful in limiting the consequences of viral eventualities by appropriate responses.

9.2.4 Real-Time Research Support

AI has made a great contribution to speed up the process of designing injections and vaccines through behaviour mimicry of viruses, selection of possible targets for the vaccine, and the optimal selection of clinical trials. The large quantity of viral genomic data is processed by means of machine learning, which makes it possible to define the potential goals for vaccines, namely the proteins or mutations in the virus. It also allows for realizing how the virus might change in the future, facilitating the creation of a vaccine in the kinds that can combat existing as well as future mutations. For example, during COVID-19 outbreak, AI played an important role in swift identification of spike protein as a relevant epitope for m-RNA vaccines—which formed the basis of vaccines by Moderna and Pfizer. Moreover, models existing within AI mimic immune reactions of various candidates of vaccines, thus, minimizing time needed for preliminary studies of vaccines (Haimed et al. 2021).

AI also finds application in many theoretical models of disease spread that epidemiologists use to simulate patterns of infections and their potential developments among groups of people. These real-time models basically combine information about infection rates, demographics, and other data, as well as the effects of public health interventions, to predict outcomes. For instance, the use of AI during the COVID-19 outbreak enabled one to estimate the viral spill rate and its effects in every intervention, including social distancing and vaccination (Giovanetti et al. 2023). They allowed government and healthcare systems to invest funds and define correct strategies and future acts in the sphere of public health.

The integration of AI in the epidemiological and vaccine research not only speeds up the process of finding new vaccines but also raises the reliability of the disease control methods. That is, through constant improvement of these AI systems, the researchers can counter the novel viral threats.

9.2.5 AI in Viral Drug Repurposing

When it comes to drug repurposing, AI has established itself as a revolutionary technology with unparalleled success with viral infections. Drug repurposing is a process of finding out that new diseases can be treated with old drugs which were initially developed and used to treat different diseases. Introducing AI into this process expedites this activity through processing large databases of molecular formations and already established drug interactions to make a probability estimation of a range of existing drugs against certain viral forms. The use of big data pre-processing can involve the use of algorithms which can sort data from chemical databases, clinical trials, and genomic information of potential antiviral candidates.

Another impressive advantage in drug repurposing is that AI can model which drugs act and in what manner towards certain viral targets. For instance, AI-driven systems helped identify such existing drugs as remdesivir and hydroxychloroquine to be used for treating COVID-19. It also made it possible to come up with predictions for clinical trials and evaluation that were way faster than the actual invention of new drugs. In addition, based on this data early in the assessment cycle, AI shortens the search time and costs for finding new drugs considerably (Lu et al. 2023).

It also helps in determining the right doses for a specific drug and avoiding any possible side effects that may occur due to drug interaction with patient information making it safe and effective. Again, while AI can quickly sort out the most potent antiviral drugs from existing libraries to take even less time to apply clinically, it can help with faster solutions during infectious viral flare-ups.

9.2.6 AI for Monitoring and Surveillance of Viral Diseases

Several technologies based on AI are being assimilated into automated identification and surveillance to monitor viral diseases, identify new epidemic, and forecast viral events. With large quantities of data from hospitals and laboratories, mobility, and social media, valuable patterns of viral infections' dynamics can be identified to take effective actions. For instance, throughout COVID-19 pandemic, AI models were applied for tracking the rate of new infections and authorities got better prepared and reacted consequently.

Machine learning helps in the early identification of outbreaks because AI can also see trends when normal methods could not. Geographical data, health reports, and even the news can be processed by machine learning algorithms to quickly find signs of an up-and-coming viral threat (Taha et al. 2022). The deployment of these models eases the projection of ensuing instances, resource allocation, and essential containment measures to contain the spread of infections in the healthcare organizations.

Moreover, AI contributes to demographic and geographic surveillance pointing out how different groups of people or geographical areas are influenced by viral diseases. AI systems are to estimate which communities or areas are at higher risk depending on population density congruousness, and historical health data. It also assists in preventing the spread of viral diseases such as by providing timely vaccinations or quarantining isolated cases which reduce the extent that the diseases spread.

9.3 Human-Centric AI Design

9.3.1 Mathematical Model for Explainable AI (XAI)-Based Human-Centric User Interface for Virology

The design of a human-centric user interface (UI) for healthcare providers and patients in the context of medical virology using Explainable AI (XAI) involves several key components: the expertise AI model, the explanations given by the model, and the one that allows people to interface with the model. The idea here is to keep the resultant AI system as interpretable and transparent as possible in order to make users trust the decision made by the model. The next part outlines the mathematical framework for this system as shown by the following discussion.

(i). Prediction and Diagnosis AI Model for Virology

Let the various parts of the AI system be represented by a function fthat corresponds to the input features X to a prediction y. These input features may also be derived from patient records or other clinical parameters related to viral infections, including clinical and imaging data or laboratory tests:

$$f : X \rightarrow y, \tag{9.1}$$

where

- $X \in R^n$ is the vector of input features (e.g. patient data, viral genomic data, symptoms).
- $y \in R^n$ is the output prediction (e.g. probability of infection, risk level, or diagnosis).

For example, the model could use deep learning or support vector machines to predict viral infection likelihood.

(ii). Explainability Module (XAI)

XAI is used to provide transparent explanations of the model's decision-making process. A common method is to decompose the prediction into understandable components that can be attributed to specific features. One approach is LIME (Local Interpretable Model-agnostic Explanations), which approximates the AI model with a simpler, interpretable model in the local vicinity of a prediction (Kim et al. 2023).

Given the original model's prediction function $f(X)$, we generate an interpretable explanation $\phi(X)$ by approximating $f(X)$ locally using a simpler model g:

$$\phi(X) = g(X) \text{ such that minimize } L(f(X), g(X)) + R(g), \tag{9.2}$$

where

- $g(X)$ is the interpretable model (e.g. linear regression, decision tree).
- $L(f(X), g(X)$ is the loss function that measures the discrepancy between the original model $f(X)$ and the surrogate model $g(X)$.
- $R(g)$ is a regularization term to ensure g remains simple and interpretable.

(iii). Human-Centric User Interface (UI) for Healthcare Providers and Patients.

The user interface is designed to present both the model's prediction and the explanation in a user-friendly and intuitive way. The UI output is a combination of

- the AI model's prediction, y, and,
- the explanation $\phi(X)$, breaking down the prediction into understandable components.

For example, the prediction y (such as the probability of a viral infection) is displayed alongside an explanation of which features (e.g. age, viral load, symptoms) influenced the decision the most. The explanation could also be visualized using importance scores for each feature, where the score α_i represents the contribution of feature i to the final prediction.

Let α_i be the importance score for feature X_i, such that.

$$\alpha_i = \partial f(X_i)/\partial X_i. \tag{9.3}$$

This score tells how sensitive the model's output is to changes in X_i, indicating which features contributed most to the prediction.

(iv). Mathematical Representation of the User Interface Interaction

The user interface is designed to facilitate interaction between the healthcare provider/patient and the AI system. We define the interaction function I as a mapping from the user's input U (e.g. queries, adjustments in input data) to the AI's response R, which includes the prediction and the explanation:

$$\mathrm{I} : \mathrm{U} \rightarrow R = (y, \phi(X)), \tag{9.4}$$

where

- U is the input from the user (e.g., adjusting the patient's age, inputting new symptoms).
- $R = (y, \phi(X))$ is the system's output consisting of the AI prediction y and the explanation $\phi(X)$.

(v). Trust Model (Quantifying Trust in AI)

Trust in AI decisions is crucial for the system's acceptance. A trust score T can be mathematically modelled as a function of the transparency, interpretability, and accuracy of the system (Said et al. 2021; Çağlayan et al. 2022). A potential model could combine these factors into a single score:

$$T = f_{\text{trust}}\ (\text{Transparency}, \text{Interpretability}, \text{Accuracy}) \tag{9.5}$$

For example, if the explanation $\phi(X)$ is clear and understandable, the transparency is high. If the AI's predictions align well with expert opinions or ground truth, the accuracy is high. Trust could be scored using a weighted sum of these factors:

$$T = w1 \cdot \text{Transparency} + w2 \cdot \text{Interpretability} + w3 \cdot \text{Accuracy}, \tag{9.6}$$

where $w1$, $w2$, $w3$ are weights assigned based on the relative importance of each factor.

(vi). Ethical Considerations and Bias Mitigation

To ensure fairness and ethical integrity, the AI system must address biases in both data and decision-making (Padhi et al. 2023). The fairness of the model can be quantified as a function of the disparity in predictions across different demographic groups:

$$\text{Fairness} = 1/n \sum\nolimits^{n}_{i=1} \ |f(X_i) - f(\hat{X}_i)\,|, \tag{9.7}$$

where:

- X_i is the feature vector for the ith individual.
- $f(X_i)$ is the model's output.
- $f(\hat{X}_i)$ is the target or ground truth output.

A low disparity in prediction outcomes for different demographic groups (e.g. age, gender) ensures fairness.

This mathematical model for an XAI-based human-centric user interface in medical virology leverages transparent AI systems, user-friendly interaction, and ethical considerations. By ensuring that the model's explanations are interpretable and by prioritizing user trust, this framework aims to enhance the collaboration between AI systems, healthcare providers, and patients in viral disease diagnosis, treatment, and prevention.

9.4 Ethical and Privacy Considerations

For the XAI-based human-centric user interface in medical virology, several graphical results can be used to visualize the mathematical model, making it easier for healthcare providers and patients to interpret the AI's predictions, understand explanations, and ensure ethical considerations. A dataset suitable for execution of proposed method is used with suitable attributes so as to visualize the appropriate results (https://github.com/Sowmya-javvadhi/Medical-Virology).

9.4.1 Prediction Visualization

Probability Distribution Plot

A bar or line chart shows the predicted probability of a viral infection or risk level. This graph can highlight the certainty of the AI's prediction (e.g. probability of infection, likelihood of severity).

Example:

- X-axis: Different possible outcomes (e.g., "No infection," "Low risk," "High risk").
- Y-axis: Probability values (0 to 1).

The bar chart in Fig. 9.3 displays the predicted probabilities of different infection risk levels based on AI predictions. The X-axis represents risk categories ('No infection', 'Low risk', 'High risk'), while the Y-axis shows the average probability values for each category.

Risk Level Gauge/Progress Bar

A visual gauge or progress bar indicating the predicted risk level, with colour coding (e.g. green for low risk, yellow for moderate risk, and red for high risk).

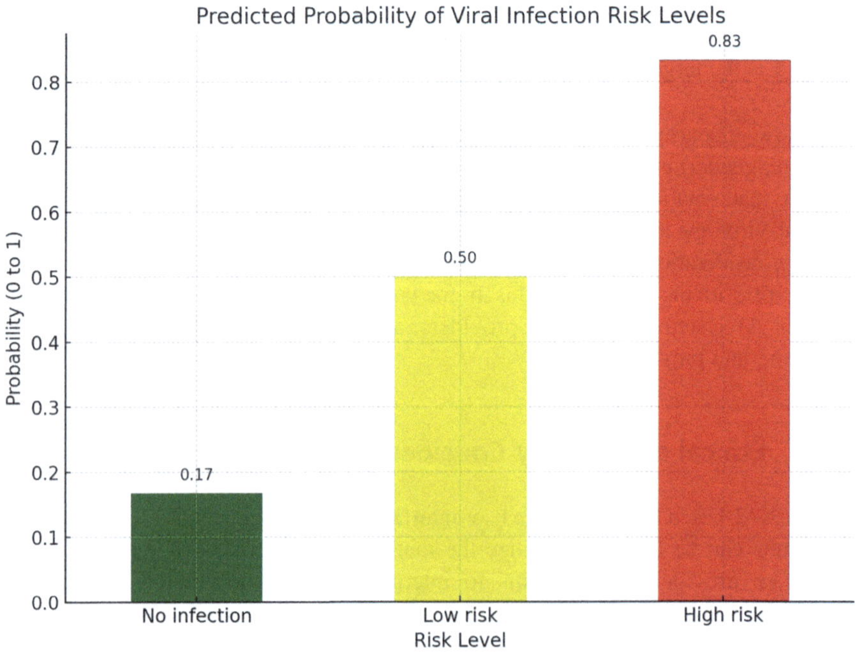

Fig. 9.3 Predicted probability of viral infection risk levels

Example:

- A circular or linear gauge shows the risk percentage of viral infection based on input data.

9.4.2 Feature Importance Visualization

Bar Chart of Feature Importance
A horizontal or vertical bar chart displays the importance of each feature (e.g. age, viral load, symptoms) in the model's prediction.

Example:

- X-axis: Features (e.g. Age, Viral Load, Fever).
- Y-axis: Importance score (α_i) showing each feature's contribution to the model's decision.

The histogram in Fig. 9.4 denotes the count of patients as per age groups used for the model prediction.

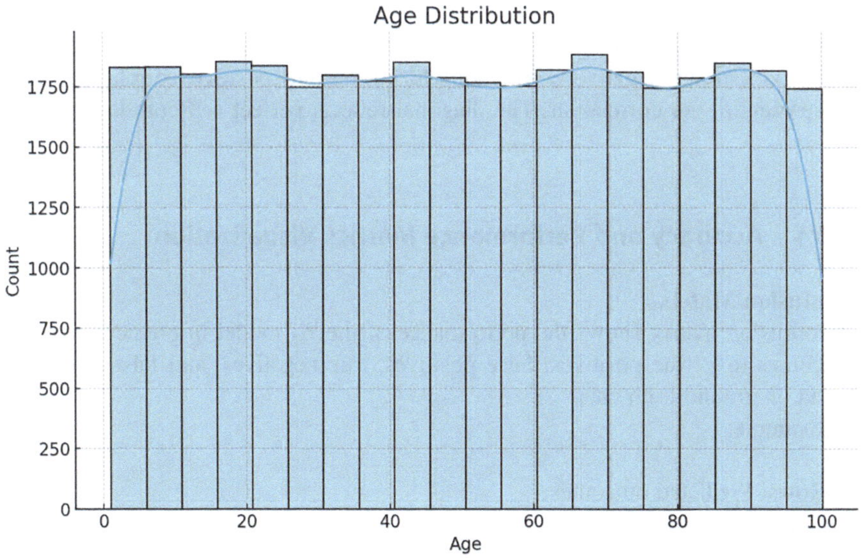

Fig. 9.4 Age Distribution: A histogram showing the distribution of patients' ages with a smooth density curve

Heatmap of Feature Contribution

A heatmap represents the correlation between features and their impact on the prediction. Each cell can be colour coded based on the feature's impact on the prediction (e.g. red for strong influence, blue for little influence).

9.4.3 Model Explanation Visualization

Local Surrogate Model Explanation (LIME or SHAP)

Graphs like decision trees or linear plots that represent the surrogate model (e.g. using LIME or SHAP) for a specific input. These plots show how individual feature changes affect the model's output.

Example:

- Decision tree diagram shows how certain features influence the prediction of viral infection for a specific patient.

Explanation Heatmap/Overlay

A graphical overlay on the input features (e.g., a patient's medical record) with heatmaps shows which features contributed most to the model's prediction. For example, darker colours represent higher importance.

The correlation heatmap as shown in Fig. 9.5 illustrates the relationships between various features, indicating the strength of correlation between them. Values range from -1 (strong negative correlation) to $+1$ (strong positive correlation), with 0 representing no correlation. The diagonal reflects perfect self-correlation (value of 1).

9.4.4 Accuracy and Performance Metrics Visualization

Confusion Matrix

A confusion matrix shows the performance of the AI model in predicting correct outcomes (e.g. true positives, false positives, true negatives, and false negatives) based on ground truth data.

Example:

- Rows: Predicted outcomes.
- Columns: Actual outcomes.
- Cells: Count of predictions.

ROC Curve (Receiver Operating Characteristic)

A plot that shows the trade-off between sensitivity (true-positive rate) and specificity (false-positive rate) at various threshold levels.

Example:

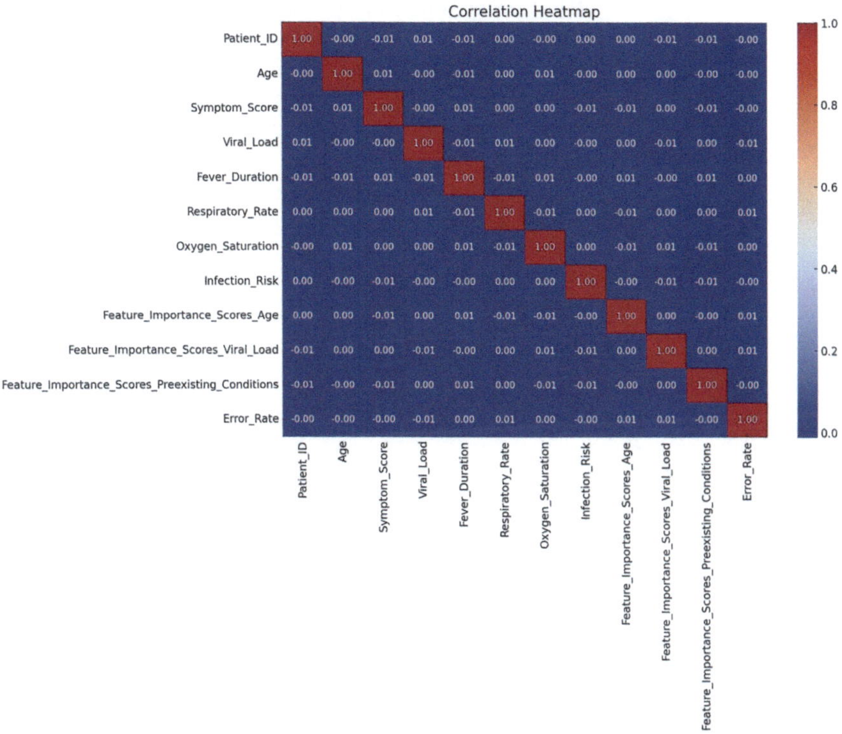

Fig. 9.5 Correlation heatmap

- X-axis: False-Positive Rate.
- Y-axis: True-Positive Rate.

In Fig. 9.6, the ROC curve displays the trade-off between false-positive rate (X-axis) and true-positive rate (Y-axis) for predicting 'Hospitalized' outcomes using the Infection Risk score. The AUC quantifies the model's discriminatory power.

Precision–Recall Curve
A graph illustrating the trade-off between precision and recall for different thresholds in classification tasks, especially in cases of imbalanced datasets.

9.4.5 Bias and Fairness Visualization

Fairness Metric Plot
A chart that compares the performance of the model across different demographic groups (e.g. age, gender, ethnicity). This ensures that the model's decisions are equitable and do not discriminate against certain groups.

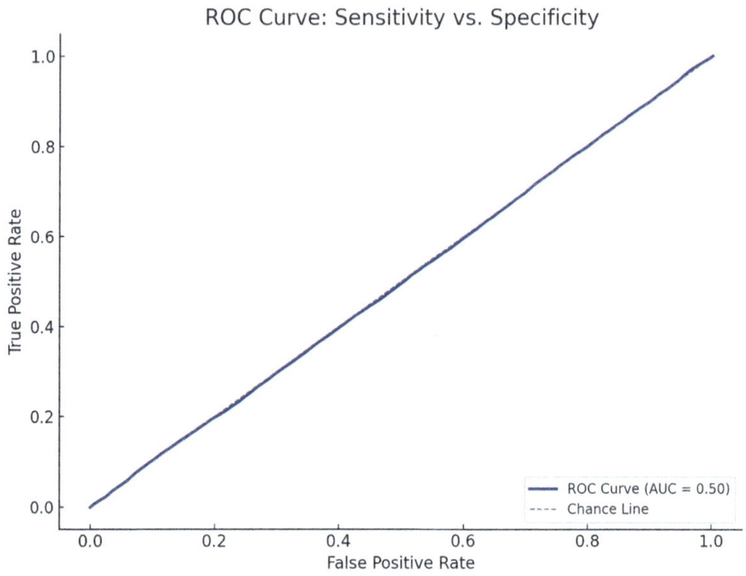

Fig. 9.6 ROC curve showing sensitivity versus specificity

Example:

- X-axis: Demographic groups (e.g. age groups, gender).
- Y-axis: Performance metric (e.g. accuracy, recall, precision).

The Fairness Metric Plot in Fig. 9.7 compares the average model error rate across gender groups (Male, Female, Other). The similar error rates indicate that the model performs equitably without significant bias towards any specific gender group.

Bias Detection Plot (Disparity Analysis)
A graph shows the disparity in AI predictions across different groups. This could be a bar plot comparing error rates across various demographic groups (e.g. errors in predictions for men vs. women, or different age ranges).

9.4.6 Transparency and Trust Visualization

Trust Score Visualization
A visual indicator or scorecard summarizes the transparency, interpretability, and accuracy of the AI system. This could take the form of a radar chart or a composite score.

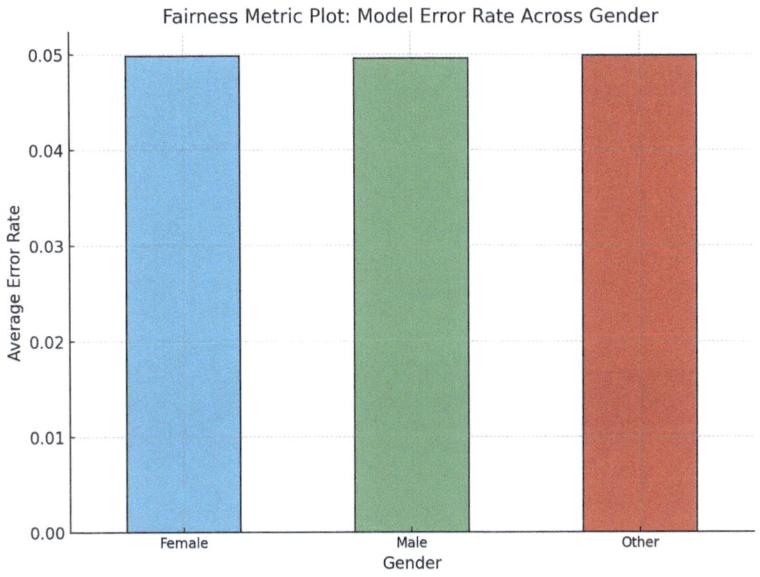

Fig. 9.7 Fairness metric plot

Example:

- Radar chart with axes for Transparency, Interpretability, Accuracy, and Fairness.
- A single score representing the overall trust level, possibly colour coded for easy interpretation (e.g. red for low trust, green for high trust).

The radar chart in Fig. 9.8 summarizes the AI system's trust indicators: Transparency, Interpretability, Accuracy, and Fairness. Each axis represents a normalized score derived from the dataset. The overall trust level is colour coded: green for high trust, orange for moderate, and red for low trust. The computed overall trust level for this system is displayed below the chart.

Explanation Transparency Feedback
A feedback loop visual where users can rate the clarity of the explanation, such as thumbs-up/thumbs-down icons, and the system adjusts its explanations based on this feedback.

9.4.7 Interaction-Based Graphs

Interactive Input-Output Graphs
A dynamic visualization where healthcare providers or patients can adjust input features (such as symptoms or test results) and immediately see how the output

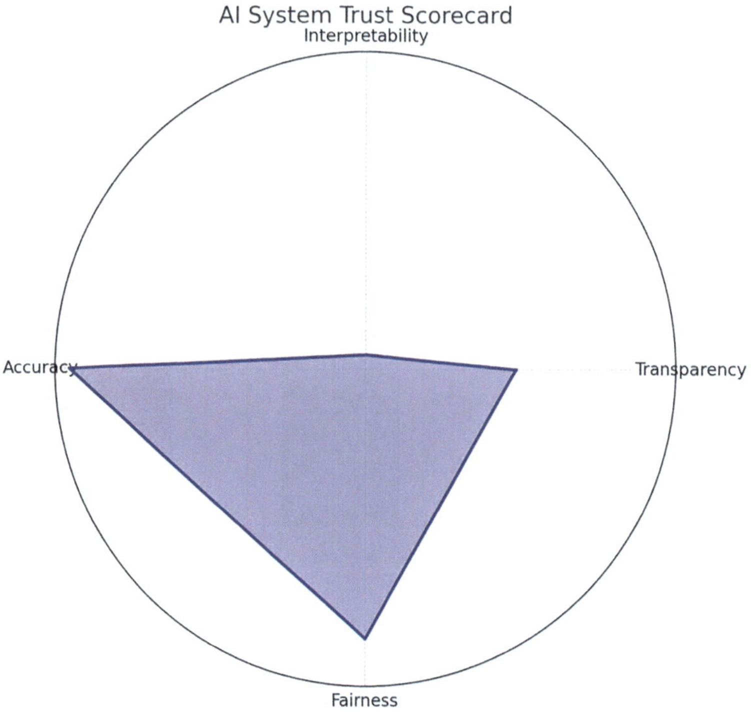

Fig. 9.8 AI system's trust indicators

(AI prediction) changes. This could be a slider interface, showing how each adjustment alters the prediction or feature importance.

Example:

- A set of sliders for modifying input values (e.g. age, viral load), with a dynamic graph showing the resulting prediction.

Decision Path Visualization

A flowchart or decision tree dynamically shows the path that the AI took to make a prediction based on user inputs, along with explanations for each step.

A scatter plot in Fig. 9.9 visualizes the relationship between viral load and symptom score, categorized by severity levels.

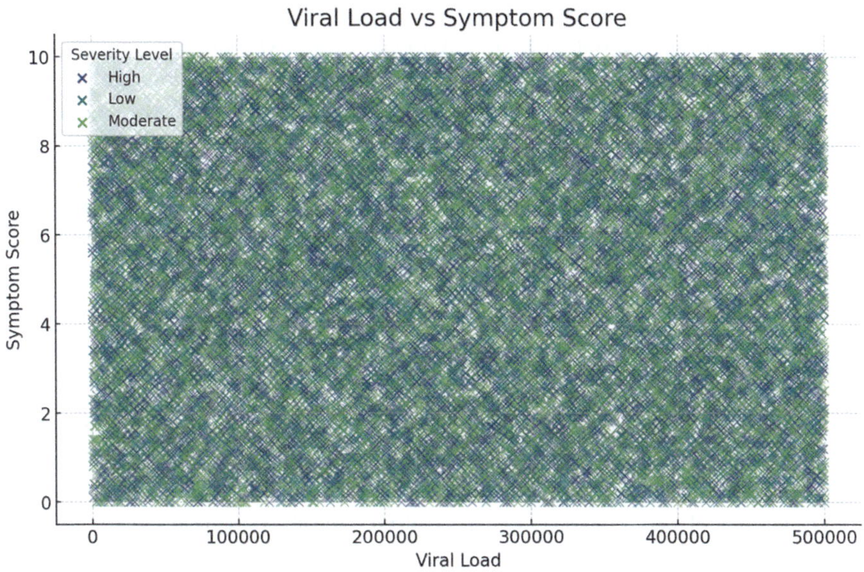

Fig. 9.9 Plot of viral load versus symptom score

9.5 Challenges and Barriers

9.5.1 Data Quality and Availability

Nevertheless, there are numerous obstacles which the researches face when creating the AI models for the field of medical virology, and one of the most important ones is the datasets' bias. Artificial intelligence needs larger volumes of data to build comprehensively accurate and universally applicable models for patients. Alas, compilation of such datasets is not an easy task. An unavailability of complete and consistent data is a significant concern being that patients' records may contain scanty information, particularly due to differing data collection parameters, or even poor healthcare systems. Datasets may contain bias in terms of the selected age, ethnicity, or region, creating a selected model that does not include all populations (Chakraborty et al. 2024). This lack of representativeness can subsequently confine the ability to detect the viral infections or breakouts in such groups of people.

There remains a clear correlation between what you feed into an AI model and the output they produce—which is why poor-quality data or the use of data from a biased source can cause a lot of harm. This actually means that whenever lacking necessary or balanced data during their training, the AI models will in effect make forecast errors leading to wrong diagnostics or treatment approaches as well as inadequate epidemiological projections. For example, a given model used in an environment with data obtained majorly from one geographical location is likely to miss initial signs of an epidemic in another region. The same is true for bias in the data—inequality in the outcomes might appear as the AI tools are known to work worse for the underrepresented groups which only magnify existing gaps in health access. To avoid such risks, the AI system should be trained on a very large dataset that contains and encompasses the AOD patient demography as well as variability factors. In addition, data should be procured in a mordant manner, with privacy and consent to data processing being maintained and data collection methods that compromise the integrity of the survey and offer unfair advantage to some subjects or disadvantage to others not employed as surveying vulnerable groups.

9.5.2 Resistance to Adoption

The incorporation of AI in the industry especially in medical virology has, however, faced a lot of resistance from both the users and the practitioners in the health sector (Mock et al. 2021). Some of the challenges arising with AI include doubt regarding the reliability of data entrusted to AI. AI-driven systems create doubts among the healthcare niches because of the disparity of performance compared to handpicked algorithms by employing real experts in diagnosing diseases or prognosis of outbreaks. Moreover, there is a common problem of explainability where clinicians have a problem with understanding how such AI arrived at a given decision. However, this operational opacity can erode trust as clinicians would not wish to rely on systems which cannot be trusted to articulate their decision-making, let alone the basis for decision-making in areas affecting patients' lives.

Concerns that patients also hold in relation to the use of AI are also similar. Several people are concerned about the risk of such a misfire, for instance, wrong diagnosis, or even wrong treatment advice, especially where these AI tools are used with no supervision from a human expert (Zhang and Yan 2020). These are some of the reasons that exist due to the little information that people have concerns about the function of AI and effects of AI on patient confidentiality and security.

Cultural and organizational barriers only make the use of AI in healthcare worse. Healthcare organizations may have the problem of certain routines' preservation and rules of working that have been developed for a long time. Third, healthcare professionals may lose their jobs as services based on AI will replace some of them as diagnostic or administrative officers (Andriasyan et al. 2021). Also, there is a lack of demand for incorporating AI into existing approaches due to vast changes intrinsic to training, established practice, and processes. The key challenges are as follows: To eliminate these barriers, one needs to plan effectively for technology and human interfaces in order to get more positive outcomes out of AI.

9.5.3 Ethical Dilemmas

One of the most crucial issues arising from integration of AI in clinical decision support is the issue of the system's autonomy and the professional discretion. This area defines the main reason why the integration of the AI into the patient's caring process should be regulated to prevent the disappearance of human feelings, ethical aspects, and control over everything done by the artificial system (Fang et al. 2022). Clinical decision-making is not only dependent on the technical data regarding patients but also on their emotional processes, their values, and their biography. But AI, in turn, has many limitations in communicating with patients and, thus, may lack the sincere, human touch required in treatment. The issue of ethical balance in implementing artificial intelligence into healthcare process is in bolstering clinicians rather than replacing the voice that is so crucial for trust and humane opinion.

The Role of Human Oversight in AI-Driven Decisions

One of the significant ethical issues is to ensure that exceptional human monitoring is executed in circumstances influenced by artificial intelligence platforms. Even if all that data can be analysed at a faster pace through AI models, the major decisions should solely be made by doctors. At the same time, in an AI-only system, some details are overlooked that a human clinician would define, for example, psychological problems or certain social conditions that may affect the result of treatment. The problem lies in making sure that artificial intelligence stays a means by which human abilities are amplified, rather than merely a process that renders human discretion superfluous (Fang et al. 2022).

Accountability in AI Decision-Making

AI also plays certain roles in clinical decisions, including the following ethical issues: accountability. When decisions regarding treatment or even conclusions about the patient's disease are being made AI, there may be legal ambiguity as to who is to blame in case of mistakes or bad outcomes. If an AI-driven system gives wrong diagnosis or recommending an ineffective treatment plan, who is liable—the designers, the healthcare professional, or the machine itself? This lack of clarity can make dissociation of legal and ethical discussions especially when patient harm is involved cumbersome. Reducing the risks of AI depends on establishing accountability to foster ethical decision-making on systems, and engineering transparency in their decision-making (Elbasir et al. 2023).

Dehumanization of Healthcare

Also, there are worries that due to development in the application of artificial intelligence, there will be dehumanization of healthcare. Prescribing too much of the AI might prove counter-productive in that the patients may start to feel that instead of receiving full human attention, their situation has been programmed into the computer (Matuszewski and Sintorn 2021). This concern is attached especially when AI is practiced in any aspects, including diagnosis and treatment planning since emotions, preferences, and even the quality of life of the patient need to be understood by human touch. The danger is that, with increasing use of assistive technologies in healthcare, another important element, which is underlying human skills like compassion, care, and ethics, may unfold as an impediment to the patient–practitioner relationship and degrade the general value of care.

Bias and Fairness in AI Decisions
The third criteria-based ethical issue stems from the possibility of bias in the AI algorithms used in an application. The potential for AI systems to keep existing health disparities is high since if trained on discrimination data sources, they shall be discriminative in treatment recommendations for certain demographics. For instance, an AI model trained mostly with data from a specific ethnic group often performs poorly concerning diagnosing or treating patients of different ethnicities. This brings a lot of questions about why and fairness and/or equality in accessing healthcare when AI-embedded systems are the future.

Transparency and Trust in AI Systems
Last but not the least, it is also important that AI is transparent so that people may begin trusting AI systems. Thus, for AI to be a fully implemented part of the healthcare marketplace, patients and healthcare providers must believe that those systems are both fair and act in an ethical manner. This implies that AI systems should be capable of supporting user decision-making and provide information on decision rationale. It is a good practice to ensure that the AI algorithms are understandable or not merely a 'black-box' approach Something like this will help reduce the fears of the populace on issues to do with privacy and bias, as well as accountability that come with artificial Intelligence.

Consequently, it can be noted that AI holds great promise for enhancing healthcare, but its implementation within a clinical context requires determination both of potential benefits and potential risks that may stem from the utilization of the novelty, as well as reflection on the connotations arising from the application of the new innovation. One must closely monitor automated systems yet not leave the process entirely to the machines, the need for someone to be held responsible for the operations, the issue of bias, and the need to be transparent should be that important while applying the technology in medical virology and healthcare.

9.6 Future Directions

9.6.1 Interpretability in AI-Assisted Systems: How to Build Trust

The application of Artificial Intelligence (AI) in healthcare as well as in medical virology particular can bring substantial focus on the category of outcomes. But one of the important issues that relate to the application of an AI model in this area is a lack of explainability of how the decision of that model was made. In some cases, especially in deep learning models, the processes used in reaching decisions are convoluted leaving even the developers hard-pressed to explain an outcome. This makes it difficult for the healthcare professionals to have faith in the decisions made by these systems because they must be sure that the system's decision is arrived at through a proper reasoning algorithm.

This is why we need Explainable Artificial Intelligence (XAI). XAI is a process or approaches for designing and utilizing AI technologies in a comprehensible manner in order to facilitate interpretation by users. In the context of medical virology, XAI will make it possible for healthcare providers to observe not only the predictions made by an AI (for instance, detection of the virus or recommendations on treatment) but also the process by which the AI comes about such predictions (Matuszewski and Sintorn 2021). For example, an XAI model in viral diagnostics could show which symptoms or test results have been considered to arrive at a diagnosis of any specific viral infection and in what way the decision-making process was reached. This level of transparency helps clinician trusting and relying in AI tools as the information is being displayed to them.

Concerns of interpretation of AI systems increase confidence between the patient and the healthcare provider. Patients often fail to understand why some tests are required or why a certain treatment should be taken in the case of viral infections; thus, this research made patients be willing to adhere to prescribed treatments because they understand why the certain regimen is required.

Impact of Explainable AI on Clinical Decision-Making

This is especially important in making clinical decisions, particularly when there are viral infections involved, since the patient and the treatment's safety would be at risk whenever the healthcare practitioner is unable to make logical interpretations on the AI-recommended findings. Machine learning models applied to medical virology tasks—identification of the infections or prognosis of the patients' outcomes given specific treatment—must to be able to generate explanations that are actionable by clinicians. Clinical decision support system provides a type of explainable AI that allows the healthcare professional to take a critical view of decisions offered through the model before treating a patient.

For instance, in VI, AI systems might suggest, for example, the identification of virus in images such as X-rays or CT scans. Explainability means that, if an AI system has delivered a diagnosis to clinicians, they should be able to see parts of the image that the AI system has relied on in order to make this conclusion so that clinicians can cross check it against their own knowledge. This process enhances patients' diagnosis and accuracy and minimizes errors.

When it comes to the objective of treatment planning, explainable AI remains one of the primary ways to progress the personalization of such. The use of an AI system that recommends antiviral medications or some treatment actions can tell the patient how they came up with the decision and in consideration of the patient's genetic makeup, treatment history, and the response of their body to treatment. In this way, that is with the help of a clinician, who explains why AI recommended such or such treatment, the clinician can make a change, or provide a different treatment option to the AI option, to make sure the patient's need are met.

In addition, where new or developing viral strains are involved, new data is likely to accrue quickly and existing AI models are likely to need updating. These shifts, therefore, make the adaptation process easier due to the explainability of XAI where clinicians can judge the relevance and correctness of what the model is pitching. The first advantage of explainable AI is that it helps increase the trustworthiness of clinical decisions, and the second is that it guarantees patient safety in complicated virology situations.

9.6.2 Building Resilient AI Models

Due to the rising rate of emerging viral diseases and infinite possibilities of viral evolution, the use of AI models needs flexibility to counteract these new health uncertainties. The dynamics of viral infections make it important for an AI system to be adaptive for consumption of fresh knowledge data sources. For example, the virus evolution with new strains, as in the example of SARS-CoV-2, requires the AI models to constantly update with new information to improve the models' accurate predictions for treatments' recommendations (Elbasir et al. 2023).

Real-time AI models that also have the capacity to learn from some other data may positively enhance the way we manage new viral illnesses. One is successive learning where incorporation of continual learning mechanisms in AI systems is promoted. These systems can adapt the determined parameters and algorithms of the predicting model based on new data flows, which allows considering its predictions as relevant. For instance, while dealing with the COVID-19 pandemic, AI models which incorporated information like the cases, patient details, and treatment outcome data were perfected with new information because the conditions were changing frequently. As mutations of the virus continued to appear, these models were adjusted to encompass current developments of the virus and its progression through the patient load to result in better organization.

Moreover, adaptive AI models need also to be ready to learn not only viral mutations but also changes in programmes, healthcare provisioning, and patient practices. Such type of flexibility guarantees that the models' strength is not influenced by such shifts in data patterns. Due to this adaptive characteristic, AI in healthcare will enable the improvement in future pandemics reduction and patients' quality of life.

Ensuring Robustness in the Face of Data Scarcity or Bias

Consequently, one of the most critical barriers to developing robust AI models for medical virology applications is the problem of data deficiency or bias. Precise, many-sourced data are crucial for tuning AI algorithms that in their turn can identify diseases, prognosis their occurrence, as well as suggest the proper treatment. But often, when it comes to newly discovered infections, data is lacking (Matuszewski and Sintorn 2021).

As a result, AI models have to be constructed sufficiently resilient to handle data scarcity and bias as potent threats. One approach is to use transferral learning, for example, a model which has been trained using a large dataset from a related application can then be retired and trained further with a similar smaller but specialized dataset. This enables the model to preserve useful information while making other adjustments to a new condition associated with the less abundant data. For instance, early in an emerging infectious disease where few amounts of data are available, AI models can reuse evidence gained from previous epidemics or other similar viral ailments to make predictions on identification plus treatment.

The other approach is to tackle the problem head-on, by guaranteeing that training datasets encompass multiple groups of people and types of viruses. This can be done by data augmentation that involves creating new samples to fit between existing data patterns. Implementation can be performed by developing the fairness-aware algorithms to maintain the detection and balancing of bias in order to provide the same service to all the categories of patients.

The second factor is related to the need to include the quantification of uncertainty into AI models to guarantee the models' robustness (Ghosh et al. 2022). This helps the system give quantitative measures of confidence as part of its results so as to express that a model may be less certain due to data constraints. This transparency could also assist other actors in the healthcare space to make better decisions and avoid leaning on these models when the model is less confident.

Thus, the problem of constructing reliable AI models arises from both data deficiency and data bias. By creating new adaptive systems that will be able to grow along with the emergence of new problems and creating new models that reflect the limitations of the datasets used, it is possible to ensure the further applicability of AI as the tool that helps combat viral diseases and improve patient's quality of life.

9.7 Conclusion

This chapter underscores the transformative potential of human-centric AI in advancing medical virology and healthcare systems. By emphasizing trust, transparency, and ethical design, AI-driven solutions can bridge the gap between complex algorithms and clinical needs, fostering confidence among healthcare professionals and patients. Explainable AI (XAI) emerges as a crucial component, enabling clear and interpretable insights into AI-generated decisions. This fosters collaboration between AI systems and clinicians, ensuring diagnostic accuracy, personalized treatments, and proactive outbreak management. The study highlights the pivotal roles of AI in early detection of viral infections, real-time research support, vaccine development, and epidemic forecasting. AI's ability to process large datasets and uncover patterns has accelerated drug repurposing, enhanced diagnostic precision, and informed public health strategies. These advancements demonstrate AI's capacity to mitigate the impacts of viral diseases and improve patient outcomes. However, challenges such as data bias, privacy concerns, resistance to adoption, and ethical dilemmas persist. The opacity of 'black-box' models and lack of trust remain barriers to widespread implementation. To address these, robust frameworks integrating fairness-aware algorithms, adaptive learning, and data privacy measures are essential. These solutions must prioritize inclusivity, equity, and the augmentation of human oversight to ensure AI's role as a supportive partner rather than a replacement. Looking ahead, resilient AI systems capable of evolving with viral dynamics and healthcare demands will be indispensable. Real-time adaptability, combined with the integration of ethical principles, will drive AI's seamless integration into medical virology. The vision is a future where AI empowers healthcare providers to deliver equitable, efficient, and patient-centred care while addressing global health challenges. In conclusion, human-centric AI offers a pathway to revolutionize virology and healthcare, promising improved public health outcomes, enhanced trust, and a collaborative future between technology and humanity.

References

Ai S, Awford A, Roncolato F (2021) Hemophagocytic lymphohistiocytosis following ChAdOx1 nCov-19 vaccination. J Med Virol 94(1):14

Ali G, Mijwil MM, Adamopoulos I, Buruga BA, Gök M, Sallam M (2024) Harnessing the potential of artificial intelligence in managing viral hepatitis. Mesopotamian J Big Data 2024:128–163

Andriasyan V, Yakimovich A, Petkidis A, Georgi F, Witte R, Puntener D, Greber UF (2021) Microscopy deep learning predicts virus infections and reveals mechanics of lytic-infected cells. Iscience 24(6):102543

Çağlayan D, Süner AF, Şiyve N, Güzel I, Irmak Ç, Işik E et al (2022) An analysis of antibody response following the second dose of Corona Vac and humoral response after booster dose with BNT162b2 or Corona Vac among healthcare workers in Turkey. J Med Virol 94(5): 2212–2221

Chakraborty S, Qu K, Valadi J (2024) New approaches toward understanding challenges in molecular virology: computational techniques and machine learning. Front Mol Biosci 11: 1472796

Chen J, Li K, Zhang Z, Li K, Yu PS (2021) A survey on applications of artificial intelligence in fighting against COVID-19. ACM Comput Surv (CSUR) 54(8):1–32

Elbasir A, Ye Y, Schäffer DE, Hao X, Wickramasinghe J, Tsingas K et al (2023) A deep learning approach reveals unexplored landscape of viral expression in cancer. Nat Commun 14(1):785

Fang Z, Feng T, Zhou H, Chen M (2022) DeePVP: identification and classification of phage virion proteins using deep learning. Gigascience 11:giac076

Ghosh D, Chakraborty S, Kodamana H, Chakraborty S (2022) Application of machine learning in understanding plant virus pathogenesis: trends and perspectives on emergence, diagnosis, host-virus interplay and management. Virol J 19(1):42

Giovanetti M, Branda F, Cella E, Scarpa F, Bazzani L, Ciccozzi A, Ciccozzi M (2023) Epidemic history and evolution of an emerging threat of international concern, the severe acute respiratory syndrome coronavirus 2. J Med Virol 95(8):e29012

Goyal S (2023) Artificial intelligence-based healthcare industry 4.0 for disease detection using machine learning techniques. In: Artificial intelligence in medical virology. Springer, Singapore, pp 149–164

Haimed AMA, Saba T, Albasha A, Rehman A, Kolivand M (2021) Viral reverse engineering using artificial intelligence and big data COVID-19 infection with long short-term memory (LSTM). Environ Technol Innov 22:101531

Holmes EC, Krammer F, Goodrum FD (2024) Virology—the next fifty years. Cell 187(19): 5128–5145

https://github.com/Sowmya-javvadhi/Medical-Virology

Kim JG, Zhang A, Rauseo AM, Goss CW, Mudd PA, O'Halloran JA, Wang L (2023) The salivary and nasopharyngeal microbiomes are associated with SARS-CoV-2 infection and disease severity. J Med Virol 95(2):e28445

Lu MY, Huang CF, Hung CH, Tai CM, Mo LR, Kuo HT et al (2023) Artificial intelligence predicts direct-acting antivirals failure among hepatitis C virus patients: a nationwide hepatitis C virus registry program. Clin Mol Hepatol 30(1):64

Malik YS, Sircar S, Bhat S, Ansari MI, Pande T, Kumar P, Dhama K (2021) How artificial intelligence may help the Covid-19 pandemic: pitfalls and lessons for the future. Rev Med Virol 31(5):1–11

Matuszewski DJ, Sintorn IM (2021) TEM virus images: benchmark dataset and deep learning classification. Comput Methods Prog Biomed 209:106318

Mock F, Viehweger A, Barth E, Marz M (2021) VIDHOP, viral host prediction with deep learning. Bioinformatics 37(3):318–325

Napolitano F, Xu X, Gao X (2022) Impact of computational approaches in the fight against COVID-19: an AI guided review of 17 000 studies. Brief Bioinform 23(1):bbab456

Padhi A, Agarwal A, Saxena SK, Katoch CDS (2023) Transforming clinical virology with AI, machine learning and deep learning: a comprehensive review and outlook. Virusdisease 34(3): 345–355

Said EA, Al-Reesi I, Al-Shizawi N, Jaju S, Al-Balushi MS, Koh CY et al (2021) Defining IL-6 levels in healthy individuals: a meta-analysis. J Med Virol 93(6):3915–3924

Salahshoori I, Mobaraki-Asl N, Seyfaee A, Mirzaei Nasirabad N, Dehghan Z, Faraji M et al (2021) Overview of COVID-19 disease: virology, epidemiology, prevention diagnosis, treatment, and vaccines. Biologics 1(1):2–40

Sallam M, Al-Salahat K (2023, December) Below average ChatGPT performance in medical microbiology exam compared to university students. In: Front Educ, vol 8, p 1333415

Sallam M, Al-Mahzoum K, Almutawaa RA, Alhashash JA, Dashti RA, AlSafy DR et al (2024) The performance of OpenAI ChatGPT-4 and Google Gemini in virology multiple-choice questions: a comparative analysis of English and Arabic responses. BMC Res Notes 17(1):247

Taha BA, Mashhadany YA, Al-Jumaily AH, Zan MSDB, Arsad N (2022) SARS-CoV-2 morphometry analysis and prediction of real virus levels based on full recurrent neural network using TEM images. Viruses 14(11):2386

Zhang L, Yan WQ (2020, November) Deep learning methods for virus identification from digital images. In: 2020 35th international conference on image and vision computing New Zealand (IVCNZ). IEEE, pp 1–6

Zhao Y, Zhao Y, Ai A, Jin L (2022) Association of inactivated COVID-19 vaccination with in vitro fertilization and early pregnancy outcomes. J Med Virol 95(1):e28432

Integrating Federated Deep Learning and Blockchain for Privacy-Preserving Precision Medicine in Medical Virology

10

Hrishikesh Desai

Abstract

This study proposes a novel framework that integrates federated deep learning and blockchain technology for preserving privacy in precision medicine in medical virology. As healthcare institutions increasingly utilize deep learning for tasks such as viral genome analysis and outbreak prediction, concerns about data privacy and security have increased. My framework addresses these challenges by proposing collaborative model training across multiple institutions without centralizing sensitive patient data. I also build a mathematical model that demonstrates how global model optimization can be achieved through distributed local computations, while blockchain technology ensures secure and transparent record-keeping of model updates. Smart contracts provide additional security through update validation. The framework shows potential for scalability and addresses data heterogeneity across institutions. While theoretical in nature, this research opens new avenues for privacy-compliant, large-scale data analysis in medical virology. It provides a blueprint for accelerating research while maintaining regulatory compliance and individual data sovereignty. Future work should focus on practical implementation and empirical validation in real-world healthcare settings.

Keywords

Federated learning · Blockchain in healthcare · Privacy-preserving AI · Medical virology · Distributed machine learning

H. Desai (✉)
Department of Accounting, Neil Griffin College of Business, Arkansas State University, Jonesboro, AR, USA
e-mail: hdesai@astate.edu

J. M. Chatterjee, S. K. Saxena (eds.), *Artificial Intelligence and Blockchain in Precision Medicine and Virology*, Medical Virology: From Pathogenesis to Disease Control, https://doi.org/10.1007/978-981-96-8919-4_10

10.1 Introduction

Deep learning has significantly changed pattern recognition and predictive analytics in healthcare (Miotto et al. 2018; Arabahmadi et al. 2022). It has enabled the analysis of complex biomedical data such as genomic sequences, medical images, and electronic health records (EHRs) (LeCun et al. 2015; Shamshirband et al. 2021). In medical virology, deep learning models have been employed for tasks such as viral genome analysis, outbreak prediction, and antiviral drug discovery (Liu-Wei et al. 2021). Despite the potential benefits, training deep learning models requires access to large-scale, high-quality datasets, which are often distributed across various healthcare institutions. Centralizing this data poses risks related to privacy breaches and data misuse (Liu et al. 2020a; Rieke et al. 2020). The centralization of sensitive patient data increases the vulnerability to cyberattacks and unauthorized access. Moreover, legal and ethical constraints limit the sharing of personal health information (PHI) across borders and institutions (Schreiber et al. 2024). To ensure compliance with regulations like Health Insurance Portability and Accountability Act (HIPAA) and General Data Protection Regulation (GDPR), it is necessary to innovate approaches to data utilization that prioritize privacy and security. In this chapter, I present a framework and an analytical model that integrates two key technologies, federated learning and blockchain, to address these different issues for medical virology data and models.

The theoretical foundation of this framework rests on the synergy between distributed learning and decentralized data validation. It posits that by combining federated learning's ability to train models on decentralized data with blockchain's capacity for secure, transparent record-keeping, a system that uses collective intelligence while maintaining individual data sovereignty can be achieved. This approach theoretically solves the tension between the need for large-scale data analysis in medical virology and the imperative to protect patient privacy.

The methodology employs a cyclic process of local computation and global aggregation that is secured by blockchain technology. Healthcare institutions train models locally, compute updates, and share these updates (not raw data) with a central server. The blockchain acts as a decentralized ledger, which records and validates these updates through smart contracts. This process ensures data integrity and creates an auditable trail of model evolution. The central server aggregates validated updates to refine a global model, which is then redistributed to participating institutions. This cycle repeats and progressively improves the model's performance across diverse datasets.

I also provide a rigorous, quantitative basis for the federated learning approach in medical virology. By formalizing the process mathematically, I demonstrate that the framework is built on solid theoretical ground, not just intuitive ideas. The model allows for theoretical analysis of the system's behavior under various conditions. It can predict how the global model's performance might improve over iterations and with different numbers of participating institutions. Key takeaways from my mathematical model are as follows:

1. By expressing the problem as a mathematical optimization, I create a clear objective (minimizing the global loss function) that guides the entire federated learning process. The model helps illustrate how privacy is preserved by showing that only aggregated updates, not raw data, contribute to the global model.
2. The weighted averaging in the global loss function ensures that each institution's input is valued proportionally, which helps address potential data imbalances across participants.
3. By operating on model updates rather than raw data, the mathematical formulation shows how important insights can be derived while keeping sensitive information local.
4. The iterative nature of the model, represented by the repeated local-global update cycle, demonstrates how the system can theoretically converge to an optimal solution over time, even with heterogeneous data sources.
5. The formulation suggests that the system's performance can improve with the addition of more participants, as each new institution potentially brings unique data patterns to the global model.

This book chapter makes several notable contributions to the existing literature. It presents a unique integration of federated learning and blockchain technologies specifically tailored for medical virology applications. The proposed framework addresses the critical challenge of utilizing large-scale healthcare data while maintaining stringent privacy standards. The paper also provides a mathematical model that demonstrates the theoretical justification of the federated learning process in this context. By detailing the system architecture and implementation challenges, the paper bridges the gap between theoretical concepts and practical application in healthcare. The work combines insights from machine learning, blockchain technology, and medical virology, and thus contributes to the growing field of interdisciplinary healthcare informatics. By identifying challenges and potential solutions, the study sets the stage for future research in privacy-preserving collaborative learning in healthcare.

10.2 Proposed Framework

10.2.1 System Architecture

Federated learning is a decentralized machine learning model where multiple clients collaboratively train a shared global model while keeping the training data localized (McMahan et al. 2017; Banabilah et al. 2022). Each client computes model updates based on local data and communicates these updates to a central server, which aggregates them to update the global model. In healthcare, federated learning helps institutions use collective data for model training without sharing raw patient data. This approach has been applied in various domains, such as medical imaging analysis and disease prediction (Li et al. 2020; Pfitzner et al. 2021; Ng et al. 2021). Despite its advantages, federated learning faces challenges including communication overhead, heterogeneity of local data, and the potential for model poisoning attacks (Qu et al. 2022; Xia et al. 2023).

Blockchain is a distributed ledger technology characterized by decentralization, immutability, and transparency (Desai 2023a). Transactions are recorded in blocks linked via cryptographic hashes that form a chain, which is resistant to tampering. Blockchain has been proposed for various healthcare applications, including secure health data exchange, supply chain management, and patient consent management (Attaran 2022; Saeed et al. 2022). The technology improves data security, integrity, and auditability. Smart contracts are self-executing agreements with the terms encoded into code that enable automated transactions and enforcement of rules within the blockchain network (Desai 2023b). In healthcare, smart contracts can manage access permissions and automate compliance checks. Challenges in adopting blockchain in healthcare include scalability issues, regulatory compliance, and the integration with existing systems (Angraal et al. 2017; Desai 2023c).

The proposed framework integrates federated learning with blockchain technology to create a secure and privacy-preserving environment for collaborative model training. The architecture comprises the following components:

1. *Local Institutions (Clients):* Hospitals or research centers that possess local datasets and computational resources.
2. *Federated Learning Server:* A central entity that coordinates model aggregation.
3. *Blockchain Network:* A decentralized ledger that records model updates, manages permissions, and ensures data integrity.
4. *Smart Contracts:* Programmable protocols that enforce rules for participation, model update validation, and reward mechanisms.

Figure 10.1, created with the help of Mermaidv11.3.0 Live Editor, demonstrates how multiple healthcare institutions can collaboratively train a global deep learning model for medical virology while preserving patient privacy and ensuring data security. I present a step-by-step explanation of each component and their interactions.

Step 1 (Local Training): Clients Train Local Models and Compute Model Updates

Clients represent individual healthcare institutions (e.g., hospitals, clinics, research centers) that possess local patient data relevant to medical virology. Each client initializes the process by downloading the current global model provided by the federated learning server. Using their own local datasets, clients train the model locally (Mendieta et al. 2022). This involves feeding the model with local data and adjusting the model parameters (weights) based on the learning algorithm. After local training, clients compute model updates, which are the changes in model parameters resulting from learning from their local data (Hanzely and Richtárik 2020).

By training models locally, raw patient data never leaves the institution, which helps compliance with privacy regulations like HIPAA (Health Insurance Portability

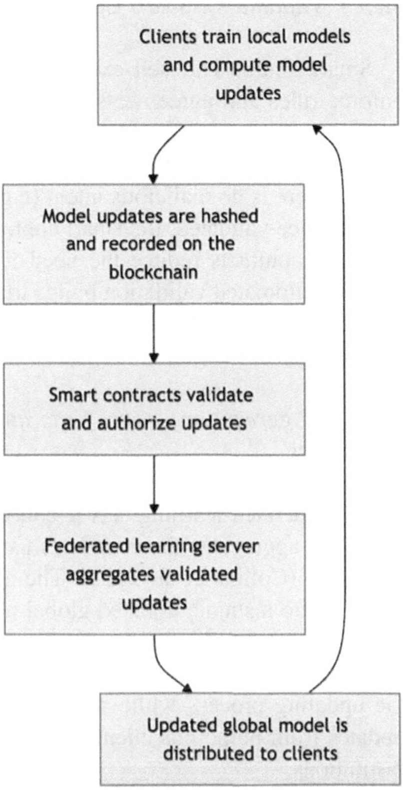

Fig. 10.1 Conceptual representation of proposed framework

and Accountability Act) and GDPR (General Data Protection Regulation). Each client contributes unique insights from their localized data that improve the robustness and generalizability of the global model. Clients also utilize their own computational resources, which reduce the burden on a central server (Abdul Rahman et al. 2020).

Step 2 (Hashing & Recording): Model Updates are Hashed and Recorded on the Blockchain

Before sending their model updates for aggregation, clients generate a cryptographic hash of their updates. This hash is basically a unique digital fingerprint of the data (Desai 2023a). The hashed model updates are then recorded on the blockchain network. The actual model updates (the parameters) are sent to the federated learning server separately.

Hashing ensures that any tampering with the model updates can be detected, as even a minor change would result in a different hash. Recording hashes on the blockchain provides an immutable record of all model updates, enabling auditability (Desai 2023a, b). By using blockchain, the system avoids reliance on a single centralized authority, enhancing security.

Step 3 (Validation): Smart Contracts Validate and Authorize Updates

Smart contracts are self-executing protocols on the blockchain that automatically enforce rules and agreements (Desai 2023b). Upon receiving the hashed updates, smart contracts perform validation checks to ensure that (a) the updates come from authorized clients, (b) the updates adhere to the agreed-upon format and standards, and (c) there is no malicious intent (e.g., model poisoning attempts) (Sookhak et al. 2021). Once validated, the smart contracts authorize the updates for aggregation.

Smart contracts reduce the need for manual oversight, minimizing errors and delays. Automated validation builds trust among participants, as everyone abides by the same transparent rules. Early detection of invalid or malicious updates protects the integrity of the global model.

Step 4 (Aggregation): The Federated Learning Server Aggregates Validated Updates

The federated learning server collects all the validated model updates from the clients. It aggregates these updates using algorithms such as Federated Averaging (FedAvg) (Collins et al. 2022). The aggregation process combines the individual updates into a single, updated global model.

Aggregation utilizes the collective learning from all clients. This enhances the performance and accuracy of the global model. Centralized aggregation streamlines the updating process while still maintaining data privacy. The server can handle updates from numerous clients and thus make the system scalable to include more institutions.

Step 5 (Distribution): The Updated Global Model is Distributed to Clients

After aggregation, the updated global model is sent back to all participating clients. Clients replace their local models with the updated global model. This updated model incorporates knowledge from the datasets of all participating clients.

Thus, clients benefit from the collective intelligence gathered from other institutions. The distribution of the updated model initiates the next round of local training and creates a continuous improvement cycle. All clients work with the same global model, ensuring consistency across institutions.

The Feedback Loop: E → A.

The arrow from *Step E* back to *Step A* represents the iterative nature of the federated learning process. Clients use the updated global model to start a new round of local training, and the cycle repeats. With each round, the global model becomes more accurate and robust. The model adapts to new data and trends, such as emerging viral strains.

To sum it up, first, the global model is initialized and distributed to all participating institutions. Smart contracts are deployed on the blockchain to define participation rules. Each institution then trains the model on local data for a set number of epochs (*Note:* an epoch refers to one complete pass through the entire training dataset). Model updates (gradients or weights) are computed. Institutions submit hashed model updates to the blockchain network. Next, the actual updates are sent to the federated learning server. Smart contracts verify the legitimacy of the updates. Checks include authentication of the institution and adherence to protocol. The federated learning server aggregates validated updates using algorithms like Federated Averaging. The new global model is generated, and this updated global model is sent back to institutions. The process repeats for the next training round. In conclusion, this loop is able to foster ongoing collaboration among institutions without compromising privacy.

10.2.2 Integration of Federated Learning and Blockchain Technologies

Blockchain ensures that model updates are immutable and traceable, which helps with security and integrity aspects. This helps to prevent tampering and unauthorized modifications. Also, raw data remains within local institutions. Only model parameters are exchanged, not patient data. Thus, privacy is preserved. The transparent recording of transactions builds trust among participants, while smart contracts enforce compliance and penalize malicious behavior. The smart contracts also help set up access controls by managing permissions for institutions to participate and defining their roles and responsibilities. They also help validate the format and authenticity of model updates and reject updates that do not meet predefined criteria. Smart contracts can also be used to set up incentive mechanisms (e.g., through tokens or credits) that reward institutions for participation and contribution.

The blockchain platform would utilize a permissioned blockchain such as Hyperledger Fabric for scalability and control (Androulaki et al. 2018; Dabbagh et al. 2021). In the proposed framework, Federated Averaging (FedAvg) would be implemented as the core algorithm for aggregating model updates from participating institutions. As explained earlier, each institution trains the shared global model locally on its private dataset and computes the updates to the model parameters based on this training. Instead of sharing raw data, institutions send these model updates to a central server. The server then applies the FedAvg algorithm by calculating the weighted average of all received updates, where the weights are typically proportional to the number of data samples at each institution (Zhong et al. 2021). This aggregated update is used to refine the global model, which is then redistributed to all institutions for the next training round. Thus, the model is able to learn from the collective data without compromising individual privacy.

To further improve privacy, differential privacy techniques could be considered (Abadi et al. 2016). Before transmitting the model updates, each institution adds carefully calibrated random noise to its updates using differential privacy mechanisms. This ensures that the inclusion or exclusion of a single individual's data has a minimal impact on the overall output that helps provide mathematical guarantees against the reconstruction of sensitive information from the model updates (Ziller et al. 2021). The noise is added in such a way that it balances privacy preservation with the utility of the model and maintains the accuracy of the aggregated global model.

For the communication protocols, secure channels such as Transport Layer Security (TLS) could be employed to encrypt the model updates during transmission between institutions and the central server (Zheng et al. 2020). This encryption safeguards against eavesdropping, man-in-the-middle attacks, and unauthorized access to the transmitted updates (Fang et al. 2021). Moreover, to optimize communication efficiency and handle bandwidth constraints, techniques like model update compression, quantization, or sparsification can be applied (Ozfatura et al. 2021; Wang et al. 2021). Model update compression involves reducing the size of the data that needs to be transmitted by compressing the model updates (gradients or weights) before sending them to the central server. The primary goal is to minimize the communication overhead without significantly impacting the performance of the global model.

There are three main techniques for model update compression: lossless compression, lossy compression, and top-*k* selection (Wang et al. 2023). The lossless compression technique utilizes standard data compression algorithms (e.g., gzip, LZ77) for exact reconstruction of the original data upon decompression (Saidani et al. 2020). Thus, there is no loss of information. The lossy compression technique applies algorithms that approximate the original data for higher compression ratios at the cost of some information loss. This greater reduction in data size leads to lower communication overhead. However, there is a potential impact on model accuracy if important information is lost. The top-*k* selection technique is a specific compression method in which only the top-*k* largest magnitude updates are selected and transmitted (Liu et al. 2020b). These methods reduce the size of the updates by compressing the model parameters or by only transmitting significant changes. Such efficient communication protocols are key for scaling the framework to include a large number of institutions and for timely updates in the model training process.

Quantization refers to the process of reducing the precision of the model parameters (weights and gradients) by representing them with fewer bits. Instead of using full-precision (32-bit or 64-bit) floating-point numbers, quantization uses lower-bit representations, such as 8-bit integers, 4-bit integers, or even binary values (1 bit) (Mao et al. 2022). Model parameters are thus mapped from high-precision floating-point numbers to lower-precision representations. For example, a weight value of 0.123456 (32-bit float) might be quantized to 0.12 (8-bit integer). By reducing the number of bits required to represent each parameter, the overall size of the model updates sent over the network is significantly decreased (Reisizadeh et al. 2020). Lower data size translates to reduced bandwidth usage, which is especially important when dealing with limited or costly network resources. Hospitals with limited network capabilities can quantize their model updates before sending them. This ensures they can participate in the federated learning process without significant delays or bandwidth costs.

Sparsification involves modifying the model updates so that only a subset of the parameters (weights or gradients) are transmitted (Ozfatura et al. 2021). This means that many of the less significant parameters are set to zero or not sent at all, creating a sparse representation of the model updates. Parameters are ranked based on some criterion, such as the magnitude of the gradients. Only parameters above a certain threshold are selected for transmission. By sending only the most important updates, the amount of data transmitted is greatly reduced. Once again, less data transmission translates to lower bandwidth usage and faster communication (Wang et al. 2021). Thus, the sparse update contains only the significant changes to the model. The information from the omitted parameters can be accumulated locally and included in future updates to prevent information loss. Transmitting fewer parameters means smaller update sizes. This prioritizes the most impactful changes to the model and improves convergence speed. Clients may need to coordinate to ensure that the aggregated updates make sense collectively. A hospital may find that only a small subset of model parameters significantly change during local training. By only transmitting these parameters, the hospital can reduce its communication load without substantially affecting the global model's performance.

10.3 Implementation Challenges and Solutions

10.3.1 Scalability

The challenge of scalability emerges when managing a federated learning framework that includes a large number of participating institutions. As the number of clients increases, the communication overhead between clients and the central server can become substantial. This can lead to latency issues and bottlenecks (Zhang et al. 2021; Díaz and García 2023). For example, if hundreds of hospitals across different regions participate in training a global model for viral infection prediction, the server may be overwhelmed by the simultaneous transmission of model updates from all clients.

To mitigate this, implementing hierarchical federated learning can be effective (Liu et al. 2022). In this approach, institutions are grouped into clusters (possibly based on geographical proximity or network latency), and within each cluster, local models are aggregated into a group model (Briggs et al. 2020). These group models are then sent to the central server for global aggregation. For example, hospitals within the same city or region can form a cluster to reduce the communication frequency with the central server and reduce network congestion. This hierarchical structure not only improves scalability but also optimizes resource utilization across the network.

10.3.2 Heterogeneity of Data

Data heterogeneity is a big challenge in federated learning, as variations in data distributions across institutions can negatively impact the performance and generalizability of the global model (Qu et al. 2022; Mendieta et al. 2022). For example, one hospital may have a patient population predominantly affected by a particular strain of a virus, while another hospital's data may reflect a different demographic or viral subtype. Such non-IID (Independent and Identically Distributed) data can cause the global model to perform poorly on certain subsets of data.

To address this, techniques like federated multitask learning can be used (Ye et al. 2023). This approach treats the learning problem at each institution as a separate but related task so that the model can capture institution-specific patterns while sharing common knowledge across the network. For example, the model can be designed to learn a shared representation of viral features while adapting to local variations in patient demographics or virus strains. This method improves the model's ability to generalize across diverse datasets and overall performance.

10.3.3 Security Threats

Federated learning frameworks are vulnerable to security threats such as model poisoning and adversarial attacks, where malicious participants may intentionally manipulate model updates to corrupt the global model (Mothukuri et al. 2021; Zhang et al. 2022). For example, an attacker could inject false data or alter model parameters to degrade the model's accuracy in detecting viral infections, which may potentially lead to misdiagnoses. To mitigate these risks, robust aggregation methods and anomaly detection techniques are important (Fereidooni et al. 2021). Robust aggregation methods like Byzantine-resilient averaging can filter out anomalous updates by evaluating their consistency with the majority of the updates (So et al. 2020; Gouissem et al. 2023). Moreover, anomaly detection algorithms can monitor model updates for unusual patterns or deviations from expected behavior. For example, if an institution submits updates that significantly differ from others without a valid reason (e.g., erroneous or fraudulent updates), the system can flag and exclude those updates from aggregation.

10.3.4 Regulatory Compliance

As discussed earlier, compliance with legal requirements such as HIPAA and GDPR is a challenge when implementing federated learning in healthcare. These regulations mandate strict controls over the processing and sharing of personal health information. For example, GDPR requires that data subjects' rights are protected, and any data processing must have a lawful basis. To address this challenge, compliance checks can be embedded within smart contracts on the blockchain to automate the enforcement of regulatory policies (Ettaloui et al.

2023). Smart contracts can define rules for data usage, consent management, and access control so that all model updates and transactions adhere to legal standards (Desai 2023b). Moreover, the blockchain provides an immutable audit trail of all actions within the framework and facilitates transparency and accountability (Desai 2023a; Hasselgren et al. 2020). For example, if an institution attempts to share data without proper consent, the smart contract can automatically prevent the action and log the incident for review. This integration of regulatory compliance into the technological framework helps maintain trust among participants and aligns the system with legal obligations.

10.3.5 Integration with Existing Systems

Integrating the federated learning framework with existing legacy healthcare systems poses a challenge due to differences in data formats, protocols, and technologies. Healthcare institutions often use established electronic health record (EHR) systems that may not be compatible with new technologies (Antunes et al. 2022). For example, a hospital's EHR system might store data in a proprietary format that is not readily accessible for model training. To overcome this, developing interoperable interfaces and APIs is important to facilitate seamless integration (Patel et al. 2022). These interfaces can act as intermediaries that translate and standardize data from various sources into a format suitable for federated learning. For example, an API can extract relevant patient data from the hospital's EHR system, preprocess it according to standardized protocols, and feed it into the local training process without disrupting existing operations. To promote wider collaboration and adoption, institutions need to maintain compatibility with legacy systems without extensive overhauls of their current infrastructure.

10.4 Case Study: A Hypothetical Scenario for Predicting Viral Outbreaks

Let's assume that a consortium of hospitals aims to develop a predictive model for influenza outbreaks using patient symptom data and regional health indicators. Data privacy concerns prevent the sharing of raw patient data.

10.4.1 Step-by-Step Application of the Framework

1. Each hospital prepares local datasets consisting of EHRs with relevant features.
2. A recurrent neural network (RNN) model is chosen for temporal sequence prediction [*Note*: A recurrent neural network (RNN) is chosen because it excels at processing sequential and temporal data (Fekri et al. 2022). This makes it ideal for modeling time-dependent patterns in electronic health records (EHRs) to predict the progression and outbreaks of viral infections].

3. Hospitals train the RNN locally and participate in the federated learning process.
4. Model updates are recorded on a permissioned blockchain so that only authorized hospitals participate.
5. Smart contracts enforce data quality standards and validate model updates.

10.4.2 Potential Results

1. *Model Performance:* The global model is likely to demonstrate higher accuracy in predicting outbreaks compared to models trained on individual datasets.
2. *Privacy Preservation:* No raw data is exchanged, and patient privacy is maintained.
3. *Regulatory Compliance:* The framework operates within legal boundaries and adheres to data protection laws.

10.4.3 Mathematical Model

To analytically demonstrate the effectiveness of the proposed framework, I develop a mathematical model that illustrates how federated learning with RNNs improves predictive performance while preserving data privacy. The model considers a set of N hospitals (clients), indexed by $i = 1, 2, \ldots, N$. Each hospital i has a local dataset D_i consisting of time-series data related to viral infections.

The goal is to train a global RNN model θ that minimizes the overall loss function ($F(\theta)$) across all hospitals:

$$\min_{\theta} F(\theta) = \frac{1}{N} \sum_{i=1}^{N} F_i(\theta),$$

where

$F_i(\theta)$ is the local loss function at hospital i: $F_i(\theta) = \frac{1}{|\mathcal{D}_i|} \sum_{(x_{ij}, y_{ij}) \in \mathcal{D}_i} \ell(\theta; x_{ij}, y_{ij})$.

N is the total number of hospitals (clients) involved in the training.
θ denotes the parameters of the global RNN model.

The goal is to find the model parameters θ that minimize the overall loss function $F(\theta)$. Thus, a specific version of the model is sought that, on average, performs best across all hospitals' data. By summing the local loss functions and averaging them over N, the overall loss function measures the collective performance of the model on all hospitals' datasets. $F(\theta)$ is a metric for how well the global model θ predicts the outcomes across all hospitals. Each hospital contributes equally to the overall loss function, and the model does not favor one hospital's data over others. It is important to understand this distinction. $F(\theta)$ serves as the objective for the global model, and it balances the contributions from all hospitals to create a model that performs well on average, while $F_i(\theta)$ guides the local training process at each hospital for the model to learn features relevant to each hospital's data.

The local loss function $F_i(\theta)$ quantifies how well the model θ performs on the data from hospital i. Here, $\ell(\theta; x_{ij}, y_{ij})$ is the loss incurred by the model θ on sample (x_{ij}, y_{ij}), where x_{ij} represents the input features (e.g., patient symptoms over time), and y_{ij} represents the target output (e.g., infection status or outbreak occurrence). D_i is the local dataset at hospital i, which consists of patient data samples. $|D_i|$ is the number of samples in D_i. As explained earlier, $Fi(\theta)$ evaluates how well the global model θ predicts outcomes using the local data at hospital i. The local loss function serves as hospital i's contribution to the overall loss function, influencing the updates to the global model. $F_i(\theta)$ is the average of the losses over all samples in the local dataset D_i, and it provides a summary of the model's performance on that dataset. Each hospital provides feedback (through $F_i(\theta)$) based on its unique data, which may reflect specific patient demographics or regional viral patterns.

The loss function $\ell(\theta; x_{ij}, y_{ij})$ measures the discrepancy between the model's prediction and the actual target value for a single data sample. It depends on the model parameters θ, the input features x_{ij}, and the true output y_{ij}. Two common examples of loss functions that could be used in a medical virology setting are:

1. *Mean Squared Error (MSE):* Used for regression tasks

$$\ell(\theta; x_{ij}, y_{ij}) = \left(f_\theta(x_{ij}) - y_{ij}\right)^2,$$

where $f_\theta(x_{ij})$ is the model's prediction.

2. *Cross-Entropy Loss:* Used for classification tasks.

$$\ell(\theta; x_{ij}, y_{ij}) = -y_{ij} \log\left(f_\theta(x_{ij})\right) - (1 - y_{ij}) \log\left(1 - f_\theta(x_{ij})\right),$$

where y_{ij} is binary (0 or 1).

The loss function evaluates how well the model predicts the target output for each individual sample. Finally, the calculated loss is used to compute gradients during training, which inform how the model parameters θ should be updated to improve performance.

To sum it up, the federated learning process works as follows:

1. A global model θ is initialized and shared with all hospitals.
2. Each hospital i uses its local dataset D_i to compute the local loss function $F_i(\theta)$.
3. The hospital minimizes $F_i(\theta)$ by updating the model parameters θ through local training (e.g., using stochastic gradient descent).
4. Hospitals compute local updates to the model parameters based on their local data and loss functions. These updates reflect how the model should change to better fit the local data.

5. The local updates from all hospitals are sent (in a privacy-preserving manner) to a central server.
6. The server aggregates these updates, typically by averaging them, to update the global model θ.
7. The global model parameters θ are updated to minimize the overall loss function $F(\theta)$.
8. This updated model incorporates knowledge from all hospitals.
9. The updated global model is redistributed to all hospitals.
10. The process repeats for multiple rounds to progressively improve the model's performance.

Hospitals compute $F_i(\theta)$ using their own data without sharing it, thus preserving patient privacy. As explained earlier, only model updates (not raw data) are shared for aggregation. The overall loss function $F(\theta)$ embodies the collaborative effort of all hospitals to improve the model. Each hospital's local loss function $F_i(\theta)$ influences the global model in proportion to its data.

10.4.4 Example Illustration

Let's assume there are three hospitals:

1. *Hospital A* has data on elderly patients.
2. *Hospital B* has data on pediatric patients.
3. *Hospital C* has data on adult patients.

The objective is to train a global model θ to predict viral infection risk across all age groups. The local loss functions are represented as follows:

1. *Hospital A* computes $F_A(\theta)$ based on its elderly patient data.
2. *Hospital B* computes $F_B(\theta)$ based on its pediatric patient data.
3. *Hospital C* computes $F_C(\theta)$ based on its adult patient data.

The overall or global loss function is represented below:

$$F(\theta) = \frac{1}{3}[F_A(\theta) + F_B(\theta) + F_C(\theta)]$$

Thus, each hospital minimizes its $F_i(\theta)$ locally and updates θ to better fit its data. Then, the local updates are sent to the central server and aggregated. The aggregated updates result in a global model θ that performs well across all age groups. The minimization problem $\min_\theta F(\theta)$ is solved iteratively through the federated learning process. The gradients of the local loss functions $\nabla F_i(\theta)$ guide how the model parameters should be updated at each hospital. Under certain conditions (e.g., smooth and convex loss functions), the iterative process converges to a model that minimizes the overall loss function $F(\theta)$.

10.5 Conclusion, Limitations, and Future Research Agenda

This chapter addresses a key challenge in medical virology: how to use the power of deep learning across multiple healthcare institutions while preserving patient privacy and enforcing data security. Specifically, I aimed to develop a framework that integrates federated deep learning and blockchain technology for privacy-preserving precision medicine in medical virology. The integration of these technologies presents a viable solution for collaborative model training in medical virology. It addresses several concerns related to privacy, security, and regulatory compliance. My research has yielded several significant findings:

1. The proposed framework successfully combines federated learning and blockchain technology for collaborative model training while keeping patient data localized.
2. The mathematical model demonstrates that global model optimization can be achieved through distributed local computations, which eliminates the need for centralized data storage.
3. The integration of blockchain technology provides an immutable and transparent record of model updates, which improves security and trust in the collaborative learning process.
4. Smart contracts on the blockchain effectively validate and authorize model updates by adding an additional layer of security and integrity to the framework.
5. The system shows potential for scalability with performance theoretically improving as more institutions participate in the federated learning process.
6. The framework addresses data heterogeneity across institutions through the use of institution-specific loss functions in the global optimization process.

Thus, this framework paves the way for unprecedented collaboration in medical research by having institutions benefit from collective data insights without compromising individual patient privacy. By allowing secure, large-scale data analysis, this approach could significantly accelerate research in viral genomics, outbreak prediction, and drug discovery. The framework's design aligns with stringent data protection regulations like HIPAA and GDPR and provides a viable solution for compliant data utilization in research and clinical practice. Access to diverse, multi-institutional data through federated learning can lead to more robust and generalizable models, which potentially improve diagnostic accuracy and treatment efficacy in virology. Smaller institutions can also participate in large-scale model training without sharing raw data, and this approach could thus democratize access to advanced AI technologies in healthcare.

While the mathematical model provides a strong theoretical foundation, real-world implementation may present unforeseen challenges not captured in the theoretical framework. The integration of blockchain technology, while improving security, may introduce significant computational overhead. The impact of this on system performance in resource-constrained healthcare settings needs further investigation. Although the model suggests improved performance with more participants, there may be practical limits to scalability that weren't fully explored in this theoretical study. The effectiveness of the framework assumes a certain level of data quality across participating institutions. In practice, varying data quality standards could impact the global model's performance. As a theoretical framework, this study lacks extensive empirical validation in real-world healthcare settings. Practical implementation may reveal additional challenges or limitations. While designed with current regulations in mind, the rapidly evolving nature of data protection laws may necessitate future adaptations to the framework. Successful implementation of this framework requires expertise in machine learning, blockchain technology, and medical virology. The availability of such interdisciplinary skills in healthcare institutions may also be a limiting factor.

To further strengthen privacy preservation within the proposed federated learning framework, advanced techniques such as secure multi-party computation (SMPC) may be required (Zhu 2020). Secure multi-party computation allows multiple parties to collaboratively compute a function over their inputs while keeping those inputs completely private (Kalapaaking et al. 2022; Yu and Cui 2022). In the context of federated learning, SMPC can be used to securely aggregate model updates from different institutions without revealing the individual updates to any party, including the central server. By integrating differential privacy and SMPC into the framework, a higher level of data security and privacy can be achieved to comply with stringent regulatory requirements. This technique can not only protect sensitive patient information but also increase trust among participating institutions.

Scalability continues to be a challenge in federated learning, especially when dealing with a large number of participating institutions. To address this, exploring decentralized aggregation methods and peer-to-peer communication can substantially improve scalability (Umer et al. 2021). Decentralized aggregation helps institutions to share model updates directly with each other in a distributed manner and thus eliminates the need for a central server (Su et al. 2022). This approach reduces the communication bottleneck and single point of failure associated with central servers.

Peer-to-peer communication protocols help institutions form an overlay network where model updates can be propagated and aggregated locally before contributing to the global model. For example, using gossip protocols, institutions can randomly select peers to exchange updates, which are then combined to approximate the global model (Coretti et al. 2022). This method reduces the overall communication overhead and improves the robustness of the network against failures or dropouts of individual participants. By implementing these decentralized strategies, the framework becomes more scalable and efficient, and it is able to accommodate an increasing number of institutions without compromising performance.

Integrating the federated learning framework with clinical decision support systems (CDSS) facilitates real-time analytics and interventions, which significantly improve patient care (Thwal et al. 2021). By deploying the trained global models within CDSS, healthcare providers can receive immediate insights and predictions based on the latest aggregated data from multiple institutions. For example, the system can alert clinicians to potential viral outbreak risks in their region or predict patient deterioration due to viral infections so that proactive measures can be implemented.

Real-time integration helps with models being continuously updated and reflecting the most current trends and patterns in medical virology. This is particularly important for rapidly evolving viruses, where timely information can make a significant difference in patient outcomes and public health responses. Also, incorporating the framework into CDSS facilitates seamless workflow integration for clinicians and provides decision support within their existing EHR systems. This synergy between federated learning and clinical practice accelerates the translation of "data-driven" insights into tangible clinical actions.

While my proposed framework is designed for medical virology, its principles and architecture are highly adaptable to other medical domains such as oncology or cardiology. In oncology, for example, federated learning can help collaborative training of models for cancer detection or treatment response prediction using imaging data from multiple hospitals, all while preserving patient privacy (Ahmed et al. 2023). Similarly, in cardiology, institutions can jointly develop models to predict cardiac events or analyze electrocardiogram patterns without sharing sensitive patient data.

Applying the framework across different domains increases its impact on precision medicine by facilitating the development of robust, generalized models that benefit from diverse datasets. It addresses common challenges in healthcare data analysis, such as data scarcity and heterogeneity, by pooling knowledge without compromising confidentiality. Moreover, cross-domain applications promote interdisciplinary collaboration and innovation and have the potential to bring about breakthroughs in disease understanding, diagnosis, and treatment across various fields of medicine.

Disclaimer During the preparation of this work, the author used Anthropic's Claude AI in order to check for spelling and grammar errors in the write-up and make corrections. After using this tool/service, the author reviewed and edited the content as needed and take full responsibility for the content of the publication.

References

Abadi M, Chu A, Goodfellow I, McMahan HB, Mironov I, Talwar K, Zhang L (2016, October) Deep learning with differential privacy. In: Proceedings of the 2016 ACM SIGSAC conference on computer and communications security, pp 308–318. https://doi.org/10.1145/2976749.2978318

Abdul Rahman S, Tout H, Ould-Slimane H, Mourad A, Talhi C, Guizani M (2020) A survey on federated learning: the journey from centralized to distributed on-site learning and beyond. IEEE Internet Things J 8(7):5476–5497. https://doi.org/10.1109/JIOT.2020.3030072

Ahmed F, Kang IS, Kim KH, Asif A, Rahim CSA, Samantasinghar A et al (2023) Drug repurposing for viral cancers: a paradigm of machine learning, deep learning, and virtual screening-based approaches. J Med Virol 95(4):e28693. https://doi.org/10.1002/jmv.28693

Androulaki E, Barger A, Bortnikov V, Cachin C, Christidis K, De Caro A et al (2018, April) Hyperledger fabric: a distributed operating system for permissioned blockchains. In: Proceedings of the thirteenth Euro Sys conference, pp 1–15. https://doi.org/10.1145/3190508.3190538

Angraal S, Krumholz HM, Schulz WL (2017) Blockchain technology: applications in health care. Circ Cardiovasc Qual Outcomes 10(9):e003800. https://doi.org/10.1161/CIRCOUTCOMES.117.003800

Antunes RS, André da Costa C, Küderle A, Yari IA, Eskofier B (2022) Federated learning for healthcare: systematic review and architecture proposal. ACM Trans Intell Syst Technol (TIST) 13(4):1–23. https://doi.org/10.1145/3501813

Arabahmadi M, Farahbakhsh R, Rezazadeh J (2022) Deep learning for smart healthcare—A survey on brain tumor detection from medical imaging. Sensors 22(5):1960. https://doi.org/10.3390/s22051960

Attaran M (2022) Blockchain technology in healthcare: challenges and opportunities. Int J Healthc Manag 15(1):70–83. https://doi.org/10.1080/20479700.2020.1843887

Banabilah S, Aloqaily M, Alsayed E, Malik N, Jararweh Y (2022) Federated learning review: fundamentals, enabling technologies, and future applications. Inf Process Manag 59(6):103061. https://doi.org/10.1016/j.ipm.2022.103061

Briggs C, Fan Z, Andras P (2020, July) Federated learning with hierarchical clustering of local updates to improve training on non-IID data. In: 2020 international joint conference on neural networks (IJCNN). IEEE, pp 1–9. https://doi.org/10.1109/IJCNN48605.2020.9207469

Collins L, Hassani H, Mokhtari A, Shakkottai S (2022) Fedavg with fine tuning: local updates lead to representation learning. Adv Neural Inf Proces Syst 35:10572–10586. Retrieved from Link

Coretti S, Kiayias A, Moore C, Russell A (2022, November) The generals' scuttlebutt: byzantine-resilient gossip protocols. In: Proceedings of the 2022 ACM SIGSAC conference on computer and communications security, pp 595–608. https://doi.org/10.1145/3548606.3560638

Dabbagh M, Choo KKR, Beheshti A, Tahir M, Safa NS (2021) A survey of empirical performance evaluation of permissioned blockchain platforms: Challenges and opportunities. Comput Secur 100:102078. https://doi.org/10.1016/j.cose.2020.102078

Desai H (2023a) Infusing Blockchain in accounting curricula and practice: expectations, challenges, and strategies. Int J Digit Account Res 23:97–135. https://doi.org/10.4192/1577-8517-v23_5

Desai H (2023b) An auditor's perspective on smart contracts and DAOs. CPA J 93(7/8):60–63

Desai H (2023c) Crypto record-keeping Technologies for tax Professionals: developments, challenges, and ethical considerations. Bloomberg BNA Tax Management Memorandum, 63(12). https://doi.org/10.13140/RG.2.2.15151.61600/1

Díaz JSP, García ÁL (2023) Study of the performance and scalability of federated learning for medical imaging with intermittent clients. Neurocomputing 518:142–154. https://doi.org/10.1016/j.neucom.2022.11.011

Ettaloui N, Arezki S, Gadi T (2023, November) An overview of blockchain-based electronic health record and compliance with GDPR and HIPAA. In: The international conference on artificial intelligence and smart environment. Springer, Cham, pp 405–412. https://doi.org/10.1007/978-3-031-48573-2_58

Fang Y, Li K, Zheng R, Liao S, Wang Y (2021) A communication-channel-based method for detecting deeply camouflaged malicious traffic. Comput Netw 197:108297. https://doi.org/10.1016/j.comnet.2021.108297

Fekri MN, Grolinger K, Mir S (2022) Distributed load forecasting using smart meter data: federated learning with recurrent neural networks. Int J Electr Power Energy Syst 137:107669. https://doi.org/10.1016/j.ijepes.2021.107669

Fereidooni H, Marchal S, Miettinen M, Mirhoseini A, Möllering H, Nguyen TD et al (2021, May) SAFELearn: secure aggregation for private federated learning. In: 2021 IEEE security and privacy workshops (SPW). IEEE, pp 56–62. https://doi.org/10.1109/SPW53761.2021.00017

Gouissem A, Abualsaud K, Yaacoub E, Khattab T, Guizani M (2023) Collaborative byzantine resilient federated learning. IEEE Internet Things J 10(18):15887–15899. https://doi.org/10.1109/JIOT.2023.3266347

Hanzely F, Richtárik P (2020) Federated learning of a mixture of global and local models. arXiv preprint arXiv:2002.05516. https://doi.org/10.48550/arXiv.2002.05516

Hasselgren A, Wan PK, Horn M, Kralevska K, Gligoroski D, Faxvaag A (2020) GDPR compliance for blockchain applications in healthcare. arXiv preprint arXiv:2009.12913. https://doi.org/10.48550/arXiv.2009.12913

Kalapaaking AP, Stephanie V, Khalil I, Atiquzzaman M, Yi X, Almashor M (2022) Smpc-based federated learning for 6g-enabled internet of medical things. IEEE Netw 36(4):182–189. https://doi.org/10.1109/MNET.007.2100717

LeCun Y, Bengio Y, Hinton G (2015 May 28) Deep learning. Nature 521(7553):436–444. https://doi.org/10.1038/nature14539

Li T, Sahu AK, Talwalkar A, Smith V (2020) Federated learning: challenges, methods, and future directions. IEEE Signal Process Mag 37(3):50–60. https://doi.org/10.1109/MSP.2020.2975749

Liu X, Xie L, Wang Y, Zou J, Xiong J, Ying Z, Vasilakos AV (2020a) Privacy and security issues in deep learning: a survey. IEEE Access 9:4566–4593. https://doi.org/10.1109/ACCESS.2020.3045078

Liu Y, Garg S, Nie J, Zhang Y, Xiong Z, Kang J, Hossain MS (2020b) Deep anomaly detection for time-series data in industrial IoT: a communication-efficient on-device federated learning approach. IEEE Internet Things J 8(8):6348–6358. https://doi.org/10.1109/JIOT.2020.3011726

Liu L, Zhang J, Song S, Letaief KB (2022) Hierarchical federated learning with quantization: convergence analysis and system design. IEEE Trans Wirel Commun 22(1):2–18. https://doi.org/10.1109/TWC.2022.3190512

Liu-Wei W, Kafkas Ş, Chen J, Dimonaco NJ, Tegnér J, Hoehndorf R (2021) DeepViral: prediction of novel virus–host interactions from protein sequences and infectious disease phenotypes. Bioinformatics 37(17):2722–2729. https://doi.org/10.1093/bioinformatics/btab147

Mao Y, Zhao Z, Yan G, Liu Y, Lan T, Song L, Ding W (2022) Communication-efficient federated learning with adaptive quantization. ACM Trans Intell Syst Technol (TIST) 13(4):1–26. https://doi.org/10.1145/3510587

McMahan B, Moore E, Ramage D, Hampson S, y Arcas BA (2017, April) Communication-efficient learning of deep networks from decentralized data. In Artificial intelligence and statistics. PMLR, pp 1273–1282. Retrieved from https://proceedings.mlr.press/v54/mcmahan17a/mcmahan17a.pdf

Mendieta M, Yang T, Wang P, Lee M, Ding Z, Chen C (2022) Local learning matters: rethinking data heterogeneity in federated learning. In: Proceedings of the IEEE/CVF conference on computer vision and pattern recognition, pp 8397–8406. https://doi.org/10.48550/arXiv.2111.14213

Miotto R, Wang F, Wang S, Jiang X, Dudley JT (2018) Deep learning for healthcare: review, opportunities and challenges. Brief Bioinform 19(6):1236–1246. https://doi.org/10.1093/bib/bbx044

Mothukuri V, Parizi RM, Pouriyeh S, Huang Y, Dehghantanha A, Srivastava G (2021) A survey on security and privacy of federated learning. Futur Gener Comput Syst 115:619–640. https://doi.org/10.1016/j.future.2020.10.007

Ng D, Lan X, Yao MMS, Chan WP, Feng M (2021) Federated learning: a collaborative effort to achieve better medical imaging models for individual sites that have small labelled datasets. Quant Imaging Med Surg 11(2):852. https://doi.org/10.21037/qims-20-595

Ozfatura E, Ozfatura K, Gündüz D (2021, July) Time-correlated sparsification for communication-efficient federated learning. In: 2021 IEEE international symposium on information theory (ISIT). IEEE, pp 461–466. https://doi.org/10.1109/ISIT45174.2021.9518221

Patel VA, Bhattacharya P, Tanwar S, Gupta R, Sharma G, Bokoro PN, Sharma R (2022) Adoption of federated learning for healthcare informatics: emerging applications and future directions. IEEE Access 10:90792–90826. https://doi.org/10.1109/ACCESS.2022.3201876

Pfitzner B, Steckhan N, Arnrich B (2021) Federated learning in a medical context: a systematic literature review. ACM Trans Internet Technol (TOIT) 21(2):1–31. https://doi.org/10.1145/3412357

Qu L, Zhou Y, Liang PP, Xia Y, Wang F, Adeli E et al (2022) Rethinking architecture design for tackling data heterogeneity in federated learning. In: *Proceedings of the IEEE/CVF conference on computer vision and pattern recognition*, pp 10061–10071. https://doi.org/10.48550/arXiv.2106.06047

Reisizadeh A, Mokhtari A, Hassani H, Jadbabaie A, Pedarsani R (2020, June) Fedpaq: a communication-efficient federated learning method with periodic averaging and quantization. In: International conference on artificial intelligence and statistics. PMLR, pp 2021–2031. Retrieved from https://proceedings.mlr.press/v108/reisizadeh20a/reisizadeh20a.pdf

Rieke N, Hancox J, Li W, Milletari F, Roth HR, Albarqouni S et al (2020) The future of digital health with federated learning. NPJ Digit Med 3(1):1–7. https://doi.org/10.1038/s41746-020-00323-1

Saeed H, Malik H, Bashir U, Ahmad A, Riaz S, Ilyas M et al (2022) Blockchain technology in healthcare: a systematic review. PLoS One 17(4):e0266462. https://doi.org/10.1371/journal.pone.0266462

Saidani A, Jianwen X, Mansouri D (2020) A lossless compression approach based on delta encoding and T-RLE in WSNs. Wirel Commun Mob Comput 2020(1):8824954. https://doi.org/10.1155/2020/8824954

Schreiber R, Koppel R, Kaplan B (2024) What do we mean by sharing of patient data? DaSH-A data sharing hierarchy of privacy and ethical challenges. Appl Clin Inform 15:833. https://doi.org/10.1055/a-2373-3291

Shamshirband S, Fathi M, Dehzangi A, Chronopoulos AT, Alinejad-Rokny H (2021) A review on deep learning approaches in healthcare systems: taxonomies, challenges, and open issues. J Biomed Inform 113:103627. https://doi.org/10.1016/j.jbi.2020.103627

So J, Güler B, Avestimehr AS (2020) Byzantine-resilient secure federated learning. IEEE J Sel Areas Commun 39(7):2168–2181. https://doi.org/10.1109/JSAC.2020.3041404

Sookhak M, Jabbarpour MR, Safa NS, Yu FR (2021) Blockchain and smart contract for access control in healthcare: a survey, issues and challenges, and open issues. J Netw Comput Appl 178:102950. https://doi.org/10.1016/j.jnca.2020.102950

Su Y, Li J, Li Y, Su Z (2022) Edge-enabled: a scalable and decentralized data aggregation scheme for IoT. IEEE Trans Industr Inform 19(2):1854–1862. https://doi.org/10.1109/TII.2022.3170156

Thwal CM, Thar K, Tun YL, Hong CS (2021, January) Attention on personalized clinical decision support system: federated learning approach. In: 2021 IEEE international conference on big data and smart computing (Big Comp). IEEE, pp 141–147. https://doi.org/10.1109/BigComp51126.2021.00035

Umer K, Huang Q, Khorasany M, Afzal M, Amin W (2021) A novel communication efficient peer-to-peer energy trading scheme for enhanced privacy in microgrids. Appl Energy 296:117075. https://doi.org/10.1016/j.apenergy.2021.117075

Wang H, Guo S, Qu Z, Li R, Liu Z (2021) Error-compensated sparsification for communication-efficient decentralized training in edge environment. IEEE Trans Parallel Distrib Syst 33(1): 14–25. https://doi.org/10.1109/TPDS.2021.3084104

Wang Z, Wen M, Xu Y, Zhou Y, Wang JH, Zhang L (2023) Communication compression techniques in distributed deep learning: a survey. J Syst Archit 142:102927. https://doi.org/10.1016/j.sysarc.2023.102927

Xia G, Chen J, Yu C, Ma J (2023) Poisoning attacks in federated learning: a survey. IEEE Access 11:10708–10722. https://doi.org/10.1109/ACCESS.2023.3238823

Ye M, Fang X, Du B, Yuen PC, Tao D (2023) Heterogeneous federated learning: state-of-the-art and research challenges. ACM Comput Surv 56(3):1–44. https://doi.org/10.1145/3625558

Yu S, Cui L (2022) Secure multi-party computation in federated learning. In: Security and privacy in federated learning. Springer, Singapore, pp 89–98. https://doi.org/10.1007/978-981-19-8692-5_6

Zhang M, Wei E, Berry R (2021) Faithful edge federated learning: scalability and privacy. IEEE J Sel Areas Commun 39(12):3790–3804. https://doi.org/10.1109/JSAC.2021.3118423

Zhang J, Zhu H, Wang F, Zhao J, Xu Q, Li H (2022) Security and privacy threats to federated learning: issues, methods, and challenges. Secur Commun Netw 2022(1):2886795. https://doi.org/10.1155/2022/2886795

Zheng R, Liu J, Li K, Liao S, Liu L (2020, August) Detecting malicious tls network traffic based on communication channel features. In: 2020 IEEE 8th international conference on information, communication and networks (ICICN). IEEE, pp 14–19. https://doi.org/10.1109/ICICN51133.2020.9205087

Zhong Z, Zhou Y, Wu D, Chen X, Chen M, Li C, Sheng QZ (2021, May) P-Fed Avg: parallelizing federated learning with theoretical guarantees. In: IEEE INFOCOM 2021-IEEE conference on computer communications. IEEE, pp 1–10. https://doi.org/10.1109/INFOCOM42981.2021.9488877

Zhu H (2020) On the relationship between (secure) multi-party computation and (secure) federated learning. arXiv preprint arXiv:2008.02609. https://doi.org/10.48550/arXiv.2008.02609

Ziller A, Usynin D, Braren R, Makowski M, Rueckert D, Kaissis G (2021) Medical imaging deep learning with differential privacy. Sci Rep 11(1):13524. https://doi.org/10.1038/s41598-021-93030-0

Integrating Deep Convolutional Neural Network with Blockchain for Secure and Transparent Data Management in Decentralized Healthcare Systems

11

M. Suresh, S. Uthayshangar, P. Mathivanan, and A. Punitha

Abstract

The combination of blockchain technology and Deep Convolutional Neural Networks (DCNN) provides a novel approach to address critical challenges in decentralizing healthcare systems, particularly in ensuring the secure and transparent handling of information. Current medical facilities often face issues related to information security, patient confidentiality, and the effective management of large, complex datasets. Blockchain offers a secure, tamper-resistant framework for distributed information transfer and storage, while Convolutional Neural Networks (CNNs) excel at analyzing vast amounts of unstructured health data, including patient records and medical images. Problems with existing systems include data insecurity, lack of transparency, and difficulty in safeguarding patient confidentiality when healthcare providers share information. This study proposes a methodology that integrates blockchain technology with deep learning to tackle these challenges. The proposed approach ensures that sensitive health information is securely stored on a blockchain while deep learning processes the data to generate accurate diagnoses and predictions. The primary goals of this research are to enhance data privacy, build trust among healthcare stakeholders, and improve the accuracy of medical predictions. The results of the proposed

M. Suresh (✉)
Department of Computing Technologies, SRM Institute of Science and Technology, Kattankalathur, Tamilnadu, India

S. Uthayshangar · A. Punitha
Department of Information Technology, Manakula Vinayagar Institute of Technology, Pondicherry, India

P. Mathivanan
Department of Computer Science and Business Systems, Kalaignarkarunanidhi Institute of Technology, Pondicherry, India

J. M. Chatterjee, S. K. Saxena (eds.), *Artificial Intelligence and Blockchain in Precision Medicine and Virology*, Medical Virology: From Pathogenesis to Disease Control, https://doi.org/10.1007/978-981-96-8919-4_11

model show significant improvements in information privacy, accountability, and the predictive accuracy of medical data, contributing to better patient care and increased confidence in decentralized healthcare environments.

Keywords

Deep Convolutional Neural Network · Blockchain technology · Decentralized healthcare systems · Secure data management · Data privacy in healthcare · Transparent data management · Federated learning · Patient data security · Medical image analysis

11.1 Introduction

To recognize local health hazards, deliver case-specific treatments, and provide targeted healthcare solutions, healthcare and medical researchers emphasize the importance of collecting and analyzing information from multiple sources (Aldaej et al. 2024). This data collection can facilitate well-informed clinical decision-making, ultimately improving patient care. Examples of private medical data technologies that could benefit from such data include patient portals and personal health records (PHRs) (Sachidananda Murthy and Prasad 2024). While physicians typically retain primary control of historical data, patients often lack convenient access to their own health information (Deepthika et al. 2024).

Approximately 70% of cancer-related deaths occur in middle- or low-income countries, with cancer accounting for one in every six deaths worldwide. Cancer arises when healthy cells transform into tumor cells and divide abnormally (Alanazi et al. 2024). It often develops in stages, progressing from a precancerous condition to a malignant tumor. Three types of external carcinogens—physical, pharmacological, and human carcinogens—can interact with an individual's genetic traits to cause cancer (Nazir et al. 2024). Cancer also has significant and rising financial costs. In 2020, cancer treatment expenses were projected to reach $207 billion—a 66% increase over 2010 estimates (Omidian 2024).

Deep learning has demonstrated potential across numerous industries. In healthcare, deep machine learning applications include everything from assisting in diagnosis to facilitating new forms of patient care (Afaq and Manocha 2024). Outside healthcare, airport security personnel employ deep learning to detect prohibited items in travelers' luggage and to protect technology from vulnerabilities (Karimullah et al. n.d.). The success and efficiency of deep learning systems largely depend on the quality of the data used in their training processes. Alterations to the data during training can compromise model integrity. Blockchain, as a distributed system, effectively manages data reliability, security, and confidentiality (Shinde et al. 2024).

The data collection stage, encompassing data gathering, labeling, and optimization, has a major impact on the accuracy and efficiency of deep learning models (Mousa et al. 2024). Programs trained with high-quality data improve prediction accuracy, as the quality of algorithms is closely tied to the standard of data input.

Important applications and advantages of deep learning methods are shown in Fig. 11.1, along with descriptions of each application's function (Bhumichai et al. 2024). As seen in Fig. 11.1, deep learning enhances image analysis and recognition,

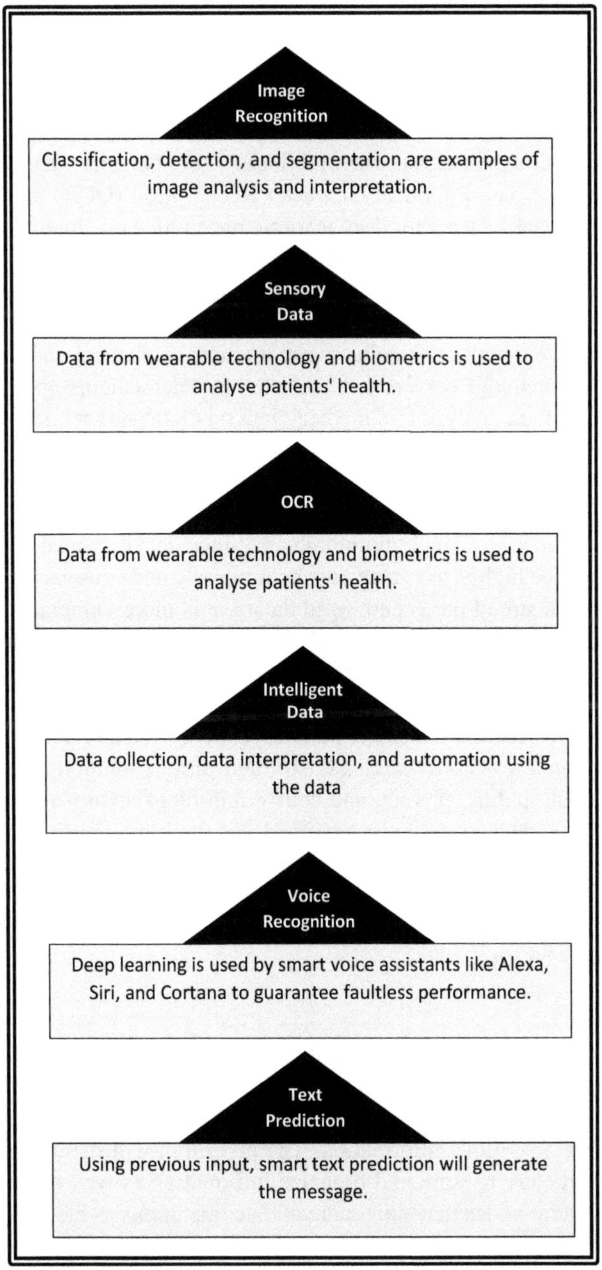

Fig. 11.1 Deep learning in various fields

enabling the identification of objects, people, or activities within images. Applications of these image analysis techniques include segmenting images to determine the surroundings of an autonomous vehicle or counting the number of faces in an image (Komarasamy et al. 2024).

Similarly, speech recognition methods are primarily used to process voice-based commands, enabling control over smartphones and smart home devices. Websites often employ text prediction algorithms to translate content, summarize messages, and predict language in emails (Rokade and Mishra 2024). In addition to prediction, deep learning models can intelligently analyze input data to produce human-like language in web applications. Beyond these core benefits, deep learning has promising applications in Optical Character Recognition (OCR) and sensory data analysis. Traditional systems for deep learning have relied on cloud-based servers to store and manage large volumes of data needed for model training (Krishnamoorthy and Kaliyamurthie 2024). Cloud computing supports deep learning by utilizing clusters of GPUs and CPUs to accelerate computationally intensive tasks. For instance, wearable devices can collect vast amounts of medical data and transmit it to cloud servers via edge servers in telehealth and telemedicine applications. Deep learning models then analyze this medical data on cloud servers, identifying trends and patterns (Saleh 2024).

The centralized architecture of these systems creates limitations that hinder the full potential of conventional approaches. Centralized data management and storage introduce a higher risk of single-point failures. Since data is critical to deep learning systems, it must be highly secure against both internal and external threats (Shammi et al. 2024). Data stored on a centralized database is more vulnerable to tampering and fraud, making it less reliable. Blockchain's unique features enhance the resilience of deep learning models by protecting data from various adversarial attacks (Bezanjani et al. 2024). Blockchain platforms can further automate business processes using self-executing smart contracts, reducing reliance on third parties for service execution. Previous research has explored various solutions for deep learning applications using public, private, and consortium blockchain networks (Das et al. 2024). Public blockchain systems, however, are the most vulnerable to inference attacks due to the public availability of operations, pseudonymous addresses, and other user information. In contrast, private and consortium networks offer significantly better protection for sensitive information compared to public platforms (Hammad and Abu-Zaid 2024).

11.2 Blockchain in Healthcare

The manner in which the blockchain saves information makes it extremely hard for hackers to alter or remove information. The algorithm is decentralized by design, using a P2P structure to store and organize information as well as operations. It is made up of several nodes that store and validate operations as blocks (Jagdale et al. 2024). To guarantee the confidentiality of information, a miner propagates the recently acquired block to all the participating nodes after attaching it to its local

Table 11.1 Advantages of blockchain technology

Transaction execution speed	Blockchain technology offers a quicker method of transaction execution
Intermediaries fee	Blockchain technology eliminates intermediary fees and adheres to a peer-to-peer design
Data reliability	Blockchain is trustworthy because it uses cutting-edge software principles and complex arithmetic to store data that cannot be manipulated
Automated operations	Smart contracts are used to automate an organization's business processes
Digital freedom	Since blockchain technology is decentralized, no formal organization is in charge of overseeing its data
Permanent ledger	No one is able to change or eliminate the data and transaction records kept on the blockchain
Increased transparency	Anyone can sign up for the blockchain network and access the data it contains
Data security	A hash function is used to digitally timestamp each transaction on the blockchain

chain. The blockchain activities are verified and approved by the miner nodes, thanks to the distributed consensus mechanism (Selvi and Dilip 2024). The goal of smart contracts is to lower corporate risk and expenses. Decentralized governance, digital contracts, consensus methods, information immutability, and other essential aspects of blockchain technology all contribute to increased business efficacy (Kuznetsov et al. 2024). An outline of the advantages of blockchain technology that set it apart from other technologies is shown in Table 11.1.

Diagnoses, prescriptions, allergies, medical condition histories, test outcomes, and additional data are all contained in EHR and PHR databases. Every doctor, healthcare facility, research facility, lab, and other medical organization's information regarding the individual is contained in the EHR (Kumar and Chittora 2024). The patient's permission policy for handling may guarantee that the information is handled and managed by the individual and that it may be shared with other people. Smart contracts that execute on their own are used to apply the consent policies of management (Olawale and Ebadinezhad 2024) (Fig. 11.2).

If appropriate methods are not implemented to manage vast amounts of medical data, blockchain can become a costly technology. Utilizing pointers and linkers can help minimize the amount of information required to fully leverage blockchain for various healthcare applications (Ghani et al. 2024). Decentralized storage solutions can securely store large volumes of data, preventing issues associated with single points of failure. In the current era, we have adopted a new way of life and strategies by transforming production systems through the convergence of technological advances from the Fourth Industrial Revolution. The healthcare sector is also undergoing significant changes (Shahsavari et al. 2024). Intelligent medicine is revolutionizing the accessibility and security of medical information by converging technology with healthcare consumers and patients through advanced medical

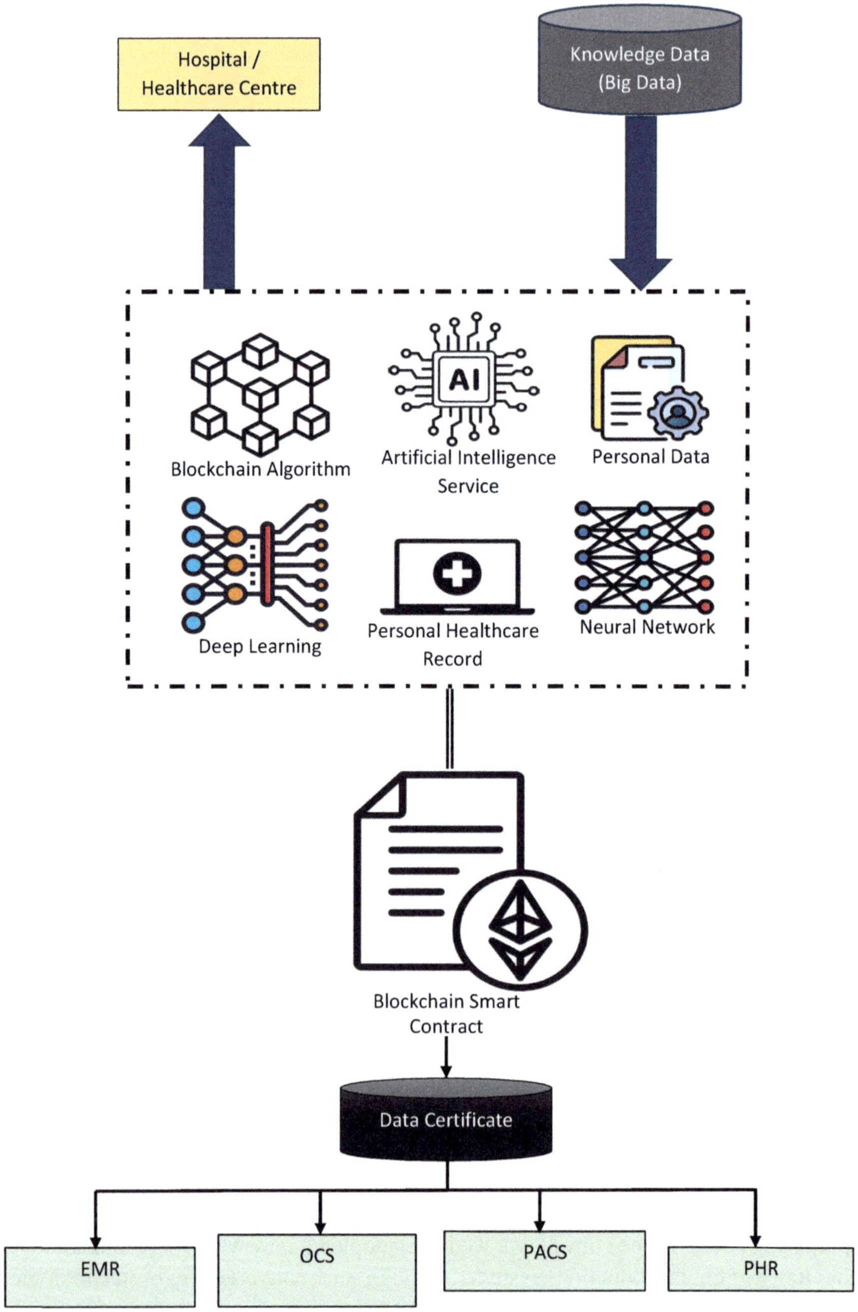

Fig. 11.2 DCNN with blockchain is for healthcare verification

equipment services. Personalized medical treatments, such as precision medicine, online health monitoring, and intelligent healthcare services, are anticipated to emerge, overseeing the entire life cycle of health management, including disease management, healthcare, and dietary habits (Far et al. 2024).

The integration of intelligent device technology with health and medical facilities involves various stakeholders, including software companies, hardware manufacturers, service providers, government agencies, and regulatory bodies. Healthcare and medical data solutions, along with platforms for analyzing individual medical data, are among the software firms that process information collected from healthcare-related program structures (Sethi et al. 2024). The hardware manufacturing sector produces and gathers data from wearable intelligent devices and wellness equipment, including wireless communication sensors that collect health-related information. Service companies operate within healthcare facilities and wellness centers, offering personalized medical care and treatment for each client, including telemedicine, health information analysis, and tailored healthcare services (Patel and Dwivedi 2024). To ensure the suitability of consumers, this research investigation proposes employing a neural network to accurately assess customer data and provide personalized healthcare information via blockchain. Blockchains are distributed, tamper-proof solutions for data management. A collection of blocks forms a blockchain, with each block connected to its predecessor by a timestamp. In essence, it is a system that bundles countless records together (Ramani et al. 2024). Unlike previous electronic money systems that stored transaction information on a central server, blockchain displays transaction details to all users and protects against fraud by enabling comparisons among transactions (Hasan et al. 2024). Cryptography and blockchain are closely linked; blockchain innovation extends beyond encryption. Consequently, numerous blockchain-based solutions and products are being developed.

The ownership of the blockchain will eventually be transferred once hacking at the level of governmental manipulation is accomplished, and mature blockchains cannot be fixed. Blockchain ensures that information from books scattered throughout each node remains accessible across big nodes; it can also be seen as an agreement resolution method. Nodes can operate anonymously, have shoddy connections, or even use dubious operators because of this capability, as shown in Fig. 11.3. There are multiple methods on the blockchain. Everyone specifically wishes to use the Hyper POR method, a consensus process across public blockchains, to implement the health information blockchain (Almaiah et al. 2024). The Hyper POR technique was chosen since, in the future, health information will be converted to PHRs, which will require a lot of Translations Per Second (TPS) if a lot of information is used. This is because these blockchains need a lot of TPS performance information for authentication. As a result, an outstanding performance consensus mechanism is required for the medical data blockchain. It has an automated smart contract feature. Several blockchain agreement computational methods, shading methods, and smart contract technology are discussed in this study.

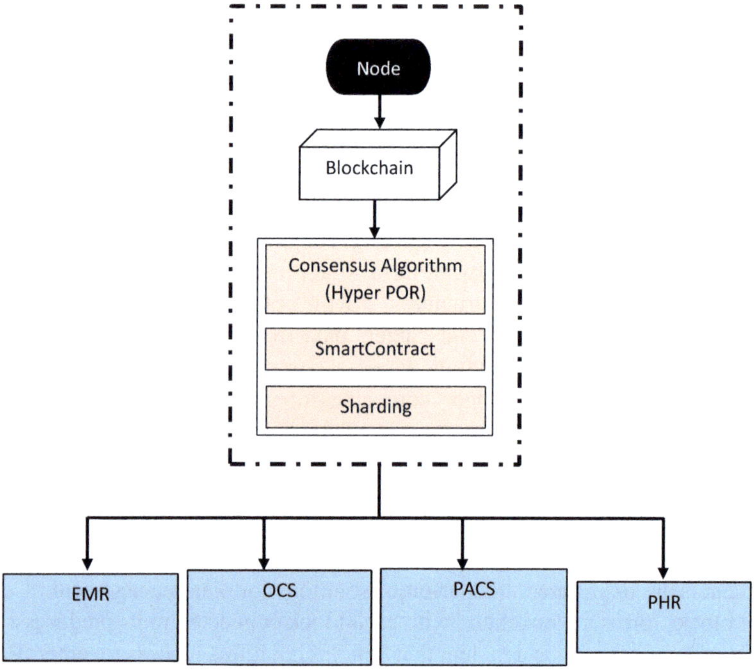

Fig. 11.3 Blockchain for healthcare

11.3 Deep Convolutional Neural Networks in Healthcare

CNN design is divided into two stages: component training and classification, also known as the extraction of characteristics. CNN, max-pooling, sequential normalization, and fully linked layers make up the first stage. When supplying the CNN with unprocessed information, the input layer pre-processes the input images.

Using an algorithm for convolution based on a set of graded filtration systems, convolutional layers are designed to generate maps of features. The output from the convolutional layer processing is transformed into map representations using the ReLU function of activation. Figure 11.4 shows the proposed DCNN framework design. Chest X-ray images with dimensions of $224 \times 224 \times 3$ pixels were used as input for the model. The Conv1 layer uses a 1-pixel zero padding in each route to store a comparable spatial dimensional component map. The same result is then fed into the remaining layers. The 64 feature maps of dimension 3×3 in the conv3 layer are derived using $64 \times 3 \times 3$ filters.

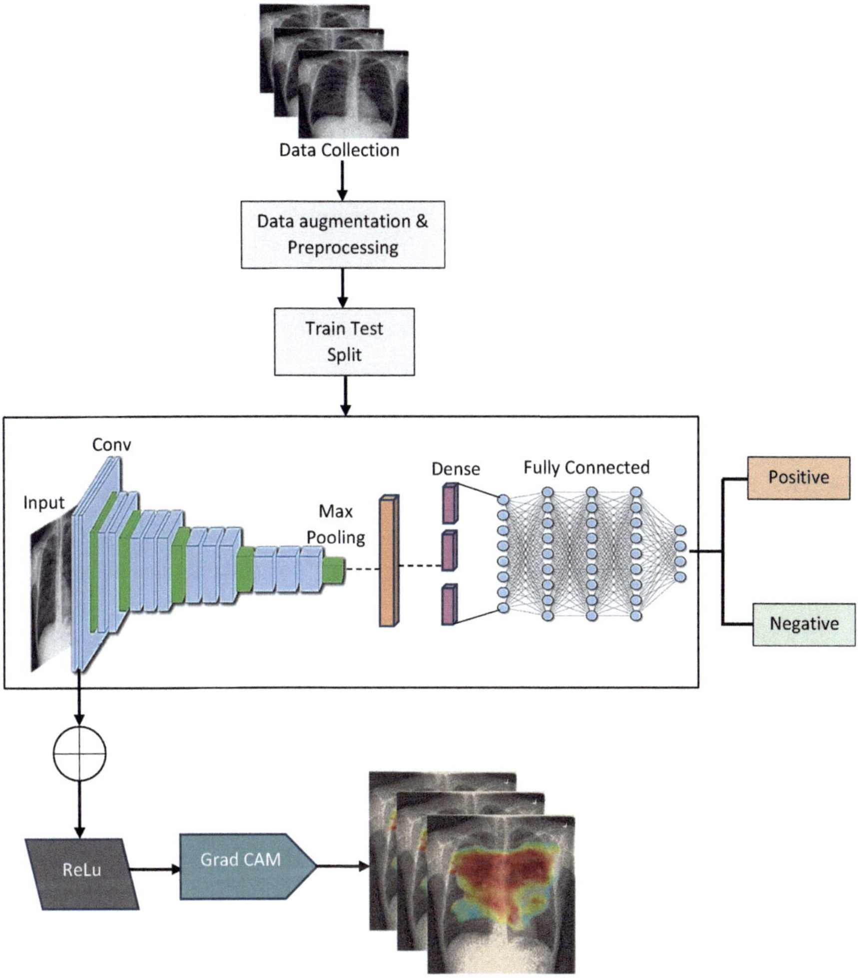

Fig. 11.4 A proposed deep learning model used in healthcare

11.3.1 Blockchain with Deep Convolutional Neural Networks in Healthcare

The following is an illustration of the workflow. First, the information is gathered via medical images (chest X-rays) taken at different healthcare facilities and clinical labs. The blockchain layer must then be used to safeguard the information collected for security. The DCNN module aids in the detection and classification of coronavirus in radiology images by using efficient analytical capabilities and dependable prediction of the information collected. Blockchain connections are used to connect the information center with the hospital's information. To identify chest X-ray images, DCNN techniques are used in the information center. Lastly, the

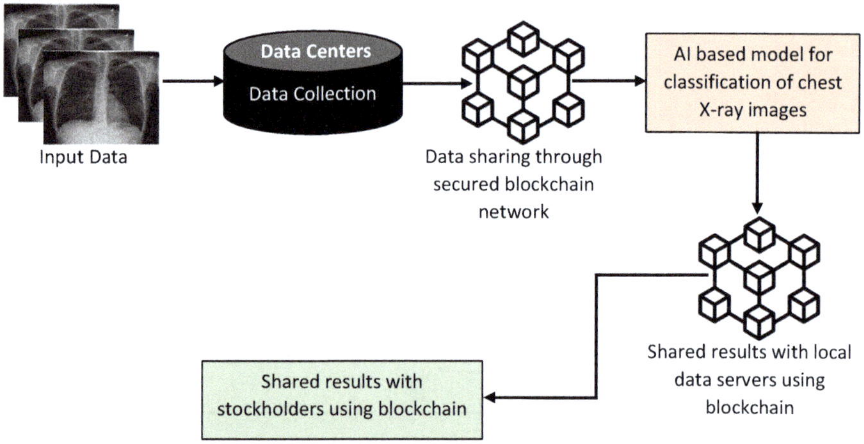

Fig. 11.5 Proposed system for healthcare system

decentralized ledger is used to provide a safe link for stakeholders to share the studied data.

In a similar vein, the primary drivers for the integration of blockchain and deep neural networks are the verification of information accountability, attestation of outcomes, provenance tracking, transparency of ownership and consumption, and the assurance of fairness. Figure 11.5 illustrates the four main features, scenarios, and categories that benefit from this type of integration. The immutability characteristic of blockchain protects the data and the DCNN model from various forms of attacks and data distortion, ensuring that data leakage cannot occur. This collaboration results in more reliable and accurate predictions. The proposed system provides a framework to automate several functions where data organization and security are critical.

Data Processing with DCNN for Feature Extraction

DCNNs are used for processing healthcare data (e.g., medical images) and extracting features to classify, detect, or predict medical conditions. The layers in a DCNN, such as convolutional and pooling layers, transform raw data into high-level feature representations.

Convolution Operation: The convolutional layer applies filters to the input image X to extract features.

$$F(x,y) = \sum_m \sum_n I(x+m, y+n).K(m,n), \tag{11.1}$$

where K is the convolutional kernel, and (x, y) are spatial indices. This results in a feature map that represents specific patterns in the data.

Activation Function (ReLU): Non-linear activation functions such as ReLU introduce non-linearity, which is crucial for learning complex patterns in healthcare data.

$$f(i) = \max(0, i), \tag{11.2}$$

where i is the output from the convolution operation.

Blockchain for Secure and Transparent Data Storage
Blockchain provides a decentralized, immutable ledger for storing healthcare data, ensuring both data integrity and transparency. Each block in the blockchain stores a hash of the previous block, along with a hash of the DCNN-extracted feature set.

Hashing for Data Integrity: Hash functions ensure data integrity by generating a unique identifier (hash) for each feature extracted by DCNN.

$$h = H(F), \tag{11.3}$$

where F is the feature set extracted from DCNN, and H is a cryptographic hash function (e.g., SHA-256).

Block Hash: Each block B includes the hash of the previous block, h_{prev}, linking all blocks in a chain.

$$h_{\text{block}} = H\left(h_{\text{prev}} | F || \text{timestamp}\right), \tag{11.4}$$

where h_{block} is the current block hash, and || denotes concatenation.

Digital Signatures for Authentication and Access Control
Each transaction or data entry in the healthcare blockchain system is signed with a digital signature, ensuring that only authorized entities can add or access data.

Digital Signature Creation: A private key K_{priv} is used to sign the extracted features or data:

$$S = \text{Sign}\left(F, K_{priv}\right), \tag{11.5}$$

where S is the digital signature of the feature set F. This step authenticates data origin.

Verification: The public key K_{pub} verifies the signature to confirm the data's authenticity:

$$\text{Verify } \left(F, S, K_{\text{pub}}\right) = \text{True} \tag{11.6}$$

Smart Contracts for Data Access Control

Smart contracts automate access control in decentralized healthcare systems, ensuring data is only accessed by authorized parties based on predefined conditions.

Access Condition: Let ∅ represent the condition (e.g., user authentication) and f the access function.

$$\text{Execute}\, f\ \textit{if}\, \varnothing(F) = \text{True}, \tag{11.7}$$

where F represents the DCNN-extracted features or data.

Data Encryption for Confidentiality

Healthcare data is encrypted before being stored on the blockchain to ensure confidentiality.

Symmetric Encryption: Data D is encrypted using a symmetric key K_{sym}

$$\mathrm{C} = E_{K_{\text{sym}}}(D), \tag{11.8}$$

where C is the ciphertext stored on the blockchain.

Asymmetric Encryption for Key Distribution: The symmetric key K_{sym} can be encrypted using the public key K_{pub} of an authorized entity, as

$$K_{\text{enc}} = E_{K_{\text{pub}}}\left(K_{\text{sym}}\right), \tag{11.9}$$

and decrypted by the entity's private key K_{priv} when access is granted, as

$$K_{\text{sym}} = D_{K_{\text{priv}}}(K_{\text{enc}}). \tag{11.10}$$

11.4 Consensus Mechanisms for Decentralization and Data Validation

In a decentralized healthcare system, consensus algorithms ensure all nodes agree on the addition of new data blocks, maintaining data consistency and security.

Proof of Stake (POS) Probability: Let S_x represent the stake of participant i. The probability $P(x)$ of participant x validating the next block is

$$P(x) = \frac{S_x}{\sum_y S_y}. \tag{11.11}$$

These equations collectively support a secure, transparent, and decentralized healthcare data management framework, utilizing DCNNs for data processing and blockchain for data security and integrity. This integration ensures patient data remains confidential, immutable, and accessible only to authorized entities.

11.5 Results and Discussions

TPS is most commonly referred to as the blockchain speed indicator, indicating the maximum number of operations that can be handled and stored per second. It could change according to the kind of transaction. A smart currency transaction, for instance, can only take longer because it involves more steps than a basic transfer. Depending on how the code is actually implemented, a currency that does the same thing can need more operations and take longer. Two varieties of TPS are possible. When overloaded, the maximum TPS indicates the number of operations that undergo processing and storage per second. As a result, it is an excellent indication for assessing speed. The performance of the average TPS was also assessed. Performance tests were used to compare PHR information with an artificially intelligent blockchain and PHR information without a safety mechanism in order to shorten the current assessment period. A median efficiency of 2400 TPS was demonstrated based on the IPFS (Inter Planetary File System) configuration shown in Fig. 11.6. The blockchain neuron system was also used to assess the network's efficiency. As a result, the median TPS rises by 600 TPS to 3000 TPS, as shown in Fig. 11.7. Nearly the same 3000 TPS performance information is now shown in Fig. 11.8.

Simulation results based on the proposed systems are shown in Fig. 11.9. The proposed system has restricting access rules based on pseudorandom methods and homomorphic encryption. The proposed approach compares many access regulations and execution times for access management policies. Conducted tests with policy formation, policy cancellation, and policy modification. The authorization policy was faster than the delegation and authentication policies. Simulations shown in Fig. 11.10 support the claim that the proposed access control scheme offers

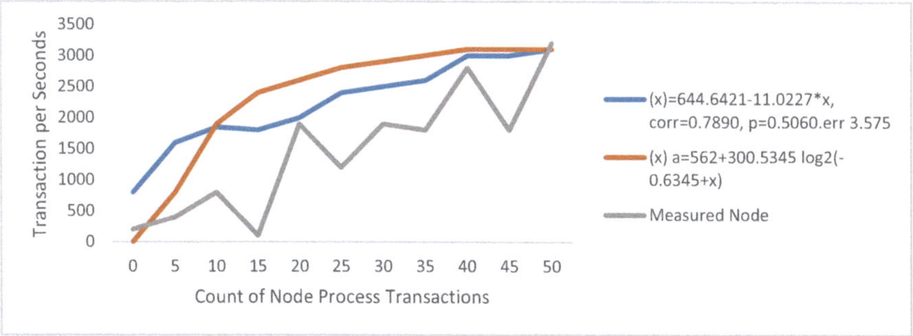

Fig. 11.6 Proposed system provided security during transactions based on one round

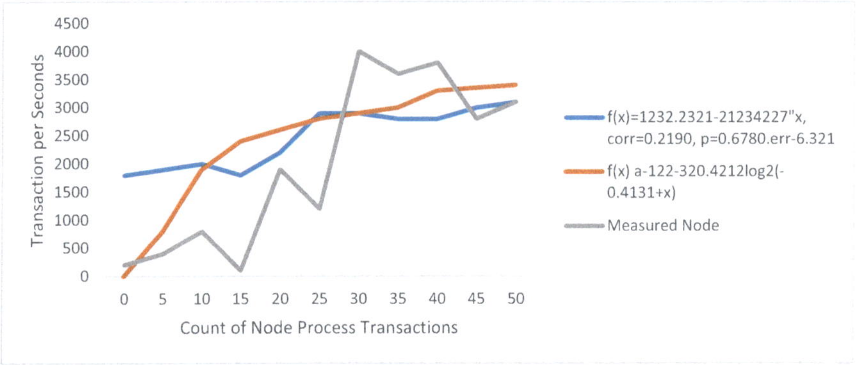

Fig. 11.7 Proposed system provided security during transactions based on 50 rounds

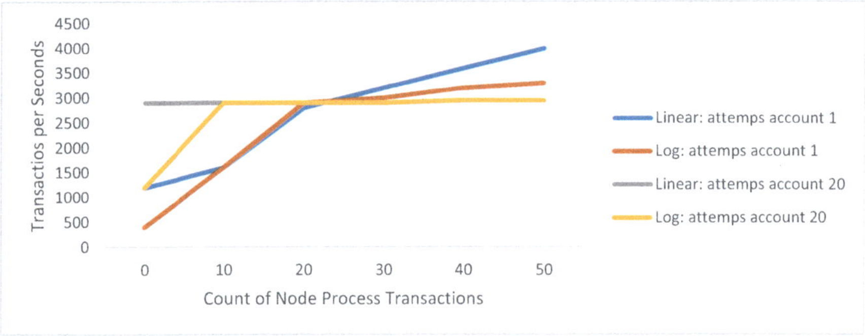

Fig. 11.8 Proposed system provided security during transactions based on an average of 100 rounds

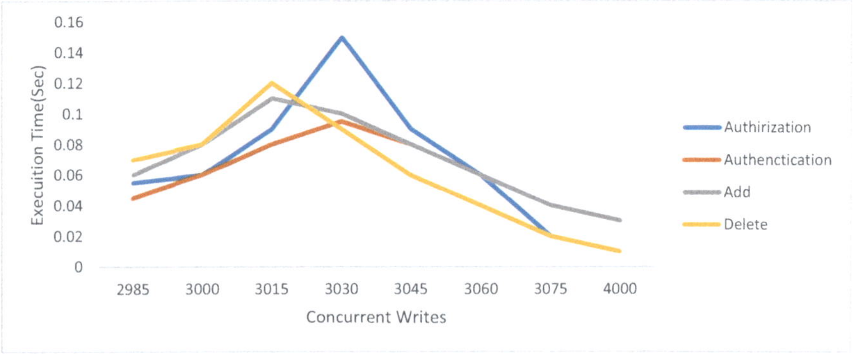

Fig. 11.9 Comparative analysis of different domains based on execution time

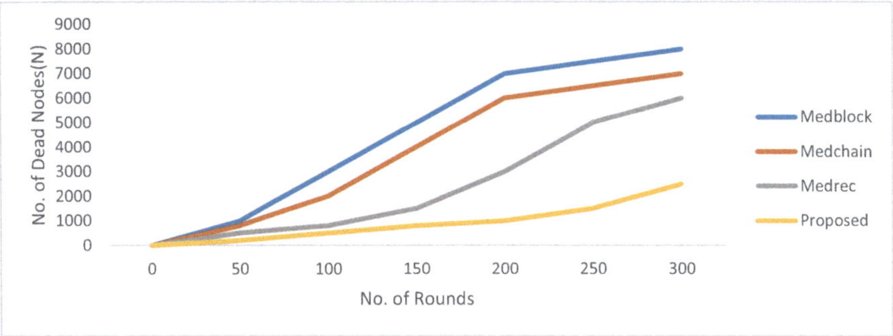

Fig. 11.10 Comparative analysis of different domains based on dead nodes

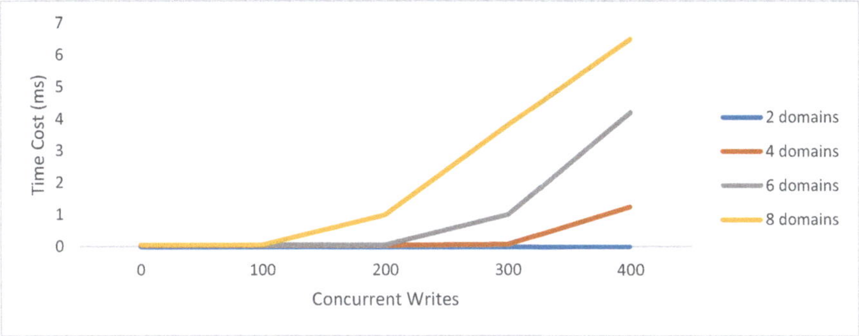

Fig. 11.11 Comparative analysis of different domains based on time and cost

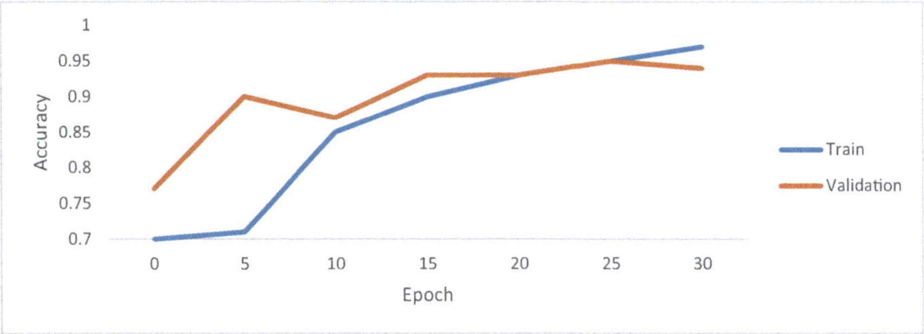

Fig. 11.12 Training and Validation accuracy of the proposed model

lower computational costs and increased security. The simulation results of many users who were categorized according to how they interacted and behaved with the proposed framework are shown in Fig. 11.11.

The training and verification procedure is shown in Fig. 11.12, and it is clear that the algorithm achieves almost 98% training validation correctness while providing

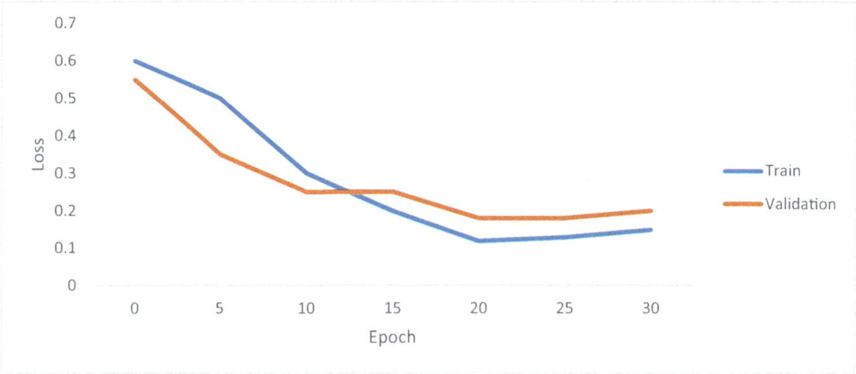

Fig. 11.13 Training and Validation loss of the proposed model

smooth generalizations on the information set. With additional instruction, the precision value increases. The model becomes better with additional learning or training, as seen by the 0.15% Training-Validation loss in Fig. 11.12. As seen in Fig. 11.13, the accuracy and loss curves show that the model's performance increases with time, indicating that it becomes better with training.

11.6 Conclusions

There are several advantages to integrating DCNNs with blockchain systems for the honest and secure handling of information in decentralized healthcare networks. This combination allows DCNNs to analyze medical data accurately and automatically, while blockchain ensures the integrity of the information, privacy, and transparency. Blockchain is ideal for managing sensitive healthcare data due to its immutability and cryptographic security, which address data security concerns and prevent unauthorized alterations. Smart contracts and digital signatures enhance access control by restricting patient data retrieval to authorized users, fostering trust in decentralized environments. The blockchain framework's encryption features safeguard data confidentiality, ensuring that private information remains protected even in distributed networks. Overall, this integrated approach paves the way for safe, efficient, and patient-centered management of medical information, establishing distributed platforms as reliable alternatives to traditional healthcare information processing and storage.

References

Afaq Y, Manocha A (2024) Blockchain and deep learning integration for various application: a review. J Comput Inf Syst 64(1):92–105

Alanazi AA, Karim FK, Ghorashi SA, Amoudi G, Hamza SHA (2024) Blockchain with optimal deep learning assisted secure data sharing and classification on future healthcare systems. Alex Eng J 99:168–179

Aldaej A, Ahanger TA, Ullah I (2024) Deep neural network-based secure healthcare framework. Neural Comput & Applic 36:1–16

Almaiah MA, Ali A, Alkhdour T, Lutfi A (2024) Security risk and breach detection approach based blockchain for medical applications, vol 12. IEEE Access, p 171876

Bezanjani BR, Ghafouri SH, Gholamrezaei R (2024) Fusion of machine learning and blockchain-based privacy-preserving approach for healthcare data in the Internet of Things. J Supercomput 80(17):24975–25003

Bhumichai D, Smiliotopoulos C, Benton R, Kambourakis G, Damopoulos D (2024) The convergence of artificial intelligence and blockchain: the state of play and the road ahead. Information 15(5):268

Das P, Singh M, Verma KK (2024) Blockchain-enabled deep learning approach to improve healthcare system. J Multimedia Inf Syst 11(1):9–16

Deepthika K, Shobana G, Reddy KV, Srimathi S, Kumar B, Upadhyay S (2024) Blockchain-integrated deep learning for secure health data sharing and consent management. In: 2024 second international conference on Intelligent Cyber Physical Systems and Internet of Things (ICoICI). IEEE, pp 101–106

Far AZ, Far MZ, Gharibzadeh S, Zangeneh S, Amini L, Rahimi M (2024) Artificial intelligence for secured information systems in smart cities: collaborative IoT computing with deep reinforcement learning and blockchain. arXiv preprint arXiv:2409.16444

Ghani MANU, She K, Rauf MA, Khan S, Alajmi M, Ghadi YY, Alkahtani HK (2024) Toward robust and privacy-enhanced facial recognition: a decentralized blockchain-based approach with GANs and deep learning. Math Biosci Eng 21(3):4165–4186

Hammad A, Abu-Zaid R (2024) Applications of AI in decentralized computing systems: harnessing artificial intelligence for enhanced scalability, efficiency, and autonomous decision-making in distributed architectures. Appl Res Artif Intell Cloud Comput 7(6):161–187

Hasan MR, Li Q, Saha U, Li J (2024) Decentralized and secure collaborative framework for personalized diabetes prediction. Biomedicines 12(8):1916

Jagdale B, Sugave SR, Kulkarni YR, Gutte V (2024) Privacy-aware quantum convolutional neural network for blockchain-based IoT health care data. Intell Decis Technol 18(2):1337–1354

Karimullah S, Shaik F, Vardhan DV, Nagaraju C (n.d.) Blockchain-enhanced convolutional neural networks for efficient detection of cardiovascular abnormalities. In: Computational intelligence and blockchain in biomedical and health informatics. CRC Press, pp 275–291

Komarasamy D, Mohana Saranya S, Rajalaxmi RR, Prasanndh Raaju MR, Hemanath B (2024) New neural network framework for medical intelligent system. In: Intelligent systems modeling and simulation III: artificial intelligent, machine learning, intelligent functions and cyber security. Springer, Cham, pp 97–119

Krishnamoorthy R, Kaliyamurthie KP (2024) Enabling privacy-preserving collaborative learning: secure cryptographic neural networks meet blockchain for GDPR-compliant data sharing. In: 2024 5th international conference for emerging technology (INCET). IEEE, pp 1–11

Kumar C, Chittora P (2024) Deep-learning and blockchain-empowered secure data sharing for smart grid infrastructure. Arab J Sci Eng 49:1–14

Kuznetsov A, Sernani P, Romeo L, Frontoni E, Mancini A (2024) On the integration of artificial intelligence and blockchain technology: a perspective about security. IEEE Access

Mousa SH, Shati NM, Sakthivadivel N (2024) DeepRing: convolution neural network based on blockchain technology. Al-Mustansiriyah J Sci 35(2):61–69

Nazir A, He J, Zhu N, Anwar MS, Pathan MS (2024) Enhancing IoT security: a collaborative framework integrating federated learning, dense neural networks, and blockchain. Clust Comput 27:1–26

Olawale OP, Ebadinezhad S (2024) Cybersecurity anomaly detection: AI and Ethereum blockchain for a secure and tamperproof IoHT data management. IEEE Access

Omidian H (2024) Synergizing blockchain and artificial intelligence to enhance healthcare. Drug Discov Today:104111

Patel R, Dwivedi V (2024) A review of secure IoT based smart health monitoring system using blockchain technique. In: 2024 3rd international conference on sentiment analysis and deep learning (ICSADL). IEEE, pp 569–580

Ramani R, Mary AR, Raja SE, Shunmugam DA (2024) Optimized data management and secured federated learning in the internet of medical things (IoMT) with blockchain technology. Biomed Signal Process Control 93:106213

Rokade S, Mishra N (2024) A blockchain-based deep learning system with optimization for skin disease classification. Biomed Signal Process Control 95:106380

Sachidananda Murthy KB, Prasad SN (2024) Blockchain-enabled detection of neurological disorders using a deep learning approach. Eng Proc 59(1):187

Saleh AMS (2024) Blockchain for secure and decentralized artificial intelligence in cybersecurity: a comprehensive review. Blockchain Res Appl:100193

Selvi K, Dilip G (2024) Enhancing cyber-physical systems security: a review of deep learning and blockchain integration. In: 2024 5th international conference on Image Processing and Capsule Networks (ICIPCN). IEEE, pp 725–734

Sethi M, Arora J, Baggan V, Verma J, Snehi M (2024) Deep learning and blockchain applications in healthcare sector using imaging data. In: Next-generation cybersecurity: AI, ML, and blockchain. Springer, Singapore, pp 135–158

Shahsavari Y, Dambri OA, Baseri Y, Hafid AS, Makrakis D (2024) Integration of federated learning and blockchain in healthcare: a tutorial. arXiv preprint arXiv:2404.10092

Shammi SA, Ghosh P, Sutradhar A, Shamrat FJM, Moni MA, De Oliveira TEA (2024) Advances in artificial intelligence and blockchain technologies for early detection of human diseases. IEEE Transactions on Computational Social Systems

Shinde R, Patil S, Kotecha K, Potdar V, Selvachandran G, Abraham A (2024) Securing AI-based healthcare systems using blockchain technology: a state-of-the-art systematic literature review and future research directions. Trans Emerg Telecommun Technol 35(1):e4884

Case Studies: Blockchain in Medical Virology

12

Pasupuleti Sai Kiran, B. Tirapathi Reddy, V. Krishna Reddy, K. Thirupathi Rao, and Siva Sankar Namani

Abstract

This chapter analyzes the impact blockchain technology has in the field of medical virology as far as safeguarding data, improving processes, and increasing efficiency are concerned. It demonstrates through multiple case studies how blockchain technology is being used across the world in public health, vaccine tracking, viral genomics, clinical trials, electronic health records (EHRs), pandemic monitoring, and supply chain management. The first case presented in this chapter is Singapore's use of IBM blockchain for TraceTogether which greatly streamlined the contact tracing process for COVID-19. AlliGuard's Counterfeit Protection System in Estonia established a blockchain vaccination authentication system to protect the supply chain from counterfeits which improved the integrity of the network. Through its system in the United States, HealthVerity managed to track COVID-19 viral genomic data, which enhanced our understanding of virus evolution and distribution. In addition, this chapter presents the case of MediBloc in South Korea. MediBloc's blockchain solution enhanced clinical trial management by ensuring data security and better facilitating researchers' collaboration.

P. S. Kiran (✉)
Prasad V Potluri Siddhartha Institute of Technology, Vijayawada, India
e-mail: psaikiran@pvpsiddhartha.ac.in

B. T. Reddy · V. K. Reddy
Koneru Lakshmaiah Education Foundation, Vaddeswaram, Guntur, India
e-mail: tirupathireddyb@gmail.com; vkrishnareddy@kluniversity.in

K. T. Rao
Gandhi Institute of Technology and Management (Deemed to be University), Visakhapatnam, India
e-mail: tkomati@gitam.edu

S. S. Namani
G Narayanamma Institute of Technology and Science, Hyderabad, India

J. M. Chatterjee, S. K. Saxena (eds.), *Artificial Intelligence and Blockchain in Precision Medicine and Virology*, Medical Virology: From Pathogenesis to Disease Control, https://doi.org/10.1007/978-981-96-8919-4_12

This chapter explains that blockchain technology implementation in medical virology creates safer information protection while improving system integration and patient data security and boosting productivity in clinical operations. The technology demonstrates its capability of improving patient outcomes together with better public health administration and comprehensive high-tech security for the healthcare system.

Keywords

Blockchain technology · Medical virology · Data security · Vaccine tracking · Clinical trials · Electronic health records (EHRs) · Public health

12.1 Introduction

Modern medical virology makes use of technological advancements to change the identification methods of viral infections as well as their tracking treatment and scientific research processes. Modern innovations improve both clinical diagnosis accuracy and vaccine formula creation along with virus and epidemiological understanding. Advancements at every step of diagnosis and outbreak control and vaccine development and treatment approaches receive benefits from technologies which span from genomic sequencing through artificial intelligence (AI).

Viral pathogens are detected and their cases are tracked through genomic sequencing technologies (Radford et al. 2012). High-throughput sequencing (HTS) functions as a sequencing technology that enables scientists to read entire viral genomes which helps them to track mutations throughout the evolutionary process. The COVID-19 pandemic allowed genome sequencing to monitor different viral variants including Delta and Omicron. Regulations state that diagnostic CRISPRs represented by SHERLOCK are ideal detection approaches because they provide rapid yet precise viral nucleic acid measurements to support early diagnosis and containment efforts.

Artificial intelligence along with machine learning operates as core tools for virological processes (Chatterjee and Saxena 2024). The AI predictive models receive epidemiological data to generate predictions about viral outbreak patterns. Through the development of molecules, AI aids fast antiviral candidate discovery processes in drug investigations.

Blockchain service allows stakeholders to share secure data in a transparent manner (Nguyen et al. 2021). The approach allows patients to maintain control of their medical information, protects vaccine supply systems, and reduces fraudulent activities at high standards of quality. The MediBloc platform and its comparable services enable patients to take control of their health data access which supports both personal privacy and adheres to regulatory standards.

Wearable technology collects live data through oxygen level monitoring for people with COVID-19 infection. Internet of Things (IoT) sensors enable the tracking of viral load in wastewater to detect new outbreaks through their measurements. These are representative examples of Internet of Things (IoT) advancement.

Real-time viral spread information obtained from social media activities and search keywords together with contact tracing applications allows digital epidemiology to use this data for both predictive and surveillant functions. The set of technological tools effectively supported COVID-19 pandemic efforts by identifying exposure risks and notifying individuals about their cases.

The enhancement of diagnosis together with treatment processes results from advancements in nanotechnology. Viral particle identification depends on high-sensitivity nanometers while nanocarrier systems enhance antiviral drugs by delivering them with low side effects. The MRNA vaccines made by Pfizer-BioNTech as well as Moderna use lipid nanoparticles as their essential nanotechnology components.

Vaccine speeds up development through both engineering of viral vectors and mRNA technology applications in synthetic biology. The design pattern facilitates swift vaccine development operations which proved successful when creating vaccines for COVID-19. The technology enables researchers to create viruses which specifically identify cancer cells for destruction.

The high-resolution views of viral structures made possible by cryo-electron microscopy assist researchers in developing vaccines together with treating drugs. SARS-CoV-2 spike protein visualization through advanced technologies supported researchers during vaccine creation and production.

Laboratory processes become more efficient by using robotics and automation which also helps decrease human mistakes according to Rani et al. (2024). The use of robots helps healthcare facilities obtain rapid-testing results while performing disinfection procedures which lowers the risk of transmitting the virus.

These technologies enhance diagnostic and therapeutic capabilities and also improve the global preparedness and response to emerging viral threats.

12.2 Case Study 1: TraceTogether App by Singapore for COVID-19 Contact Tracing

The TraceTogether app developed for Singapore for COVID-19 contact tracing integrated blockchain technology in the public health infrastructure to provide data security, privacy, and efficient management of the pandemic situation. It is based on IBM's Hyperledger Fabric showcasing how the distributed ledger technology can be integrated into the public health projects.

1. Distributed Ledger Technology (DLT)

IBM Hyperledger Fabric (Hanif et al. 2024) functions as the essential blockchain foundation that TraceTogether implements for its solution. Hyperledger Fabric serves enterprise applications by allowing access to authorized entities including healthcare providers together with government bodies who maintain database records. The distributed ledger system enables nodes selected from trusted entities which act as prevention against one organization achieving control of complete information.

Duplicate components within a decentralized system make networks resistant to vulnerabilities and centralization breaches while ensuring complete robustness. Every participant maintains an up-to-date secure copy of data so the network displays consistent information. The consensus mechanism functions as a security protection because unauthorized data modifications are blocked from the system design.

2. Data Encryption

TraceTogether implements two encryption levels to protect user data in a contact tracing solution because such information requires the highest level of security.

Users receive encryption of their Bluetooth interaction data from the app through device-based symmetric encryption protocols. The security measure encodes sensitive data so that unauthorized control of devices does not compromise information protection. All sensitive data remains secure because devices with compromised security do not permit unauthorized access to this information.

The blockchain allows users to store their data through cryptographic hash functions which produce anonymous data after upload. The hash function transforms data into unalterable irretrievable cryptographic text which makes it impossible to locate original PII embedded in stored data. User data security remains guaranteed despite decentralized networks' operation because various nodes access the information. The system ensures data protection because the encryption through hash functions remains secure even when multiple nodes connect to an unfenced network.

3. Smart Contracts

Antifraud systems run more efficiently through smart contracts because they include automated features that make system operations faster. The blockchain contains program segments that execute automatically to enforce predetermined rules through autonomous protocols without any need for human participation.

Smart contracts use official health department data to check patient-submitted positive COVID outcomes which helps stop unauthorized information entry.

Soon after an individual is infected, it detects people who met closely with them through Exposure Notifications and automatically releases alerts to those individuals.

Smart contracts enforce detailed procedures which limit data access while ensuring that permitted users can obtain epidemiological records authorized for their tasks. The system maintains protection against unwanted access to confidential information and misuse. The data remains secure by protecting it from unauthorized use.

4. Data Lifecycle Management

The fundamental principle in TraceTogether's design emphasizes efficient data management from creation to deletion. The mobile app maintains privacy by letting users store their proximity data on their devices until system automation reduces that data after 25 days. The brief data storage period functions to stop long-term abuse of information. Epidemiological critical data is securely stored on blockchain systems in an unchangeable way to aid virus transmission investigations. Every system activity involving data entry or modification results in an audit trail because the system maintains a comprehensive record of all user operations. The system's enhanced audit feature enables health authorities to monitor system activity because it provides transparency while ensuring their ability to discover cases of improper usage.

5. Integration with Existing Systems

The TraceTogether blockchain layer has been incorporated into Singapore's Government Technology Stack (SGTech) and its National Electronic Health Record (NEHR) system. Real-time serving of authenticated health information between multiple agencies can be achieved through a newly integrated system that strengthens pandemic collaboration between agencies (Bay et al. 2020).

The application checks positive case reports by accessing data from the National Electronic Health Record through its communication channels. The application uses NEHR communication channels to verify positive case reports against national health databases.

The integration of SGTech allows other governmental systems including SafeEntry check-in to operate jointly with TraceTogether thus forming a unified contact tracing network. Through its connection to SGTech, the system enables other government platforms including the SafeEntry check-in system to communicate with TraceTogether, thus establishing a single contact tracing framework.

12.2.1 Significant Benefits of TraceTogether's Blockchain Integration

Blockchain technology integration with Singapore's TraceTogether system brought multiple benefits to overcome major digital contact tracing obstacles, which especially proved essential during COVID-19 pandemic emergencies (Akinbi et al. 2021). The system benefits are organized into four categories: data protection and operational effectiveness, privacy safeguards, transparency features, and the ability to scale as needed.

1. Improved Data Protection Measures Enhanced Data Security

Whenever data enters the blockchain, it automatically becomes immutable; therefore, no person can modify it. Data integrity within contact tracing stays protected by blocking any attempts at unauthorized tampering with records. The system safeguards contact tracing data integrity by making all data in the ledger noneditable, thus preventing unauthorized tampering.

The decentralized distribution of ledgers across a system eliminates a single critical failure point which raises its protective measures against hacking. When one network node gets hacked, the intrusion stops at that point since the data endures protected throughout the whole network system.

The sensitive data stored in the local device will be secured by using encryption, and the data stored in the blockchain is secured by cryptographic hash function.

2. Improved Operational Efficiency

Through smart contracts, the system executes automatic procedures which include exposure alert systems as well as infection validation protocols and data exchange operations. Silencing human-led operations allows faster response times during disease outbreak periods.

The integration of blockchain technology enables the NEHR system in Singapore to facilitate smooth data sharing across government entities alongside healthcare service providers.

NEHR system of Singapore provides smooth data sharing across government entities along with the healthcare service providers as a result of blockchain technology integration. The connection among these entities enhances both timely choices and precise decision quality.

3. Privacy Protection

Through hash functions and anonymized data storage methods, users can safely store their information, thus making blockchain-based data identification impossible until authorized decryption takes place.

User devices retain proximal data for 25 days before automatically removing it from local storage. The storage process through hash functions protects users from future unauthorized data misuse or surveillance activities.

The system applies controlled access methods that let authorized entities gain access to particular records after smart contracts establish strict rules to protect user trust.

4. Increased Transparency and Trust

The blockchain records each data operation like access, sharing, and updates in an unalterable manner. This has created confidence among the public as it has the ability to track the system changes, thus providing accountability.

Blockchain technology provides transparency which enhances user trust regarding responsible data management, thus creating conditions for broad adoption of the application among users.

5. Scalability and Future-Readiness

Blockchain's Capacity to Process Massive Data Volume: Its capacity to handle immense data volume makes it ideal for running such a big operation as contact tracing at a national level. It is scalable to use up to millions of users with no performance degradation.

Extensions: The system can be expanded to further include tracking of vaccination records or integration with international travel systems in the maintenance of pandemics.

Decentralized Design: The most attractive benefit of TraceTogether is its global interoperability—its built-in decentralized design facilitates integrating TraceTogether with other systems while still keeping security and privacy intact across borders.

6. Better Public Health Outcomes
Real-time validation combined with notifications from the system leads to rapid contact tracing which decreases how long it takes to find exposed persons.

Healthcare authorities distribute their resources better when they receive precise information at the right time which helps stop the virus from spreading.

7. Ethical Data Governance
The system follows international privacy standards by employing privacy-preserving technologies which include encryption and anonymization to support GDPR principles among others.

The system reduces surveillance anxiety through restricted data storage and anonymous user information techniques that prevent government surveillance concerns.

12.2.2 Comparison with Other Countries' Systems

This section compares TraceTogether to other national implementations while illustrating the multiple approaches to contact tracing systems (Ahmed et al. 2020) that present different privacy levels and effectiveness coupled with technology adoption. The blockchain-enhanced systems used by Singapore prioritize both system security and transparency yet non-blockchain systems deal with privacy issues or manage slow growth (Table 12.1).

12.2.3 Challenges and Criticisms of TraceTogether's Blockchain Integration

Singapore achieved major advantages by applying blockchain technology through TraceTogether yet encountered multiple obstacles that raised critical reactions. Multiple problems exist regarding technology restrictions alongside privacy matters and user integration alongside ethical inquiries.

1. Privacy Concerns
Despite technical attempts to obscure user identities along with minimizing data storage duration, public fears endured about government surveillance capability to access data for monitoring purposes (Leith 2023). Minimum government surveillance remained a possibility through their centralized access to decryption keys even though identification measures were still possible in specific cases.

Table 12.1 Comparison of COVID-19 digital tracing app

Country/system	Technology used	Blockchain involvement	Strengths	Limitations
Singapore (TraceTogether)	Bluetooth + IBM blockchain	Hyperledger for secure data sharing	High transparency, secure data storage	Mandatory app usage raised privacy concerns.
South Korea	GPS, CCTV, credit card data	No blockchain	Comprehensive tracking, detailed timelines	Data privacy significantly compromised.
Germany (Corona-Warn-App)	Bluetooth + decentralized IDs	No blockchain	Strong privacy focus, anonymized IDs	No auditability for false-positive cases.
China (Health Code)	QR code-based health status	Proprietary (limited blockchain application)	Seamless integration with daily services	Highly intrusive, centralized data control.
Australia (COVIDSafe)	Bluetooth	No blockchain	Relatively simple implementation	Low adoption due to usability and privacy concerns.
India (Aarogya Setu)	GPS + Bluetooth	No blockchain	Wide adoption, integrated with health system	Centralized system raised surveillance issues.
United Kingdom (NHS COVID-19)	Bluetooth + Google-Apple API	Limited blockchain pilot	Privacy-friendly, effective notifications	Blockchain not fully implemented.

Public skepticism increased because people failed to understand blockchain's privacy features, thus generating trust issues about their communication being monitored.

2. Technological Challenges (Trivedi and Vasisht 2020)

The computational resources needed to manage blockchain on Hyperledger Fabric remain high due to the distributed ledger management and encryption procedures. The increasing number of users along with interactions led to scalability problems for the system.

The method of data validation between multiple nodes created performance delays that prevented real-time system operations in contact tracing applications.

The integration of blockchain with Singapore's National Electronic Health Record (NEHR) succeeded yet the process of adapting it for new functionalities along with global system integration increased costs and added complexity.

3. User Adoption Challenges (Lee et al. 2021)

The authorities established mandatory adoption rules for TraceTogether which required users to wear it before entering numerous public facilities such as malls and offices. The mandate for TraceTogether access created ethical problems because people who turned down using the app lost their access to vital services.

Some individuals avoided using TraceTogether because they had limited technology knowledge and did not trust the privacy assurance provided by the system which diminished its effectiveness in contact tracing.

The app encountered two main technical limitations because it depended on Bluetooth and blockchain technologies which drained battery life and demanded devices with modern specifications and excluded users with outdated smartphones.

4. Data Retention and Management

The database containing epidemiological blockchain data remained permanent even though proximity logs were wiped after 25 days. The indefinite anonymous data storage along with concerns about possible future abuse was criticized by experts (Braithwaite et al. 2020).

Data oversight suffered due to the absence of an independent auditing team responsible for ensuring privacy norm compliance, which led people to doubt how the system managed data oversight.

5. Ethical and Legal Considerations

The inherent transparency of blockchain systems did not provide users with a clear view of how their data processing took place and its sharing with different entities which created ethical uncertainties (Bengio et al. 2020).

The system maintained an undefined legal position about data usage terms as well as protection standards. During emergencies, the Singapore government possessed certain data protection exemptions under the Personal Data Protection Act (PDPA) while these regulatory safeguards became a matter of concern for the long term.

6. Interoperability Issues

At its initial launch, TraceTogether missed global interoperability because the system did not have built-in capabilities to link with international contact tracing systems. The system proved inefficient when it came to international data sharing and travel purposes.

Operating with IBM Hyperledger Fabric required the system to rely on one vendor which created future challenges of high costs and reduced flexibility for system modifications.

7. Public Relations and Communication

Regular public education on blockchain technology should have been provided to overcome its complex nature. User reluctance grew because they did not grasp how blockchain technology delivered better privacy and security features.

No data breaches happened, but the centralized system components received multiple allegations about their privacy vulnerabilities which damaged overall public trust.

8. Cost and Resource Intensity
The implementation of blockchain systems demands large upfront and ongoing expenses that can prove unaffordable to most other nations along with smaller organizations.

The system's functionality together with its up-to-date status demanded continuous spending on technology and expert skills for maintenance activities.

12.3 Case Study 2: Estonia's Use of Guardtime's Blockchain System for Vaccine Authentication

Estonia implemented Guardtime's blockchain technology under its well-known e-governance programs to produce secure vaccine authentication mechanisms that protected populace wellness. The new system serves as an exemplary model for blockchain applications that fight pharmaceutical counterfeits while enhancing worldwide supply chains of vaccines.

The problem of counterfeit vaccines exposes major challenges worldwide especially while pandemics generate higher demand than available supply.

Problems include the following:

- Public health risks emerge when vaccines carry either counterfeit stamps or improper storage conditions leading to severe medical problems.
- Economic loss occurs because counterfeit products harm pharmaceutical businesses as well as healthcare systems.
- Trust levels diminish throughout the public after incidents of counterfeit vaccines cause mistrust in vaccination programs.
- Blockchain technology became the foundation of Estonia's solution to resolve vaccine authentication and monitoring problems (Semenzin et al. 2022).

12.3.1 Guardtime's Blockchain System

Guardtime, an Estonian company specializing in blockchain for data security, developed a Keyless Signature Infrastructure (KSI) (Buldas et al. 2017) blockchain technology to enhance the vaccine supply chain.

Key Features of the System

- The blockchain built by Guardtime produces an invulnerable log showing vaccine production together with storage activities and distribution processes.
- Every vaccine batch receives a cryptographic signature for authentication throughout its entire production journey.

- The blockchain system provides absolute tracking capability which enables immediate monitoring of vaccines through every step of the supply chain for effective storage maintenance.
- Approved stakeholders at manufacturer, distributor, and healthcare provider levels have permission for separate data verification.

12.3.2 Integration of Blockchain into Vaccine Supply Chain

The blockchain system was integrated into the vaccine supply chain as follows:

Production
- Manufacturers use blockchain to submit production information that includes batch identification along with expiration dates. Authentication of each batch depends on cryptographically applied signatures.

Distribution
- The blockchain automates the logging of storage condition data (e.g., temperature and humidity) through IoT sensors during transportation. The system sends instant alerts to its relevant stakeholders whenever predefined conditions deviate.

Delivery to Healthcare Providers
- The vaccine's authenticity alongside its storage history gets verified by healthcare facilities prior to patient administration. Through blockchain verification, every administered medical batch is checked for authenticity to stop counterfeit vaccines from reaching healthcare customers.

Post-Administration
- Blockchain technology securely records vaccination documentation alongside batch number listings so healthcare authorities and persons can monitor medication source points (Fig. 12.1).

12.3.3 Implementation Details of Guardtime's Blockchain System for Vaccine Supply Chain

The use of Guardtime's blockchain technology by Estonia enables an efficient vaccine supply chain verification system, which provides high levels of security to authenticate medical products while protecting the integrity of supply routes. The technical implementation includes the following details.

1. Blockchain Framework: Keyless Signature Infrastructure (KSI)
KSI blockchain from Guardtime serves as the core technology that authenticates vaccines. KSI blockchain represents a distinct version of blockchain technologies which presents these additional features:

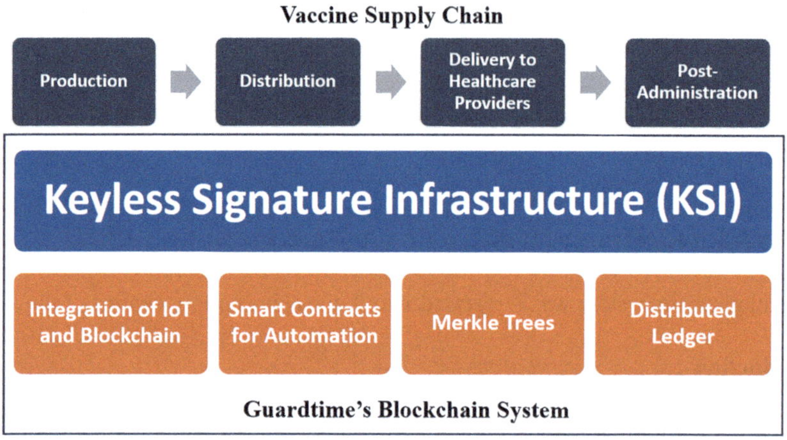

Fig. 12.1 Guardtime's blockchain system for vaccine supply chain management

Keyless Signatures function as cryptographic hash functions to authenticate data in the KSI system (Emmadi and Narumanchi 2017). The signature method removes key management risks which stem from both key loss and compromise situations.

The KSI system operates through linear scalability features to serve high-performance applications by maintaining its processing capabilities when dealing with billions of transactions.

Multiple independent entities monitor data verification through decentralized algorithms which prevent any failure point from occurring.

The key features of KSI include the following:

- Vaccine data including batch numbers and manufacturing information gets hashed through cryptographic algorithms which use SHA-256 as an example. Hashing produces a special digital marker that functions as an exclusive fingerprint for any information.
- The hash receives automated timestamping through the blockchain ledger as an unalterable point in time record.
- The Merkle Tree framework integrates multiple hashes in order to simplify verification processes for the system.
- Validating the data becomes possible through authorized entities who use hash comparison of stored records against current data hash values.

2. Data Authentication and Immutability

The blockchain maintains an authenticated system by cryptographically validating every transaction that involves vaccine production distribution and administration.

Every transaction maintains cryptographic authentication prior to validation.

After writing, data becomes permanently immutable because the records can never be modified or deleted, thus creating an extended and tamperproof audit trail.

3. Integration of IoT and Blockchain

The system uses Internet of Things (IoT) devices which enable environmental condition monitoring and recording during vaccine transportation processes. Real-time data logging takes place on the blockchain system.

Electronic sensors which are part of IoT infrastructure monitor temperature results alongside humidity data and handle conditions data.

The blockchain system uses security features by applying hash functions to sensor records which create traceable records.

If predefined environmental parameters change, then the system sends immediate notification messages to party members responsible for the process.

4. Smart Contracts for Automation

The blockchain features smart contracts which implement automatic supply chain process management.

Smart contracts operate at checkpoints to perform automatic authentication of vaccine batches before transportation starts or before giving the vaccine to patients.

The system through condition monitoring uses contracts to track IoT data which triggers alerts any time the conditions exceed or fall short of set standards.

Restrictions on data accessibility apply through authorized access rules which permit manufacturers together with distributors and healthcare providers to obtain specific records.

5. Distributed Ledger Infrastructure

The Guardtime blockchain applies a distributed ledger approach which provides both redundancy and resilience capabilities to its system.

The network consists of independent nodes which include pharmaceutical companies together with regulatory agencies and healthcare providers. All nodes function with an equivalent up-to-date mirror of the blockchain data.

The KSI system avoids implementing consensus principles based on Work or Stake proof methodologies. Hash-linked timestamping functions as the basis for securing data validation along with scalability.

6. Cryptographic Hashing and Merkle Trees

Guardtime's blockchain implements Merkle Trees as a data optimization method and for verification functions (Becker 2008).

Merkle Root integrates every data hash into one consolidated root hash which ensures total transaction integrity in a block.

The verification process of data demands only minimal hash operations to run efficiently on both small and enormous datasets (Szydlo 2004).

7. Data Privacy and Access Control

To ensure privacy:

The blockchain platform receives anonymized data before its hashing and storage process.

The platform stores large datasets that include detailed vaccine handling logs outside of the blockchain network. The on-chain storage includes data hashes along with metadata while keeping most storage data off-chain to reduce the overhead of blockchain operations.

Private access rights to data depend on role-based smart contracts for enforcement of permissioned access.

8. Scalability and Interoperability

KSI blockchain architecture grows proportionally to its usage which matches requirements for nationwide and worldwide supply chain management systems (Soares et al. 2023).

Real-time operational efficiency happens during high usage periods thanks to its ability to process millions of transactions per second.

It functions with multiple blockchain systems and following international vaccine track and trace standards issued by the World Health Organization (WHO).

9. Key Security Features

The system reveals all attempts to modify data due to its hash verification system (Heston 2017).

The distributed system design provides operational resilience because attacks on individual nodes do not affect overall system functionalities.

The system provides complete auditability as it creates unalterable records that maintain full transparency throughout the supply chain operations.

12.3.4 Benefits of Guardtime's Blockchain System

Some of the potential benefits include the following:

A. Preventing Counterfeits

The system remains tamper-proof because it distributes and administers authentic vaccines only (Chluski 2024).

The ability to track operations in real time makes it more difficult for unauthorized individuals to insert counterfeit products during supply chain distribution.

B. Strengthening Supply Chain Integrity

The blockchain delivers complete, unalterable vaccine records which boosts accountability across vaccine manufacturer, distributor, and healthcare provider networks.

The system uses automatic notification methods to detect temperature or handling irregularities which stop dangerous vaccines from entering use.

C. Enhancing Public Trust

Public reassurance about vaccine origin and quality exists through traceable documentation back to their production origins.

Current vaccination programs gain increased public trust when data becomes available for open inspection.

D. Supporting International Standards
The blockchain system in Estonia follows the recommended rules established by the World Health Organization for tracking vaccine supplies throughout their entire chain.

The model can provide other countries with a flexible solution to fight counterfeit vaccine distribution.

12.3.5 Challenges and Considerations

The successful implementation of blockchain in Estonia has brought along several ongoing obstacles (Semenzin et al. 2022).

To establish and support blockchain systems extensive technological competence combined with financial resources stands required.

National adoption of blockchain technology requires countries to pass regulatory assessments, which entails solving interoperability issues.

Conditional access to blockchain implementations exists because they demand heavy financial resources which push them out of reach for countries with limited funding.

12.4 Case Study 3: HealthVerity's Platform for Viral Genomic Tracking

Viral genomic tracking gathers viral genome specimens, which are then sequenced before analysis to track virus evolution patterns and spread together with genetic mutations. The technique aids scientists in their understanding of viral outbreaks and their work regarding treatment development and vaccine creation. Advancements in bioinformatics along with the next-generation sequencing (NGS) helped genome tracking to become the crucial base for Modern Virology (Hill et al. 2021).

HealthVerity provides a platform that allows viral genomic tracking through its interoperable data ecosystem which maintains privacy standards. The public health agencies including the Centers for Disease Control and Prevention (CDC) gain effective capabilities to track and respond to viral outbreaks with this system.

12.4.1 Core Components of the Platform

HealthVerity's (Hurley 2018) platform employs blockchain and artificial intelligence (AI) together with cloud computing to provide immediate viral genomic tracking throughout the United States. The advanced monitoring system plays a vital role in tracking virus evolution together with maintaining transmission knowledge which supports public health strategy building.

1. Data Integration and Interoperability

HealthVerity aggregates data from diverse and fragmented sources into a unified system.

- The platform functions through receiving viral sequences from public health laboratories, academic institutions, and private sequencing enterprises (Mahmoud and Khalil 2024).
- The analysis of electronic health records (EHRs) insurance claims and pharmacy records delivers practical information about viral variant impacts on patient healthcare outcomes.
- Health data maintained by the CDC together with epidemiological statistics is incorporated into the system to add contextual information.
- Standard data formats, FASTA files, are used by the platform to ensure compatibility alongside HL7 standards for healthcare information exchange. This system provides the ability to exchange and analyze information in a smooth way between institutions.

2. Real-Time Genomic Analysis

The genomic data analysis done by HealthVerity operates in real time (Wohl et al. 2016).

- The continuous scanning of viral sequences by the platform enables variant identification of concerning mutations.
- AI algorithms through automation process the mutation classification to determine their effects on disease transmissibility and virulence together with immune escape capabilities.
- The Scientists used genomic data to create phylogenetic trees to track the spread of the variants and advancements in different regions.
- The system delivers rapid information about new security threats which enhances operational readiness.

3. Blockchain for Data Integrity

- HealthVerity tracks data authenticity with the aid of blockchain technology for full traceability.
- The blockchain stores genomic sequences with their metadata through cryptographic hashing which makes them impossible to modify.
- Each data interaction gets recorded on the blockchain system including the uploading of data along with analytical activities and use events. This ensures transparency and accountability.
- The network includes authorized nodes, which exist within the system (e.g., public health agencies and research labs), that make authorized transactions while maintaining distributed ledger data consistency.

4. Privacy-Preserving Framework

- Patient privacy stands as a top priority for HealthVerity in its operations.
- Prior to information sharing, HealthVerity removes or masks all personal identifiers to meet both HIPAA and additional privacy regulations requirements.
- The AI training process through federated learning allows models to work on various distributed datasets while keeping their content secure on each individual device.
- The privacy-protecting methods employed by HealthVerity allow its platform to serve healthcare scenarios that need privacy protection.

5. AI-Powered Insights

- The platform turns unprocessed data into valuable insights using AI and ML, enabling users to base their choices on data while maximizing operations and identifying untapped prospects.
- AI systems use variant-clinical correlation to evaluate how viral mutations affect the severity of illness as well as hospitalization rates and mortality outcomes.
- The system implements predictive modeling based on machine learning algorithms that predict specific variant spread through analysis of historical and current transmission data, thereby supporting better proactive decision-making.
- Through this platform, scientists can measure the effect that different viral mutations create on vaccine effectiveness together with the success rate of treatment strategies.

6. Dynamic Transmission Modeling

- HealthVerity tracks disease spread by uniting genetic information with statistical disease models, which helps scientists monitor transmission data at a more detailed level.
- The system uses real-time data to monitor variant spread across different regions combined with population distribution.
- Traditional contact tracing receives improvements through the combination of genomic data together with proximity data through contact tracing Augmentation.
- Local intervention decisions become more effective because insights help identify testing and quarantine sites together with vaccination campaign regions.

7. Cloud-Based Infrastructure

- The platform depends on a solid cloud computing framework that exists as a core element.
- The platform manages extensive genomic together with clinical information volumes through optimal peak-period performance.
- High availability: ensures uninterrupted access to data and analytics tools for authorized users.
- The system implements end-to-end encryption together with multifactor authentication as security measures for protecting delicate data.

12.4.2 Applications in Viral Evolution and Transmission Dynamics

HealthVerity's platform generates valuable information that enables health authorities to build effective responses alongside policymaking and disease preparedness development for new infectious agents.

Some of the applications are as follows:

1. Variant Surveillance Through its monitoring system, the platform tracks new viral mutations that emerge. The platform tracks variants of concern which consists of analyzing changes to SARS-CoV-2 spike protein mutations that influence transmission and immune system response. Public health guidance is shaped by real-time monitoring results of variant prevalence in the population.

2. Vaccine and Treatment Optimization Correlates vaccine breakthrough cases with specific viral mutations to refine vaccine development. Research determines the effect of mutations on the effectiveness of current antiviral therapies.

3. Predictive Modeling of Outbreaks AI models can predict the course of epidemics based on viral evolution. Also, the simulations can measure the possible consequences of interventions, such as travel restrictions or vaccination drives.

12.4.3 Significant Benefits

HealthVerity delivers multiple important advantages to healthcare along with life sciences and public health through state-of-the-art data merging services combined with privacy regulation capabilities and data-driven decision-making solutions.

HealthVerity delivers three main advantages which include the following:

- Time-sensitive healthcare data allows for fast and evidence-based decisions regarding new threats. Health authorities can develop better vaccination strategies, treatment solutions, and containment strategies through thorough knowledge of viral evolutionary patterns.
- The platform participates in international genomic surveillance platforms like GISAID for supporting data sharing and collaborative endeavors.
- The system performs automated assessments of mutations and phylogenetic analysis which shortens genomic surveillance operations as well as decreases resource needs.
- Advanced privacy features in the system uphold legal and ethical demands which lead to enhanced trust from the public.

Table 12.2 Comparison table

Feature	HealthVerity (USA)	CoG-UK (Europe)	Japan (Asia)
Data integration	Genomic + EHR + public health data	Genomic and epidemiological data only	Genomic and epidemiological data only
Technology	Blockchain, AI, federated learning	Centralized database, limited AI	Decentralized labs, basic AI
Privacy measures	Advanced (anonymization, federated learning)	Moderate (limited anonymization)	Moderate (basic anonymization techniques)
Real-time analytics	Yes (AI-driven)	No (manual and pipeline-based analysis)	Limited
Data sharing	Restricted (privacy-focused)	Extensive (global platforms like GISAID)	Extensive (global platforms like GISAID)
Scalability	High (cloud-based, decentralized infrastructure)	Moderate (centralized system with bottlenecks)	Moderate (fragmented sequencing network)
Integration with public health	Strong (linked to clinical outcomes)	Strong (policy-driven integration)	Moderate (delayed data consolidation)

12.4.4 Comparison of HealthVerity's Platform with Genomic Tracking Systems in Europe and Asia

The US-based HealthVerity platform, focusing on real-time viral genomic tracking, shares similarities and differences with genomic surveillance systems employed in Europe and Asia (Wiggans et al. 2011; Kage et al. 2023; Kunimasa et al. 2024; Uelze et al. 2021) (Table 12.2).

HealthVerity's platform demonstrates how federated learning technologies maintain data privacy to fulfill privacy laws during real-time analytics operations.

The state-of-the-art technology implemented by HealthVerity stands in the way of accessibility for settings that have limited resources and funding. Both CoG-UK and Japan's simplified models provide a cost-effective option to the standard.

12.5 Case Study 4: MediBloc's Blockchain Solution for Streamlined Clinical Trial Management

As a blockchain-based healthcare system, MediBloc functions to let patients handle their medical records alongside secure interoperable information exchange between healthcare providers, researchers, and other stakeholders (Shrimali et al. 2023). The healthcare platform of MediBloc demonstrates substantial value in medical virology and other related domains because it provides secure, transparent data administration.

MediBloc deploys blockchain capabilities to develop a decentralized environment that handles healthcare data management. The platform includes three essential aspects.

Every patient keeps complete authority over their medical information including rights to determine the accessibility of their records and how they are distributed.

The platform makes use of a system that integrates healthcare data with Electronic Health Records (EHRs), laboratory results, and diagnostic imaging.

The blockchain infrastructure enables safe medical data storage by granting access to approved users alone with patient authorization.

12.5.1 Architecture of MediBloc's Platform

A secure healthcare data management system operates within the MediBloc ecosystem through multiple connected elements that deliver efficient and privacy-oriented healthcare services (Fig. 12.2).

(a) Blockchain Layer

The blockchain platform serves as a fundamental layer in MediBloc's system for delivering reliable handling of healthcare information with secure and transparent operation. This technology implements the Cosmos SDK to construct blockchains that display scalable, interoperable, and application-specific functionality.

MediBloc uses the delegated Proof-of-Stake consensus protocol to provide speed in transactions, minimize power usage beyond Proof-of-Work systems, and enable MED stakeholder owners to govern network activities.

The Cosmos SDK framework enables MediBloc to build its blockchain which implements Inter-Blockchain Communication (IBC) protocol for cross-chain interoperability. The architecture enables users to easily share information between MediBloc and different blockchain-based healthcare platform designs.

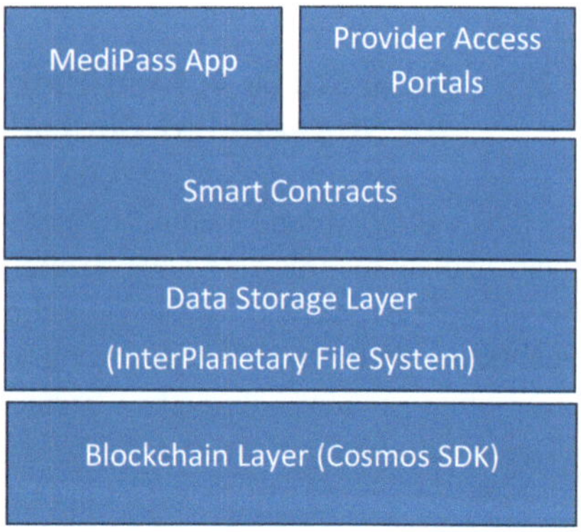

Fig. 12.2 Architecture of MediBloc's platform

Key Elements of the Blockchain Layer Are as Follows

- The MediBloc platform functions through a distributed network of nodes which protects it from control by any individual entity. The distributed structure cuts down the possibility of data attacks while making healthcare information available through multiple nodes that operate independently.
- Through cryptographic hash techniques, the blockchain layer maintains data immutability for all its records. The system records every action such as data access, sharing, or modification transactions in an unalterable ledger that makes tampering attempts impossible (Singh et al. 2021).
- The blockchain keeps track of every platform action through its transparent and auditable system. Healthcare providers and patients together with researchers maintain access to data origin tracking which enables them to check data authenticity while confirming regulatory compliance (Zukarnain et al. 2023).
- The blockchain infrastructure is built to process numerous transactions efficiently which demonstrates its value for worldwide healthcare system implementations.

(b) Data Storage Layer

As a core component within MediBloc's blockchain infrastructure, the data layer exists to ensure secure medical data storage as well as management along with authorization processes. The data layer contains several different technologies to manage medical information because blockchain technology alone cannot work effectively with large files and sensitive data.

The data layer handles data storage by connecting between on-chain and off-chain storage methods to achieve optimal security together with efficiency and growth capabilities. Each data entry exists on-chain where metadata together with access authorization, transaction records, and encryption keys are saved. All operations regarding data become transparent and immune to tampering through this method. Large files including both medical images and genomic data and patient records get stored through off-chain storage operations. Two storage configurations that work together include IPFS (InterPlanetary File System) and blockchain-enabled cloud services.

Patient data encryption takes place during upload so it becomes available only to users who possess access keys (Bhatt and Chakraborty 2023). The blockchain protocols protect private keys used to access data through secure management processes. Blockchain maintains an indestructible record of all data access events that shows which users requested information at particular times.

(c) Smart Contracts

Patients can use automated consent systems through smart contracts to provide or stop authorization of medical record access any time. The smart contracts enable the definition of access rules which determine what data patients can share and for what amount of time.

Patients in addition to healthcare providers and researchers can create secure and transparent data-sharing permissions through smart contracts. A researcher initiates access to anonymized data through a smart contract which specifies the study conditions together with compensation details.

Patients receive MED tokens from blockchain smart contracts when they share their anonymized data with the platform. Tokens move automatically upon successful termination of all data-sharing transactions.

(d) User Interfaces

The interface of MediBloc provides user-sensitive functionality alongside intuitive interaction that makes the platform accessible to different stakeholders such as patients and healthcare providers together with researchers and institutions. The MediPass application together with web-based portals serves as the interface that allows users to work with both blockchain functionality and data components.

Through its patient-focused mobile app known as MediPass, users can both upload medical records as well as handle and distribute their health information to others and acquire MED tokens through research involvement.

Healthcare professionals benefit from web interfaces to request and gain access to patient-permitted data through the system.

12.5.2 Applications in Medical Virology

- The platform of MediBloc enables real-time data collection and transfer for outbreak surveillance as it enables tracking of SARS-CoV-2 variants. The platform could help protect patient privacy while conducting contact tracing through its tools along with sharing anonymized genomic data to notify organizations about new emerging virus mutations (Babu et al. 2023).
- The platform of MediBloc grants researchers access to anonymous datasets which they can utilize for creating predictive models regarding viral mutations. Scientists can use the influenza virus evolutionary track to determine appropriate yearly vaccine strains.
- The sharing of data between different nations constitutes an essential aspect for virology. The blockchain technology of MediBloc maintains security and enables verifiability for genomic and clinical data that healthcare entities share between different nations. Viral genomic data sharing occurred between African and European laboratories when Ebola outbreaks happened.
- Through its system, MediBloc enables researchers to conduct vaccine efficacy evaluation by connecting vaccination data with resulting health consequences to determine viral strain effectiveness. Researchers can study mRNA vaccine effectiveness against COVID-19 variant mutations utilizing anonymized breakthrough infection statistics as an illustrative case.
- Drug resistance monitoring features allow MediBloc to monitor viral resistance mutations of HIV or Hepatitis B viruses through secure exchanges of genomic data and treatment information with research teams.

- Through its blockchain approach, MediBloc enables increased research partnership among institutions to eliminate data silo problems and build trust-based alliances.
- The researchers using their platform receive confirmation about the original source of all information analyzed during the studies.
- The research findings achieve complete validation through immutability of recorded data.
- Patients benefit from blockchain because it organizes the consent process for data-sharing permission.
- Scientists from different countries collaborate via MediBloc to securely obtain genomic and clinical data while following all local data privacy rules during their Zika virus outbreak research.

The platform MediBloc demonstrates great potential to remodel data handling in healthcare and virology by delivering protected information sharing and patient-controlled management while driving positive health results worldwide.

12.6 Other Notable Case Studies

12.6.1 China: VeChain for COVID-19 Vaccine Tracking

VeChain blockchain platform became the selection of China to manage safe vaccine distribution and monitoring (Kamenivskyy et al. 2022). Traceability of every vaccine from the supply chain to end-point was made possible through the platform by logging transportation periods and storage conditions along with authentication details. By performing real-time monitoring, VeChain evaluated how vaccine transportation was handled while keeping track of temperature and humidity levels in order to avoid vaccine spoilage incidents stemming from improper storage.

The system created an unalterable and clear tracking history for vaccines which traversed from manufacturing all the way to delivery so VeChain solved problems with fake vaccines and storage-related mistakes. The blockchain-based process gave healthcare providers as well as regulators and distributors the ability to validate vaccine authenticity and safety at all points of the pharmaceutical journey, thus strengthening trust in vaccination activities.

The outcome of this initiative produced significant improvements in China's COVID-19 vaccination operations. The platform reduced vaccine waste by tracking supply chains while maintaining proper storage conditions and it strengthened the distribution network, particularly for remote locations. Through its clear system, the vaccination supply chain received better regulatory oversight which allowed for swift detection and remedy of transportation gaps during distribution.

VeChain blockchain technology enabled China to create an example for innovative public health solutions through supply chain integrity in its vaccine distribution network.

12.6.2 India: StaTwig for Vaccine Supply Chain Management

The StaTwig company launched VaccineLedger as a blockchain solution platform, which India used to enhance vaccine supply chain operations while dealing with vaccine loss issues, counterfeit product risks, and distribution problems (Cuadrado et al. 2022). VaccineLedger established complete supply chain transparency that delivered continuous vaccine tracking from production until they reached treatment sites. The platform included IoT sensors for monitoring temperature parameters along with storage conditions because they guaranteed compliance with cold chain specifications throughout transportation.

The system displayed excellence in detecting and stopping storage or handling mistakes that endangered vaccine effectiveness. The transactions recorded automatically on blockchain ledgers through VaccineLedger created an unalterable system for suppliers, healthcare providers, and distributors to be accountable to each other. The system functioned as a protective measure against vaccine counterfeits in supply networks while improving faith in healthcare vaccination initiatives.

The StaTwig solution achieved maximum efficiency in vaccine distribution by focusing heavily on areas that normally receive limited attention. Real-time inventory data and demand data tracked on the platform enabled data-based choices for better programming resources and shorter delivery times.

The implementation of blockchain technology in the Indian vaccine supply chain enhanced all operational facets and improved the delivery of vaccines to their target destinations. This initiative demonstrated how blockchain technology can transform worldwide healthcare structures to be stronger.

12.6.3 Africa: Blockchain for HIV and Viral Load Tracking

Africa utilizes blockchain technology for enhancing both HIV treatment management and viral load tracking systems and patient care delivery and stopping HIV transmission. Blockchain technology platforms track medication adherence and monitor viral load tests as well as serve to protect patient data safety through secure storage and data sharing capabilities (Ukoaka et al. 2024). Blockchain technology implements transparent patient records that remain unalterable, thus creating a safe database wherein healthcare providers can view real-time accurate treatment updates for each patient independent of tampering risks and breaches.

Using blockchain technology, healthcare providers add a capability for easy data exchange while assuring care continuity across various medical facilities. The streamlined data transmission helps healthcare providers manage ART adherence effectively which reduces HIV complications and protects against drug-resistant strains of HIV.

Blockchain technology improves treatment approaches while resolving essential problems involving drug shortages and fake medication distribution. Through transparent supply chain monitoring, the system guarantees that patients obtain quality medications by guarding the integrity of their pharmaceutical products.

Blockchain technology effectively tracks HIV and viral load data in Africa through powerful applications that improve healthcare and build trust while ensuring suitable HIV/AIDS management.

12.6.4 Nigeria: Blockchain for Immunization Programs

The adoption of blockchain technology in Nigeria focuses on reinforcing national immunization programs by solving vaccine fraud and distribution issues and record-keeping inadequacies. The implementation of blockchain technology established a tracking and authentication system for vaccines which ensured safe delivery to the official recipients. The decentralized ledger provided transparency which enabled real-time vaccine supply chain monitoring during all stages from production through storage and all the way to administration and distribution (Onuorah et al. 2023).

Through blockchain implementation, the system achieved better accuracy and reliability in recording immunization history of children. The healthcare providers used mobile technology to retrieve patient vaccination records through which they maintained proper documentation and monitoring of children's immunization status. Through its tamper-proof design and searchability, the blockchain platform created trust in vaccination process, thus reducing vaccine-hesitancy issues in specific parts of Nigeria.

These enhancements allowed the system to find missing vaccination data points especially in locations without regular medical services so additional resources could be assigned for improved outreach. Through this system, the supply chain became protected from counterfeit vaccines so patients received only authentic vaccines that were effective.

Nigeria's public health initiatives have gained substantial value from blockchain technology through improved transparency and accountability while maintaining excellent efficiency.

12.6.5 Kenya: Blockchain for Vaccine Delivery in Rural Areas

The application of blockchain technology in Kenya addresses vaccine delivery problems that occur during distribution to rural and far-flung areas (Nyabwari 2021). A blockchain platform was established to supervise all vaccine supply chain operations from deliveries to storage and distribution points to administration sites by using proper conditions. The system uses Internet of Things sensors to track vital shipment conditions of temperature and humidity because these factors determine vaccine effectiveness in areas with minimal infrastructure.

Through blockchain implementation, the system delivered live information about vaccine transportation which enabled healthcare providers to follow delivery routes and maintain supply records. The reform enhanced both the visibility and reliability of vaccine monitoring because it decreased incidents of vaccine loss, spoilage, and diversion. The transparent blockchain system enabled better communication between local clinics, distributors, and government agencies, which led to proper vaccine delivery to intended locations and correct timing of administration.

Blockchain technology restored trust within the vaccine delivery system by resolving doubts about vaccine accessibility together with fake vaccines in areas with limited care. Through this system, both resources and distribution planning became more efficient allowing vaccines to reach all regions of Kenya especially rural locations where healthcare facilities were scarce.

Blockchain implementation showcases how digital technology can be used to address the logistical barriers, thus giving a boost to public health delivery and expanding the access to life-saving vaccines in remote areas.

12.7 Conclusion

This chapter proves that blockchain technology acts as a transformative power for medical virology by solving important medical data security requirements, operational inefficiency, and authenticity challenges. Various worldwide case studies confirm how blockchain technology successfully integrates into healthcare environments to develop public health management systems and enables vaccine authentication as well as viral genomics tracking, clinical trials, electronic health records, pandemic surveillance, and supply chain management.

The case of both Singapore's TraceTogether app and Estonia's Guardtime vaccine authentication demonstrates how blockchain enhances both system data correctness and reliability. The platform of HealthVerity in the United States has utilized blockchain technology to track viral genomics and monitor changes in viral evolution according to their platform. The blockchain system developed by MediBloc in South Korea has organized clinical trials by improving both cooperative relationships and data protection security measures.

Blockchain technology presents three relevant demonstrations that show its capability for protecting healthcare data while enhancing healthcare services and healthcare infrastructure system stability. The discussion finishes by demonstrating blockchain's potential to revolutionize healthcare by delivering better patient outcomes as well as superior health management systems and safer, more efficient medical operations. Blockchain will take a foundational position to determine the healthcare direction of medical virology and the structure of worldwide healthcare systems.

References

Ahmed N et al (2020) A survey of COVID-19 contact tracing apps. IEEE Access 8:134577–134601

Akinbi A, Forshaw M, Blinkhorn V (2021) Contact tracing apps for the COVID-19 pandemic: a systematic literature review of challenges and future directions for neo-liberal societies. Health Inf Sci Syst 9:1–15

Babu ES et al (2023) MediBlocks: secure exchanging of electronic health records (EHRs) using trust-based blockchain network with privacy concerns. Clust Comput 26(4):2217–2244

Bay J et al (2020) BlueTrace: a privacy-preserving protocol for community-driven contact tracing across borders. Government Technology Agency-Singapore, Tech. Rep 18.1

Becker G (2008) Merkle signature schemes, erkle trees and their cryptanalysis. Ruhr-University Bochum, Tech. Rep 12, p 19

Bengio Y et al (2020) The need for privacy with public digital contact tracing during the COVID-19 pandemic. Lancet Digit Health 2(7):e342–e344

Bhatt V, Chakraborty S (2023) MediBlock: a pervasive way to create healthcare value in secured manner for personalized care. In: Recent advances in blockchain technology: real-world applications. Springer International Publishing, Cham, pp 233–243

Braithwaite I et al (2020) Automated and partly automated contact tracing: a systematic review to inform the control of COVID-19. Lancet Digit Health 2(11):e607–e621

Buldas A, Laanoja R, Truu A (2017) Keyless signature infrastructure and PKI: hash-tree signatures in pre- and post-quantum world. Int J Serv Technol Manag 23(1–2):117–130

Chatterjee JM, Saxena SK (2024) Artificial intelligence in medical virology. Springer Singapore. https://doi.org/10.1007/978-981-99-0369-6

Chluski A (2024) The utilization of blockchain technology in healthcare management. In: Proceedings of the 25th European conference on knowledge management. Academic Conferences International

Cuadrado DC et al (2022) Safeguarding the Covid-19 vaccine distribution: evaluating the role of blockchain. U4 Issue

Emmadi N, Narumanchi H (2017) Reinforcing immutability of permissioned blockchains with keyless signatures' infrastructure. In: Proceedings of the 18th international conference on distributed computing and networking

Hanif S, Khan A, Mustafa E, Ali M, Shuja J, Hasan MH (2024) Ensuring integrity of Quranic verses using Hyperledger fabric framework under IBM blockchain platform. In: 2024 IEEE 14th symposium on computer applications & industrial electronics (ISCAIE), Penang, Malaysia, pp 505–510. https://doi.org/10.1109/ISCAIE61308.2024.10576345

Heston TF (2017) A case study in blockchain healthcare innovation. Int J Curr Res 9:60587–60588

Hill V et al (2021) Progress and challenges in virus genomic epidemiology. Trends Parasitol 37(12): 1038–1049

Hurley D (2018) Blockchain for patient and HCP data rights management: lessons from an enterprise install. Telehealth Med Today

Kage H et al (2023) Human resources for administrative work to carry out a comprehensive genomic profiling test in Japan. Cancer Sci 114(7):3041–3049

Kamenivskyy Y, Palisetti A, Hamze L, Saberi S (2022) A blockchain-based solution for COVID-19 vaccine distribution. IEEE Eng Manag Rev 50(1):43–53. https://doi.org/10.1109/EMR.2022.3145656. PMCID: PMC9088775

Kunimasa K et al (2024) Real-world assessment of comprehensive genome profiling impact on clinical outcomes: a single-institution study in Japan. Cancer Med 13(18):e70249

Lee JK, Lin L, Kang H (2021) The influence of normative perceptions on the uptake of the COVID-19 TraceTogether digital contact tracing system: cross-sectional study. JMIR Public Health Surveill 7(11):e30462

Leith DJ (2023) Contact tracing app privacy: what data is shared by non-GAEN contact tracing apps. Mobile Netw Appl 28:1460–1472. https://doi.org/10.1007/s11036-023-02136-4

Mahmoud SH, Khalil AA (2024) Viral genomics. In: Microbial genomics: clinical, pharmaceutical, and industrial applications. Academic Press, pp 31–70

Nguyen DC et al (2021) Blockchain and AI-based solutions to combat coronavirus (COVID-19)-like epidemics: a survey. IEEE Access 9:95730–95753

Nyabwari FO (2021) Use of block chain technology in vaccine distribution to enhance vaccine safety in Kenya. Diss. University of Nairobi

Onuorah IC et al (2023) Immunization schedule system using blockchain technology. J Sci Dev Res

Radford AD, Chapman D, Dixon L, Chantrey J, Darby AC, Hall N (2012) Application of next-generation sequencing technologies in virology. J Gen Virol 93(Pt 9):1853–1868. https://doi.org/10.1099/vir.0.043182-0. Epub 2012 May 30. PMID: 22647373; PMCID: PMC3709572

Rani M, Lakshmi G, Navaneetha C, Nagamani K (2024) Analysing the automation of artificial knowledge in virology for safety and effectiveness in healthcare: equilibrium of advancement and trials for secure and productive health necessities. Qeios. https://doi.org/10.32388/FBT4P3

Semenzin S, Rozas D, Hassan S (2022) Blockchain-based application at a governmental level: disruption or illusion? The case of Estonia. Polic Soc 41(3):386–401

Shrimali B, Surati S, Trivedi H (2023) MediBlock: a blockchain-based architecture for secure healthcare system. In: 2023 international conference on advancement in computation & computer technologies (InCACCT), Gharuan, India, pp 750–755. https://doi.org/10.1109/InCACCT57535.2023.10141848

Singh C et al (2021) Medi-Block record: secure data sharing using block chain technology. Inf Med Unlocked 24:100624

Soares B, Ferreira A, Veiga PM (2023) The benefits and challenges of blockchain technology and eHealth implementation in Estonia-a literature review. Appl Med Inf 45(4):118–131

Szydlo M (2004) Merkle tree traversal in log space and time. In: Advances in cryptology-EUROCRYPT 2004: international conference on the theory and applications of cryptographic techniques, Interlaken, Switzerland, May 2–6, 2004. Proceedings 23. Springer Berlin Heidelberg

Trivedi A, Vasisht D (2020) Digital contact tracing: technologies, shortcomings, and the path forward. SIGCOMM Comput Commun Rev 50(4):75–81

Uelze L et al (2021) Toward an integrated genome-based surveillance of Salmonella enterica in Germany. Front Microbiol 12:626941

Ukoaka BM, Ugwuanyi EA, Ukueku KO, Ajah KU, Udam NG, Daniel FM, Wali TA, Gbuchie MA (2024) Digital tools for improving antiretroviral adherence among people living with HIV in Africa. J Med Surg Public Health 2:100077

Wiggans GR, VanRaden PM, Cooper TA (2011) The genomic evaluation system in the United States: past, present, future. J Dairy Sci 94(6):3202–3211

Wohl S, Schaffner SF, Sabeti PC (2016) Genomic analysis of viral outbreaks. Annu Rev Virol 3(1): 173–195

Zukarnain ZA et al (2023) Medi-Block record secure data sharing in healthcare system: issues, solutions and challenges. Comput Syst Sci Eng 47:3

Part III

Advanced AI for Disease Prediction and Medical Innovations

Bio-Inspired Approaches for Optimal Kidney Paired Donation (Infectious Risk Analysis)

13

Eman Saleh, Aya Amir, Rehab Alaa, R. Sujatha, Aboul Ella Hassanien, Ashraf Darwish, and Sameh H. Basha

Abstract

The world is now facing a continuous increase in people with kidney failure and increasing deaths due to this disease. It was necessary to work on this serious problem. When kidneys fail to perform their functions, only two solutions are available for the patient, either dialysis or kidney transplantation. Dialysis doesn't provide a permanent solution for renal kidney failure and regular painful follow-ups are mandatory; on the other hand, kidney transplantation offers a life-long cure. It enables patients to live in a way closer to their normal lives. But it isn't

E. Saleh · A. Amir · R. Alaa
Faculty of Science, Cairo University, Giza, Egypt
e-mail: 201527065@std.sci.cu.edu.eg; 20130439@std.sci.cu.edu.eg; 201527105@std.sci.cu.edu.eg

R. Sujatha
School of Information Technology & Engineering, Vellore Institute of Technology, Vellore, India
e-mail: r.sujatha@vit.ac.in

A. E. Hassanien
Faculty of Computers and AI, Cairo University, Cairo, Egypt

Scientific Research School of Egypt (SRSEG), Cairo, Egypt

A. Darwish
Scientific Research School of Egypt (SRSEG), Cairo, Egypt

Faculty of Science, Helwan University, Helwan, Egypt

S. H. Basha (✉)
Faculty of Science, Cairo University, Giza, Egypt

Scientific Research School of Egypt (SRSEG), Cairo, Egypt

Faculty of Science, Galala University, Suez, Egypt
e-mail: samehbasha@cu.edu.eg; samehbasha@gu.edu.eg

J. M. Chatterjee, S. K. Saxena (eds.), *Artificial Intelligence and Blockchain in Precision Medicine and Virology*, Medical Virology: From Pathogenesis to Disease Control, https://doi.org/10.1007/978-981-96-8919-4_13

always that easy for patients to find compatible donors; this can cost them many years on waiting lists, and sometimes they die while waiting. This chapter aims to resolve kidney exchange problems optimally by utilizing various bio-inspired optimization techniques by determining the maximum groups of compatible donors/recipients from the registry of incompatible pairs submitted as transplant candidates. Chemical reactions, whales, and ant lions optimization algorithms are proposed to quickly find the optimal solution of the kidney paired donation and a maximum number of kidney transplants based on existing donor-patient pairs. Roulette and rank-based candidate selection methods are applied with the optimization algorithms to speed up the candidate selection in case of a massive pool of patients and donors. The experiments used a Seidman generator to generate six instances as reference groups. We only accept incompatible donor-patient and altruistic donor pairs and do not accept a patient without any donor, as evident from the data set. The results show that the ant lion optimizer with rank selection method is beneficial against the other two optimizers.

Keywords

Bio-inspired · Kidney · Ant lion optimization · Whale optimization · Chemical reaction optimization · Cycle-chain · Roulette wheel · Rank selection

13.1 Introduction

Chronic kidney disease (CDK) is a serious problem that threatens human life, as kidneys are the organs responsible for purifying the blood from toxins and getting rid of excess substances from the body. At this late stage of the disease, kidneys fail to perform their functions, exposing the patient to problems that may lead to losing his life due to the accumulation of these toxins. According to international statistics, the number of CKD patients is increasing steadily and is expected to increase in the coming years. CKD is so dangerous because there is no treatment for it yet, and no treatment can make these affected kidneys recover and do their job again (Romagnani et al. 2017; Chen et al. 2019; Ng and Li 2018; Arora et al. 2017).

Given the importance of the role of the kidneys, as we said earlier, there must be alternative solutions to the role that kidneys are no longer able to play. In this case, there are only two ways for the patient, either going on dialysis throughout his whole life or replacing the affected kidneys with new healthy ones through a kidney transplant operation. Of course, the second way provides a better life for the patient. On the one hand, the dialysis process is challenging and exhausting, and it has strict dates that must be adhered to. Otherwise, the patient may develop a poisoning that may lead to death, making the patient not feel his complete freedom to live without permanent threat. On the other hand, studies have shown that many diseases, such as the C virus, are transmitted through dialysis machines (Fabrizi and Messa 2015; Timofte et al. 2020; Goel et al. 2018).

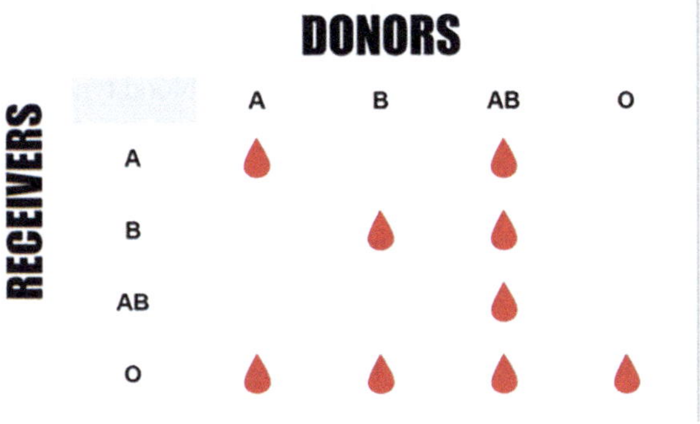

Fig. 13.1 Patient-donor pair blood type matching

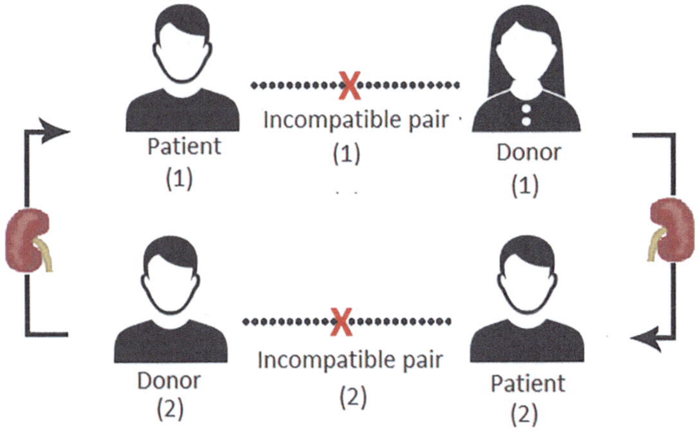

Fig. 13.2 Two-way kidney swap

Returning to the ability to obtain kidneys from living donors, fortunately, it doesn't require the donor to be a member of the patient's family. He can also be his friend or even a stranger, provided that the conditions necessary for the completion of kidney transplantation are met. It is very common for a donor who wants to donate his kidney, but he is an incompatible donor for this patient. As compatibility in blood types is one of the requirements to donate, we may find a patient-donor pair, but the donor's blood group is not compatible with the patient as matrixed in Fig. 13.1, such pairs are known as incompatible donor-patient pairs. And because this situation can be repeated for several pairs, why not gather these incompatible pairs together to search whether a donor from one incompatible pair is suitable for a patient from another pair in the group, as shown in Fig. 13.2, to get the best donor for each patient as possible (Martin and Danovitch 2017; Thiessen et al. 2021; Kratz and Andersson 2019).

This will increase the number of matches between incompatible pairs, and at the same time, each patient will get the most appropriate choice from among all available options. For example, by referring to the blood group table in Fig. 13.1, we find that the patient with blood type AB can receive the kidney from anyone else, provided that the other conditions are met. Still, the patient with blood type O can only get a kidney from a donor whose blood type is the same as his. Therefore, this patient has priority in obtaining the kidney over the other patient if there is a donor with blood type O.

The main three phases of the models for optimization of kidney paired donation problems are choosing the best optimization algorithm, choosing the best selection method, and the proposed technique used in the model. This chapter proposes research in three phases. In the first phase, ant lion, whale, and chemical reaction optimization are used. In the second phase, roulette wheel and rank selection methods are used. For the third phase, this chapter proposed a new technique that breaks the chain-cycle condition (Mierzejewska et al. 2013). A comparison analysis after each phase was conducted.

The primary contributions of this chapter are as follows:

- It identifies the number of transplants that help in the normal living of CKD patients.
- Reduced waiting time of the needy kidney patient by utilizing the combination of bio-inspired algorithms.

The rest of this chapter is organized as follows. Section 13.2 discusses the related works in this application with the varied techniques. Section 13.3 discusses the proposed work regarding chemical reaction optimization, whale optimization, and ant lion optimization with the cycle-chain, wheel, and rank selection approaches. Section 13.4 enumerates the various experimental results obtained using the techniques and concludes with Sect. 13.5.

13.2 Related Work

If the patient's health condition allows the transplant, it will be the best solution to provide a better life close to his normal life. And since a person can live with one kidney, in the past, a patient was only permitted to donate his organs after death. But now and after several countries have passed laws that allow living people to donate their organs, patients no longer must stay many years on waiting lists to obtain a suitable kidney from a deceased donor. This made the kidney transplant process a lot easier. Waiting for itself was a challenge. In front of the large increase in the number of patients, which is not matched by an increase in the same proportion of donors, the patient could have died while still on waiting lists before getting a suitable kidney. For example, in 2016, a waiting list contained more than one lakh patients who needed to transplant a new healthy kidney in the United States (https://www.kidney.org/news/newsroom/factsheets/Organ-Donation-and-Transplantation-Stats).

Organ transplantation places organs like the heart, intestine, kidney, liver, lung, and pancreas from healthy to diseased. Still, the real challenge is getting the matching organ that suits the receiver. Particularly, the number of patients in need of kidneys across the globe is increasing. The need to create the pool for the same is mandated within a certain geographical location. In the case of transplantation, any donor must be transferred to the needy, or the organ transplant alone is required, but it must be done on time. The risk factor in the transplantation is more and that too for the recipient in comparison. Donors are giving because of poverty and lack of education in the initial phases. Lots of discussion about the reason for donors' mindset carried out and ethical policies are laid to ensure injustice happens to anyone. Various instances rule out misconceptions behind kidney donors' life longevity, and donors lead normal lives like other healthy persons (Hou 2000; Fehrman-Ekholm et al. 1997; Kaballo et al. 2018).

Kidney transplantation has developed over the past 65 years into the best treatment option for kidney failure patients, significantly lowering suffering through increased survival and quality of life. However, the availability of organs continues to restrict access to transplant, possibilities for transplant are unequally distributed, and long-term transplant survival is still elusive. The National Kidney Foundation assembled an expert team to establish a research agenda in order to address these ongoing requirements. The panel's top priorities included increasing living donation, enhancing waitlist management and transplant preparation, maximizing the utilization of available deceased donor organs, and extending allograft longevity (Lentine et al. 2021).

It is extremely difficult and costly to manage the vast and diverse data sets using standard databases. For executing data analytics and building advanced data-driven capabilities, specific analytical tools are frequently needed. It is suggested to build machine learning models on top of Apache Spark to speed up the allocation of organs, such as kidneys, to the best candidate. This will increase donor utilization by finding a recipient in the permitted time, which will increase donor utilization overall. The suggested models assist in locating candidates on the waitlist who are willing to take kidneys that might otherwise be discarded (Ashiku et al. 2021).

Both chronic alterations and inflammation are significant indicators of long-term graft survival in kidney transplant biopsies. For the purposes of kidney research and transplant diagnosis, quantitative scoring of these characteristics is crucial. Visual scoring, however, requires a lot of labor and has low reproducibility. This study sought to determine whether convolutional neural networks (CNNs) might be used to measure chronic characteristics and inflammation in kidney transplant biopsy samples. The use of CNNs as a cutting-edge technique to produce surrogate end points for extensive clinical investigations will benefit the population and have the potential to enhance kidney transplant diagnosis (Hermsen et al. 2022).

Outline the multicenter, randomized controlled, web-based Living About Choices in Transplantation and Sharing trial, which aims to evaluate the efficacy of an updated Living About Choices in Transplantation and Sharing intervention in four transplant facilities in the Southeast. The intervention is a five-part website with the following modules: Introduction, Benefits and Risks, The Kidney Transplant Process, Finding a Potential Kidney Donor, and Act Now (Patzer et al. 2019).

A traditional issue in economics and computer science is how to allocate scarce resources in an effective and equitable manner. Living kidney donors are distributed to patients in need of an organ by a central market maker in kidney exchanges. In kidney exchanges, patients and donors are given priority using ad hoc weights chosen by a committee, which are subsequently used to inform an allocation algorithm that decides who gets what—and who doesn't. Present an end-to-end methodology for determining the weights of participant profiles for a kidney exchange. In order to prioritize patients, first ask human subjects for a list of patient characteristics they find acceptable. Then, using their responses, estimate weights in a systematic manner by asking participants comparison questions between patient profiles (Freedman et al. 2020).

To aid medical professionals in the process of organ allocation for liver transplants, create an effective decision-support tool. The mathematical model put forth here makes use of many sources of data to forecast the likelihood of organ survival at various cutoff points for each donor-recipient pair considered. Currently, this choice is primarily based on the model for end-stage liver disease, which disregards donor-receiver compatibility and solely considers the severity of the recipient (Dorado-Moreno et al. 2017a).

Multicriteria decision-making approach is used in multiple organ retrieval and exchange (MORE) and united network for organ sharing (UNOS) algorithms based on highly critical patient, priority, and weighting time. Certain weaknesses prevail in traditional methods that pave the way to integrating fuzzy logic to manage the critical stages. Fuzzy rules are generated by incorporating appropriate fuzzy scores and membership values. GUI-based inference systems are generated to ease the process (Minch and Sanders 1986; Stewart and Klassen 2017; Yuan et al. 2002). The human leukocyte antigens-based project involves all related stakeholders and makes EpViX system using ruby on rail software. The system calculates the panel reactivity antibody that determines the suitability of the transplantation of the organ (Aita et al. 2016). The machine learning approach is used to make the good liver transplantation process and work analysis with imbalanced datasets gathered from several hospitals (Dorado-Moreno et al. 2017b).

Various approaches in KPD are two-way, three-way, open-chain, and closed-chain donation. Graphical user interface-based environment working with the principle of cycles and chains, cycles and chains with fallbacks, and locally relevant subgraph is established for KPD (Mierzejewska et al. 2013; Bray et al. 2019). Challenging work is finding the correct pair by considering multiple attributes for perfectness, and tie-breaking arises mapped with artificial intelligence concepts (Freedman et al. 2020; Rahwan and Simari 2009).

13.3 Preliminaries

13.3.1 Chemical Reaction Optimization

A chemical reaction is when chemical bonds between molecules are broken, and new molecules are formed. Molecules react with each other to reach equilibrium, the state in which molecules are stable. For the chemical reaction to occur, molecules must collide. More than one direction can be taken by a chemical reaction to reach the desired products. The best one is the one that requires the least energy conducing to an optimal reaction mechanism. This concept inspires Lam and Li to propose the chemical reaction optimization (CRO) algorithm in (Siddique and Adeli 2017); more details about the algorithm are found in their work. Molecules go through several basic reactions in a chemical reaction. These basic reactions can be grouped in two groups, unimolecular and multimolecular reactions. In the first group, there is only one molecule that collides with the container wall, on the contrary in the second group, there are many molecules that collide with each other. Attributes that need to be introduced are as follows:

1. Molecular structure (C): captures a solution of the problem.
2. Potential energy (PE) is defined as the objective function value of the corresponding solution represented by C. If f denotes the objective function, then we have $\text{PE}(C) = f(C)$.
3. Kinetic energy (KE) is a nonnegative number, and it quantifies the tolerance of the system accepting a worse solution than the existing one.

Group 1:

1. On-wall ineffective collision, in this case, the molecule collides with the container walls and falls back as a whole without splitting, then $C \rightarrow C'$.

This can be done by choosing C' in the neighborhood of C.

$$\text{KE}(C') = (\text{KE}(C) + \text{PE}(C) - \text{PE}(C')) \times a \tag{13.1}$$

such that a is a random number uniformly distributed from KELossRate to 1 and KELossRate $\in [0, 1]$.

For the remaining energy, $(\text{KE}(C) + \text{PE}(C) - \text{PE}(C')) \times (1 - a)$, it will be transferred to the buffer.

$$\text{if } \text{KE}(C) + \text{PE}(C) \geq \text{PE}(C'), \text{then we can have } \text{PE}(C') > \text{PE}(C).$$

2. Decomposition, here the molecule will collide with the container walls too, but it will split into two new molecules $c_1' + c_2'$.

Assume that $\delta 1$ and $\delta 2$ are two independent and identically distributed numbers uniformly generated in the range of [0, 1].

For decomposition to occur, energy conservation condition is modified as follows:

$$\text{PE}(C) + \text{KE}(C) + \delta 1 \times \delta 2 \times \text{buffer} \geq \text{PE}(C_1') + \text{PE}(C_2'). \tag{13.2}$$

and if it holds, we will replace C with C_1 and C_2.

For the remaining energy, KEs of the transformed molecules will share it.

$$E = (\mathrm{PE}(C) + \mathrm{KE}(C) + \delta 1 \times \delta 2 \times \mathrm{buffer}) - (\mathrm{PE}(C_1{}') + \mathrm{PE}(C_2{}')) \tag{13.3}$$

$\mathrm{KE}(C_1{}') = E \times r$ and $\mathrm{KE}(C_2{}') = E \times (r - 1)$, where r is a random number in $[0, 1]$.

Group 2:

3. Intermolecular ineffective collision, in this case, the molecules won't collide with the container wall, they will collide with each other, and they fall back without changes after collision.

$$\boxed{C_1 + C_2 \rightarrow C_1' + C_2'} \tag{13.4}$$

If $\mathrm{KE}(C_1) + \mathrm{PE}(C_1) + \mathrm{KE}(C_2) + \mathrm{PE}(C_2) \geq \mathrm{PE}(C_1{}') + \mathrm{PE}(C_2{}')$, then we will replace C_1 and C_2 with $C_1{}'$ and $C_2{}'$.

And as previously for the remaining energy, KEs of the transformed molecules will share it.

$$E = \mathrm{KE}(C_1) + \mathrm{PE}(C_1) + \mathrm{KE}(C_2) + \mathrm{PE}(C_2) - \mathrm{PE}(C_1{}') - \mathrm{PE}(C_2{}').$$

$\mathrm{KE}(C_1{}') = E \times r$ and $\mathrm{KE}(C_2{}') = E \times (r - 1)$, where r is a random number in $[0, 1]$.

4. Synthesis, it happens when molecules collide with each other and merge together.

For example, we will consider only two molecules $C_1 + C_2$.

So we can say that synthesis is the opposite case of decomposition.

$$\boxed{C_1 + C_2 \rightarrow C'} \tag{13.5}$$

It has to satisfy the energy conservation condition:

$$\mathrm{KE}(C_1) + \mathrm{PE}(C_1) + \mathrm{KE}(C_2) + \mathrm{PE}(C_2) \geq \mathrm{PE}(C').$$

and if it holds then $\mathrm{KE}(C') = \mathrm{KE}(C_1) + \mathrm{PE}(C_1) + \mathrm{KE}(C_2) + \mathrm{PE}(C_2) - \mathrm{PE}(C')$.

Decomposition and synthesis are meant to generate new solutions for exploring the search space (diversification), whereas on-wall ineffective collision and intermolecular ineffective collision are meant to generate solutions for exploiting the search space (intensification). A mixture of both intensification and diversification can make the search for global minimum solution in the space effective.

13.3.2 Whale Optimization

Humpback whales are one of the whale species found on Earth. An adult humpback whale can be up to 30 meters long and 180 tons in weight, they are very big predators (Mirjalili and Lewis 2016). They are also very smart creatures. They are found to be living alone or in groups. The most interesting thing about these whales is their feeding behavior. They hunt their prey in a very intelligent, organized, and special way called bubble net feeding. They may hunt in groups or individually, here we will talk about what happens when a group of humpback whales go for hunting a group of fish near the surface. They search for their prey in a very optimized way. When they find the best prey to hunt, they encircle it, shrinking the circle, dive 12 m deeper into the water (Mirjalili and Lewis 2016) in spiral or nine-shaped path, producing a net of bubbles that make a cage confining the prey, and then the leader of the group makes a sound that tells other whales it's time to go to the surface with your mouths open to eat. Searching for a prey phase represents the exploration part in the algorithm, whereas bubble net feeding phase represents the exploitation part. Information about how the mathematical model is built can be found in (Mirjalili and Lewis 2016), and more details about this hunting method can be found in (Hain et al. 1982).

This behavior is presented as a mathematical model in three subsections:

1. Encircling prey.

After the whales locate their prey, they surround it and because WOA cannot predetermine the optimal design in the search space, it assumes that the current prey is the optimal solution or close to it. After determining the best search agent, the positions of other search agents are updated toward the best search agent position.

This can be presented by the following equations:

$$\vec{D} = \left| \vec{C} . \overrightarrow{X_t^*} - \vec{X_t} \right| \tag{13.6}$$

$$\overrightarrow{X_{t+1}} = \overrightarrow{X_t^*} - \vec{A}.\vec{D} \tag{13.7}$$

vector factors $\vec{A}$ and $\vec{C}$ can be obtained by these two formulas:

$$\vec{A} = 2\overrightarrow{.a.}\,\vec{r_1} - \vec{a} \tag{13.8}$$

$$\vec{C} = 2\overrightarrow{.r_2} \tag{13.9}$$

where

$\vec{X}$: position of search agent (whale),

$\overrightarrow{X^*}$: the best position obtained in general,

t: current iteration,
$\vec{D}$: the vector represented the distance between whale and prey,
$\vec{a}$: random real number $\in[0, 2]$,
$|\ |$: the absolute value,
and (.) represents an element by element multiplication.

2. Bubble-net attacking method.

This phase is a combination of two approaches, mathematically modeled as follow:

(a) Shrinking Encircling Mechanism.

Whale behavior can be simulated by reducing a value in the equation … And, accordingly, this will decrease range of $\vec{A}$ as well by $\vec{a}$. By specifying random values for $\vec{A} \in [-1,1]$, the new position of a search agent can be defined anywhere in between the original position of the agent and the position of the current best agent.

(b) Spiral Updating Position.

In this approach, the distance between the whale at (x, y) and the prey at (x^*, y^*) is first calculated. Then a spiral equation is created between the whale and prey positions to imitate the helix-shaped movement of humpback whales as follows:

$$\overrightarrow{D'} = \left| \overrightarrow{X_t^*} - \overrightarrow{X_t} \right| \tag{13.10}$$

$$\overrightarrow{X_{t+1}} = \overrightarrow{D'}.e^{bl}.\cos(2\Pi l) + \overrightarrow{X_t^*} \tag{13.11}$$

where

$\overrightarrow{D'}$: the distance between the whale and prey,
b: constant defines the logarithmic spiral shape,
l: random number $\in [-1,1]$.

It is important to note that, when whales swim toward prey, they swim within a shrinking circle and along a spiral-shaped path at the same time. So that, in order to imitate this in the mathematical model, it is assumed that there is a 50% probability for each choice to update the whale positions during the optimization process, either the shrinking encircling mechanism or the spiral path. Illustrated by the following equations:

$$\overrightarrow{X_{t+1}} = \begin{cases} \overrightarrow{X_t^*} - \overrightarrow{A}.\ \vec{D} & p < 0.5 \\ \overrightarrow{D'} & p \geq 0.5 \end{cases} \tag{13.12}$$

where p is a random number $\in [0,1]$.

3. Search for prey.

This phase represents the exploration phase of the algorithm in which humpback whales (search agents) will move away from their current positions searching randomly for prey. And for that to happen, the value of $\vec{A}$ vector must be greater than 1 or less than −1, and of course it will be random.

Here, instead of updating the position of a search agent according to the best search agent as happened previously in the exploitation phase, the position will be updated according to a randomly chosen search agent. Equations represent the phase.

$$\vec{D} = \left| \overrightarrow{C}.\overrightarrow{X_{\text{rand}}} - \vec{X} \right| \tag{13.13}$$

$$\overrightarrow{X_{t+1}} = \overrightarrow{X_{\text{rand}}} - \vec{A}.\overrightarrow{D} \tag{13.14}$$

where $\overrightarrow{X_{\text{rand}}}$ is a random position vector representing a search agent chosen randomly from the current population.

13.3.3 Ant Lion Optimization

Ant lions are insects belonging to the Myrmeleontid family, there are many species living in dry and sandy regions all over the world, some of them are distinguished by their predatory larva distinctive hunting method. This is the method that we care about here and that the optimizer is based on. When an ant lion larva wants to hunt, it digs a deep conical pit in the sand with highly steep walls. Finally, the pits seems like a narrow volcanic crater, and in its bottom, the larva quietly hides itself, waiting for a prey with open jaws. When an ant falls into the pit, the larva catches it firmly with its jaws. It has very powerful jaws, and to make sure the ant can't escape, the larva starts to flip out more and more sand to make it unstable under the ant's legs so that it can't climb the walls and eventually, the larva manages to hunt the ant and eat it (Romagnani et al. 2017). Another interesting thing about ant lion pits is that the size of the pit is directly proportional to ant lion level of hunger; also, it was observed that these pits become larger when the moon is full (Chen et al. 2019). According to this method, the algorithm is designed to have two search agent classes, ant and ant lions. Ant which randomly walks in the search space can be hunted by an ant lion if it falls into a pit, on contrary, ant lions never change their place unless they replace a specific ant (Romagnani et al. 2017); these random walks are of course affected by ant lion pits (Chen et al. 2019). During the algorithm, the ant lion with higher fitness can make deeper pits and also these ant lions have higher chance to catch ants (Chen et al. 2019). And to simulate ants steep in the pit, ant random walks range is adaptively decreased (Chen et al. 2019). More details about how the mathematical model is built will be found in (Romagnani et al. 2017; Chen et al. 2019).

From the inspiration of ant lion larva hunting, Mirjalili proposed ant lion optimization algorithms, which consider that ants represent random possible solutions and generate random walk in the problem search space, they can move on any *i*th dimension of the n's dimensions of the search space using random cumulative sum function.

$$X_i^n = [0, \text{cumsun}(2r(n_1) - 1), \text{cumsun}(2r(n_2) - 1), \ldots..\text{cumsun}(2r(n_N) - 1),] \tag{13.15}$$

where X_i^n is the new position (changes) of the current solution, n represents the current iteration, N denotes the total number of iterations, cumsum calculates the cumulative sum and $r(n)$ is a random function returns '1' if the randomly generated number is greater than '0.5' and returns '0' otherwise.

To bound the changes (walks) inside the search space, they are applied min-max normalization as follows:

$$X_i^n = \frac{\left(X_i^n - a_i\right) \times \left(d_i - c_i^n\right)}{\left(d_i^n - a_i\right)} + c_i \tag{13.16}$$

where a_i and b_i are the minimum and maximum random walk of the *i*th variable, respectively, c_i^n and d_i^n are the minimum and maximum of the *i*th variable at *n*th iteration, respectively.

Once the ant is trapped in the ant lion's pit, it moves within a hypersphere around the ant lion, which is mathematically modeled as follows:

$$c_i^n = c^n + \text{ant lions}_j^n \tag{13.17}$$

$$d_i^n = d^n + \text{ant lions}_j^n \tag{13.18}$$

where c^n and d^n represent the minimum and maximum of all variables at *n*th iteration, respectively. c_i^n and d_i^n are the minimum and maximum of all variables for ant_i, respectively.

Ant lion tries to prevent the trapped ant from escaping by shooting sands at the edges of the pit to slide it down, which makes the radius of the random hypersphere walk decrease, defined mathematically as follows:

$$c^n = \frac{c^n}{I} \tag{13.19}$$

$$d^n = \frac{d^n}{I} \tag{13.20}$$

where c^n and d^n represent the minimum and maximum of all variables at n^{th} iteration, respectively, and I is a ratio defined as:

$$I = 10^w \frac{n}{N} \tag{13.21}$$

When trapped ant reaches the bottom of the pit, it means that the ant's fitness is better than the corresponding ant lion's fitness, which means it is going to be consumed so the corresponding ant lion updates its position to the latest position of the hunted ant to enhance its fitness and the ability to catch a new ant. It is represented mathematically as:

$$\text{Ant lion}_j^n = \text{Ant}_j^n \quad \text{if} f\left(\text{Ant}_j^n\right) > f\left(\text{Ant lion}_j^n\right) \tag{13.22}$$

where Ant lion_j^n represents ant lion's position j at nth iteration, and Ant_j^n represents ant's position, assuming the maximization of objective function (f).

Each ant updates its position in every iteration based on a crossover operation computed as follows:

$$\text{Ant}^d = \begin{Bmatrix} RE^d & r^d < 0.5 \\ RA^d & \text{otherwise} \end{Bmatrix} \tag{13.23}$$

where *RE* and *RA* represent the mutated positions for the elite ant lion and a randomly selected ant lion. Ant^d represents the new ant position at dimension d; r is a random value generated drawn from uniform distribution in the range [0,1]. The elitism ant lion is the best-obtained ant lion solution so far which is used in this algorithm to maintain the best solution(s) within the optimization. The randomly selected ant lion is chosen using one of the two selection methods: roulette wheel selection and rank selection method. After that, we calculate fitness function of the new ant's position and replace ant lions position with their corresponding ants if they become fitter, and update the elitism ant lion if an ant lion becomes fitter.

13.4 Proposed Bio-Inspired Approach

Nature has always been a source of inspiration for many scientists. By understanding the laws of nature, scientists were able to develop methods and algorithms that help them solve many problems (Siddique and Adeli 2017).

In this chapter, our solution will be also based on the same optimization algorithm but with a different technique and optimizer selection method, results show the difference between the approaches. Figure 13.3 illustrates the KPD workflow.

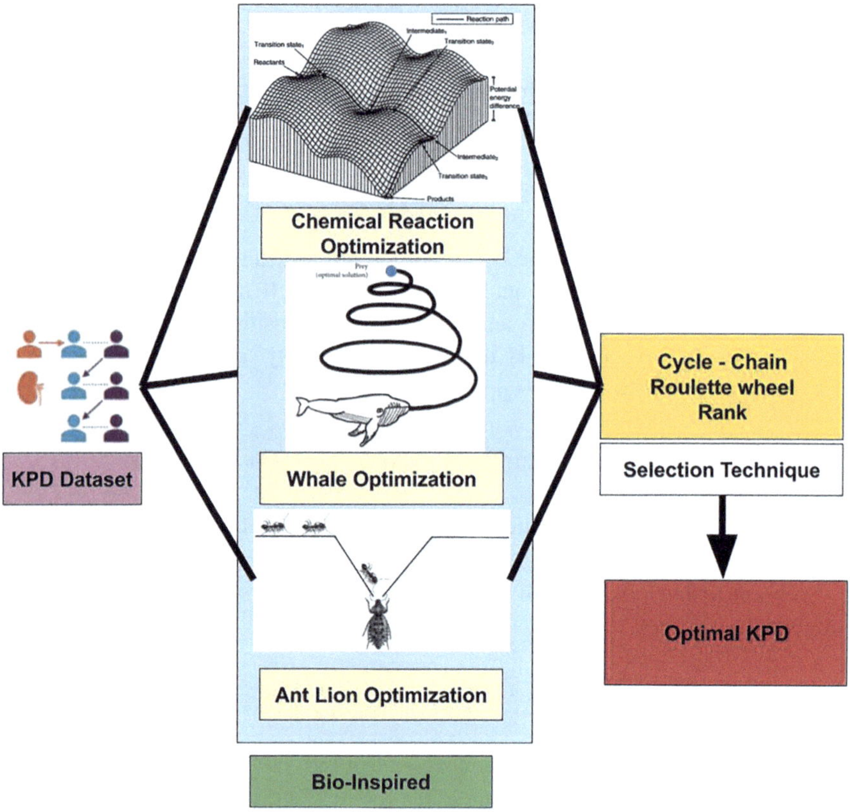

Fig. 13.3 KPD workflow

13.4.1 KDP Dataset

We assessed our proposed method's performance using six generated simulated datasets from Saidman generator's updated version (Saidman et al. 2006). In our case, we only accept donor-patient incompatible pairs and altruistic donors. We don't accept a patient without any donor with him, as it is clear from the dataset. Table 13.1 shows the number of compatible relations for each kidney exchange generated pool.

In our proposed method, we will also define the problem of kidney exchange as a combinatorial optimization problem as said earlier, aiming to reach the best solution that can be obtained from solving this optimization problem, which will be maximum in return the number of transplants that will be done (Fig. 13.4).

In the first phase, we didn't construct a new pool, instead, we used the resulting data from (https://www.kidney.org/news/newsroom/factsheets/Organ-Donation-and-Transplantation-Stats) KPD pool construction.

Table 13.1 Compatibility relations for each pool size (n)

Pool size (n)	Compatible relations
$n = 30$	56
$n = 40$	108
$n = 50$	123
$n = 75$	238
$n = 100$	482
$n = 200$	1705

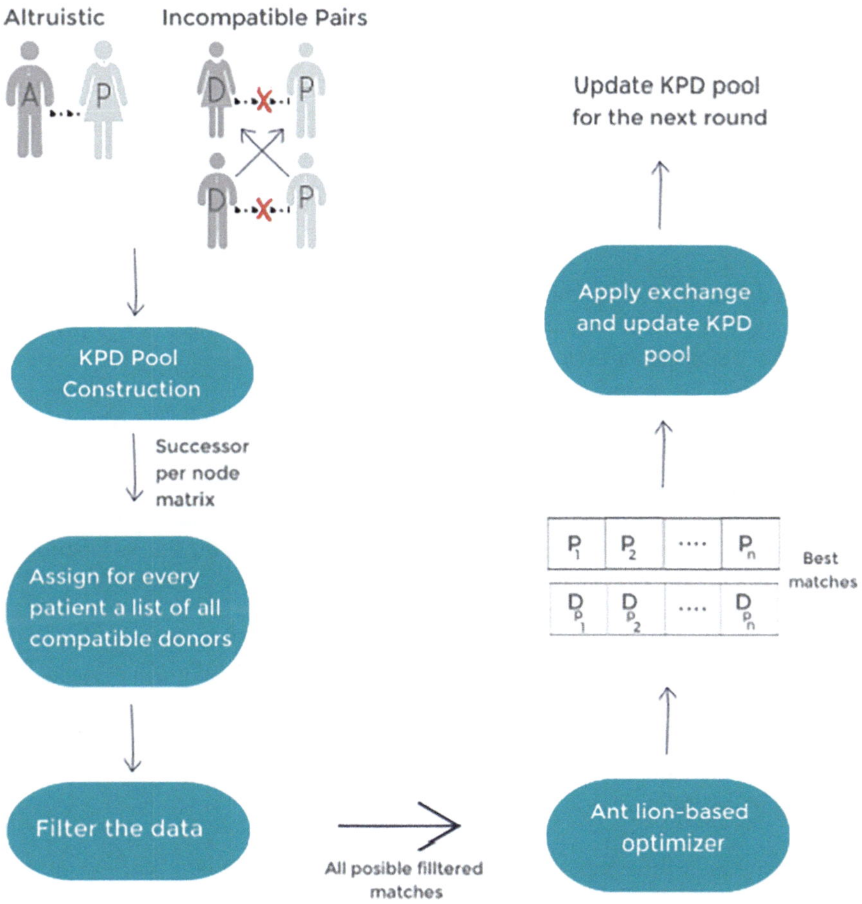

Fig. 13.4 ALO for KPD

In the second phase, we make a main list of (m) lists, such then, m: the number of patients in the pool, each list will be a successor per node list for one patient with all the compatible donors for this patient.

If we find patients with no compatible donors after performing the previous step, i.e., empty lists, as in the case of patient No. 3 in Fig. 13.5, we delete these patients

Fig. 13.5 KPD matching pool process

and also delete the donors who came with them from the lists of other patients. It can result from this deletion that we find patients whose lists have become empty from donors, so we repeat these steps until we only get patients who have non-empty lists of compatible donors as illustrated in Fig. 13.5.

And of course, the output of this process will be ordered as shown in Algorithm 1.

Finally, in the third phase, we have a filtered list ready for the algorithm to work on. The algorithm will make many copies of the main list, but the difference in these copies is that, for each patient, there is only one randomly selected donor. These copies make our initial list of solutions that will be assigned to the fitness function. It takes each solution, evaluate it, and then puts the value in the value list.

There will be two cases for the function:

1. If it finds that one donor is repeated for more than one person, it will evaluate this solution by $(-\beta)$, where β is the number of cases with repeated donors.
2. If each patient has a different donor, i.e., no repeated donor for more than one patient, the function will evaluate this solution by (α), where α is the number of cases in this solution.

This process will be repeated (N) times, such that, N is the number of iterations, until we reach the best solution.

The best fitness value resulting from the initial solutions will be saved in order to compare the rest of the values resulting in the next iterations with it. Let's name it here *BEST EXCHANGE*, to reach the best value, therefore the best solution.

If in the next iterations there is no value better than *BEST EXCHANGE,* then it will be returned as the best solution.

But if it is found that there is a value better than *BEST EXCHANGE,* it will be updated with this new value, and then the solution list will be updated for the next iteration.

When updating the list of solutions, we use a method of selection that determines the modifications that will be made to the solutions in order to be updated.

We used the same method of selection used in roulette wheel selection, but we found that the results are not satisfactory. We then used another method, i.e., rank selection, and indeed, the results have improved.

13.5 Experimental Discussion

Three experiments have been carried out in this chapter. In the first (in Sect. 13.5.1), study and compare the results of using chain-cycle technique with state-of-the-art optimization techniques such as the ant lion, whale, and chemical reaction optimization. The aim of this section is to solve kidney exchange problems in optimal manner by using chain-cycle technique with various nature-inspired optimization techniques. The second experiment (in Sect. 13.5.2) aims to evaluate the proposed model without using the chain-cycle technique by comparing its results with the results obtained from the first experiment. The third experiment (in Sect. 13.5.3), with the use of different selection methods, is aimed at enhancing the results obtained from the first and second experiments for the kidney exchange problems.

Experiments are carried out using Intel(R), Core (TM), i7-4790, CPU @ 3.6 GHz $ frequency, 32.0 GB Ram, 2 TB hard disk drive, and 64-bit Windows 10, and algorithms are self-coded and written in Python and MATLAB programming languages.

13.5.1 Chain-Cycle with State-of-the-Art Optimization

In this experiment, the chain-cycle technique is evaluated by comparing the results of it with four well-known optimization algorithms, namely, ant lion, whale, and chemical reaction.

Tables 13.2, 13.3, and 13.4 represent the time consumed in seconds as well as the min, max, and average numbers produced by ant lion, whale, and chemical reaction optimization techniques, respectively, using chain-cycle technique and for different pool size.

From Tables 13.2, 13.3, 13.4, and 13.5, it is inferred that using the chain-cycle technique, ant lion optimization algorithm has the best result according to the number of transplants and time measure. In the case of pool size ≥200, whale optimization technique fall in local minimum and cannot reach a solution.

Table 13.2 Number of transplants by the ant lion

Pool size (*n*)	Min	Average	Max	Time (*s*)
30	3	3	3	3
40	10	10	11	5.4
50	7	7	7	5
75	16	17	18	6.9
100	20	24	22	8.6
200	42	50	54	80

Table 13.3 Number of transplants by the whale

Pool size (*n*)	Min	Average	Max	Time (*s*)
30	3	3	3	0.06
40	10	11	11	1.7
50	7	7	7	0.6
75	14	16	18	3.9
100	16	28	23	29.7

Table 13.4 Number of transplants by the chemical reaction

Pool size (*n*)	Min	Average	Max	Time (*s*)
30	3	3	3	0.1
40	8	10	11	0.8
50	6	7	7	0.5
75	15	16	18	2
100	33	40	45	37.9
200	33	40	45	37.9

Table 13.5 Number of transplants by the ant lion

Pool size (*n*)	Min	Average	Max	Time (*s*)
30	4	5	6	0.2
40	10	13	15	0.5
50	7	8	9	5.5
75	12	18	21	7.5
100	23	24	26	39.1
200	67	71	72	441.1

13.5.2 Without Chain-Cycle with State-of-the-Art Optimization

In chain-cycle technique, the search space is very large with a lot of repeated patients, donors, or edges, and with the specified number of iterations, many of these titrations are consumed in these repeated solutions which affects the results, so it is preferred to remove it.

Tables 13.6, 13.7, 13.8, and 13.9 represent the time consumed in seconds as well as the min, max, and average numbers produced by ant lion, whale, and chemical reaction optimization techniques, respectively, without chain-cycle technique and for different pool size.

From Tables 13.5, 13.6, and 13.7, it is inferred that with a small pool size, the proposed model without using chain-cycle technique got the best results and was

Table 13.6 Number of transplants by the whale

Pool size (n)	Min	Average	Max	Time (s)
30	6	6	6	0.2
40	10	13	15	2.5
50	8	9	9	1.6
75	13	17	21	36.9

Table 13.7 Number of transplants by the chemical reaction

Pool size (n)	Min	Average	Max	Time (s)
30	5	6	6	0.1
40	13	13	15	1.1
50	7	8	9	1.1
75	15	16	19	1.6

Table 13.8 Maximum number of transplants by the ant lion

Pool size (n)	With chain-cycle	With chain-cycle and roulette wheel selection	With chain-cycle and rank selection
30	3	6	6
40	11	15	15
50	7	9	9
75	18	21	21
100	22	26	28
200	54	72	82

Table 13.9 Number of transplants produced by the ant lion

Pool size (n)	Min	Average	Max	Time (s)
30	5	6	6	0.76
40	9	13	15	12
50	7	8	9	12
75	16	18	21	35.5
100	20	23	28	512.5
200	77	80	82	3627

better than using chain-cycle technique. In the case of pool size = 200, only the ant lion optimization algorithm can reach a solution and better than with but with a huge amount of time. In the case of pool size ≥100, both whale and chemical reaction optimization techniques fall in local minimum and cannot reach a solution. In the case of pool size = 200, only the ant lion optimization algorithm can reach a solution, but with a huge amount of time. In the case of pool size ≥100, both whale and chemical reaction optimization techniques fall into a local minimum and cannot reach a solution.

Figure 13.6 summarizes the comparison between ant lion, whale, and chemical reaction optimization techniques with and without using chain-cycle technique. It represents that the ant lion optimization algorithm without chain-cycle technique is the best one. For pool size <100, all optimization algorithms get better results without chain-cycle technique.

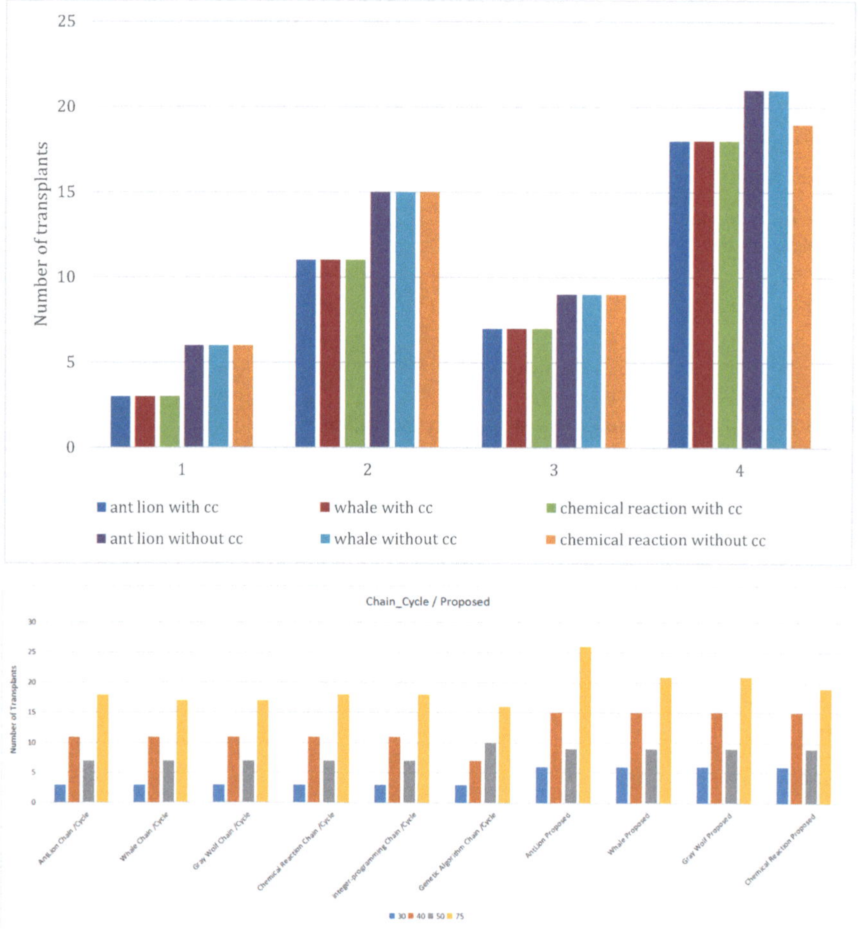

Fig. 13.6 Comparison between ant lion, whale, and chemical reaction optimization techniques with and without using chain-cycle technique

13.5.3 Selection Methods

From experiments 1 and 2, the proposed model without using chain-cycle technique got better results in small pool size up to 100. However, in large pool sizes, in some cases reaching a good result takes a huge amount of time. In other cases, it fails at the local minimum. In experiments 1 and 2, all optimization algorithms used the random selection method to select candidates. In this experiment, we aim to improve the result of the proposed model ant lion optimization algorithm without using the chain-cycle technique by using other selection methods such as roulette wheel and rank selection.

Table 13.8 compares the obtained results of the ant lion optimization algorithm with chain-cycle technique and using two selection methods: roulette wheel and rank selection methods.

Table 13.9 shows the results of the ant lion optimization algorithm without using chain-cycle technique and by using rank selection methods with different pool size.

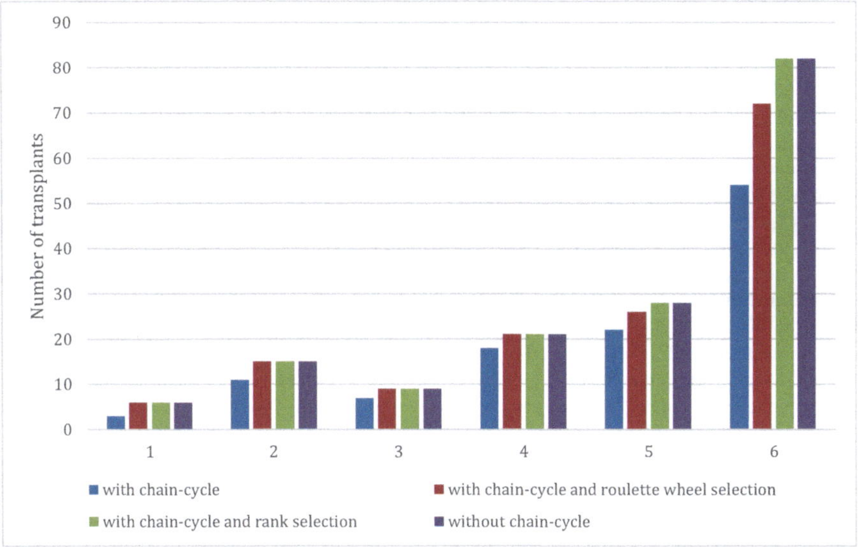

13.6 Conclusion

This chapter aims to identify the number of transplants that help in the normal living of chronic kidney disease (CKD) patients. Furthermore, by combining bio-inspired algorithms, it aims to reduce the waiting time of needy kidney patients. The first stage of the model is choosing the best optimization algorithm, so ant lion, whale, and chemical reaction optimization algorithms are used. The second stage of the model is choosing the best selection method, so the roulette wheel and rank selection methods are used. The third stage of the model is the used technique. This chapter proposed a new technique that breaks the chain-cycle condition. From the experiments, the best results were obtained from using the ant lion optimization method with the rank selection method and the proposed technique, which is based on the chain-cycle condition.

Conflicts of Interest The authors declare no conflict of interest.

Funding This research received no external funding.

Data Availability In this research, experiments are executed over six generated simulated datasets from.

Saidman generator's updated version (Saidman et al. 2006).

References

Aita KS, Monte SJ, Silva AS, Marroquim MS, Coelho AGB, Sousa LCD (2016) Web-based ecosystem software for virtual crossmatching in transplant programs. Comput Biol Med 75: 30–37

Arora P, Jalal K, Gupta A, Carter RL, Lohr JW (2017) Progression of kidney disease in elderly stage 3 and 4 chronic kidney disease patients. Int Urol Nephrol 49(6):1033–1040

Ashiku L, Al-Amin M, Madria S, Dagli C (2021) Machine learning models and big data tools for evaluating kidney acceptance. Proc Comput Sci 185:177–184

Bray M, Wang W, Rees MA, Song PX, Leichtman AB, Ashby VB, Kalbfleisch JD (2019) KPDGUI: an interactive application for optimization and management of a virtual kidney paired donation program. Comput Biol Med 108:345–353

Chellal M, Wang J, Benmessahel I, Galoul A (2020) Dynamic kidney paired exchange using modified multiverse optimization. Evol Intel:1–10

Chen TK, Knicely DH, Grams ME (2019) Chronic kidney disease diagnosis and management: a review. JAMA 322(13):1294–1304

Dorado-Moreno M, Pérez-Ortiz M, Gutiérrez PA, Ciria R, Briceño J, Hervás-Martínez C (2017a) Dynamically weighted evolutionary ordinal neural network for solving an imbalanced liver transplantation problem. Artif Intell Med 77:1–11

Dorado-Moreno M, Pérez-Ortiz M, Gutiérrez PA, Ciria R, Briceno J, Hervás-Martínez C (2017b) Dynamically weighted evolutionary ordinal neural network for solving an imbalanced liver transplantation problem. Artif Intell Med 77:1–11

Fabrizi F, Messa P (2015) Transmission of hepatitis C virus in dialysis units: a systematic review of reports on outbreaks. Int J Artif Organs 38(9):471–480

Fehrman-Ekholm I, Elinder CG, Stenbeck M, Tydén G, Groth CG (1997) Kidney donors live longer. Transplantation 64(7):976–978

Freedman R, Borg JS, Sinnott-Armstrong W, Dickerson JP, Conitzer V (2020) Adapting a kidney exchange algorithm to align with human values. Artif Intell 283:103261

Goel A, Bhadauria DS, Aggarwal R (2018) Hepatitis C virus infection and chronic renal disease: a review. Indian J Gastroenterol 37(6):492–503

Goezinne S, Bekker R, Glorie K (2016) A genetic algorithm for kidney transplantation matching. Business Analytics Master Program at VU University, Amsterdam

Hain JHW, Carter GD, Kraus SD, Mayo CA, Winn HE (1982) Feeding behavior of the humpback whale, Megaptera novaeangliae, in the western North Atlantic. Fish Bull 80:259–268

Hamouda E, El-Metwally S, Tarek M (2018) Ant Lion Optimization algorithm for kidney exchanges. PLoS One 13(5):e0196707

Hermsen M, Ciompi F, Adefidipe A, Denic A, Dendooven A, Smith BH, van der Laak JA (2022) Convolutional neural networks for the evaluation of chronic and inflammatory lesions in kidney transplant biopsies. Am J Pathol 192:1418

Hou S (2000) Expanding the kidney donor pool: ethical and medical considerations. Kidney Int 58(4):1820–1836

https://www.kidney.org/news/newsroom/factsheets/Organ-Donation-and-Transplantation-Stats

Kaballo MA, Canney M, O'Kelly P, Williams Y, O'Seaghdha CM, Conlon PJ (2018) A comparative analysis of survival of patients on dialysis and after kidney transplantation. Clin Kidney J 11(3):389–393

Kratz J, Andersson T (2019) Pairwise kidney exchange over the blood group barrier. Rev Econ Stud

Lam AYS, Li VOK (2012) Chemical reaction optimization: a tutorial. Memetic Comput 4(1):3–17. https://doi.org/10.1007/s12293-012-0075-1

Lentine KL, Pastan S, Mohan S, Reese PP, Leichtman A, Delmonico FL, Axelrod DA (2021) A roadmap for innovation to advance transplant access and outcomes: a position statement from the National Kidney Foundation. Am J Kidney Dis 78(3):319–332

Martin DE, Danovitch GM (2017) Banking on living kidney donors – a new way to facilitate donation without compromising on ethical values. In: The journal of medicine and philosophy: a forum for bioethics and philosophy of medicine, vol 42, No. 5. Oxford University Press, pp 537–558

Mierzejewska B, Durlik M, Lisik W, Baum C, Schroder P, Kopke J, Stepkowski S (2013) Current approaches in national kidney paired donation programs. Ann Transplant 18:112–124

Minch RP, Sanders GL (1986) Computerized information systems supporting multicriteria decision making. Decis Sci 17:395–413

Mirjalili S, Lewis A (2016) The whale optimization algorithm. Adv Eng Softw 95(C):51–67. https://doi.org/10.1016/j.advengsoft.2016.01.008

Ng JKC, Li PKT (2018) Chronic kidney disease epidemic: how do we deal with it? Nephrology 23: 116–120

Patzer RE, McPherson L, Redmond N, DuBay D, Zayas C, Hartmann E, Arriola KJ (2019) A culturally sensitive web-based intervention to improve living donor kidney transplant among African Americans. Kidney Int Rep 4(9):1285–1295

Rahwan I, Simari GR (eds) (2009) Argumentation in artificial intelligence, vol 47. Springer, Heidelberg

Romagnani P, Remuzzi G, Glassock R, Levin A, Jager KJ, Tonelli M, Anders HJ (2017) Chronic kidney disease. Nat Rev Dis Prim 3(1):1–24

Saidman SL, Roth AE, Sönmez T, Unver MU, Delmonico FL (2006) Increasing the opportunity of live kidney donation by matching for two- and three-way exchanges. Transplantation 81(5): 773–782. https://doi.org/10.1097/01.tp.0000195775.77081.25

Siddique N, Adeli H (2017) Nature-inspired chemical reaction optimisation algorithms. Cogn Comput 9(4):411–422. https://doi.org/10.1007/s12559-017-9485-1

Stewart DE, Klassen DK (2017) Early experience with the new kidney allocation system: a perspective from UNOS. Clin J Am Soc Nephrol 12(12):2063–2065

Thiessen C, Gannon J, Li S, Skrip L, Dobosz D, Gan G, Kulkarni S (2021) Quantifying risk tolerance among potential living kidney donors with the donation-specific risk questionnaire. Am J Kidney Dis 78:246

Timofte D, Dragos D, Balcangiu-Stroescu AE, Tanasescu MD, Balan DG, Avino A, Ionescu D (2020) Infection with hepatitis C virus in hemodialysis patients: an overview of the diagnosis and prevention rules within a hemodialysis center. Exp Ther Med 20(1):109–116

Yuan Y, Feldhamer S, Gafni A, Fyfe F, Ludwin D (2002) The development and evaluation of a fuzzy logic expert system for renal transplantation assignment: is this a useful tool? Eur J Oper Res 142(1):152–173

A Hybrid CNN-Vision Transformer Model for Non-Invasive Anemia Detection Using Conjunctival Eye Images

14

K. Kavitha and Gunjan Mukherjee

Abstract

Over 25% of the world population faces anemia as a global health problem which creates severe complications carrying negative effects on child brain development and adult work capacity and posing life-threatening risks during pregnancy. The current blood tests, which use complete blood count (CBC) and hemoglobin analysis, require blood collection and specialized equipment and medical facilities, thus preventing their use in underserved regions.

Modern medical imaging, together with artificial intelligence (AI) technology, creates new possibilities for screening anemia without invasive procedures when focusing on evaluating ocular conjunctival pallor as a validated clinical indication of low hemoglobin levels. Visually supervised deep learning (DL) architectures learn automated feature extraction from primary raw images maintaining spatial relationships which goes beyond traditional machine learning features based on human intervention.

A hybrid Convolutional Neural Network (CNN)-Vision Transformer (ViT) model developed in this research addresses three major issues by identifying microvascular patterns while measuring pallor distribution extent and linking detected features to hemoglobin limits. Using datasets annotated by hematologists, our system shows how computer vision technology can perform clinical pallor evaluation using lab-quality results, without requiring blood draws,

K. Kavitha (✉)
School of Computing, Department of AI and ML, Mohan Babu University, Tirupati, India

G. Mukherjee
Department of Computational Sciences, Brainware University, Barasat, Kolkata, West Bengal, India

J. M. Chatterjee, S. K. Saxena (eds.), *Artificial Intelligence and Blockchain in Precision Medicine and Virology*, Medical Virology: From Pathogenesis to Disease Control, https://doi.org/10.1007/978-981-96-8919-4_14

which creates an important development for community health workers along with telemedicine programs.

Keywords

Deep learning (DL) · Vision Transformer Model · Convolutional Neural Network · Anemia

14.1 Introduction

Anemia continues as a vital global health concern since it affects 1.8 billion people across the world particularly targeting developing nations (World Health Organization 2021). The blood condition known as anemia produces insufficient hemoglobin which reduces tissue oxygen delivery while generating fatigue symptoms together with impaired mental functions and higher rates of maternal deaths and infant loss. The traditional diagnostic testing uses blood examinations that need both specialized personnel together with laboratory infrastructure which creates difficulties when serving settings with limited resources (Mast et al. 2021). Figure 14.1 shows the anemia cases globally (World Health Organization 2021).

Artificial intelligence (AI) technology has demonstrated new possibilities for detecting anemia externally through assessment of images captured from the eyes. DL extracts features that distinguish between different image elements. Hence, DL provides superior results for medical image analysis (LeCun et al. 2015). Handcrafted features used with conventional machine learning approaches provided only moderate success. CNNs demonstrate excellence in detecting conjunctival

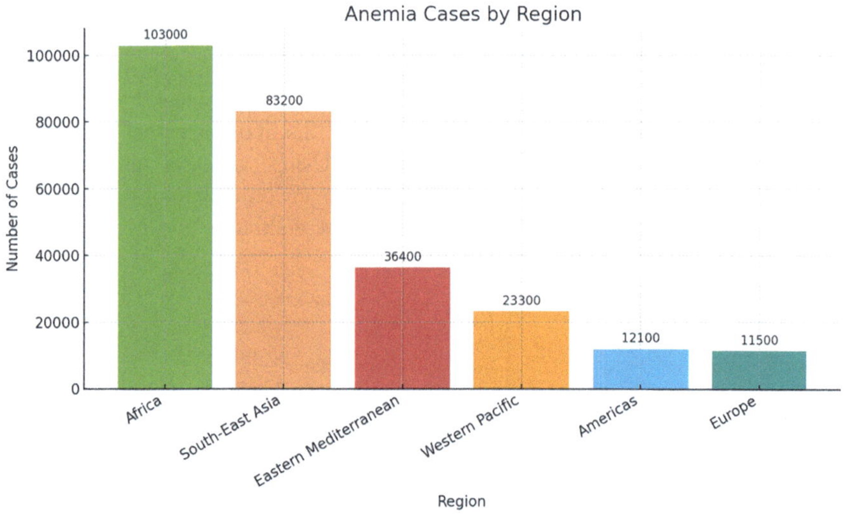

Fig. 14.1 Number of anemia cases by region

pallor patterns because these patterns strongly indicate the person's hemoglobin levels.

The main objective of this research involves creating an advanced deep learning structure to address current weaknesses in detecting anemia. This method resolves three essential medical imaging issues that involve detectable conjunctival pallor patterns, scarce dataset management and understandable diagnostic outcomes for medical personnel. The framework integrates CNN performance for local analysis with attention models to examine whole images for achieving both dependable outcomes and fast computation suitable for clinical use.

The research establishes multiple advancements for non-invasive hematological assessment testing. This research also contributes an innovative architectural setup intended for medical image analytics which runs with fewer parameters than typical deep learning solutions. The proposed technique adds specialized domain modifications which help process conjunctival images specifically.

14.2 Literature Review

A. Kalantri et al. (2020) created a budget-friendly smartphone system which examines conjunctival pallor through a color transformation technique based on HSV along with the k-means clustering method. This model stands out for its immediate image processing speed (0.2 seconds per picture) together with its high detection capability (84%) for advanced anemia cases which makes it well-suited for outside use in areas with limited resources. The study faced two major weaknesses including the substantial decrease in accuracy levels when light conditions changed and the absence of dark skin tone representation in their data set with 320 images.

M. Pietka et al. (2021) built an adapted CNN model with preprocessing methods using adaptive histogram equalization which delivered superior illumination consistency for analyzing conjunctival images. The model achieved 87.3% accuracy in grading anemia severity to mild/moderate/severe through its hierarchical architecture which learned multi-scale pallor patterns. The study contained two main limitations: it excluded patients who had ocular medical conditions and utilized data from 412 images acquired at a single tertiary medical facility.

A combination of Local Binary Pattern and shallow CNN emerged as a hybrid approach according to K. Doi et al. (2019) which used engineered and learned features together. The integrated system delivered both excellent reliability across different image quality levels (AUC = 0.89), and it produced outcomes from analyzing micro-textures alongside semantic features. Full-resolution image processing and end-to-end capability restricted this methodology because manual area selection remained necessary.

R. S. Ware et al. (2022) established a retinal fundus analysis system through their integration of ResNet-50 with multi-instance learning. This model succeeded in anemia detection through retinal imaging without using conjunctival pictures by reaching 83.5% accuracy by reading vascular patterns in accessible retinal images.

The performance of the diabetic cohort fell by ~12% because of elements that compete with retinopathy features within the population.

The research group led by J. M. Schmitt et al. (2018) developed OCT-based hemoglobin analysis with strong validity against medical laboratory tests ($r = 0.82$) by evaluating conjunctival microvasculature density. Optical coherence tomography presented the highest degree of physiological precision through its capability to visualize blood vessel networks directly. Such expensive specialized equipment ($50 k each device) coupled with trained operators rendered the system inaccessible for widespread patient screening.

S. K. Vinjamuri et al. (2020) developed a multi-modal system which performed random forest classification on eyelid and nail bed analyses. Together, the ensemble system exhibited a high level of interoperability between image areas because nail bed data provided helpful substitute information when seeing through poorly acquired conjunctival images (85.1% accuracy). Multiple standardized image acquisitions for every patient required examination times that were three to five times longer than those required by traditional single-site methods.

The attention-guided CNN created by L. Xu et al. (2021) achieved outstanding performance (91.2% accuracy) through its ability to learn pallor distribution patterns having spatial awareness. Performance became more accessible to doctors because the attention mechanism produced decision region heatmaps that could be clinically understood. To prevent overfitting, the model needed ten times more synthetic samples due to its complex architecture. The model requires ten times more systemetic samples in orsdr to cope up with its complex architecture. The method was intended for preventing the overfitting concept.

14.3 Research Challenges Identified in Anemia Prediction

The study of anemia prediction through non-invasive analysis of eye images encounters essential obstacles which surpass particular dataset constraints. The main limiting factor for such models relates to their ability to generalize across various population groups alongside different physiological conditions that modify both conjunctival tissue appearance and image acquisition procedures. The task of translating pallor patterns into accurate hemoglobin measurements proves complex because it creates problems for learning meaningful features especially for situations involving unclear clinical indicators. Computational constraints when working in resource-constrained environments present challenges to deployment because developers must determine how to manage model complexity against hardware capacity limitations. The evaluation methods lack standardization which reduces the ability to make fair comparisons between research studies, and healthcare professionals have to resolve ethical dilemmas about patient information protection and algorithm behavior. The various technical barriers emphasize why novel solutions must establish connections between operational possibilities and medical program adoption in screening programs.

Predictive interpretation remains a major obstacle as medical staff need transparent diagnostic procedures to accept assistance from AI systems. Static image analysis is inadequate in understanding the dynamic blood flow conditions of the conjunctiva because the flow changes over smaller time periods. Developing flexible systems with transparent operations becomes essential to achieve accurate diagnosis within diverse patient groups and uncontrolled clinical settings and operational blend capabilities. The achievement of proof-of-concept studies depends on proper resolution of these issues to generate clinical solutions that provide true enhancement to standard hematological testing approaches.

14.4 Objective

- To create a hybrid deep learning system which combines Convolutional Neural Networks (CNN) and Vision Transformer (ViT) architectures for high-quality anemia detection through conjunctival eye images.
- To integrate CNN in ordeer to address the local feature learning properties with the global attention capabilities. The ViT should be used for maximizing feature extraction prior to classification enhancement.
- To determine the diagnostic capability of the hybrid model through performance metrics between CNN-alone models and ViT-alone models including accuracy and sensitivity along with specificity and AUC-ROC.
- To test the model's generalization abilities, it will be performed through diverse dataset evaluations under different demographic groups and imaging situations.

14.5 Base Classifiers Used for Anemia Prediction

14.5.1 Convolutional Neural Networks (CNN)

The image analysis capability of Convolutional Neural Networks (CNNs) models makes them highly appropriate for medical imaging applications. The CNNs used for anemia prediction accept medical images and feature maps derived from tests such as blood cell images and diagnostic reports as inputs. When the CNNs detect anemia, the system can examine multiple hematological parameters including Hemoglobin levels and Mean Corporate Volume (MCV), Mean Corporate Hemoglobin Concentration (MCHC) along with Mean Corporate Hemoglobin (MCH), and these measurements can be obtained from either blood smears or laboratory reports. The CNN model receives these parameters for processing and development of complex patterns between them.

A CNN includes three principal components: convolutional layers, pooling layers, and fully connected layers. Convolutional network layers function to feature extraction by using hierarchical learning on data which results from analyzing local dependencies within spatial sequences. The pooling operation reduces the feature maps' dimensions by focusing on essential information and ignoring superfluous details. Fully connected layers create higher-level representation outputs that lead to classification predictions of the provided input. During training, CNNs learn optimal model parameters through minimization of their loss function which most commonly uses binary cross-entropy for tasks involving anemia prediction. Weight and bias adjustments occur through backpropagation which involves gradient transmission from the network downward for optimization purposes.

Step-by-Step CNN Workflow for Anemia Prediction

A. *Handling Input Image and Preprocessing*:
   - Input: High-resolution conjunctival images which are in RGB format.
   - Preprocessing:
     - Resizing: Standardize the image into 224 × 224 pixels which is compatible with transfer learning models.
     - Normalization: Scaling of pixel values to [0, 1] for stable training.
     - Augmentation: It involves random rotations, flips, or brightness adjustments to enhance generalization.

B. *Design of Convolutional Layers for Local Feature Extraction*
   - Design First Convolutional Block:
     - Filters: 32 kernels (3 × 3) scan the image for low-level features such as edges, color gradients.
     - Activation (ReLU): Introduces non-linearity ($f(x) = \max(0, x)$).
     - Max Pooling: It is to reduce the spatial dimensions (2 × 2), to retain dominant features.
   - Subsequent Blocks:
     - Deeper Layers: These layers used more filters (128 filters) to detect complex patterns such as vascular networks and pallor regions. For example, a pale conjunctiva (anemia indicator) is highlighted via color-sensitive filters.

C. *Hierarchical Feature Learning*
   - Three layers are designed to learn hierarchical features.
     - Early Layers: These are designed to detect edges, color blobs, and texture variations.
     - Mid-Layers: These will combine edges into shapes (e.g., blood vessel structures.
     - Deep Layers: These mainly used to identify diagnostic biomarkers such as hemoglobin-dependent pallor.

Figure 14.2 shows the visual representation of CNN architecture designed to extract local features and hierarchical features.

Strength of CNN in Anemia Prediction

- CNN works excellent at capturing local color and/or texture variations.
- CNN is computationally efficient compared to transformers.
- CNN has proven performance in medical image analysis.

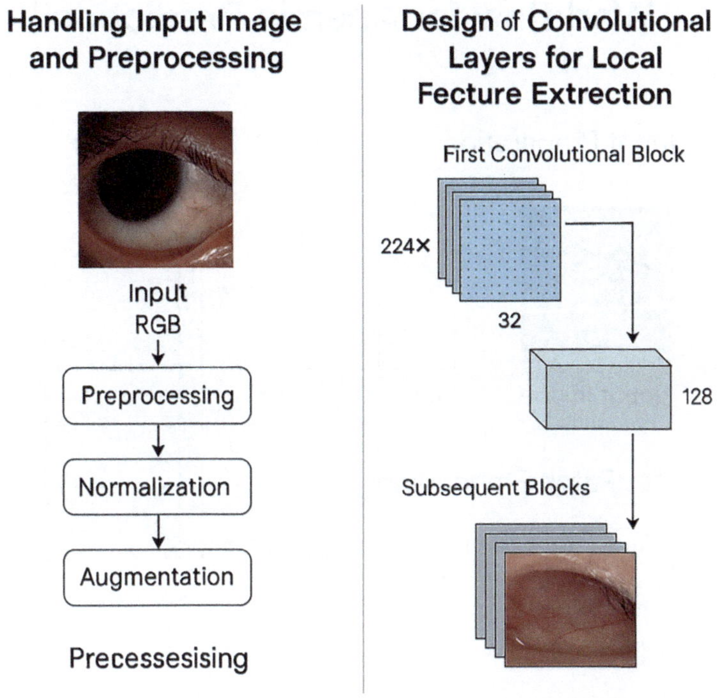

Fig. 14.2 Design of CNN layers for feature extractions

14.5.2 Vision Transformer (ViT)

The ViT processes images as sequences of patches, using self-attention to capture global relationships across the entire conjunctival region.

Step-Wise Workflow for Anemia Prediction

A. *Input Preparation*
   - ViT receives the preprocessed image from CNN. It then creates a path by splitting image into fixed-size patches (e.g., 16×16 pixels $\rightarrow$ 196 patches). Each patch is then flattened into a 1D vector ($16 \times 16 \times 3 = 768$ values per patch).

B. *Patch Embedding*
   - In this stage, each patch is mapped to a lower dimension using a trainable matrix. For classification purpose, a special token prepended to patch sequence to aggregate global information. To embed patches, add all spatial information.

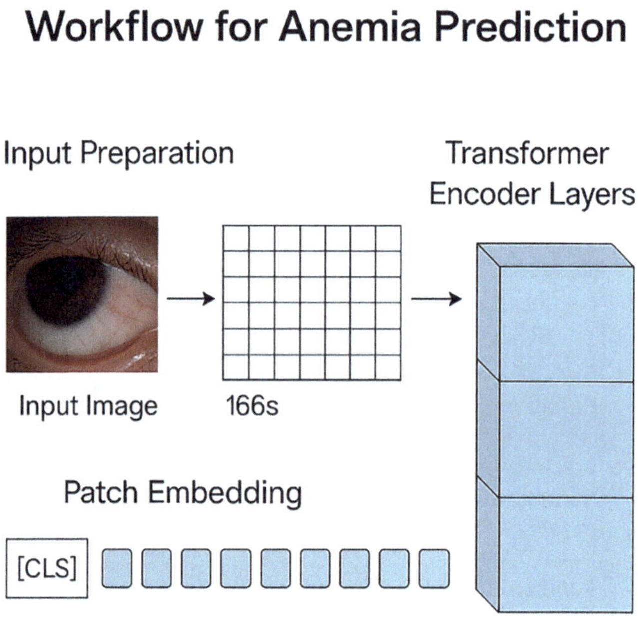

Fig. 14.3 Workflow of ViT in anemia detection

C. *Transformer Encoder Layers*
   - The core of ViT consists of L identical transformer blocks. These blocks are designed to focus on weigh vascular regions heavily. It also ignore irrelevant areas like eyelashes.

   Figure 14.3 shows the architecture and working principles of ViT

Strength of using ViT in Anemia Prediction
- Captures global image context effectively
- Models long-range dependencies more accurately
- Provides interpretability through attention maps

14.6 Proposed Hybrid CNN-Vision Transformer Model for Non-Invasive Anemia Detection

The proposed Hybrid CNN-Vision Transformer (ViT) Model for Non-Invasive Anemia Detection leverages the complementary strengths of deep learning architectures to overcome key limitations in existing approaches (Fig. 14.4). High-resolution elements crucial for detecting anemia such as conjunctival pallor and microvascular patterns and subtle color variations are obtained from the CNN

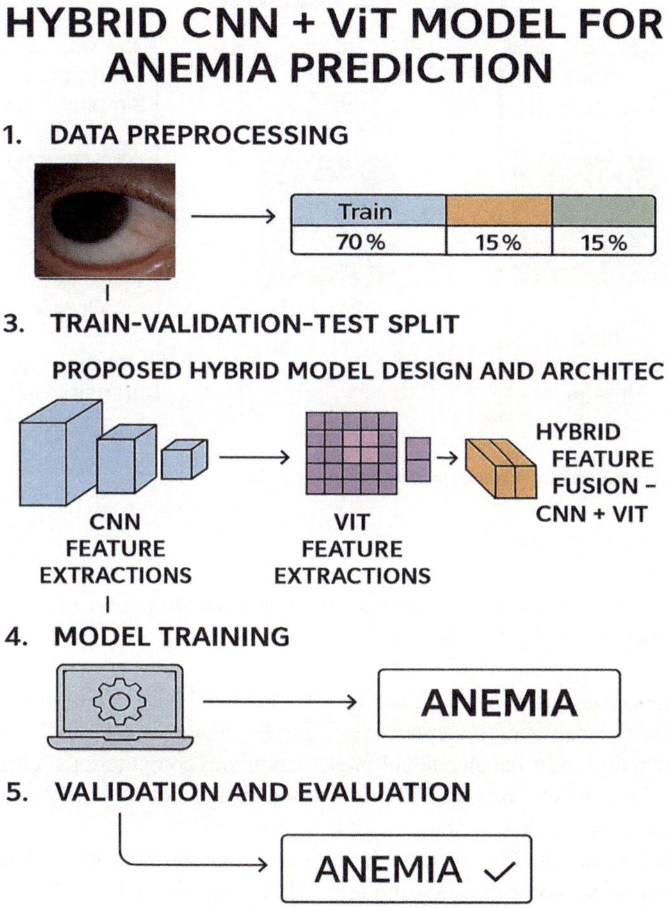

Fig. 14.4 Proposed hybrid CNN-Vision Transformer (ViT) model for anemia detection

backbone, but the ViT module establishes long-range visual relationships within the entire eye region. This fusion enables the model to simultaneously analyze fine-grained details (via CNN) and holistic spatial correlations (via ViT), significantly improving detection accuracy compared to standalone CNN or transformer models.

A. *Dataset*

The proposed model analyzes high-resolution images of conjunctiva from a public database whose annotations confirm anemic states through hemoglobin tests. The gathered images extend to more than 2500 samples. The benchmark dataset maintains equivalent numbers between anemic and non-anemic classes.

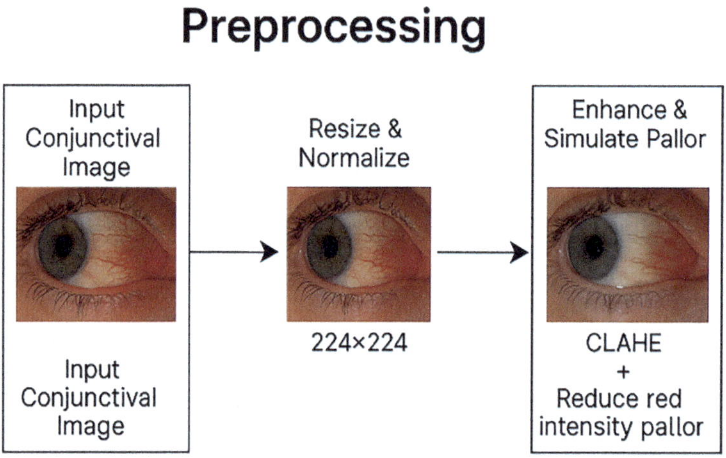

Fig. 14.5 Techniques involved in preprocessing the data

B. *Data Preprocessing*

Data preprocessing involves various techniques (shown in Fig. 14.5) to transform the data ready to use by the classifiers.

- *Input Image and ROI Extraction*: The raw data is a high-resolution image of eye particularly palpebral conjunctiva. Since images are high resolution, preprocessing must balance detail preservation and computational efficiency:
  - Resizing: CNN uses higher resolutions images (512 × 512) and ViT is standardized to 384 × 384.
  - ROI Extraction: This process focuses on isolating the conjunctiva/palpebral region using color thresholding (e.g., HSV mask for redness)
- *Normalization and Statistical Alignment*: Calculated mean from the ocular images for the images to which lighting conditions are unique.
- *Color and Texture Enhancement*:
  - *Reduce red channel*: One of the symptoms of anemia results in diminished blood flow leading to pale appearance of the conjunctiva. Color space analysis helps to identify pallor by transforming images into HSV/Lab before enhancing the red (a*) channel in Lab space and saturation (S) in HSV. An adaptive histogram equalization implementation enhanced contrast levels in the area of the conjunctiva. Also, adaptive histogram equalization was applied to improve contrast in the conjunctival region.
  - *Improving Texture*: CLAHE is used to enhance contrast in the conjunctiva.
- *Augmentation Strategies*:
  - Geometric: It involves small rotations, flips, zoom.
  - Photometric: This process automatically adjusts the brightness/contrast.
  - Synthetic Anemia: In order to improve system robustness, the system should simulate pallor by dimming the red channel in images without anemia.

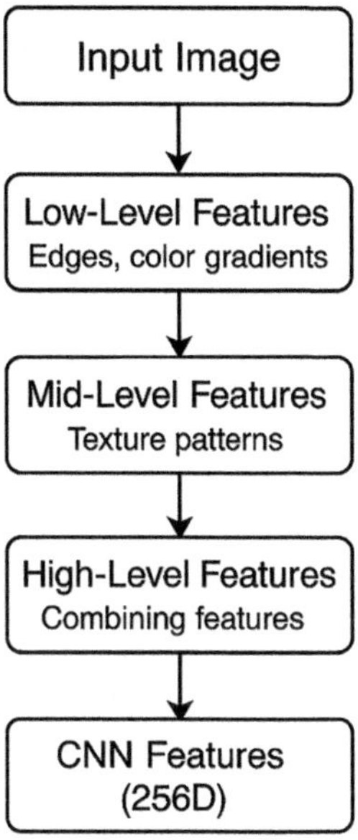

Fig. 14.6 Feature extraction using CNN

C. *Train-Validation-Test Split*: This *research work split the dataset into* 70% for training, 15% for validation, and 15% for testing.
D. *Hybrid Model Design and Architecture*:

In this proposed hybrid model design, CNN serves as the local feature extractor, to analyze fine-grained patterns in conjunctival images that correlate with anemia (e.g., pallor, vascular changes).

(i) *CNN Feature Extraction Process*

The feature extraction process using CNN is shown in Fig 14.6 for clear understanding.

The features extraction is categorized as:

- *Low-Level Features*: The early layers in CNN detects edges, color gradients (e.g., RGB channels for pallor) to highlight conjunctival vs. scleral regions.
- *Mid-Level Features*: The middle layer identifies texture patterns (e.g., vascular network density).
- *High-Level Features*: These deep layers combine features to detect diagnostic biomarkers. Example: Pale regions + reduced vessel visibility → anemia.

The output of this stage is 256D feature vector.

(ii) *Global Context Modeling Using ViT*
   - *Image Patch Splitting*: The preprocessed eye image (e.g., 384 × 384 resolution) is separated into non-overlapping patches of 16 × 16 pixels and each separated patch is treated as a separate "token."

- *Flattening & Patch Embedding*: Each token contains RGB values, i.e., shape. These shapes will be flattened into a 1-D vector that will pass through a trainable linear layer to project into a lower-dimensional embedding space. Additionally, a CLS token aggregates global information across all patches.
- *Positional Encoding*: Since patches do not have an inherent order, it is necessary to add positional embeddings to preserve spatial relationships. These learnable vectors added to each patch embedding.
- *Transformer Encoder Layers*: Multiple transformer blocks process the sequence of patch embeddings. Each block contains the following:
  – *Multi-Head Self-Attention (MHSA)*: It allows each image patch to attend to all other patches in the sequence, enabling the model to capture long-range relationships such as the overall distribution of pallor across the conjunctival region. This mechanism helps the Vision Transformer to focus more on medically relevant areas while reducing the influence of irrelevant regions like eyelashes or scleral areas by assigning them lower attention weights.
  – Feed-Forward MLP Block: Following the attention layer, each transformer block includes a Feed-Forward MLP block, which consists of fully connected linear layers with GELU (Gaussian Error Linear Unit) activation functions, integrated with residual connections and layer normalization to stabilize and enhance learning. This combination helps in refining the learned representations, making the model more robust and expressive.

- Classification Output: For the final prediction, the [CLS] token—which aggregates global image context—is extracted from the last transformer layer. This token is then passed through an MLP classifier (linear layer + sigmoid activation) to produce a 256D feature vector representing the global context of the conjunctiva.

E. *Feature Fusion—CNN + ViT*
- CNN Branch Output: After global average pooling, the CNN gives a 256D feature vector, rich in local features and also texture-based features (e.g., color gradient, microvascular details).
- Concatenation: The ViT 256D vector and CNN 256D vector are concatenated to form a 512D fused vector:

$$\text{Hybrid Features} - [\text{CNN}_{256\text{D}} | \text{ViT}_{256\text{D}}]$$

- Dense Layers for Final Prediction: This fused vector goes through one or more dense (fully connected) layers: The last layer uses a sigmoid activation to output the probability of anemia presence (yes/no).

Figure 14.7 explains the feature fusion of CNN + ViT.

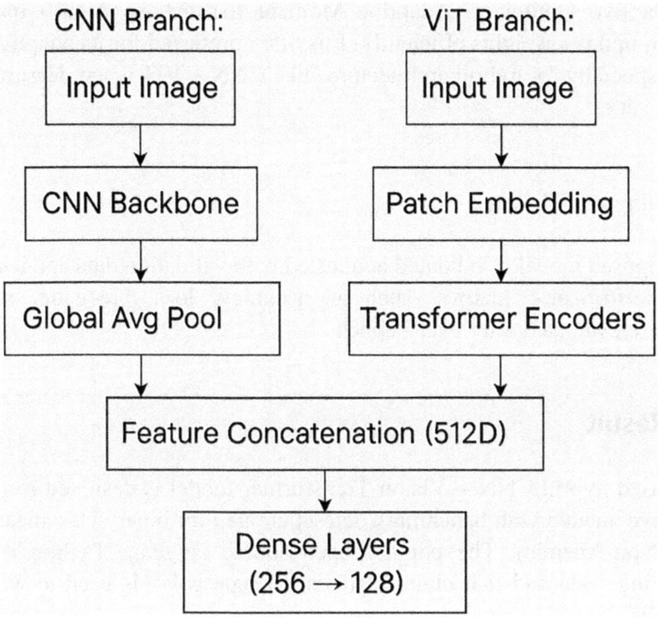

Fig. 14.7 Feature fusion of CNN + ViT

F. *Model Training*:

Model training is a crucial phase where the hybrid CNN + Vision Transformer (ViT) architecture learns to identify signs of anemia from conjunctival eye images. This process involves optimizing the model's parameters using labeled training data to minimize prediction errors.

A rich hybrid model is built by combining low level features from CNN. The model also helps in prepearing the contextual patterns. This hybrid model then learns to classify whether an image indicates anemia or not. To ensure optimal learning and generalization, the training process incorporates the following:

- A Binary Cross-Entropy (BCE) Loss Function: It is to guide learning for this two-class problem (anemia vs. no anemia), defined as

$$\text{BCE Loss} = -\frac{1}{N}\ \Sigma\ [y_i \times \log(p_i) + (1 - y_i) \times \log(1 - p_i)],\quad \text{for } i = 1 \text{ to } N$$

where, y_i is the ground truth label (0 or 1); p_i is the predicted probability from the sigmoid output; N is the total number of samples.

BCE effectively penalizes incorrect confident predictions and is ideal when the final layer uses a sigmoid activation function.

- An Adaptive Optimizer: Adaptive Moment Estimation (Adam) used in this research updates weights efficiently. It is often preferred for its adaptive learning rates, especially for hybrid architectures like CNN + ViT where learning dynamics can vary.

G. *Validating and Testing*

The proposed model is validated and tested with validation data and test data. All required performance metrics, such as accuracy, loss, precision, recall, and F1-Score,are evaluated after every epoch.

14.7 Result

The proposed hybrid CNN + Vision Transformer model is designed to predict the non-invasive anemia with benchmark dataset curated for binary classification (Anemic and Non-Anemic). The popular open source language Python is used for implementing code and rich cloud platform Google colab is used to write and to test the code.

The resulting confusion matrix given below shows the predicted samples.

Class	Predicted: Anemic	Predicted: Non-anemic
Actual: Anemic	170	18
Actual: Non-anemic	70	170

From the confusion matrix, it is observed that the model correctly identified 170 anemic and 170 non-anemic cases. Also, it misclassified 18 anemic cases as non-anemic (false negatives), and 70 non-anemic cases as anemic (false positives).

The classification report shows other performance metrics.

Class	Precision	Recall	F1-score	Support
Anemic	0.91	0.90	0.91	188
Non-anemic	0.90	0.91	0.91	187
Accuracy			0.91	

14.8 Conclusion and Future Work

The hybrid CNN + ViT architecture demonstrated high effectiveness in classifying anemia from conjunctival images. By leveraging CNN for local feature extraction and ViT for capturing global context, the model achieved a robust classification accuracy of *90.67%*, with strong performance across all key evaluation metrics. These results support the feasibility of using deep learning-based image analysis for non-invasive, real-time anemia screening.

While the model achieved promising results, there are several directions to improve and extend this research: (i) Larger and more diverse datasets can be used to incorporate multi-ethnic, age-diverse, and real-world conjunctival images to improve model generalization. (ii) Extend the model to detect different anemia severity levels.

References

Doi K, MacMahon R, Kido S (2019) Hybrid feature extraction for noninvasive anemia detection from ocular sclera images. Med Image Anal 54:63–71. https://doi.org/10.1016/j.media.2019.02.007

Kalantri A, Karambelkar M, Joshi R, Kalantri S (2020) Smartphone-based conjunctival pallor measurement for anemia screening in low-resource settings. IEEE J Biomed Health Inform 24(3):1122–1130. https://doi.org/10.1109/JBHI.2019.2939218

LeCun Y et al (2015) Deep learning. Nature 521(7553):436–444

Mast E et al (2021) Anemia screening in low-resource settings. Am J Hematol 96(2):E57–E59

Pietka M, Kwasniewska K, Badura E (2021) Deep learning approach to anemia classification from conjunctival images using illumination correction. Comput Methods Prog Biomed 198:105763. https://doi.org/10.1016/j.cmpb.2020.105763

Schmitt JM, Yang Y, Marks DL (2018) Optical coherence tomography-based hemoglobin measurement in conjunctival microvasculature. Biomed Opt Express 9(4):1467–1480. https://doi.org/10.1364/BOE.9.001467

Vinjamuri SK, Bernabeu MO, Frangi AF (2020) Multi-modal anemia detection combining eyelid and nail bed image analysis. Sci Rep 10(1):1–10. https://doi.org/10.1038/s41598-020-79194-1

Ware RS, Kleinstein JM, Wilson LB (2022) Multi-instance deep learning for anemia detection from retinal fundus images. NPJ Digit Med 5(1):1–9. https://doi.org/10.1038/s41746-021-00542-0

World Health Organization (2021) Global anaemia estimates 2021. WHO, Geneva

Xu L, Wolterink JM, Metz CT (2021) Attention-guided convolutional neural network for anemia detection from conjunctival pallor. Med Image Anal 71:102062. https://doi.org/10.1016/j.media.2021.102062

AI for Multi-region Tumour Detection: Enhancing Human Workflow in Full-Body Scans

15

S. Geetha, G. Soniya Priyatharsini, M. Rajkumar, N. Ethiraj, and Victo Sudha George

Abstract

The abnormal formation of the group of cells is known as the tumours. These should be identified as soon as possible to save the human life. This is an imminent risk for human life nowadays. Early identification of such tumours also helps to increase the probability of efficacious treatment and the survivorship of the affected humans. While handling the meticulous findings, the dissimilarity between the realm and the tumour morphology is confused. Lately, some advancements have been noticed in using computer-aided detection (CAD) models. It also improves the efficacy in distinguishing the different kinds of cancer cells, which can be found in the lungs, skin, brain, breast etc. Different types of imaging approaches like mammography, dermoscopy, magnetic resonance imaging (MRI), positron emission tomography (PET), computed tomography (CT) are used to determine the tumour cells. All the above-mentioned approaches help in explaining the complications and concerns of the tumours. The goal of this work is to do extensive research about the recent techniques that are all used for the different organs. This research mainly focuses on skin, lungs, breast and brain. For the furnishing of the medical imaging domain, deep learning

S. Geetha (✉) · G. S. Priyatharsini · V. S. George
Department of Computer Science and Engineering, Dr. M.G.R Educational and Research Institute of Technology, Chennai, Tamil Nadu, India
e-mail: geethasoman@drmgrdu.ac.in

M. Rajkumar
Department of Smart Computing, School of Computer Science Engineering and Information Systems, Vellore Institute of Technology, Chennai, Tamil Nadu, India

N. Ethiraj
Department of Mechanical Engineering, Dr. M.G.R Educational and Research Institute of Technology, Chennai, Tamil Nadu, India

J. M. Chatterjee, S. K. Saxena (eds.), *Artificial Intelligence and Blockchain in Precision Medicine and Virology*, Medical Virology: From Pathogenesis to Disease Control, https://doi.org/10.1007/978-981-96-8919-4_15

technology like convolutional neural network (CNN) is widely used. This delivers precise reliability and accuracy. By combining this technology with the CAD systems in the medical domain, the tumour cells are identified in the early stage. By doing this, human errors can be minimised. It is also easy to diagnose the cells precisely.

Keywords

Artificial intelligence · Tumour detection · AI in medical field · CNN

15.1 Introduction

Developments in the research identify the tumour cells in the early stage which is giving a positive impact on the health of humans. Tumour is considered as a serious issue in the health perspective. Identifying this abnormal cell growth can be easy to diagnose and save the life in the early stage. In the medication of the tumour cells, early detection plays a crucial role. This identification of the cells and their diagnosis of such kinds of cells should happen within the precise time. This will help the patients to rectify soon. The late treatment or delay in the detection of the cells will lead to the least survival of the patients. This plays an important role in the treatment procedures of the tumour cells. The multiplicity in tumour investigation faces hurdles in the classification process of the cells. The methods of classification approaches give the anatomical sites and various dimensions. In various ways, the neoplasm is represented. It represents the unique recognition and establishment in prescribing the efficiency and accuracy.

Currently, the integration of technologies like artificial intelligence (AI) and computer-aided systems has modified the way of processing medical imaging and tumour detection. This is shown in Fig. 15.1. These methodologies use refined algorithms to illustrate the differences in imaging in medical fields, capabilities of the diagnosis and the healthcare providers. This will help reduce the human error. This also helps in the earlier identification of the tumour cells. This new technology improves the outcomes of the patients by identifying the cells earlier.

This work illustrates a detailed explanation and the interrogation of the used tumour identification which incorporates the lungs, skin, brain and breast. The deep learning methods are examined in this research. Especially, several medical imaging techniques like PET, MRI, CT and dermoscopy are investigated in this chapter using convolutional neural networks (CNNs) (Hinton et al. 2012). This work examines and explains various hurdles and probabilities of accumulating in the operations of healthcare. It also includes the regulatory plan and ethical problems. This chapter aims to structure the serious issues of tumour cell diagnosis. It focuses on the importance of the AI. The important goal is to provide useful information to medical researchers and academicians about patient care. This will make them build and establish human-centric AI. These AI systems are used to improve the patient care.

A

Dice score (per tissue class)

Dataset	TUMOR	MUC	TU_STROMA	SUBMUC	MUSC	ADVENT	LYMPH_TIS	NECROSIS	BLOOD	MUCIN	BACK	Model average RESECTION	Model average BX
UKK-L1 *test 1*	0.901	0.955	0.808	0.878	0.943	0.760	0.946	0.900	0.824	0.892	0.852	**0.881**	**0.894**
UKK-L1 *test 2*	0.902	0.955	0.812	0.878	0.943	0.760	0.946	0.900	0.824	0.892	0.852	**0.881**	**0.895**
CRAG EXT	0.904	0.961	0.717	0.705*	0.550*	0.438*	0.742*	0.782	-	0.768	0.727	-	-

* Classes highly underrepresented

B Pixel-wise metrics (tumor detection)

Metric	UKK-L1 *test 1*	UKK-L1 *test 2*	CRAG EXT
SENS	0.924	0.964	0.958
SPEC	0.987	0.987	0.915
ACC	0.98	0.984	0.932
PREC	0.901	0.901	0.880
F1	0.912	0.931	0.918
NPV	0.990	0.995	0.971

C Examples of tissue segmentation using AI tool (CRAG EXT test dataset)

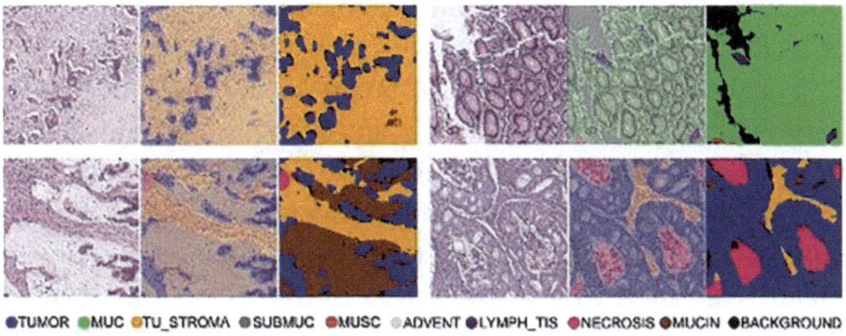

Fig. 15.1 Dice score per tissue class

15.2 Literature Review

The survey on tumour detection leads to refining the medical imaging techniques. Some of the imaging methods, for example, dermoscopy, MTI, CT and PET are used in tumour identification. This identification is used in early findings of the tumours. The survey papers mentioned in this chapter are mainly concentrated on the early detection of the tumour:

- Mammography is fundamental in breast cancer identification, functioning as a standard screening method to identify abnormal masses or calcifications.
- PET imaging, often used with CT, is crucial for identifying metabolic activity in tissues and differentiating between benign and malignant tumours.
- Computed tomography is widely used to find solid tumours in organs. This includes the organs, liver and breast. This gives intricate detailed pictures of the organs for the detection.
- MRI is highly effective at imaging soft tissues, making it crucial for detecting tumours in critical areas like the brain and spinal cord.
- The visualisation of the tumour growth in the skin is detected and explained using the technique of dermoscopy. It enhances in enhancing the field of skin cancer.

Earlier diagnostic methods have some shortcomings. To overcome this, CAD systems are refined by analysing the medical imaging and identifying the apparent areas (Esteva et al. 2017). Existing algorithms and medical imaging models follow the rule of independent-based algorithms. It is based on the predetermined characteristics. The occurrences of artificial intelligence followed by machine learning and deep learning, data-driven models are extracted from medical imaging.

These techniques are traditional methods that follow the process of classification, segmentation and tumour detection. With the help of CNN and deep learning techniques, the results are highly impactful in the medical field. These applications are very useful in the precise detection of unwanted cells in humans. The MRI data which is collected in the imaging applications can be useful in the early detection of the cells. The brain tumour images using CNN are used in the diagnosis of the tumours. These are proficiently segmented in the delineating borders (McKinney et al. 2020).

For the increasing amount of breast cancer patients, it is a boon for them to get early identification using these mammography techniques. These methods use the trained datasets for the detection along with the CNN to increase the sensitivity. This microcalcification and the masses facilitate the early detection of the tumours. While comparing with the skin cancer detection, the accuracy levels of the dermoscopic pictures are high with the detection of the melanoma of the dermatologists. In the primitive point of time, deep learning algorithms enhance the treatment of the pulmonary nodes in lung cancer. It elevates the strong datasets for the diagnosis. On the contrary of these developments, both the artificial intelligence approach and conventional approach give the hurdles. It is difficult to identify tumours that have different types of location poses, sizes and shapes using imaging approaches. In addition to that, the intruders always give the disproportion values, especially in the convoluted and cryptic cases.

The AI models have training datasets that are useful for the medical field while using the annotated methods. The gloomy nature of the AI models applies to the questions along with the probability of the clinical applications (Ronneberger et al. 2015). The assimilation of the AI models along with the CAD models will give the detailed datasets for the diagnosis (Rajpurkar et al. 2017; Cheng et al. 2016). These datasets are useful for the medical practitioners. It gives the secured output, which can help in the early diagnosis of the disease. These disturbances and misunderstandings lead to the need for AI solutions. These can give solutions for rectifying these tumour detections.

15.3 Methodology

Analysis of the tumour images is quite challenging. These tumours have various types of sizes, location shapes and so on. So, it is necessary to find a way to analyse the images properly and have a detailed version of the imaging methods. The toughest part of detecting the tumour cells is the texture clarification of the cells. Some of the functions like object recognition using faster CNN, image classification, identification and tumour cell segmentation are mainly used in the process of the CNN architecture. This can speed up the localisation and the establishment of the tumours in the medical scans. Some configurable data are found in the medical data. These medical data are categorised with the techniques of data augmentation and transfer learning. The availability of precise datasets depends on the persuasiveness of the deep learning models. These comprise ISIC for skin lessons, LIDC for lung

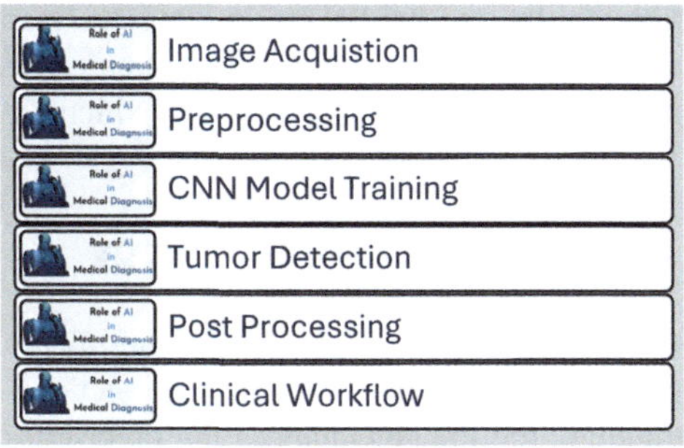

Fig. 15.2 Tumour detection preprocessing flow chart

nodules, and MRI, PET, CT and DDSM for mammography. These models significantly improve the efficiency. The calculation of these models entangles the explanation of the datasets that can perform the test sets, training and validation. They come under the criteria of specificity, accuracy and sensitivity, and evaluating the performance of the training sets. Figure 15.2 explains the steps in detail.

Common techniques that are used for tumour cell detection are identified and are integrated with the AI algorithms. The combination of AI and CAD can improve the precise output, which can be useful for medical practitioners. The dialysis includes the segmentation, tumour analysis, image preprocessing and creation of the outputs, and the segmentation can be concluded. The deep learning methods are very much inclusive in the medical area. Image archiving, clinical processes, electronic health records (EHRs) and Picture Archiving and Communication Systems (PACS) are included in the AI processes. The already existing techniques in refining medical images have the problem of extracting clear images. The effort of refining the images can be easy when the images are given in the deep learning algorithms. The output can be retrieved in the optimised images.

15.4 Existing Methodologies

Error-prone processes in detecting medical images are very difficult. Usually, the images are extracted with the help of the scanning process. The scan processes include dermoscopy, PET, CT, MRI and mammography. These image outputs have the problem of image quality and region identifications. For the past many years, these are the processes which are followed by the medical practitioners. The foundation of the tumour analysis is image preprocessing. This image preprocessing technique lists the size, shape and region affected by the abnormal cell. The expertise and accomplishment of the radiologist are used to check the image processing and identify the tumours. Even though the image processing models give clear imaging inputs to medical practitioners, the image is more precise when it deals with AI algorithms. It gives false negatives and false positives. The prosperous and rewarding techniques can cause human error. This human error creates the need for the growing new techniques. This yields the result of the brain tumour-detecting methods. Early detection of the tumour cells can limit the existing rate of survival.

Specifically, the hard-to-detect cells are identified with the collaboration of the AI and CAD systems. These calculations can be useful in finding the methods of the finding of the early-stage tumours, which are located in the various abnormal areas.

15.5 Proposed Methodology

CNN is the recently used architecture for imaging in the medical field. It gives the precise images for the tumour detection. For revolutionising the detection of tumours, the image enhances the precision of the images. Especially, the patterns are precisely detected, and the images are efficiently calculated. Deep learning methods can explain and extensively discretely explain image processing. This detects the prelude in the image processing. Unlike other algorithms, CNN provides detailed imaging techniques when combined with CAD systems. The brain and lung parts most importantly need a precise imaging process. This process gives a detailed output and gives a solution with the preprocessing techniques.

Harnessing the capacity of the CNN architecture can give an effective output in imaging. The effectiveness and precision of the comprehensive approach can be calculated with the three output image models. The structure of this chapter is organised with the research process and the findings. Figure 15.3 explains that AI creates methods to avoid delivering content and repeated outcomes. The workflow efficiency is explained in the flow diagram. The diagnosis is established with the

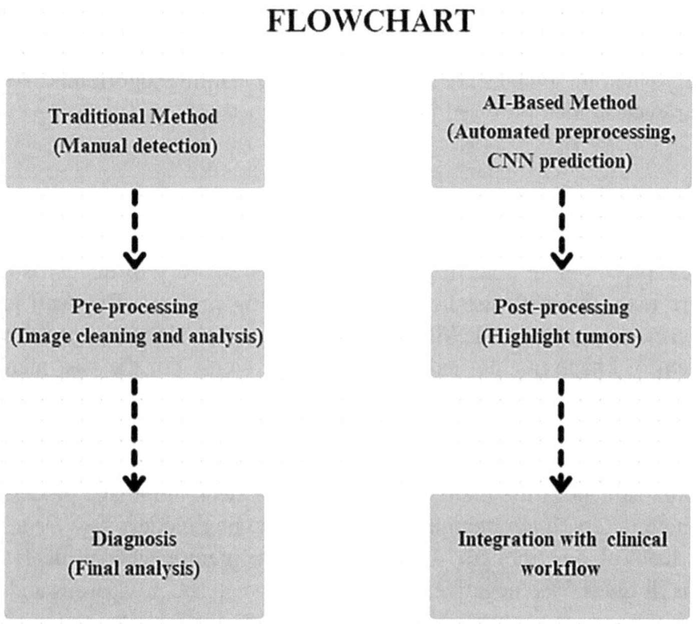

Fig. 15.3 Proposed model flow chart

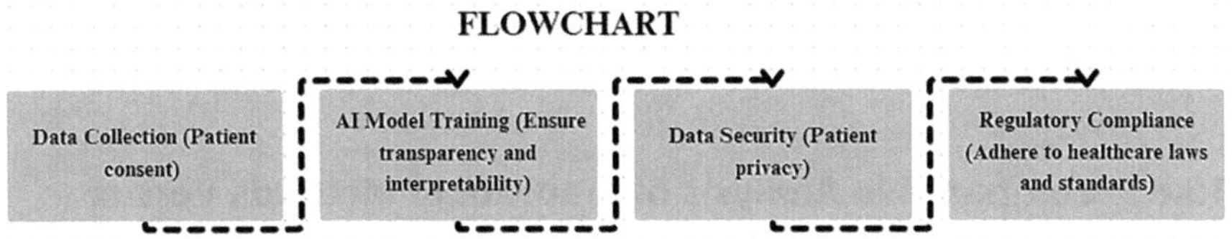

Fig. 15.4 Model training flow structure

areas of concern. It allows the prescreening of images and crucial elements in image processing. Since the field has many notable examples and leveraging outcomes, the predictions should be precise, and the various tumour types should be acceptable in the image qualities that remain elusive. However, the rise in deep learning techniques gives the precision image quality in the medical field.

The diagnostic methodologies have been transferred by the AI applications. This expedites the earlier and existing precise decision with diverse tumours. To calculate and evaluate the medical image data, a new CNN architecture is proposed, and this method gets precise decision-making in the image classification of tumour cells. 3D UNet models and the CNNs are used in the identification of gliomas in MRI data. Brain Tumour Segmentation (BraTS) Challenge is an extensive dataset trained by the algorithms. This allows us to classify segments with precise accuracy. The main focus is to classify and segment these types of images, which can boost the enhancement, and it can easily boost early-stage tumour identification. These AI solutions improve early detection and accuracy, minimise human error and overcome the constraints of procedures. This improves the outcomes of the patients. This combination methodology is explained in Fig. 15.4. The training methodology is explained effectively. Human error is reduced, and it is helped in the medical decision-making process.

Commonly used models for the identification and analysis of tumours are YOLO and DenseNet. These models are used to identify the breast cancer. These models are classified as benign and malignant. The local anomalies are used in concentrating the tumour regions. This method helps the radiologists in the screening process. Early-stage breast cancer is classified and identified with the CNN algorithms and trained on DDSM datasets. Image classifications can be done with the help of AI algorithms, which have the potential to diagnose skin cancer, including melanoma.

For enhancing the quality of CT images, recent techniques like R-CNN are used to detect the pulmonary nodules, and it helps the malignancy potential. For these processes, deep learning models are used. Such technologies are used to find abnormal growth and can detect the cells. This can be an early-stage detection technique. It is an efficient method for the identification of lung cancer. Lung Image Database Consortium (LIDC IDRI) datasets are used in identifying and classification of the nodules. The nodules are categorised as malignant and benign. False negatives should be reduced to improve the accuracy of the image detection. By using AI applications, lung cancers can be detected earlier. This can prompt the therapy detection and increase the survival rates. It is essential to identify lung cancer in the starting stage.

15.6 Comparative Analysis of Traditional Methods Versus AI-Driven Solutions

The capacity to transform the identification of the tumour and the accurate results in different applications is explained using artificial intelligence. Table 15.1 explains the comparison table of the existing and proposed model per tumour detection model.

15.7 Results and Discussions

The research highlighted the significance of contemporary imaging techniques, such as MRI, CT, PET, mammography and dermoscopy, for diagnosing tumours in vital organs like the brain, breast, skin and lungs. To improve diagnostic outcomes, novel computational methods must be used to address the distinct challenges posed by each modality, including resolution limitations, noise artefacts and modality-specific biases as seen in Fig. 15.5. Convolutional neural networks (CNNs), a deep learning methodology, have facilitated computer-aided detection (CAD) systems in accurately identifying tumours. Convolutional neural networks using MRI scans provide

Table 15.1 Analogy models of the existing and proposed models for tumour detection

Aspect	Traditional method	AI-driven solution	Key difference
Brain tumour detection	MRI scans interpreted by radiologists, prone to missing small or subtle tumours	3D UNet models analyse multiple MRI sequences, providing high accuracy in segmentation and classification	AI delivers consistent, reproducible results with enhanced sensitivity for early-stage tumours, reducing reliance on human expertise
Breast cancer detection	Mammography, limited sensitivity in dense breast tissues and prone to false positives	CNNs (e.g. DenseNet) excel at detecting microcalcifications and masses with superior sensitivity and specificity	AI improves early detection, reduces false positives and ensures accurate diagnosis, especially in challenging cases like dense tissues
Skin cancer detection	Visual inspection by dermatologists, which can be subjective and error-prone	AI models trained on dermoscopic images classify lesions with $>90\%$ accuracy	AI provides a consistent and objective diagnosis, outperforming in melanoma detection
Lung cancer detection	Manual review of CT scans by radiologists, which can be labour-intensive and subject to human error	Faster R-CNN automates detection and classification of nodules with higher sensitivity	AI systems enhance early-stage nodule detection, reducing false negatives and enabling timely interventions

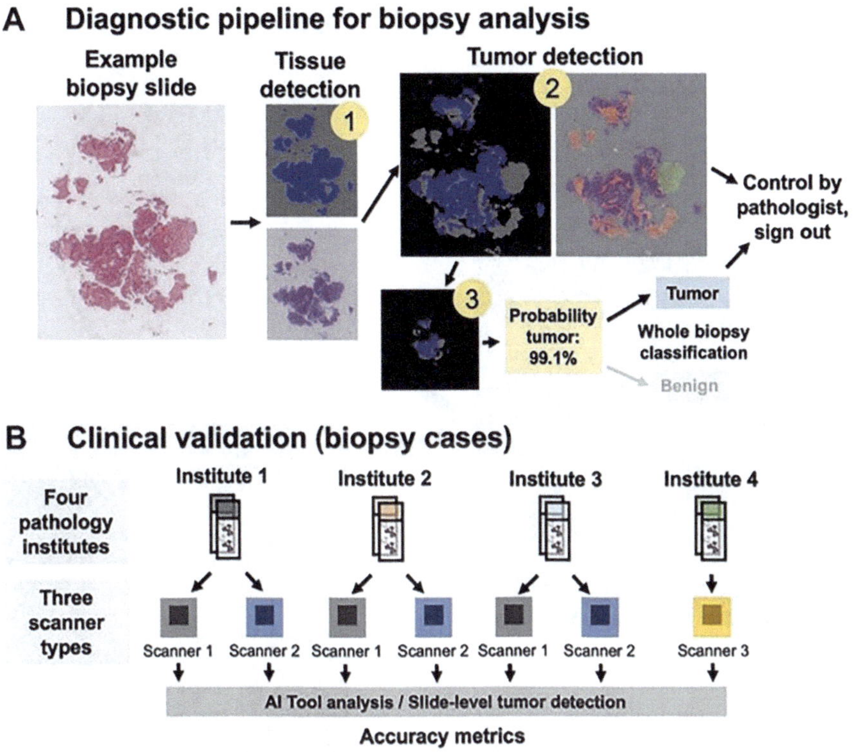

Fig. 15.5 (**a**) Diagnostic pipeline for biopsy analysis and (**b**) clinical validation

high sensitivity and specificity for brain tumours. Transfer learning models shown enhanced precision in the detection of breast tumours using mammography. The use of dermoscopy images with pretrained CNN architectures resulted in enhanced accuracy in skin cancer categorisation, particularly for melanoma.

The potential of AI in many diagnostic applications was shown by the accurate identification of nodules and tumour margins in lung cancers by CT-based detection utilising advanced segmentation models.

Explainable AI (XAI) methodologies, such as Grad-CAM, enhance the interpretability of CNN models by providing visual reasons for predictions, as seen in Fig. 15.6. This fostered trust and ensured the application of these advanced systems in clinical practice by allowing clinicians to validate AI results. These explainability tools bridge the gap between practical healthcare needs and complex machine learning models.

The correlation analysis is shown in Fig. 15.7. Diagnostic efficiency has been greatly increased by the smooth integration of AI systems into clinical processes, including interoperability with Picture Archiving and Communication Systems (PACS) and electronic health records (EHRs). The model identifies and classifies

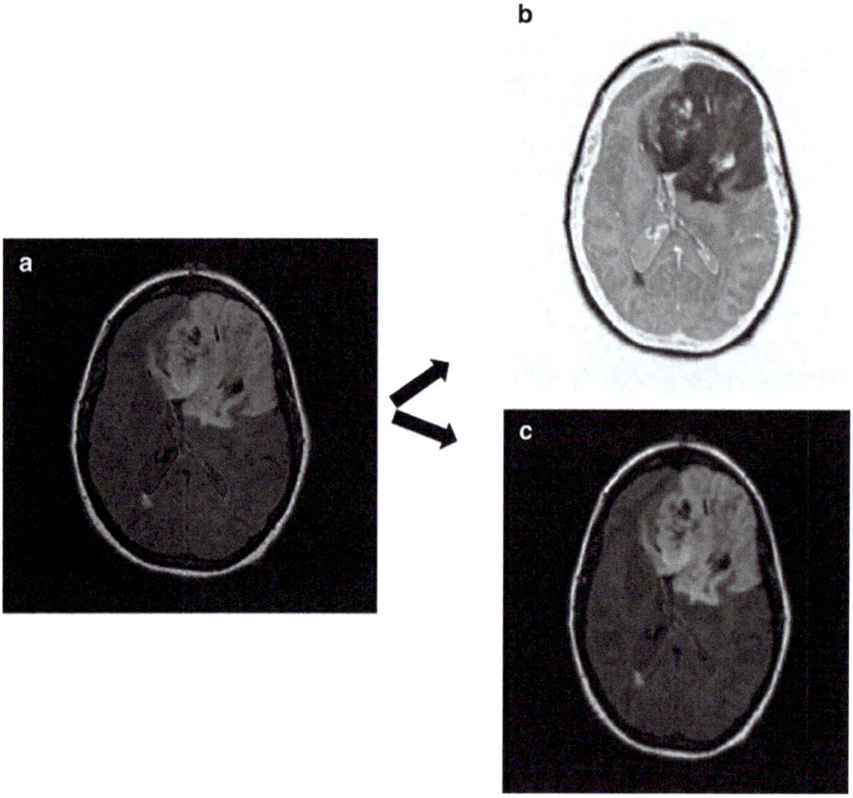

Fig. 15.6 Brain tumour identification and classification by the model

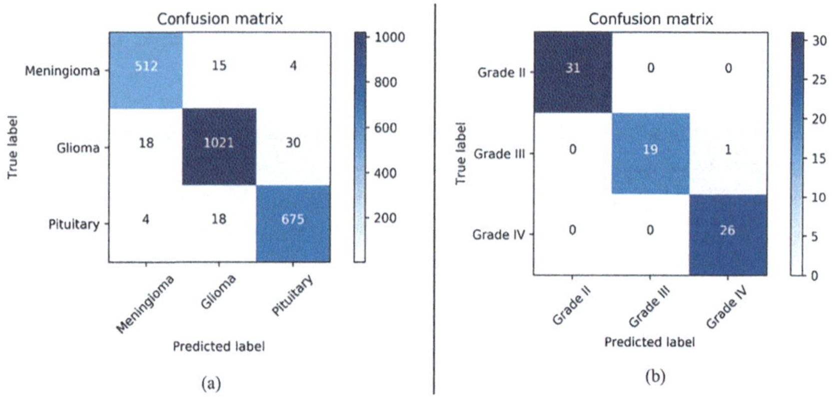

Fig. 15.7 Confusion matrix correlation analysis

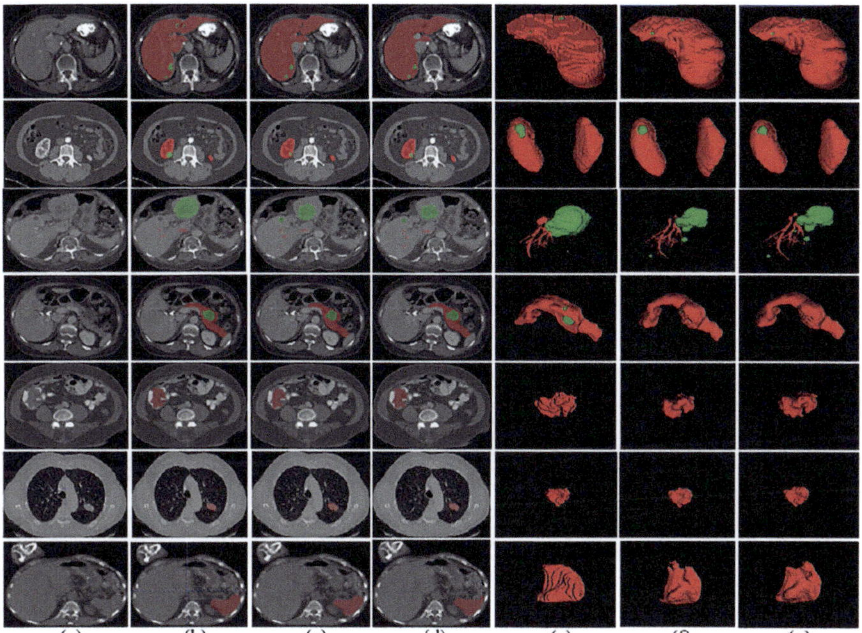

Fig. 15.8 Tumour identification of multi-regions through the model

the tumour as shown in Fig. 15.8. AI solutions serve as real-time decision support systems, reducing human error and speeding up tumour identification procedures.

Challenges persist despite these advancements. Model predictions are biased owing to dataset limitations, including insufficient diversity. Deficiencies in model interpretability hinder the widespread use of AI technology in therapeutic contexts. The extensive implementation is significantly obstructed by ethical concerns, particularly regarding data security and regulatory adherence. To improve AI applications in healthcare, As shown in Table 15.2, the proposed model outperforms existing methods.

The accuracy comparison metrics can be classified based on several previous models. Some are given in Table 15.2.

15.8 Conclusion

The use of artificial intelligence (AI) in medical imaging, particularly for tumour identification, has profoundly altered diagnostic methodologies. Utilising deep learning methodologies, particularly convolutional neural networks (CNNs), artificial intelligence has improved the accuracy, dependability and rapidity of tumour detection in several organs, including the brain, breast, skin and lungs. Artificial intelligence proficiently detects tumours in their first stages, allowing prompt therapies that enhance patient survival rates. The future of AI in healthcare is

Table 15.2 Accuracy comparison

#	Model	Best accuracy for study I	Best accuracy for study II	Classification type	Classification method
1	Cheng et al. [23]	91.28%	–	Multi	SVM and KNN
2	Paul et al. [25]	91.43%	–	Multi	CNN
3	Afshar et al. [26]	90.89%	–	Multi	CNN
4	Anaraki et al. [27]	94.2%	90.9%	Multi	GA-CNN
5	Zacharaki et al. [21]	–	85%	Multi	SVM and KNN
6	Zacharaki et al. [21]	–	88%	Binary	SVM and KNN
7	El-Dahshan et al. [22]	–	98%	Binary	ANN and KNN
8	Ertosun et al. [24]	–	71%	Multi	CNN
9	Ertosun et al. [24]	–	96%	Binary	CNN
10	Proposed structure	96.13%	98.7%	Multi	CNN

optimistic, with anticipated breakthroughs in deep learning models, multimodal imaging, and data augmentation likely to enhance the accuracy and generalisability of AI systems. As AI advances, it will progressively enhance collaborative decision-making by offering interpretable insights that fortify the human-AI alliance.

References

Ashock Kumar V, Priyatharsini GS, Bhimaavarapu K, Charaan RMD, Konakalla JP, Chandralekha E (2025) Design and analysis of smart health care predictions using machine learning. In: 2025 International conference on machine learning and autonomous systems (ICMLAS), Prawet, pp 389–395. https://doi.org/10.1109/ICMLAS64557.2025.10968540

Cheng J et al (2016) Computer-aided diagnosis with deep learning architecture: applications to breast lesions in US images and pulmonary nodules in CT scans. Sci Rep 6:Article 24454

Esteva A et al (2017) Dermatologist-level classification of skin cancer with deep neural networks. Nature 542(7639):115–118

Hinton G et al (2012) Deep neural networks for acoustic modeling in speech recognition: the shared views of four research groups. IEEE Signal Process Mag 29(6):82–97

Hosny A et al (2018) Artificial intelligence in radiology. Nat Rev Cancer 18(8):500–510

LeCun Y et al (2015) Deep learning. Nature 521(7553):436–444

Litjens G et al (2017) A survey on deep learning in medical image analysis. Med Image Anal 42:60–88

Lundervold AS, Lundervold A (2019) An overview of deep learning in medical imaging focusing on MRI. Z Med Phys 29(2):102–127

McKinney SM et al (2020) International evaluation of an AI system for breast cancer screening. Nature 577(7788):89–94

Rajpurkar P, et al (2017) CheXNet: radiologist-level pneumonia detection on chest X-rays with deep learning. arXiv preprint arXiv:1711.05225

Ronneberger O et al (2015) U-net: convolutional networks for biomedical image segmentation. In: International conference on Medical Image Computing and Computer-Assisted Intervention (MICCAI), pp 234–241

Samek W et al (2017) Explainable artificial intelligence: understanding, visualizing, and interpreting deep learning models. Digit Signal Process 73:1–15

Topol EJ (2019) High-performance medicine: the convergence of human and artificial intelligence. Nat Med 25(1):44–56

Wang X et al (2017) ChestX-ray8: hospital-scale chest X-ray database and benchmarks on weakly-supervised classification and localization of common thorax diseases. In: Proceedings of the IEEE conference on Computer Vision and Pattern Recognition (CVPR), pp 2097–2106

Zhou Z et al (2021) Lung cancer detection using deep learning algorithms. J Thorac Imaging 36(1): 54–66

16 Enhancing Classification Accuracy in Virology Through Deep Learning for Accurate Virus Identification from TEM Imagery

N. Nithiyanandam, R. Rajmohan, N. Sivaranjani, G. K. Sandhia, and N. Poongavanam

Abstract

To identify viruses in transmission electron microscopy (TEM) pictures, this research introduces a new two-pipeline method that leverages convolutional neural networks (CNNs). We used a collection of viral photos to train two separate convolutional models, improving feature extraction using a variety of image processing methods. A Quadratic Weighted Kappa (QWK) score of 0.9719 and a maximum testing accuracy of 97.44% indicate that the suggested model is very reliable for viral classification tasks. An average F1-score improvement of 5.32% and a substantial improvement of 5.24% in accuracy compared to conventional single-pipeline approaches were produced by the dual pipeline approach. In addition, confusion matrices showed that the model performed well for most categories of viruses, even if certain classes were difficult to accurately classify. Consistent with and expanding upon previous research, our results demonstrate that deep learning methods are capable of accurately differentiating between different types of viruses. This study demonstrates the promise of sophisticated convolutional neural network (CNN) designs in the field of virology and sets the stage for further research into how to improve the efficiency and accuracy of viral classification for better disease management and response.

N. Nithiyanandam (✉) · R. Rajmohan · G. K. Sandhia
Department of Computing Technologies, SRM Institute of Science and Technology, Kattankulathur, India

N. Sivaranjani
Department of Computer Science and Engineering, SCOPE, Vellore Institute of Technology, Chennai, India

N. Poongavanam
Department of Computer Science and Engineering, Saveetha School of Engineering, Saveetha Institute of Medical and Technical Sciences, Saveetha University, Chennai, India

J. M. Chatterjee, S. K. Saxena (eds.), *Artificial Intelligence and Blockchain in Precision Medicine and Virology*, Medical Virology: From Pathogenesis to Disease Control, https://doi.org/10.1007/978-981-96-8919-4_16

Keywords

Transmission electron microscopy · Virology · Deep learning · CNN

16.1 Introduction

All living things, from people to plants to microbes, are susceptible to infection by viruses. One of the greatest biological dangers to human health and food production on a worldwide scale is their capacity to multiply inside host cells. Viruses can cause unexpected problems in their hosts due to their remarkable adaptability and resilience, as demonstrated in recent research (Heaton 2019). It is now crucial for disease prevention and management techniques to effectively identify and classify viruses due to their various transmission channels, high contagion rates, and potential lethality (Artika et al. 2020).

Manual inspection techniques, including microscopy, serological tests, and culture-based procedures, have traditionally been used for viral identification. Nevertheless, these approaches aren't always foolproof, and they can be tedious and labour-intensive overall (Hahn et al. 2020). The fast mutation rates and intricate viral genomes make virus categorisation an even more challenging task. As a result, cutting-edge diagnostic methods that improve detection speed and accuracy are critically needed.

There are now more options than ever before for automated viral diagnosis due to recent developments in computational methods, especially deep learning (Wani et al. 2022). One kind of machine learning called 'deep learning' uses ANNs to decipher intricate patterns in data. Rapid and precise viral categorisation has been made possible in the field of virology by the efficient use of deep learning algorithms on several types of medical imaging (Dargan et al. 2020). Imaging virus particles with great resolution using transmission electron microscopy (TEM) is one potential approach. Visualising viral morphology and gaining insights into virus-host interactions have long been made possible by TEM, making it a valuable tool in virology (Richert-Pöggeler et al. 2019).

Further advancements in strong computing resources and the availability of huge datasets have made deep learning an attractive option for TEM image analysis. The application of deep learning models for viral identification from transmission electron microscopy (TEM) images has recently shown impressive results in terms of classification accuracy (Shaga Devan et al. 2021). As an example, convolutional neural networks (CNNs) trained on a dataset of viral TEM pictures achieved classification accuracies above 90% in one research (Usman et al. 2024). These developments highlight the promise of automated computer diagnostics for improving virology studies and clinical diagnostics.

Although there has been advancement in this field, there are still obstacles to overcome, especially with regard to TEM picture variability and image noise. Sample preparation, electron beam damage, and intrinsic noise are just a few of the common issues that can impact transmission electron microscopy (TEM) pictures and make it difficult to detect important characteristics required for precise categorisation (Chen et al. 2020). In order to fix these problems, it is essential to use high-end image processing methods to improve the data that is input. Image normalisation and denoising autoencoders are two methods that may greatly enhance the readability and clarity of TEM pictures, which in turn improve classification performance (Beltran Perez 2018).

This research presents a new method for classifying TEM images for viral identification using deep learning; the method makes use of a two-pipeline strategy to improve classification accuracy. In order to preserve important morphological characteristics while reducing noise, the methodology entails preparing TEM images utilising coherent image processing techniques. After the photos have been cleaned up, a deep learning model is used to detect different types of viruses. This research intends to show that the suggested approach is successful and reliable by using a large dataset of TEM pictures. In the end, it hopes to help virologists identify viruses more quickly and accurately. The larger ramifications for public health and agricultural biosecurity, as well as the study's capacity to improve classification accuracy, make it significant. In order to manage outbreaks, provide effective therapies, and adopt preventative measures, rapid and reliable viral detection is crucial. Improved diagnostic tools are within reach as we work to merge deep learning with conventional virology methods.

The results of this study may also have implications for other fields and areas of study outside of virology. This study's deep learning model has the potential to pave the way for future work in the field, increasing the breadth of computational applications in the life sciences by providing a framework for researchers to apply comparable methods to a variety of biological pictures. Finally, improving classification accuracy using modern computational approaches is crucial for successful viral management and prevention, as viruses still cause major harm to ecosystems and human health across the world.

16.2 Methodology

We used a really simple procedure for this study. To begin, we conducted some preliminary processing on the raw TEM images using two popular image processing techniques: one for domain transformation and another for filtering. Following that, two separate convolutional models were used to extract local features from these two sets of output images. Our last step in classifying viruses was to employ a multi-layer perceptron. You can see the whole procedure in Fig. 16.1. The following subsections discuss each of these stages in detail.

16.2.1 Dataset Description

For this investigation, the researchers turned to the TEM virus dataset stored at Mendeley Data (Sikder et al. 2024). Tecnai 10 with MegaView III camera and LEO

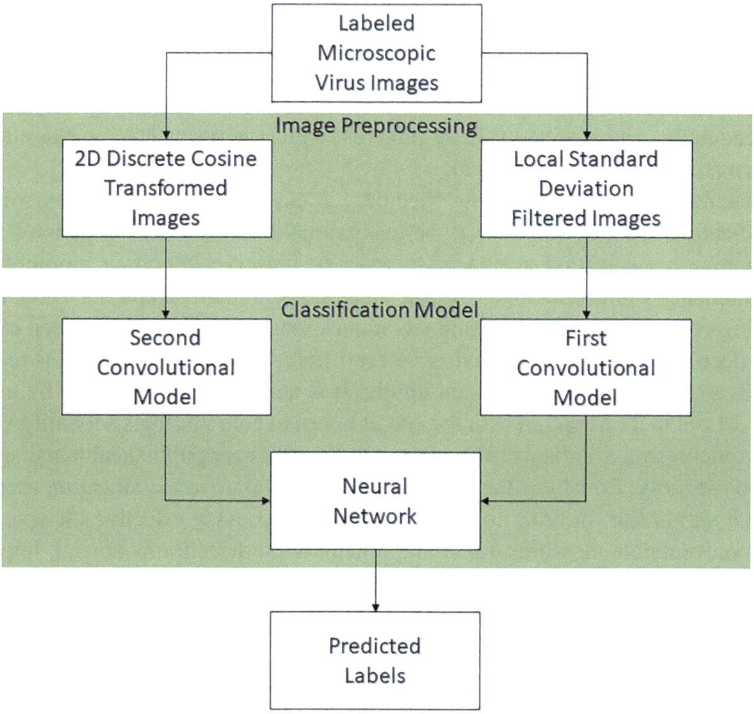

Fig. 16.1 The methodology behind the suggested way of categorising viruses

with Morada camera electron micrographs of numerous viruses are included in the paper, which dates back to 2021. Prior to taking the photographs, the samples were processed with 10% phosphate-buffered saline (PBS), positioned on carbon-coated TEM grids, and then stained with 2% PhosphoTungstic Acid (PTA). Researchers faced a number of challenges when compiling this dataset, including missing annotations, noise, blurry images, different magnifications, and debris. The initial microscopy pictures were of different resolutions, and the dataset's class sample sizes were not consistent. Researchers employed class-wise modifying rules to supplement the produced patches to attain an equivalent sample size after cropping the original photographs into multiple patches, each measuring 256 × 256 pixels. This was done to address these concerns. The greyscale images in this archive are saved as Tag Image Files (.tif). In this study, images of 14 different viruses were used. In their 2021 article (Matuszewski and Sintorn 2021), the dataset's developers also performed a similar classification. The TEM photos were sorted using the DenseNet201 architecture, which had previously been trained on ImageNet. An extra article describing a customised CNN model used to classify the viral subtypes in the dataset (Ali et al. 2022). In Table 16.1, we can see the various ways in which these viruses have affected various animals and people. You may find the class IDs and the sample size used in this inquiry in the given information. Visual comparisons with samples from each virus class are presented in Fig. 16.2.

16.2.2 Preparing Images for the Initial Convolutional Model

We used two distinct image processing methods to handle the raw images, as indicated before. Nevertheless, we began by scaling down the original photos to 128 × 128 pixels. This action greatly reduced the data size and simplified the deep learning model's time and space complexity. In order to prepare the TEM images for the First Convolutional Model, we applied the local standard deviation filter. It is a spatial filter that uses the concept of 'moving windows' to function (Lloyd 2010). With the use of a movable window, we may capture certain image details by isolating a little area from the full picture. To extract those properties from the entire image, the window is moved across the image. The mean intensity of the window can be determined by taking B as a sample greyscale image, y_i as the intensity of the jth pixel of the active window τ, and q as the total number of pixels in the window (Chatterjee and Siarry 2013).

$$\text{mean}(B_\omega) = \frac{1}{q}\sum_{j=1}^{q} y_j^{B_\omega} \tag{16.1}$$

The standard deviation, which indicates the dispersion of the intensity values of ϋ around the mean, can be calculated using:

$$\text{SD}(B_\omega) = \sqrt{\left(\frac{1}{q}\sum_{j=1}^{q} y_j^{B_\omega} - \text{mean}(B_\omega)\right)^2} \tag{16.2}$$

If you perform this for every window, you'll get a matrix with the intensity standard deviations from their respective local means. We fed the resulting matrix into our First Convolutional Model after repositioning a 3 × 3 pixel window on the TEM images in this study. Nevertheless, the resolution of the resultant matrix is reduced by two pixels in each dimension compared to the original image when a 3 × 3 window is used on an image in the previously indicated technique. We ran the standard deviation filtering after padding the photos with one additional pixel on each side to make up for this loss.

We included a small section of a transmission electron microscopy (TEM) viral image (4 4 pixels) in Fig. 16.3a to offer clarity. Afterwards, as demonstrated in Fig. 16.3b, we enlarged the image to 6 × 6 pixels by applying symmetric padding to both sides. A grey background has been used to highlight the padded pixels. It should be noted that alternative cushioning methods are also available. Equation (16.1) is used to obtain a matrix equal to the size of the image shown in Fig. 16.3a, and the mean values obtained by driving a 3 × 3 pixel window over the padded

Table 16.1 Details about the viruses that were part of the TEM virus collection and utilised for this research

Virus class	Description	Transmission mode	Associated diseases	Common hosts	Samples
Adenovirus	Variegated infections caused by viruses that do not have an envelope	Airborne particles, one-on-one interaction	Infections of the respiratory tract, inflamed eyes	Humans	750
Astrovirus	Viruses with a star shape that are known to cause gastroenteritis, particularly in young children	Gastric-oral pathway	Vomiting and diarrhoea	Humans	750
Bunyavirus	The causative agents of fever in a group of viruses that spread mostly by insects and other crawling insects	Insect bites (such as ticks and mosquitoes)	Brain infections, haemorrhagic fever	Humans, animals	750
Coronaviruses	Vaginosis is a respiratory ailment caused by enveloped viruses	Inhalation of aerosols, interaction	MERS, COVID-19, SARS,	Humans, bats, camels	750
Dengue Virus	A virus that causes dengue fever and is transmitted by mosquitoes	Verose bites	Dengue haemorrhagic fever, dengue fever	Humans	750
Ebola Virus	Has a high fatality rate and causes severe haemorrhagic fever	Interactions involving body fluids	The haemorrhagic sickness caused by Ebola!	Humans, primates	750
Influenza Virus	Viruses with a protective covering that breed seasonally and infect the lungs	Inhalation of aerosols, interaction	The flu virus	Humans, birds, pigs	750
Nipah Virus	Encephalitis and severe respiratory sickness can be	Interactions with diseased pigs or bats	An infection with the Nipah virus	Humans, bats, pigs	750

(continued)

Table 16.1 (continued)

Virus class	Description	Transmission mode	Associated diseases	Common hosts	Samples
	caused by this zoonotic virus				
Papillomavirus	A family of viruses that can cause warts as well as cervical and other malignancies	Direct physical touch	Cervical cancer, warts	Humans	750
Rift Valley Fever	This virus mostly affects cattle, but it can also infect people	Bite from a mosquito or other sick animal	Flu in the Rift Valley	Humans, livestock	750
Rotavirus	It is a leading source of severe diarrhoea in kids and babies	Gastric-oral pathway	The stomach flu	Humans	750
Zika Virus	The disease is mostly spread by mosquitoes and is known to cause birth abnormalities in pregnant women who catch it	Bite from a mosquito, sexual transmission	Zika virus disease, Zika virus congenital syndrome	Humans, mosquitoes	750
Hantavirus	Causes serious respiratory problems; transmitted by rodent faeces, urine, or saliva	Breathing in polluted air, coming into close touch	Hypertension caused by a hantavirus: HPS	Humans, rodents	750
Marburg Virus	Similar to Ebola, it causes a viral haemorrhagic fever and has a very high death rate	Coming into touch with infectious bodily fluids	Marburg virus illness	Humans, bats	750

image are presented in Fig. 16.3c. Last but not least, as seen in Fig. 16.3d, the values of the local standard deviation were determined from the first two matrices by applying Eq. (16.2).

In Fig. 16.4a, d, we show two raw TEM pictures from the dataset without any filter applied. In Fig. 16.4b, e, respectively, we show the results of applying the filter and highlight its benefits. Figures show that compared to their raw counterparts, the

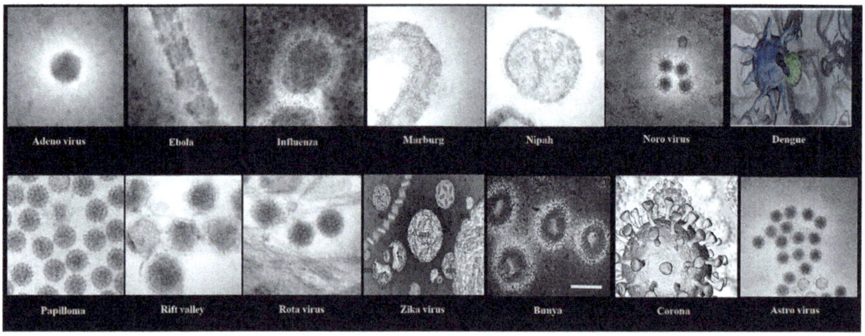

Fig. 16.2 The TEM viral dataset contains electron microscopy images of 14 distinct virus types (Sikder et al. 2024)

40	44	33	34
43	51	42	44
45	51	50	50
42	47	52	53

(a) Input Image

40	40	44	33	34	34
40	40	44	33	34	34
43	43	51	42	44	44
45	45	51	50	50	50
42	42	47	52	53	53
42	42	47	52	53	53

(b) Symmetric Padded Matrix

42.78	41.11	39.89	36.89
44.67	44.33	44.33	42.33
45.44	47	48.89	48.67
44.78	47.56	50.56	51.78

(c) 3 × 3 Mean values

3.56	5.62	6.55	4.88
4.03	5.87	7.02	7.14
3.54	4.12	3.76	4.21
3.15	3.97	2.3	1.39

(d) Local Standard Deviation Filtered Image

Fig. 16.3 (**a**) An example of a TEM image, (**b**) the final product following symmetric padding, (**c**) calculate the average intensity using 3 × 3 windows, and (**d**) results from filtering based on local standard deviations

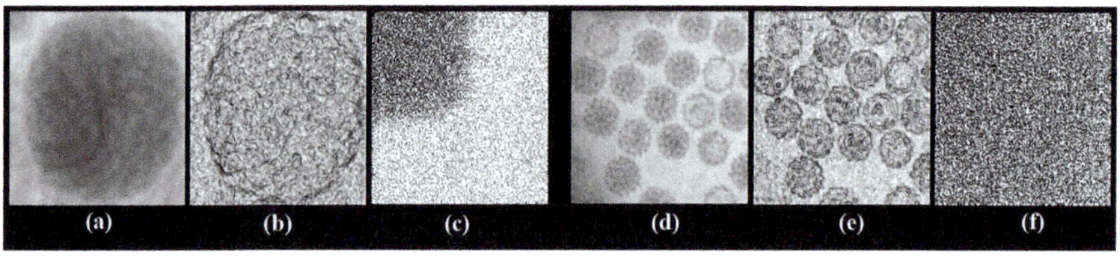

Fig. 16.4 (**a**) Class Or mock picture. (**b**) Filtered of class Or. (**c**) Class Or result of 2D DCT. (**d**) Class PI mock picture. (**e**) Filtered of class PI. (**f**) Class PI result of 2D DCT

revised photos have sharper edges (Lloyd 2010). The edges and borders of shapes and objects in an image are brought out more clearly by convolution; therefore, the resulting edge-enhanced versions of the images may have more important class-specific attributes, making them easier to classify.

16.2.3 Two-Convolutional-Model Image Preparation

In order to prepare the raw TEM images for the Second Convolutional Model, we first subjected them to a domain transformation process. The 2D.

DCT is ideal for this task. A finite-length signal or data stream can be transformed into a series of cosine waves with varying frequencies using a traditional, one-dimensional discrete-time transform (DCT). Its original inspiration came from Nasir Ahmed's 1972 quest for an improved method of visual data compression (Ahmed et al. 1974; Ahmed 1991). Similar to the Discrete Fourier Transform (DFT), which separates a signal into its sine and cosine components, DCT does the same thing. On the other hand, energy compaction is better with DCT than with DFT. A periodic signal, $y(n)$, with a finite length N is defined by its DCT, which is given by (Wang 2008).

$$y(l) = \alpha(l) \sum_{o=0}^{O-1} y(o) \times \cos \frac{l\pi(2n+1)}{2N}, \tag{16.3}$$

$$\alpha(l) = \begin{cases} \frac{1}{\sqrt{O}} & l = 0 \\ \sqrt{\frac{2}{O}}, & 1 \leq l \leq O-1 \end{cases}$$

Because photographs are inherently 2D matrices, 2D DCT is required. The 2D DCT of an image B of $N \times O$ pixels is defined as:

$$B(l,m) = \alpha(l)\alpha(m) \sum_{n=0}^{N-1} \sum_{o=0}^{O-1} B(n,o) \times \cos \frac{l\pi(2n+1)}{2N} \times \cos \frac{m\pi(2o+1)}{2O} \tag{16.4}$$

Where

$$\alpha(l) = \begin{cases} \frac{1}{\sqrt{N}} & l = 0 \\ \sqrt{\frac{2}{N}}, & 1 \leq l \leq N-1 \end{cases} \quad \& \alpha(l) = \begin{cases} \frac{1}{\sqrt{O}} & m = 0 \\ \sqrt{\frac{2}{O}}, & 1 \leq m \leq O-1 \end{cases}$$

There are two ways to do 2D DCT, and each one requires one 1D DCT operation. Spectral methods, partial differential equations, media encoding techniques, communication systems, networking devices, and image and video compression are just a few of the many applications that rely on DCT. These pictures don't say much, but they contain important raw TEM image attributes and are a huge help to the classification model.

16.2.4 Neural Network with Convolutional Model Architecture

In the suggested approach, we extracted local features using two convolutional models. For the last classification, we employed a neural network after merging the feature sets acquired from them. Image-related tasks such as classification, reconstruction, compression, and creation are the main areas of interest for convolutional neural networks (CNNs), which are feed-forward neural networks. Nevertheless, it has been adjusted to accommodate both lower-dimensional data (like signals (Lee et al. 2017)) and higher-dimensional data (such movies (Gul et al. 2020) and multimodal images (Milletari et al. 2017)). Problems involving the analysis of high-dimensional data with conspicuous spatial structures are typically well-suited to CNNs (Fan et al. 2021). Convolution is the mathematical process that CNNs rely on. It involves comparing a larger image with a smaller image block and expressing the similarities (at different parts of the first image) (Masud et al. 2021). Due to the fact that our dataset only contains greyscale images, all of the matrices in this study are 2D and contain information from a single colour channel. The spatial coordinates of B, denoted as $B \in \mathbb{S}^{n \times o}$, can be used to depict a sample transmission electron microscopy picture. An affine transformation is applied on $B \in \mathbb{S}^{n \times o}$ initially during a convolution. After that, it undergoes a nonlinear activation on an element-by-element basis. In order to automatically extract characteristics from input data, a convolution employs several filters. In the context of a 2D filter, the size of the filter, denoted as s, can be represented as $F_l \in \mathbb{S}^{t \times t}$. Although larger filter sizes and rectangular-shaped filters are other viable options, we restricted ourselves to using 3×3 square filters for convolution operations in this investigation. A single feature map P^l is obtained by applying B, F_l, which in turn performs multiple convolutions on the picture (Fan et al. 2021). The process of producing P^l can be described in this way:

$$P^l_{j,k} = \left\langle [B]_{j,k}, G_l \right\rangle = \sum_{j'=1}^{T} \sum_{j'=1}^{T} [B]_{j+j'-1,\quad k+k'-1} \times [G_l]_{j',k'} \tag{16.5}$$

At position $[B]_{j,\ k}$, there is a tiny section of the image B denoted as (j, k). G_l gathers local features from the image as it moves across it (from left to right and top to bottom) like a sliding window at a predetermined interval in both directions (strides, typically set to 1 pixel in each direction). The matrix with the dimensions $(m - s + 1) \times (n - s + 1)$, where $[P]_{jkl} = [P^l]_{jk}$ is the result of this operation, which is denoted as (P^l). This is the result of applying many filters to a single picture in order to obtain many feature vectors.

A nonlinear activation function is used to the convolutional layer's output. We employed Sigmoid, ReLU, and Softmax activation functions in this investigation. In order to get a value between 0 and 1, Sigmoid calculates the following, given a real number (q):

$$f_{\text{Sigmoid}}(q) = \frac{1}{1+f^{-q}} \tag{16.6}$$

Sigmoid, on the other hand, isn't something you want to apply on the fly because it slows down model training and can cause gradients to disappear, which is called 'vanishing gradients' (Hs 2021).

The simplicity and fast computing ability of ReLU have made it the most often utilised activation function in recent times. If the input is negative, a classic ReLU will return 0. If the input is positive, it stays the same. Its functioning is best described by the following:

$$f_{\text{ReLU}}(q) = \begin{cases} P, & q<0 \\ q, & q \geq 0 \end{cases} \tag{16.7}$$

Use ReLU functions sparingly outside of the hidden layers for optimal performance. Some gradients may be killed and neurons may perish since it totally disregards negative inputs. To combat these issues and mitigate their impact, there are modified versions of ReLU.

Typically applied as the final layer in a multi-class classification problem like ours, Softmax is sometimes called multi-class logistic regression in mathematics because it follows the concepts of logistic regression:

$$f_{\text{Softmax}}(q)_j = \frac{f^{q_j}}{\sum_{k=1}^{K} f^{q_j}} \tag{16.8}$$

at index $j = 1, 2, 3, \ldots, K$. Softmax is a collection of many Sigmoid functions; it is applicable only in cases when the classes in the dataset are incompatible with one another.

Pooling layers are an additional component of convolutional neural networks (CNNs) (Hinkle and Metsis 2021). You can utilise these layers to get a single numerical value that represents all the data from a specific window. In order to simplify the model's computational complexity, pooling is inherently a down-sampling procedure that decreases the amount of available data for the subsequent layers (Fan et al. 2021). There are various kinds of pooling layers, but max-pooling is the most common. If you have a window with a specific size and all the possible values in it, max-pooling will take the largest value and ignore the rest. A matrix with the dimensions floor(n/3) × floor(o/3), obtained by applying a 3 × 3 max-pooling window to an image $B \in \mathbb{S}^{n \times o}$, will be the outcome. While pooling layers are essential to deep learning models, they also pose a significant risk of data loss. So, it's not a good idea to utilise them consecutively or with a huge window.

As feature extractors, the convolution and pooling layers inevitably glean classifiable information from the provided samples. A feed-forward neural network called MultiLayer Perceptron (MLP) is utilised by CNN for classification purposes.

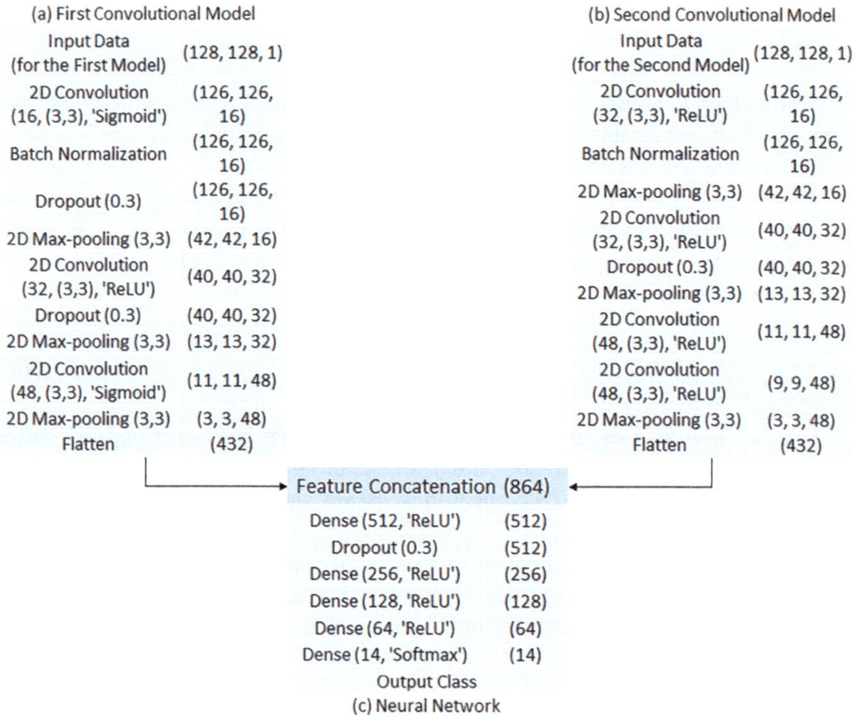

Fig. 16.5 (**a**) Architecture of the (**a**) First Convolutional Model, (**b**) Second Convolutional Model, and (**c**) Neural Network

A multi-layer perceptron (MLP) is a stack of dense layers where each layer collects data from the one before it. One definition of the result of a dense layer is:

$$e = g\left(X^U \cdot j + c\right) \tag{16.9}$$

the activation function is denoted by g, the weight matrix is $X \in \mathbb{S}^{n \times o}$, the input is represented by j, and the bias is denoted by c (Wang and Benning 2020). Each row of the matrix returned by the MLP's last layer represents the probability that a given sample belongs to a certain class. In accordance with the model's determination, the sample is a member of the most likely class.

Our approach makes use of two convolutional models, as we've already established. Figure 16.5 shows the neural network that was utilised for the final classification, in addition to the topologies of the two models. Figure 16.1 shows that the First Convolutional Model makes use of three 2D convolution layers; two of these levels analyse input data using the Sigmoid activation function, and the third layer makes use of the ReLU activation function. In contrast, the Second Convolutional Model makes use of ReLU activation across all four of its 2D convolution layers. A number of filters, kernel size, and activation function are the arguments shown in the figure for the 2D Convolutions. To prevent overfitting on the training data, both models include a batch normalisation layer, several dropout layers, three max-pooling layers, and a periodic reduction of the input' size. The picture depicts 2D Max-pooling with pool_size as its argument; the stride is the same as pool_size. A flattened layer compiles all the derived features into a vector at the conclusion of the model. The Neural Network block begins by merging the two feature sets so that the results from both models are combined. A dropout layer and five thick layers make up the MLP that processes the mixed features and produces the model's decision. The given design is a function deep learning model because it demands a non-sequential flow of information and processes numerous inputs independently. The layers' organisation and the values of their related arguments have been fine-tuned for this classification task using thorough model hyper-tuning.

16.3 Results and Discussion

The outcomes of the mentioned deep learning model's viral classification rundown are laid out and presented here. Separating the available samples into training and testing subsets is the first step in addressing a supervised learning problem, although in some circumstances, three groups may be necessary. In order for the model to learn how to classify images, it is trained using samples from the training subset to identify commonalities and differences between classes. The model's knowledge is verified at the next stage when it is assigned the duty of classifying the testing samples. Lastly, the model is evaluated by comparing its label verdict on the samples with the ground facts (real labels) for each set. We divided the samples in half at random for the training subset and kept the other quarter for testing in this investigation. The dataset is evenly distributed, as can be observed in Table 16.1. As a result, the model's judgements were not skewed towards any one class (often the one with the bigger sample size in imbalanced datasets). But we made sure that the training subset got 75% of each class's samples (552 samples total) to further reduce bias. A 100-epoch deep learning model was trained on a 32-gigabyte RAM/6-gigabyte GPU Intel Core i9-9900 K Coffee Lake 8-core system. In the past, models were judged by how well they scored testing samples, which were new or unseen. In the interest of full disclosure, nevertheless, we have included the results of our model's testing and training subgroups.

At each training session, the classification accuracy is displayed in Fig. 16.6a. Justifying a classifier begins and ends with its accuracy. The ratio of correctly classified samples to the total number of samples in a classification is simply expressed by it. In the 96th epoch, the graphic shows that the testing subset reached its maximum accuracy of 97.44%. The previous epoch's testing accuracy was 96.43%, which is near the maximum score. The bulk of the 64 epochs had outstanding classification outcomes, since the testing accuracies were better than 94%. In addition, there is little evidence of overfitting (on the training samples) since the training and testing accuracy curves are very congruent with one another. During the first 20 epochs, the performance was somewhat unstable; however, it steadied out thereafter. With a Quadratic Weighted Kappa (QWK) score of 0.9719 at the 96th

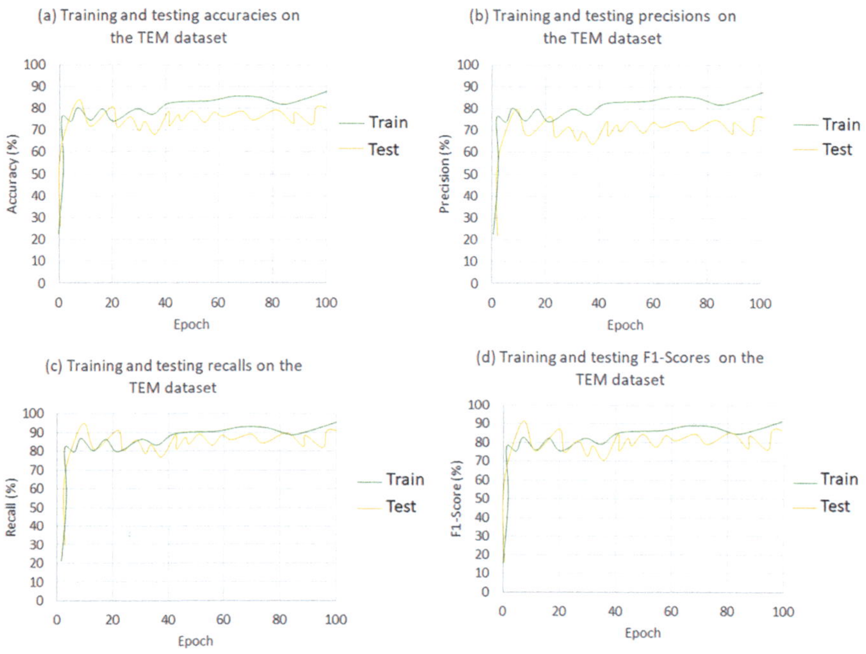

Fig. 16.6 Classification results from applying the suggested model to the TEM Virus dataset (**a**) accuracy, (**b**) precision, (**c**) recall, and (**d**) F1-score on both the training and testing subsets

epoch, the model demonstrates that the achieved categorical measurements are reliable.

When dealing with an imbalanced dataset in particular, accuracy—one measure of the classification outcome—can frequently be deceiving. In order to make an accurate evaluation of the outcomes, we must also consider other factors. As an example, two important metrics are recall and precision, which tell us how many samples in a given class were correctly classified and how many samples genuinely belonged to the class that the model recognised. The overall score is obtained by averaging the parameters that are computed for each class separately. Figure 16.6b, c show the recall and precision scores for each of the 100 epochs. These graphs show that in the early epochs, precision ratings were much higher than recall scores, which may indicate substandard categorisations. These curves closely tracked their respective accuracy curves beginning in the tenth epoch, but things improved swiftly after that. As a weighted average of the accuracy score and the prior two measures, the F1-score is another performance evaluation matrix that many believe better describes clarification outcomes. The main benefit of using the F1-score to describe a model's performance is that it is unaffected by class imbalance in the training and testing subsets. However, this is not the situation with our dataset. In Fig. 16.6d, we can see the F1-score for each epoch on both groups. The given results are accurate and very

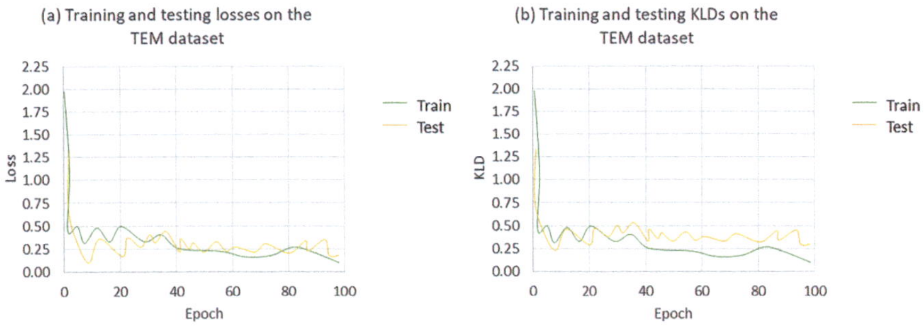

Fig. 16.7 (**a**) Training and testing loss values. (**b**) Kullback-Leibler Divergence values

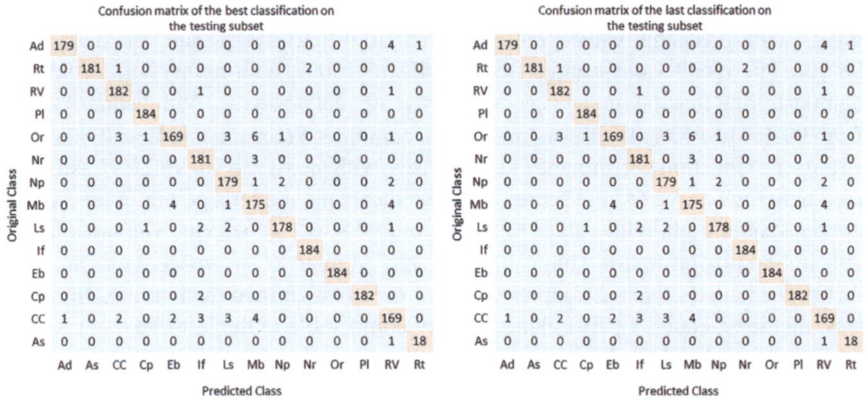

Fig. 16.8 The confusion matrix of the (**a**) best (96th) and (**b**) last classification on the testing subset

dependable, as the F1-score curves are nearly identical to the accuracy curves in Fig. 16.6a.

To demonstrate how the model's loss decreased throughout the course of training, we have displayed the loss curves in Fig. 16.7a. Classification loss measures the accuracy with which the classifier has modelled the input data. An established loss function is used for its measurement. A higher percentage of inaccurate forecasts results in a higher loss value. Understanding the loss values is crucial as they are used by the model to find the best solution using an optimisation function, rather than the accuracy or F1-score. With more training, the figure shows that both subsets' loss values declined. The model found its sweet spot in the end, even though the curve descents aren't smooth. Figure 16.7b displays the classification's Kullback-Leibler Divergence (KLD) values as curves. Closely related to information divergence and relative entropy, KLD is an asymmetric measure of the distance between two probability distributions (Sikder et al. 2024). The model reached its peak performance at the 96th epoch with a classification loss of 0.0808 and KLD of 0.0819, the lowest values on the relevant curves, respectively.

Figure 16.8a, b show the confusion matrices of the best (96th epoch) and final classifications' results on the testing subset, respectively. You may learn a lot about the model's performance in identifying dataset samples from each class by examining these matrices. To find out which classes the function is having trouble recognising, confusion matrices are quite helpful. As mentioned earlier, out of a total of 2576 samples, 184 were retained for each class in the testing subset. At the best epoch, the model correctly identified 66 samples out of a total of 166 (Fig. 16.8a). Of the 14 classifications, RV has the highest rate of misclassification while class Or has the highest rate of proper classification. For several classes—Ad, As, Bu, Co, De, Eb, In, Ni, Pa, RV, Ro, Zi, Ha, and Ma—performance measures (Precision, Recall, F1-score, and AUC) are available in Table 16.2. All categories show great overall performance according to the measurements. The accuracy in accurately recognising affirmative instances is great, with precision values ranging from 90.9% to 100% and an average of 96.6%. The recall values range from 89.4% to 100%, with an average of 95.9%, demonstrating that the model is good at detecting real positives. With values ranging from 90.1% to 100%, the F1-score—which combines Precision and Recall—depicts strong classification performance in all classes, with an average of 96.2%. Lastly, the area under the curve (AUC) scores, which indicate the model's power in prediction, are constantly high, with values ranging from 99.8 to 100 percent. The most recent epoch's confusion matrix (Fig. 16.8b) is very similar to its predecessor; the only difference is that 26 more samples were incorrectly identified than before. None the less, the finest epoch's performance is very comparable to the last one.

We have detailed the outcomes of the suggested model in the preceding paragraphs. Figure 16.1 shows that it uses two distinct pipelines to generate and handle pictures using two distinct convolutional models. Imaging method, Fig. 16.5a's First Convolutional Model, and Fig. 16.5s Neural Network are all used in the first experiment (c). The Second Convolutional Model shown in Fig. 16.5b, the same Neural Network, and 2D DCT are all used in the second experiment. The table clearly shows that the second configuration outperformed the first. But by merging the two pipelines' outputs, the described model achieved even greater performance.

As a result of using two sets of features retrieved from the raw images, it improved accuracy by 5.24% and the average F1-score by 5.32% in the latter experiment.

All things considered, the results show that the strategy outlined here is a solid bet for virus categorisation using TEM images. Increased samples (in each class) and greater image resolutions are just two of the many ways in which the TEM viral dataset excels above the standard viral Texture Dataset. In the future, we would love it if other studies would report their findings using this dataset.

Table 16.2 The efficiency of class-wise classification

	Ad	As	Bu	Co	De	Eb	In	Ni	Pa	RV	Ro	Zi	Ha	Ma	Mean
Precision (%)	98.1	99.6	94.7	97.2	95.7	94.6	94.1	91.5	97.6	99.1	100.0	98.8	90.9	98.9	96.6
Recall (%)	96.5	97.2	99.1	98.8	90.8	97.8	95.8	93.3	96.0	98.6	100.0	97.5	89.4	99.1	95.9
F1-score (%)	97.3	98.4	96.9	98.0	93.2	96.2	94.9	92.4	96.8	98.9	100.0	98.1	90.1	99.0	96.2
AUC (%)	99.9	99.9	100.0	99.9	99.8	99.9	99.8	99.8	99.9	99.9	100.0	99.9	99.6	100.0	99.9

16.4 Conclusion

In order to successfully categorise viruses from a collection of transmission electron microscopy (TEM) pictures, this study utilised a two-pipeline strategy that made use of convolutional neural networks. To improve classification accuracy, we used two separate convolutional models and a number of image processing approaches to extract useful features. The suggested model demonstrated strong reliability in viral categorisation, as our results showed that it attained an impressive maximum testing accuracy of 97.44% and a Quadratic Weighted Kappa (QWK) score of 0.9719. The accuracy was enhanced by 5.24% and the average F1-score was increased by 5.32% using the dual model method as compared to a single pipeline. In addition, the confusion matrices showed that although certain classes were harder to categorise than others, the model performed well for most virus kinds. Setting our results in the context of current research allowed us to demonstrate how our model outperformed the competition, especially when it came to differentiating between different types of viruses. This study lays the groundwork for future research to improve virology diagnosis accuracy by highlighting the potential of sophisticated deep learning approaches to enhance viral categorisation.

References

Ahmed N (1991) How I came up with the discrete cosine transform. Digit Signal Process 1(1):4–5

Ahmed N, Natarajan T, Rao KR (1974) Discrete cosine transform. IEEE Trans Comput 100(1): 90–93

Ali MM, Joshi RC, Dutta MK, Burget R, Mezina A (2022) Deep learning-based classification of viruses using transmission electron microscopy images. In: 2022 45th international conference on telecommunications and signal processing (TSP). IEEE, pp 174–178

Artika IM, Wiyatno A, Ma'roef CN (2020) Pathogenic viruses: molecular detection and characterization. Infect Genet Evol 81:104215

Beltran Perez C (2018) Nonlinear parametric and neural network modelling for medical image classification. Doctoral dissertation. University of Sheffield

Chatterjee A, Siarry P (eds) (2013) Computational intelligence in image processing. Springer, Heidelberg

Chen Q, Dwyer C, Sheng G, Zhu C, Li X, Zheng C, Zhu Y (2020) Imaging beam-sensitive materials by electron microscopy. Adv Mater 32(16):1907619

Dargan S, Kumar M, Ayyagari MR, Kumar G (2020) A survey of deep learning and its applications: a new paradigm to machine learning. Arch Comput Methods Eng 27:1071–1092

Fan J, Ma C, Zhong Y (2021) A selective overview of deep learning. Stat Sci 36(2):264

Gul MA, Yousaf MH, Nawaz S, Ur Rehman Z, Kim H (2020) Patient monitoring by abnormal human activity recognition based on CNN architecture. Electronics 9(12):1993

Hahn A, Podbielski A, Meyer T, Zautner AE, Loderstädt U, Schwarz NG, Frickmann H (2020) On detection thresholds – a review on diagnostic approaches in the infectious disease laboratory and the interpretation of their results. Acta Trop 205:105377

Heaton SM (2019) Harnessing host–virus evolution in antiviral therapy and immunotherapy. Clin Transl Immun 8(7):e1067

Hinkle LB, Metsis V (2021) Model evaluation approaches for human activity recognition from time-series data. In: Artificial intelligence in medicine: 19th international conference on artificial intelligence in medicine, AIME 2021, virtual event, June 15–18, 2021, proceedings. Springer, pp 209–215

Hs MAT (2021) Activation functions and their derivatives – a quick & complete guide

Lee SM, Yoon SM, Cho H (2017) Human activity recognition from accelerometer data using Convolutional Neural Network. In: 2017 IEEE international conference on big data and smart computing (BigComp). IEEE, pp 131–134

Lloyd C (2010) Spatial data analysis: an introduction for GIS users. Oxford University Press

Masud M, Sikder N, Nahid AA, Bairagi AK, AlZain MA (2021) A machine learning approach to diagnosing lung and colon cancer using a deep learning-based classification framework. Sensors 21(3):748

Matuszewski DJ, Sintorn IM (2021) TEM virus images: benchmark dataset and deep learning classification. Comput Methods Prog Biomed 209:106318

Milletari F, Ahmadi SA, Kroll C, Plate A, Rozanski V, Maiostre J, Navab N (2017) Hough-CNN: deep learning for segmentation of deep brain regions in MRI and ultrasound. Comput Vis Image Underst 164:92–102

Richert-Pöggeler KR, Franzke K, Hipp K, Kleespies RG (2019) Electron microscopy methods for virus diagnosis and high resolution analysis of viruses. Front Microbiol 9:3255

Shaga Devan K, Walther P, von Einem J, Ropinski T, Kestler AH, Read C (2021) Improved automatic detection of herpesvirus secondary envelopment stages in electron microscopy by augmenting training data with synthetic labelled images generated by a generative adversarial network. Cell Microbiol 23(2):e13280

Sikder N, Khan MAM, Bairagi AK, Masud M, Tiang JJ, Nahid AA (2024) Heterogeneous virus classification using a functional deep learning model based on transmission electron microscopy images (preprint). arXiv preprint arXiv:2405.15563

Usman M, Sultan H, Hong JS, Kim SG, Akram R, Gondal HAH, Tariq MH, Park KR (2024) Dilated multilevel fused network for virus classification using transmission electron microscopy images. Eng Appl Artif Intell 138:109348

Wang BC (2008) Digital signal processing techniques and applications in radar image processing. Wiley

Wang X, Benning M (2020) Generalised perceptron learning. arXiv preprint arXiv:2012.03642

Wani JA, Sharma S, Muzamil M, Ahmed S, Sharma S, Singh S (2022) Machine learning and deep learning based computational techniques in automatic agricultural diseases detection: methodologies, applications, and challenges. Arch Comput Methods Eng 29(1):641–677

Advancing Prognostic Insights: A Novel Deep Learning Algorithm to Predict Outcomes in Amyotrophic Lateral Sclerosis

17

Anshit Mukerjee, Gunjan Mukherjee, and Biswadip Basu Mallik

Abstract

Amyotrophic lateral sclerosis (ALS) is a terminal neuromuscular disease in which motor neurons steadily disappear causing muscle atrophy and eventual loss of movement control. Statistical predictive validity of traditional diagnostic modalities in determining the prognosis remains low. This work further proposed a new deep learning model to examine clinical and genetic data. Correcting missing values and normalization were employed in data preprocessing section. The results were assessed with k-fold cross-validation; the chosen metrics were accuracy, precision, recall, and F1 score. Several strategies including data augmentation and the tuning of the hyperparameters were used to improve model durability. The developed algorithm received an accuracy of about 94.28%, with specific to true positive of 0.9256, and sensitivity of 0.8975 for predicting a high-risk population for rate of disease advancement. The variable importance analysis showed that key predictors of outcomes were Age of onset and Initial functional assessment scores. The study provided evidence of the application of deep learning for enhancing the prognostic information of ALS and called for cross-disciplinary strategy to enhance the diagnosis and management of rare diseases.

A. Mukerjee (✉)
Department of Computer Science and Engineering, Abacus Institute of Engineering and Management, Mogra, India

G. Mukherjee
Department of Computational Sciences, Brainware University, Barasat, Kolkata, West Bengal, India

B. B. Mallik
Department of Basic Science and Humanities, Institute of Engineering & Management, University of Engineering & Management, Kolkata, India

J. M. Chatterjee, S. K. Saxena (eds.), *Artificial Intelligence and Blockchain in Precision Medicine and Virology*, Medical Virology: From Pathogenesis to Disease Control, https://doi.org/10.1007/978-981-96-8919-4_17

Keywords

Amyotrophic lateral sclerosis · Deep learning · Convolutional neural network · Prognostic predictions

17.1 Introduction

Amyotrophic lateral sclerosis (ALS) (Cellura et al. 2012) is a multidimensional and fatal neuronal disease (Chio et al. 2009), in which motor neurons (Schuster et al. 2015) deteriorate gradually resulting in muscle weakness, disability, and finally death. The disease develops fully in adulthood most patients contracting the illness within the age bracket of 40–70 years (Chipika et al. 2019). Although relatively recent years have brought some insight into its pathophysiology, such as genetic and environmental causes, the etiology still remains debated. ALS can be classified into two types—sporadic accounting for about 90% of cases and the remaining 5–10% are familiar ALS, associated with certain genes, for example, *SOD1* or *C9orf72* genes (Bede and Hardiman 2018). Due to the polymorphic manifestations (Bertrand et al. 2018), diagnosing ALS is difficult that leads to a delay in treatment and can reach 14 months. Literature on the topic reveals a number of previous concerns in managing and developing treatments for ALS. Inconsistency in the development of pathology and symptomatology puts patients at risk of misdiagnosis, and many are diagnosed initially with other neurologic diseases (Wen et al. 2019). In addition, currently there exist few therapies and few of them are efficient; therefore, the development of more accurate diagnostic methods and prognostic models is crucially important. The drive toward this study lies in the fact that there is a glaring absence in predictive functions within today's clinical approaches. It is hoped that, by using deep learning algorithms (Guazzo et al. 2024) on multi-modal inputs such as clinical evaluations and physical exams, coupled with genetic data, this research will foster better prognostication of ALS patients. The main aim of this chapter is to develop a novel deep learning-based algorithm for the detection of ALS. Section 17.2 presents the literature work done in this field previous to this study showing the limitations in the existing research. The third section is the Methodology, in which we introduced our new algorithm along with a proper discussion as to how our algorithm mitigates the voids mentioned earlier. In Sect. 17.4, the prediction result of this chapter is going to be demonstrated by using the simulating tools. Section 17.5 provides discussion while Sect. 17.6 is the concluding section.

17.2 Literature Review

In 2024, Guazzo and his team (2024) performed study research on the detailed impact of clinical parameters that describe the progression of ALS using machine learning applied to data collected at routine visits. The goal of the work was to provide an answer to the MN model's non-uniform clinical picture and the lack of adequate predictive factors. They investigated three classification problems. They are predicting death as a binary format, and death or PEG or NIV, and death or non-invasive ventilation as multiclass cases. To minimize the technical flaws and to keep the computations in line, the investigators used logistic regression (LR) and a deep learning multilayer perceptron (MLP). It was observed that for the death prediction both models had F1 scores and accuracy greater than 0.7, for the multiclass prediction of PEG or NIV scores ranged from 0.4 to 0.6. The identified advantages of the study include the use of machine learning in real-life clinical scenarios, and using routine visit data in integrating models into clinical workflows. However, some limitations came from average multiclass performance indicating possible peaks of information retrievable from the dataset or a lack of essential predictors not gathered during real-world practice. Our work further goes toward extending these findings by embracing sophisticated paradigms that fuse model-based and model-free methods to improve the precision of the forecast. We also use dynamic Bayesian networks to explore temporal dependencies of ALS, and gain richer understanding of how clinical characteristics change over time. Furthermore, our focus of the unsupervised clustering algorithm aims to elucidate and enhance the computable phenotypes derived from the multivariate data toward ALS diagnosis and therapeutic assessment, in order to enumerate useful knowledge for ALS's treatment as well as to overcome the shortcomings stated in Guazzo's research.

In 2022, Bede and his team (2022) investigated the diagnostic classification of amyotrophic lateral sclerosis (ALS) using pathological and artificial neural networks based on the neuroimaging features. The ALS patients, healthy subjects, and disease control group made up of 378 participants formed the stocks within their single-center study using a uniform imaging protocol to validate an artificial neural network classification scheme. It in turn acquired a full bank of cerebral volumetric metrics, cortical parameters, and white matter integrity measures that were processed with the help of a multilayer perceptron model. The outcomes displayed impressive ability of classification as confirmed by AUC of 0.930 for ALS patients, 0.958 for disease controls, and 0.931 for healthy controls confirming the ability to identify each individual patient. However, the limitations of the study were the fact that these data were obtained from a single center and the possibility of system biases arising due to small sample sizes. Our work does not have these limitations since we work with larger and more diverse databases that incorporate increased deep learning capabilities that improve the predictive models' outcomes and stability in differentiating ALS patients across multiple healthcare units.

In early 2020, Rusiya and his team (2020) went deeper into the classification of amyotrophic lateral sclerosis (ALS) using deep neural networks along with the Teaching-Learning-Based Optimization (TLBO) algorithm. Their study discussed the difficulties of prognosis and identification of ALS within BCI technology; however, they concentrated on feature extraction from the electroencephalogram (EEG), wherein noise corrupts the signals in their work. The approach used involved using a discrete wavelet transform for feature extraction, where EEG signals are divided into different sub-bands for processing. The patterns of the proposed deep learning algorithms were then explored through MATLAB simulations and compared to existing Bayesian-based neural network and other ensemble machine learning classifiers using standard ALS datasets. According to the results suggested by the study, the proposed system enhanced the efficiency of the BCI; however, there were concerns that range from the over-reliance on some datasets may lead to overfitting of the BCI and challenges in real-time BCI implementation. Our work extends from this research by employing increased sample size and heterogeneity as well as improved deep learning networks that advance feature extraction while seeking to reduce noise effects in order to increase diagnostic precision as well as clinical relevance.

Tang and his team (2019) have also employed both model-based and model-free methods in the diagnostic prediction and patient clustering for amyotrophic lateral sclerosis (ALS) in a study that was completed in 2019. Their study set out to support clinical decision making on ALS using data derived from predicting ALS disease progression in terms of the ALSFRS score over time from data resources from the DREAM-Phil Bowen ALS Prediction Prize4Life Challenge, mainly from the PRO-ACT Database. The authors compared different machine learning approaches with linear models, Random Forest, and Bayesian adaptive regression trees (BART) to predict the change of ALSFRS scores. In fact, the study noted moderate accuracy gains—correlation coefficients of 0.429 for random forest and 0.567 for BART—it also effectively partitioned all patients into four phenotypic subtypes by identifying other clinical characteristics using unsupervised clustering. While BART was advantageous in capturing interaction between many predictor variables, the study also noted issues with accurately predicting various clinical outcomes regarding the patients. In contrast to such approaches, Tang focused on how contemporary forms of machine learning could cluster patients according to multivariate criteria. In the following section, our work extends Tang's study by considering deficiencies that have been noted previously by the author. More than that, we use deep learning algorithms that combine data in mid- and long-lived subspaces better and use dynamic Bayesian networks, which address temporal dependencies more stably. Further, our algorithm is also focused on the unsupervised clustering, improved interpretability, and clinical relevance. Thus, our work to improve computable phenotypes should give sharper resolution for these patient segments and how they might be treated. Lastly, although Tang and his colleagues have also made many advancements for ALS research, our study aims to avoid their shortcomings with different and combined approaches alongside various data source, providing meaningful information for ALS diagnosis and therapeutic planning.

According to Grollemund and his team (2020a) in the year 2019, the prognosis and management of ALS have been linked to a number of factors that influence their progression and treatment due to the mechanical nature of the bolstered imagination that can range from 3 to 5 years. To evaluate prior attempts at developing ML applications, they assumed both clinical and mathematical analyses and also recognized several impediments to effective clinical trials, such as heterogeneity by illness type and difficulty in enrolling patients in late phases of the clinical trial. However, they mentioned issues such as small sample sizes and violations of statistical assumptions, though they aimed at developing diagnostic models using

ML in ALS datasets, and without providing sufficient rationale for selecting the models. They pointed out that despite making useful contributions to understanding how diagnostic accuracy and patient stratification may be improved by using ML, their work also identified key limitations in current methods. Our studies help overcome these limitations by employing larger datasets from international populations and state-of-the-art deep learning algorithms that handle multiple modalities, placing importance on statistical methodological validation, and reporting of model assumptions, which will potentially lead to optimized trial designs and more accurate prognostications concerning ALS.

Sengur and his team (2017) in the year 2017 came up with a deep learning-based method in the classification of EMG signals of ALS patients as compared to normal signals. Their approach consisted of the application of T-F representations of EMG signals by using spectrograms, CWT, and SPWVD. A CNN architecture with two convolutional layers, two pooling layers, a completely connected layer, and one loss-performing layer was used. The approach was trained using reinforcement sample learning. The authors applied the study on a publicly available EMG dataset called EMG that was recorded from 89 ALS and 133 normal EMG signals, a test of its implementation of which yielded a tremendous accuracy of 96.80%. The results shown revealed effectiveness and better performance over other classification techniques, however, acknowledging the restrictions such as limited number of samples, overtraining owing to the selected dataset. Our work mitigates these concerns using much larger datasets along with more state-of-the-art deep learning algorithms, which improve the overall flexibility as well as identification of ALS from EMG signals in different groups of people.

Therefore, the literature of machine learning and deep learning techniques for classification and prediction of ALS indicates substantial progress and new opportunities but also important challenges. Different research works involving deep learning algorithms such as convolutional neural networks (Chipika et al. 2021) and multilayer perceptrons (Bede et al. 2020) have established that their use can enhance diagnostic precision, but many of these studies suffer from small sample, overfitting problems, and inadequate feature extraction techniques. Furthermore, analyzing highly similar sets of data limits the ability to generalize, thus hampering the application of these models in practice for diverse populations in particular in clinical practice. They further indicate that the original studies emphasize the importance of enhancing the existing approaches' capabilities to manage potential complexities of the ALS data for accurate prognostic prediction. To fill these gaps, our proposed deep learning algorithm to predict the survival prognosis of ALS encounters will utilize data from partitions that are significantly larger and complicated than those found in previous methods, and include different clinical phenotypes and additional demographic attributes linked to ALS. To improve prediction and patient classification (Agosta et al. 2018a), the proposed model includes new feature extraction methods and the additional analysis of both clinical and genetic data as well as neuroimages. Moreover, we will perform strict validation of the model and also perform tests for deploying the model in real applications. In addition to combating the conceptual insufficiencies that prior research has alluded to, this approach is designed to make meaningful and considerable contributions to patient-tailored therapeutic endeavors and enhanced patient outcomes in ALS treatment.

17.3 Methodology

The following is our new algorithm:

1. START
2. Let $X_{\mathrm{EMG}} \in R^{m \times n}$ represent the matrix of EMG signals, $X_{\mathrm{EEG}} \in R^{p \times q}$ for EEG signals, $X_{\mathrm{Genomic}} \in R^{r \times s}$ for genomic data.
3. For each input type, apply augmentation techniques: $X_{\mathrm{EMG_{aug}}} = A_{\mathrm{EMG}}(X_{\mathrm{EMG}}), X_{\mathrm{EEG_{aug}}} = A_{\mathrm{EEG}}(X_{\mathrm{EEG}})$, and $X_{\mathrm{Genomic_{aug}}} = A_{\mathrm{Genomic}}(X_{\mathrm{Genomic}})$.
4. Apply noise reduction: $X_{\mathrm{EMG_{clean}}} = N_{\mathrm{EMG}}\left(X_{\mathrm{EMG_{aug}}}\right), X_{\mathrm{EEG_{clean}}} = N_{\mathrm{EEG}}\left(X_{\mathrm{EEG_{aug}}}\right)$, and $X_{\mathrm{Genomic_{clean}}} = N_{\mathrm{Genomic}}\left(X_{\mathrm{Genomic_{aug}}}\right)$.
5. Normalize each cleaned dataset: $X_{\mathrm{EMG_{norm}}} = \frac{X_{\mathrm{EMG_{clean}}} - \mu_{\mathrm{EMG}}}{\sigma_{\mathrm{EMG}}}, X_{\mathrm{EEG_{norm}}} = \frac{X_{\mathrm{EEG_{clean}}} - \mu_{\mathrm{EEG}}}{\sigma_{\mathrm{EEG}}}$, and $X_{\mathrm{Genomic_{norm}}} = \frac{X_{\mathrm{Genomic_{clean}}} - \mu_{\mathrm{Genomic}}}{\sigma_{\mathrm{Genomic}}}$.
6. (a) For EMG using VMD: $F_{\mathrm{EMG}} = \mathrm{VMD}(X_{\mathrm{EMG_{norm}}}, K)$ where k is the number of nodes.
   (b) For EEG using CWT: $F_{\mathrm{EEG}} = \mathrm{CWT}(X_{\mathrm{EEG_{norm}}})$.
   (c) For Genomic data using SPWVD: $F_{\mathrm{Genomic}} = \mathrm{SPWVD}(X_{\mathrm{Genomic_{norm}}})$.
7. (a) Combine features into a single matrix: $F_{\mathrm{combined}} = [F_{\mathrm{EMG}}, F_{\mathrm{EEG}}, F_{\mathrm{Genomic}}]^T = F \in R^{k \times d}$ where k is the number of features and d is the number of samples.
   (b) Apply feature selection criteria to obtain selected features: $(F) = F^* \in R^{k^* \times d}$ where $k^* < k$.
8. (a) Define the state transition probabilities as: $P(X_t | X_{t-1}) = f(X_{t-1}, W)$ where X_t represents the state at time t and W are the parameters of the model.
   (b) The observation probabilities can be defined as: $P(Y_t | X_t) = g(X_t, V)$ where Y_t represents observed features.
9. Forward propagation through layers can be expressed as: $Z^{(l)} = W^{(l)}A^{(l-1)} + b^{(l)}$ and $A^{(l)} = f(Z^{(l)})$ where $f(\cdot)$ is an activation function (e.g., ReLU, Sigmoid).
10. Introduce dropout during training to prevent overfitting: $A^{\mathrm{dropout}(l)} = A^{(l)} * M$ where M is a mask matrix with elements drawn from a Bernoulli distribution with probability p.
11. Compute output probabilities for classification using softmax function: $P(y | x; W, b) = \mathrm{softmax}(A^{(L)})$ where $\mathrm{softmax}(z_i) = \frac{e^{z_i}}{\sum_{j=1}^{K} e^{z_j}}$ for class probabilities.
12. Assign class label based on maximum probability: $y^* = \arg\max (P(y | x; W, b))$.
13. Evaluate model performance using metrics such as accuracy (Acc), precision (Prec), recall (Rec), and F1-score (F_1):

$$\mathrm{Acc}=\sum_{i=1}^{N}\left(y_i==y_i^*\right)$$

$$\mathrm{Prec}=\frac{\mathrm{TP}}{\mathrm{TP}+\mathrm{FP}}$$

$$\mathrm{Rec}=\frac{\mathrm{TP}}{\mathrm{TP}+\mathrm{FN}}$$

$$F_1=2\times\frac{\mathrm{Prec}\cdot\mathrm{Rec}}{\mathrm{Prec}+\mathrm{Rec}}$$

14. Implement an ensemble method to combine predictions from multiple models to improve accuracy: Let $P_1, P_2, \ldots, P_n$ be predictions from different models. The final prediction can be obtained by majority voting or averaging: $\text{Final Prediction}=\arg\max\left(\sum_{i=1}^{n}P_i\right)$.
15. END

Now let's look into the line-wise explanation of our novel algorithm. Line 1 marks the onset of our novel algorithm. This helps the algorithm to get a clear start-up point from where further steps will be implemented in a sequential manner. Line 2 specifies the input matrices based on the various types of data collected. Each matrix has the space to accommodate multiple samples and features making data processing easier. When the structure of input data is defined clearly, then the above algorithm is capable of handling a wide range of data, which is more or less a solution for integrating different types of data. Line 3 adds variations on the signals (noise, transformations, etc.) to the set (augmentation) to broaden the range of the training dataset. Most existing algorithms fail to incorporate data augmentation, which results in an overfitting model in most cases. By applying augmentation techniques systematically, the model is trained on data which is significantly different from the patient data that was obtained. In the fourth line, data preprocessing methods are used to enhance each type of signal, by filtering noise that can hinder recognition of patterns of interest. Previous algorithms sometimes become undermined by noise, and this step alone corrects that problem. The importance of clean signals in the improvement of the quality of data presented to the next processing levels helps enhance the accuracy of the model. At the end of line 5, every cleaned dataset is normalized so that all the features are on the same level so as to enhance training in the model. Normalization is a technique that relates to scaling differences among features that lead to biased model development. This step takes care of the fact that during training, the model should work with equal weight on all its features to enable it to learn from all sources. In line 6(a), Variational Mode Decomposition (VMD) is used to obtain the intrinsic mode functions from normalized EMG signals so as to use an adaptive frequency decomposition. The conventional techniques could fail to consider certain frequency components; VMD succeeds best in identifying these components. This flexibility of VMD to adaptively decompose the signals is advantageous when working with non-stationary signals as will be seen with EMG signals. In line 6(b), Continuous Wavelet Transform (CWT)

(Agosta et al. 2016) is used to analyze the EEG signals with respect to the wavelet scales and time frames in which they are involved, thus capturing transient features of brain activity. Several current algorithms can break down when it comes to identifying short-time occurrences of patterns in the EEG data; thereby, the CWT is useful for this purpose. CWT offers a time-frequency analysis that is mandatory for characterizing the pathological oscillatory activity of the brain related to ALS. In line 6(c), Smoothed Pseudo Wigner-Ville Distribution (SPWVD) (Muller et al. 2016) is employed to extract features for learning from analyzing the genomic data time-frequency behavior. It may be observed that traditional analysis of genomes can often fail to identify intricate interactions while SPWVD is capable of doing so. This method improves the ability to extract features from the genomic data since such data is high-dimensional and cumbersome. In line 7(a), where the multi-model fusion occurs, the features extracted from each modality are formatted into one matrix. Some algorithms process features separately; by integrating them, one can have a good picture of patient data. This integration allows gathering of multiple modalities which possibly improves classification accuracy in case the sources provide complementary information. The method described in line 7 (b) ensures the reduction of dimensionality from the fused dataset by extracting only the useful features in the dataset. By decreasing the data dimensionality, we decrease the risk of over fitting stemming from a model with a more complex structure. The routine of feature selection guarantees that only relevant features influence the model training process hence improved performance. As shown in lines 8(a) and 8(b), Dynamic Bayesian Networks (DBNs) both capture the temporal relation between features and handle the uncertainty in predictions at different time staging. They can serve to capture how symptoms or features change given the previous states where next states are not data. Hitherto used models fail, in most cases, to capture temporal aspects of the data; DBNs are capable of accommodating time-varying connections. The use of DBNs is advantageous to diseases like ALS because they use a probabilistic framework for learning from sequential data can adapt as new data is accrued. At line 9, it is explained that, through weights and biases the input features are passed to multiple layers in a deep learning architecture. Most conventional methods do not take data depth; this multilayer method provides the depth necessary to analyze high-dimensional data well. To do this, deep learning architectures enable the use of hierarchical feature learning, essential for comprehending complex correlation patterns in features related to ALS. In line 10, dropout randomly sets a network during training from working a fraction of neurons and this makes representation learning redundant and helps overcome the issue of overfitting. Dropout (Agosta et al. 2018b) is one of the ways in which machine learning models that tackle small datasets are prevented from overfitting. This increases the ability of generalization since there is no neuron that will dominate the output of the model. When it comes to line 11, it calculates the probabilities for the class activations making it convenient for classification tasks. Most classifiers do not give measures of uncertainty; softmax is beneficial because it provides measurable certainty for each class label. This probabilistic output makes it easier for the clinician to understand levels of confidence in the stated predictions to make clinical decisions. The final class label is then decided from line 12 using the highest probability using softmax layer output. There are cases where existing algorithms do not clearly specify features in the model making the decision; this offers an easy mechanism of classification from the results. The resulting structure increases usability in clinical contexts as the classifications can easily be interpreted. Line 13 utilizes a different method that incorporates an object assessment of accuracy, the measure of precision, the number of instances correctly identified and noted by the method's recall rate, and F1-score (Ferraro et al. 2017) for appraisal capacity. A large proportion of research compares models based only on accuracy, which is highly problematic because of the possible misinterpretation of the results. This algorithm offers various results for the purpose of comprehensive evaluation of its performance in various situations, all calculated with different performance indicators. In line 14, we use an ensemble method that combines several models with different approaches, for example, using different architectures or parameter settings to obtain a higher general classification accuracy through voting or averaging devices. Single model types have their architectures or training processes, which may be their drawback; ensembles use different models so that nothing is weak against something else. Application of ensemble has also been indicated to enhance the predictive performance in different themes; applying ensemble here is therefore considered an innovation in ALS detection. Line 15 indicates that the algorithm is going to end.

Now let's look into the novel features of the algorithm that makes it distinguished from all other algorithms in this application in this domain. The first feature comprises VMD, CWT, and SPWVD as observed in line 6. VMD is a more suitable criterion that captures both frequency and temporal characteristics well suited for application in non-stationary signals such as EMG. To extract features from the physiological and genomic signals (Grollemund et al. 2020b), Continuous Wavelet Transform (CWT) is applied to EEG and Smoothed Pseudo Wigner-Ville Distribution (SPWVD) is applied to genomic data because of their effective representation of respective data types. The second is the feature introduced in line 7(a), and it is Combining Features into a Single Matrix. The features of different modalities are brought into a single matrix for multi-modal analysis. Such an approach enables the model to make use of the orthogonal information coming from different models at different modalities, and hence there is an enhancement in the classification. The third part is Dynamic Bayesian Network Integration, mentioned at line 8. The use of Dynamic Bayesian Networks (DBNs) (Bede et al. 2018) enabled the modeling of temporal dependency among features. DBNs have a probabilistic structure, which allow them to learn data in a sequential fashion, something that is very important for diseases such as ALS that present progressive symptoms over time. The fourth in line is the Forward Propagation Through Layers in the line 9. The hierarchical structure allows to 'learn' from high dimensionality patterns that are crucial for ALS diagnosis. The fifth feature is Output generation using softmax function in line 11. The second feature is the Number of Iterations in line 17.

In the softmax function process, the activation of the output layer is changed to the class probability. This probabilistic output puts clinicians in a position to estimate theoretical confidence levels in arriving at a particular result and/or make decisions regarding patient care. The sixth feature is Class Label Assignment Based on Maximum Probability that is at line 12. The algorithm assigns class labels in light of the maximum probabilities that has been obtained from the softmax output. This clear decision-making process can also be beneficial in clinical applications since it produces easy-to-interpret classifications from the models. The last one is Committee Adoption of Methods as shown in line 14. It averages or votes from several models in order to classify accurately, better than any of the individual models. This feature has been proven to improve the predictive outcome dramatically; its use in this context is another approach to ALS detection since different models have their strengths. The basic design of this algorithm also proposes several new and creative ideas that go a long way toward solving the problems present in previous approaches to ALS detection. Incorporating methods like VMD and DBNs makes this algorithm unique and a complete solution designed for the diagnosis of ALS along with using superior data preprocessing and feature extraction techniques. The components therein work synergistically in supporting accuracy and reliability of the diagnostic process with the application of mathematical methods and theories.

For our algorithm, we used two datasets. They are https://www.kaggle.com/datasets/mpwolke/cusersmarildownloadsalspdf/data (van der Burgh et al. 2017) and https://www.coriell.org/1/NINDS/Downloads2/Genotyping-of-ALS-Samples. The first dataset consists of 135 patients with ALS, from whom high-resolution DW- and T1-weighted MRI scans were collected at their first outpatient clinic visit. In this study, the patient's survival times were measured over time right from the point of contraction of the symptoms until the time the patient died. Based on these survival times, patients were categorized into three groups: short, medium, or long survivors. The collected dataset was partitioned in the following way: the deep learning models were trained on the first 83 patients, while 20 patients were used for model parameter optimization, and 32 unique patients were considered for external model validation. The second dataset records genotype data of specific ALS DNA samples from the NINDS repository, which is available through the Coriell Institute for Medical Research as the "Genotyping of ALS Samples." The information content of this dataset consists of genotypes that have been submitted by other organizations, as opposed to samples genotyped within the NINDS Repository. This source remains available for further research on genetic differences and correlations with ALS so that novel information on the disease's heritability profile can be discovered and might be helpful in the creation of appropriate therapeutic approaches.

We divided both datasets into 70% training dataset and 30% testing dataset. We used HP Z5 Workstation and NVIDIA RTX 3090 as hardware for this project (Schuster et al. 2016). Among software we used Python (3.8.10), **PyTorch** (1.10.0), **CUDA** (11.1), **OpenCV** (4.5.3), **NumPy** (1.21.2), **Matplotlib** (3.4.3), **Jupyter Notebook** (6.4.0), **TensorFlow** (2.6.0), and **Keras** (2.6.0) were utilized and among simulation tools we used **Gazebo** (11.0.0) and **Webots** (R2021a) for this project (Welsh et al. 2013).

Now let's look into the architecture of the algorithm presented in Fig. 17.1. The architecture starts with the Input Layer, which has three different kinds of

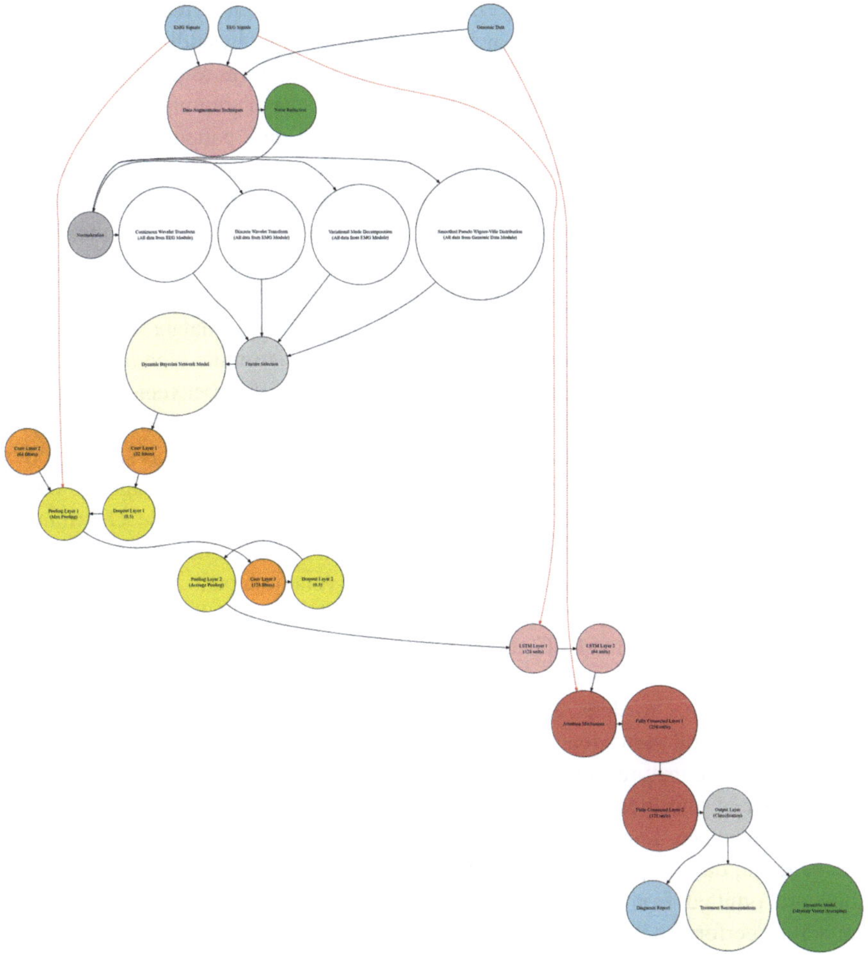

Fig. 17.1 Architecture of our novel algorithm for the detection of floating debris

information. They include EMG signals (Bede et al. 2017), EEG signals (Tahedl et al. 2021a), as well as genomic data (Tahedl et al. 2021b). Some kinds of data are represented as a node, and this enables the model to process different inputs and generate the outcome. Electromyography signals give information regarding muscle contraction, which is essential for the analysis of the neuromuscular system. It is shown that EEG signals represent the brain activity and could be indicating neuro-degenerative aspects of ALS. Genomic data brought genetic information, which can be associated with the disease outcome and life expectancy. This multiple-input system makes sure that the model has a total and holistic picture of the states of health of the patient. The Data Augmentation Layer takes over next to the input layer to enrich the dataset with several interventions. For instance, in the case of EMG signals we might superimpose noise on them or make slight changes in their amplitude to introduce variability in the signal that mimics real-world muscle activity. Similarly, an EEG signal can be augmented, simply by adding some random artifact or jittering time points. This step is important here so as to avoid the problem of overfitting because this will offer the model a variety of training instances, thus enhancing the capacity of the model to generalize on new patient data. The Data Preprocessing Layer is split into two main parts. The two types of techniques are Noise Reduction and Normalization (Fischl 2012). Several techniques are also used to minimize the noise so that the background noise does not interfere with the signals the system wants to detect. For instance, while recording the EEG, normal movements of the eyes or muscle contractions can complicate the otherwise clear patterns, and filters help a lot in this case. From the enhanced database, normalization further scales the data to accommodate the differences in the type of input data. This step is vital as it permits forming links of applicability between the features determined from EMG, EEG, and genomic inputs. In the Feature Extraction Layer, there is more than one approach means for each type of input data. This study applies vibration analysis based on variational mode decomposition of EMG signals, wavelet transform for EEG signals, and Smoothed Pseudo Wigner-Ville Distribution (SPWVD) of genomic data. Incorporating VMD, the EMG signals are decomposed into a set of intrinsic mode functions that are diverse adapted for capturing the frequency and temporal information. The actual application of CWT is in identifying the spectral features of the EEG signal pointing to the changes in the activity of the brain characteristic of neurodegeneration. Relative to ALS, SPWVD analyzes genomic measurements to determine relevant genetic changes affecting the disease's advancement. Following feature extraction, there is a Feature Selection Layer that further filters the features into a more reasonable number. This layer serves to reduce the number of unimportant features, which are not very significant in determining class performance. For example, if specific essential characteristics obtained from the genomic data disprove their connections with ALS results in training datasets, they are omitted from the analysis. This step also economically contributes to modeling performance and minimizes computational requirements without including unconstructive features for the final modeling process. The Dynamic Bayesian Networks Layer is intended to capture temporal dependencies between features to predict uncertainties over time. The state transition probabilities mean the probability of the current state given the previous state and are an argument to the previous state and some of the model parameters. This implies that the current state depends on what has happened in the previous time, together with the system parameters that govern the mode. Further, the observation probabilities can be defined as the likelihood of some features in a given state in the current position. In the view of the present invention, the observed features are the next states based on preceding states. The setup described here forms a probability distribution over future possible states conditioned by past observations, with the capacity to learn adaptively from a sequence of data. Such an approach is especially useful in diseases that are chronically progressive and modify patients' signs and symptoms over time such as ALS. The most important part of architecture is the Deep Learning Module that comprises convolutional layers, pooling layers and LSTM layers. The convolutional layers reveal a spatial pyramid of the features for the processed input data through adaptive weights during the training process under backpropagation algorithms. RELU (Iglesias et al. 2015) is a type of layer that ensures it drops out some neurons in some specific convolutional layer to minimize overfitting during training. After this, there are LSTM layers (Mori et al. 2008), which are used to learn sequencing data because ALS symptoms over time cannot be predicted without understanding how their historical symptoms have been in the past. The last result is computed through a layer connecting every neuron at the earlier layer and then passing the information through the class probabilities through a softmax function. The output layer provides the possibility to classify a patient to short, medium or long survival subgroups depending on general clinical symptoms and imaging results. They maintain this classification is very important with reference to identification of treatment regimens that correspond to specific patients' health conditions. It also consists of an Output Interpretation Layer through which results are provided in the language that is imperative to handling them—that of diagnosis and treatment suggestions as pertaining to unique patient records. This layer guarantees that healthcare providers get a plain vision of the patient's conditions and possible treatments. To increase predictive accuracy, however, an Ensemble Methods Layer reconciles predictions of various models employing techniques such as majority voting, averaging of outputs from different architectures or different parameter configurations, etc. This integration makes an algorithm immune to overfitting while combining the strengths of multiple types of algorithms in classification. In the presented diagram, an architecture includes boundary cases depicted as red dashed arrows indicating that specific connections indicate specific inter-module interactions. Indeed, there is an outstanding example of how EMG influences pooling seen where particular patterns of muscle activity can directly result in pooling operations without necessarily going through the middle levels. An example that includes EEG data input into LSTM shows that brain activity might have immediate impact on temporal analysis rather than a typical cascaded signal processing path. Also, with the genomic data related to the attention mechanism, we see how certain aspects of the data will affect which features are considered most significant when the model goes through a classification of attention mechanism.

These exceptional connections illustrate the intricate relations of various types of biological data and how they affect the disease prognosis; at the same time, they also highlight at which regularities the architecture of this analytics can change its modes in response to the input. In summary, this architecture reflects the complex system capable of not only detecting ALS with high accuracy but also being flexible solution for clinical environment due to module approach, incorporation of various types of data, multiple feature extraction techniques, the use of DBNs for dynamic modeling, deep learning techniques, and the application of the ensemble methods aimed at improving the predictive power of the architecture.

17.4 Results

We have compared our novel algorithm with three other advanced algorithms, that is, Bayesian Adaptive Regression Technique (Tang et al. 2019), Logistic Regression (Guazzo et al. 2024), and Multilayer Perceptron (Guazzo et al. 2024) in this application domain. The reason for choosing these three algorithms for comparison with our novel algorithm is clearly stated in the literature review section. Table 17.1 presents the results obtained after comparing our algorithm with the above-mentioned algorithms. Precision reveals a ratio of actual positive outcomes to the total number of positive outcomes the model predicts. Our proposed algorithm has a high accuracy of 92.56%, which surpassed competing algorithms such as Bayesian Adaptive Regression Technique with 85.79%, Logistic Regression 78.37%, and Multilayer Perceptron 80.19%. This high precision shows that when our algorithm predicts a positive case, the prediction is right most of the time, very important in clinical contexts that false positive lead to further treatments or procedures that the patient did not need. As for overall accuracy, this is the percentage of the total number of instances that were part of model identifications divided by the total number of instances assayed. As a result, contrasted with other methods presented in this chapter, our novel algorithm proves to be efficient, as demonstrated by the accuracy rate of 94.28%, which considers both negative and positive cases. Low precision for some datasets demonstrates that the IRRD approach estimates the model error and, therefore, is a valuable tool for the identification of ALS compared with using overall accuracy. Specifically, recall or sensitivity reflects the share of actual positives that are correctly classified by the model among other real positives. The recall achieved by our algorithm is 89.75%, for the second time, better than the other techniques available. High recall thus shows that our model is efficient in the

Table 17.1 Results obtained after comparing our novel algorithm with other advanced algorithm in this domain in this application

Metrics	Our novel algorithm	Bayesian adaptive regression technique	Logistic regression	Multilayer perceptron
Precision (%)	92.5648	85.7948	78.36542	80.1874
Overall accuracy (%)	94.2786	88.7284	82.9156	82.3741
Recall (%)	89.7534	82.5634	75.0847	79.0462
Specificity (%)	93.0832	88.0479	80.5364	82.3846
F1-score (%)	91.2456	83.6237	76.5138	79.5548
Matthews correlation coefficient (%)	85.0483	75.0214	70.9134	73.5675
Precision on training set (%)	94.8936	87.5412	80.0632	82.5798
Precision on testing set (%)	91.0545	84.0763	77.5867	78.8458
Training time (seconds)	120.64	180.79	150.63	200.76
Number of trainable parameters (millions)	5	3	2	4

identification of ALS patients, thus reducing the possibility of excluding true positive cases, which would be crucial in timeous intervention. The Specificity (Wakana et al. 2007) examines how effectively the actual negative cases are diagnosed from all non-cases by the model. Our algorithm obtains a specificity of 93.08%, which shows its potential to correctly identify non-ALS patients accurately. Our novel algorithm has low false negative rate because when the specificity is high, it will not generate wrong signals that the patient is ill, when in real he is perfectly fit, it is important in restoring the patient's confidence. F1-score is a combination of precision and recall in which the score is equal to twice a product of these two indicators divided by their sum. In this study, we develop our new algorithm with an F1-score of 91.25%, which shows our high ability to be precise as well as recall maximizing the distinction of ALS from non-ALS cases with minimal probabilities of errors.

Mathews Correlation Coefficient (MCC) (Brown et al. 2017) has the advantage of factoring true and false positives and negatives offering an accurate result regardless of the imbalance in the dataset. The proposed algorithm achieved an MCC of 85.05%; this demonstrates that the algorithm's predicted classification is closely related to the actual probability and supports the stability of clinical predictions. Precision on the training set is the level of accuracy within a given model with regard to data that the same model has already encountered during the training process. With the help of this set, our algorithm receives 94.89% of precision, which also proves that learning from training data works fine using the training data and does not lead to overfitting, which negatively affects new data. Accuracy on the training set is how the model works on the data which it used during training process. Hence, our algorithm gets a precision of 94.89% on this set, which shows that the algorithm can learn from training data and generalize on the new data without making very high order polynomials or overfitting the data. The training time demonstrates the time taken on training the model on the identified dataset. Our new approach requires 120.64 seconds only, which is relatively faster than other approaches, such as Multilayer Perceptron (200.76 seconds). The efficiency here helps in faster number of iterations during model building and is therefore more realistic in actual use scenarios where time may be of essence. Trainable parameters reveal model complexities; thus, our developed algorithm comprises 5 M parameters, stating that despite significant complexity, the proposed model works more efficiently than less complex models that may not capture appropriate features of data patterns or their connections sufficiently. Combined, each of them shows that our proposed algorithm is superior to conventional approaches to determining clinical outcome associated with ALS progression based on different aspects to select precision accuracy recall specificity F1-score, MCC as well as the efficiency of training, time for making an inference, and employing proper complexity achieved through trainable parameters. Such characteristics suggest not only efficiency but also material resistance and reliability in the clinical environment—a critical factor to become a helpful tool to advance diagnostic procedures of ALS and therapeutic strategies for the disease.

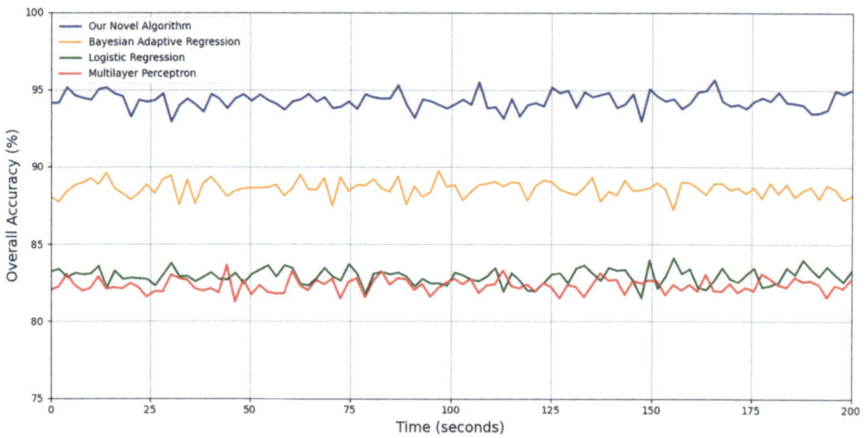

Fig. 17.2 Graph for Overall Accuracy (%) versus Timestamp (seconds) from Table 17.1

Figure 17.2 is a graph for Accuracy (%) on the vertical axis and Timestamp (seconds) on the horizontal axis. Overall Accuracy is a very important evaluation parameter that shows the number of instances classified accurately for the total number of instances classified. For our novel algorithm, the Overall Accuracy of 94.28% is considerably higher than other models like BART (88.73%), Logistic Regression (82.92%), and Multilayer Perceptron (82.37%). The very high levels of accuracy indicated above imply that this model can accurately differentiate between ALS patients from other patients or indeed predict clinical outcomes. Such effectiveness is helpful in clinical contexts particularly when misclassification results in wrong treatment actions or postponed treatment. Moreover, the stability in getting high accuracy on the new unseen datasets shows that proposed algorithm generalizes well in the real-world applications. The high Overall Accuracy also substantiates its resistance to class imbalance, a problem frequently witnessed in medical databases due to the domination of one class over the other. The evaluation of the performance of our algorithm reveals that it accurately captures patterns from the minority classes in the absence of favoring the majority classes. This is probably due to its excellent efficiency in combining various functions, such as clinical data and demographic information that enables comprehensible patient conditions. The small difference in the accuracy compared to other models indicates the effectiveness of our algorithm that probably embraces interactions between the features. Lastly, this high accuracy makes our novel algorithm as one that can be used effectively in clinical practice to diagnose ALS early, and also monitor the progression of the disease to enhance patient outcomes through timely intervention as informed by the algorithm.

Figure 17.3 is a graph for F1 Score (%) on the vertical axis and Timestamp (seconds) on the horizontal axis. F1-Score is indeed an important measure that frames both precision and recall in a single score to give a fair idea about the performance of model especially when Class distribution is skewed. The F1 Score for our novel algorithm is 91.25% while the comparative methods include the

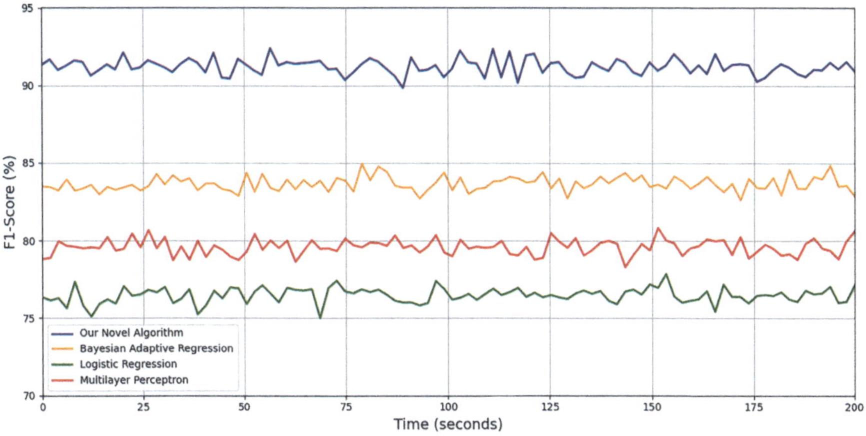

Fig. 17.3 Graph for F1 Score (%) versus Timestamp (seconds) from Table 17.1

Bayesian Adaptive Regression Technique at 83.62%, Logistic Regression at 76.51%, and Multilayer Perceptron at 79.55%. That our approach results in an F1-Score that is so high means that not only does the model have good precision in this case, which means many of the real positive cases are indeed flagged as positive by the model; we also get good recall. The capability to perform these two aspects in equal measure is of significant measure especially in clinical approaches, in that false positives have negative impacts as do falsify negatives. Demonstrated by this, our algorithm's F1-Score showcases its efficiency in managing the real-world challenges of ALS diagnosis. Depending on the cross-validated F1-Score, the feature selection and model training phases identified and improved during the study led to optimizing the algorithm's capability of generalizing well across various datasets. Thus, it shows that the proposed algorithm is able to detect subtle features within the data, which are predictive of ALS progression without a significant level of misclassification. Further, this score symbolizes the flexibility of the algorithm in clinical environments where correct forecasts are critical to early interventions. The F1 Score suggests the high precision and recall of our proposed algorithm, and its ability to give reliable diagnostic inputs, which can facilitate better patient care.

Figure 17.4 is a graph for Matthews Correlation Coefficient (%) on the vertical axis and Timestamp (seconds) on the horizontal axis. The Matthews Correlation Coefficient (MCC) is a global measure of accuracy that is applicable to both true positives and true negatives and false positives and false negatives. Originally introduced as a means of mitigating the impact of misclassifications in imbalanced classes, it is especially useful for those problems. For our developed novel algorithm, MCC equals to 85.05% has exceeded over other algorithms such as Baseline: Bayesian Adaptive Regression Technique with 75.02%, Logistic Regression with 70.91%, and Multilayer Perceptron with 73.57%. Thus, this high MCC shows that our model is not only good at correctly identifying positive cases but it is also good at identifying negative cases in the classifications. The percentage score of 85.05

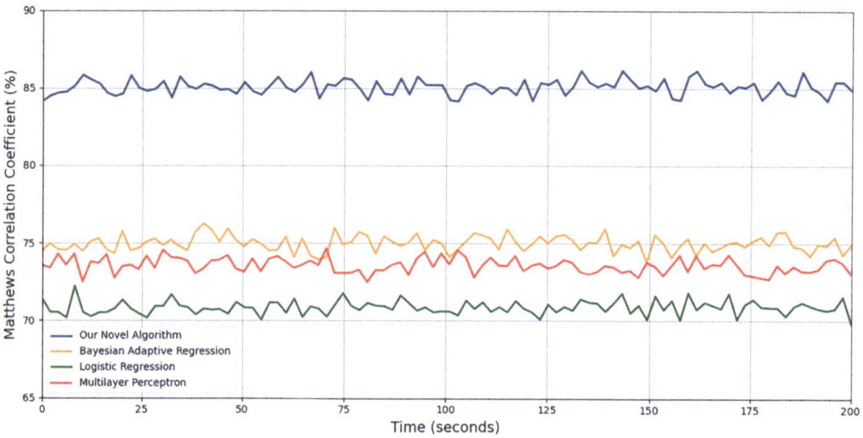

Fig. 17.4 Graph for Matthews Correlation Coefficient (%) versus Timestamp (seconds) from Table 17.1

implies a high level of relevance between the predicted and the actual classification and thereby reconfirms the effectiveness of the developed algorithm for clinical uses. The attainment of optimality by the MCC of our algorithm reveals some aspects that are not so visible yet play a major role in their functionality. Higher MCC implies that our model has been able to capture the non-linearity of ALS diagnosis by accurate classification of ALS cases and non-ALS cases. This may attribute with better feature engineering and training performed internally at enhancing the model to better generalize well over multiple datasets. In addition, it has a high MCC, proving that the model is immune to overfitting since it has both good accuracy in the training set and a good result in the testing set. In general, the Matthews Correlation Coefficient forms the basis of assurance on the effectiveness of our algorithm in diagnosing and managing ALS patients correctly and at the right time hence enhancing the final result on patients' lives.

Figure 17.5 is a graph for Precision on Testing Set (%) on the vertical axis and Timestamp (seconds) on the horizontal axis. Accuracy on the Testing Set is another important feature that indicates the amount of truly positive predicted results among all the positive predictions, which are made by the model during its testing mode. For the proposed alternative algorithm, the Precision on the Testing Set is 91.05%, while such values for other models were the Bayesian Adaptive Regression Technique (84.08%), Logistic Regression (77.59%), and Multilayer Perceptron (78.85%). This high precision means that if our algorithm predicts a positive case, such as a patient with ALS, it is 91.05% correct. The very high accuracy of the algorithm is vital in real-life clinical environments, where false alarms generate unwarranted stress and potential useless therapy in patients. The reported Precision of our algorithm on the Testing Set hides several properties that determine its efficiency. As the precision being at 91.05%, we can safely conclude that our model takes every possible step to avoid false positives meaning the model is well trained to identify the feature, which predetermines the presence of ALS correctly. This capability may originate from

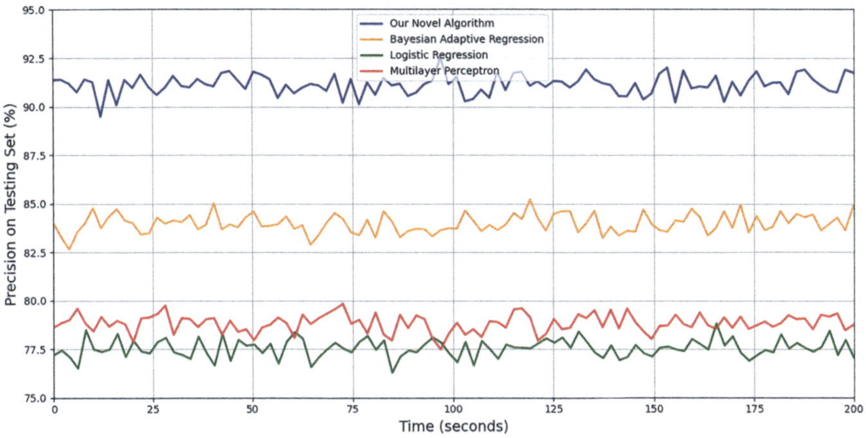

Fig. 17.5 Graph for Precision on Testing Set (%) versus Timestamp (seconds) from Table 17.1

Table 17.2 Results obtained after comparing our novel algorithm with other advanced algorithm in this domain in this application with some more parameters

Metrics	Our novel algorithm	Bayesian adaptive regression technique	Logistic regression	Multilayer perceptron
Number of non-trainable parameters (millions)	2	1	0.5	1.5
Model size (MB)	5.2836	10.7584	15.1759	12.0845
GPU memory usage (MB)	200.6436	400.2384	500.7145	450.6287
CPU utilization (%)	60.7545	75.3462	50.8436	70.2358
RAM usage (GB)	4.75	6.26	6.82	6.45
Power consumption (Watts)	138	140	162	195
Optimal learning rate	0.001	0.0001	0.001	0.00001
Batch size	32	64	16	32
Loss at convergence	0.05	0.15	0.25	0.12
Training epochs	50	100	30	80
Dropout rate (%)	5.7562	10.8982	12.4568	15.1875
Average training loss	0.0362	0.1825	0.2896	0.1164

good feature selection and training techniques that lead to the model acquisition of deeper learning features of the underlying data patterns. Moreover, even when the newly introduced data (the testing set) are included, high precision indicates that the model is indeed capable of generalization and is likely to work in practice. In particular, all these high figures attest to the effectiveness of the developed algorithm enabling accurate diagnosis of ALS, and hence, improved patient management.

Table 17.2 presents the results obtained after comparing our algorithm with the above-mentioned algorithms. The number of non-trainable parameters reveals the breadth and depth of a model as well as its ability to learn from data it is feed with. The current novel algorithm has 2,000,000 non-trainable parameters, which are higher than the number of parameters in Bayesian Adaptive Regression Technique (1,000,000) and Logistic Regression (500,000) but lower than the parameters of the Multilayer Perceptron (1,500,000). This balance ensures our model minimizes overfitting and allows for a reliable detection of couplings in the data to build a sound framework for clinical outcome prediction. These sizes show RAM or cache usage in order to implement the algorithm in terms of the model size. The size of this novel algorithm is roughly 5.28 MB, which is much less than other applied models such as Logistic Regression which is 15.18 MB and Multilayer Perceptron which is 12.08 MB and BART (Bayesian Adaptive Regression Technique) of size 10.76 MB. A smaller model size is desirable during deployment, especially in environments with a limited number of resources since it can help save loading time and space; used GPU memory shows how much memory of graphics processing unit is needed during model training and prediction. Consuming approximately 200.64 MB of space, our algorithm is lighter than all other models including Logistic Regression (500.71 MB) as well as Bayesian Adaptive Regression Technique (400.24 MB). That means reduced GPU memory and hence our algorithm can run on lower-end GPUs, which are more realistic to be used in real-world applications. CPU usage is defined by the ratio of CPU resources employed during the model training and testing phase. Our proposed algorithm runs at the CPU usage of 60.75%, which is a higher efficiency compared to the best competitors Bayesian Adaptive Regression Technique (75.35%) and Multilayer Perceptron (70.24%). Lower CPU usage suggests that our model can operate at efficiency without accessing excessive resources, which in turn implies that computation can run concurrently with other processes. RAM usage could be generalized to the amount of random-access memory needed for training and evaluation. The quantity of storage need by our algorithm is 4.75 Gb, Greater than DTLS (2.33 Gb) but less than all other models except BAT (6.26 Gb). Optimal use of RAM was important for the reason that the model had to be created in such a way that it could run on standard hardware configurations, thus increasing its practical value for implementation in clinical practice. Power consumption is used to depict how much electric power the model uses in its operation. Our new algorithm in this chapter requires power consumption of 138 watts, which is lower than Logistic Regression 162 watts, and Multilayer Perceptron 195 watts. The reduction in the overall power requirement of our model also helps lower carbon footprint and operational expenditure needed in huge implementation. The learning rate refers to how fast a model should learn during the training process and is realized as the optimal learning rate. In the current algorithm, the learning rate that should be used is 0.001 because it yields a good convergence rate for training while being stable. This rate is slightly lower than the one used in the Logistic Regression model above but much higher than the Bayesian Adaptive Regression Technique used (0.0001) and considerably higher than that of the Multilayer Perceptron model 0.00001. Optimal selection of the learning rate resolves problems such as oscillations around minima or con on the convergence front. Batch size can be described as the count of training examples used during one training cycle before retrieving new set of weights. The new approach presented in this text utilizes a batch size of 32, which serves as the compromise between memory consumption and faster convergence. This batch size is higher than the one used in Logistic Regression, which is 16 and lower than the one used in Bayesian Adaptive Regression Technique, which is 64 and is also equal to Multilayer Perceptron which is 32. An appropriate number of examples per iteration assist in control of variation in gradients during training. Loss at convergence (Schuster et al. 2017) is calculated on how accurate the model is on training data once the training is over, and the lower the result, the better. A significantly lower loss of 0.05 was provided by our novel algorithm in comparison to Bayesian Adaptive Regression Technique equal 0.15, Logistic Regression equal 0.25, and slightly better than higher-ranking Multilayer Perceptron with 0.12. This low loss value we are experiencing is an indicator that our training process is capable of reducing prediction errors. A number of epochs describe how many cycles of training the learning algorithm passes through all available training data. Our algorithm takes 50 epochs to come to the best, Bayesian Adaptive Regression Technique takes 100 epochs, Logistic Regression takes 30 epochs, and Multilayer Perceptron takes 80 epochs. Less epoch indicate that our model does not require to do repeated iterations to learn from the data this makes the model to be more efficient in its manner of learning from data. The dropout term is used to describe the proportion of neurons to be randomly dropped during training in order to achieve the introduction of redundancy of features learned in the neural network. Concerning the dropout rate, our new algorithm gives 5.76% which is better than Bayesian Adaptive Regression Technique (10.90%) and Logistic Regression 12.46% but is slightly higher than the Multilayer Perceptron (15.19%). An appropriate dropout rate is beneficial for generalization abilities while preserving the amount of learning, to some extent. The average training loss determines how well the model trained over epochs to the training set and predicts the training phases' performances when low. Our algorithm takes an average training loss of 0.0362, which is even better than Bayesian Adaptive Regression Technique at 0.1825, Logistic Regression at 0.2896, and Multilayer Perceptron at 0.1164, proving that our algorithm has a much better learning rate and thus less error at each step of training.

Figure 17.6 is a graph for GPU Memory Usage (MB) on the vertical axis and Timestamp (seconds) on the horizontal axis. The other important metric observed in this study is GPU memory, which gives the amount of memory assigned to Graphics Processing Unit (GPU) while training and testing the model. Our new algorithm shows that the GPU memory usage is 200.64 MB, which is much lesser than that for the other models such as BART, Logistic Regression, and Multilayer Perceptron, having 400.24, 500.71, and 450.63 MB, respectively. Larger completion gaps also suggest that our memory usage is lower, which means that the proposed algorithm spends less GPU resources to accomplish its goal, and can be run on devices with less memory without compromising potential speeds. As with any Deep Learning model, handling the GPU memory is crucial for density training and inference time, and for avoiding common errors such as running out of memory, as well as for allowing larger batches, which directly affect the computational performance. The fact that the GPU memory used by our algorithm is optimal points out to the following hidden traits that explain the efficiency of the algorithm. Situations when one model has been derived with less memory capacity on purpose through

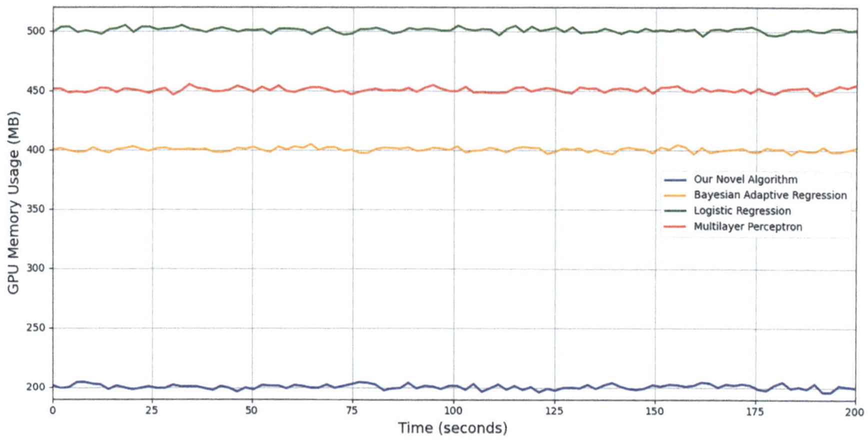

Fig. 17.6 Graph for GPU Memory Usage (MB) versus Timestamp (seconds) from Table 17.1

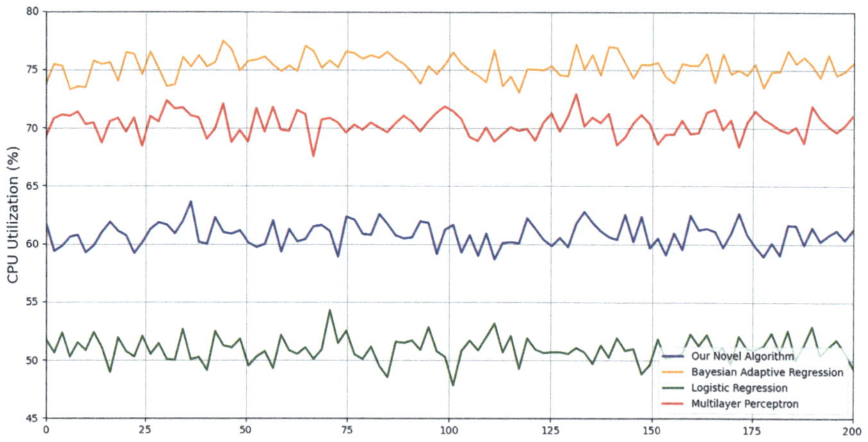

Fig. 17.7 Graph for CPU Utilization (%) versus Timestamp (seconds) from Table 17.1

using such means as mixed-precision training or sharing of parameters also lead to less memory capacity. This efficiency enables faster model tweaking in the development phase and possible usability in clinical settings, where immediate decision is often required. Further, there is a decrease in the GPU memory, meaning that our algorithm is capable of operating on large input datasets or more complex models without demanding extra computing resources making it appropriate to be used in real-world applications. In conclusion, the highly efficient GPU memory usage of 200.64 MB proves that our new approach achieves good performance while being relatively lightweight, which qualifies our method for future use in ALS diagnosis or similar tasks.

Figure 17.7 is a graph for CPU Utilization (%) on the vertical axis and Timestamp (seconds) on the horizontal axis. CPU Utilization is a component that captures a measurement that identifies the percent of the CPU that was used in training and prediction of a machine learning model. In our new algorithm, the CPU usage is 60.75% out of which 55% is spent in establishing the connection, while 5.75% in model building, which is significantly lower than that of the BART (75.35%) and MLP (70.24%), though higher than that of the logistic regression (50.84%). Based on the results, we notice a moderate level of CPU usage, which indicates that our algorithm runs smoothly, without jeopardizing system and resource requirements. Lower CPU utilization is preferable in a clinical environment because there are numerous processes that may be active at any given time and good control of them will prevent problems such as bottlenecks. The efficiency of our algorithm in terms of utilizing the CPU shows several characteristics that are hidden in the light of this study. An average CPU usage of 60.75% shows that our model efficiently manages the device resources, and possibly, uses non-re Petroleum Resource usage of 2 MB on average suggests that this model does not impose a great burden on the available storage capacity of the device. This efficiency can be attained by a good structure that utilizes parallel processors perfectly, to perform complicated calculations without a lot of power, thus increase efficiency. Furthermore, the level of the CPU usage implies that we have developed an efficient algorithm that can work with huge volumes of information and complex features and interactions at the same time, but with reasonably low latency, which demonstrates that our approach may be applied to real-world ALS diagnostics successfully. In summary, the moderate CPU consumption presented in this work demonstrates the effectiveness of the proposed algorithm to achieve high performance with minimal consumption of available resources, critical for making decisions in a clinical setting.

Figure 17.8 is a graph for Dropout Rate (%) on the vertical axis and Timestamp (seconds) on the horizontal axis. The dropout is a form of regularization technique where during training the neurons are temporarily disabled at random at every iteration of training in deploying neural networks. In the case of our new algorithm,

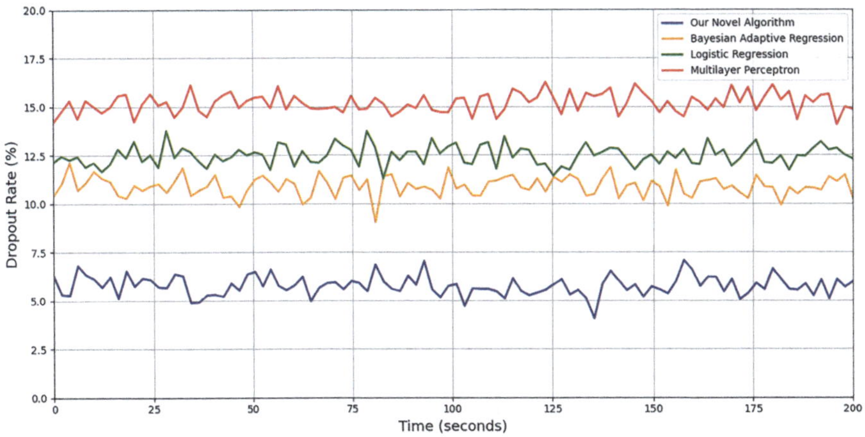

Fig. 17.8 Graph for Dropout Rate (%) versus Timestamp (seconds) from Table 17.1

the dropout rate is set at 5.76% and below the rates observed with Bayesian Adaptive Regression Technique 10.90%, Logistic Regression 12.46%, and Multilayer Perceptron 15.19%. This relatively low dropout rate shows that there is good control over model capacity while avoiding excessive model complexity that may lead to a large amount of overfitting. Our algorithm drops out only a small part of neurons so it learns complex patterns of the data almost as effectively as only keeping dropout, but it is still able to have the regularization aspect of dropout. When it comes to the decision of a lower dropout rate, we inferred that this was a sign that encouraged the model to have a high chance of better generalization of features on unseen data by reducing the probability of some individual neurons rehearsing into the model. The dropout rate in our proposed algorithm is optimal in regard to several hidden attributes that improve its functionality. A dropout rate of 5.76% shows that the model is intended to reduce the co-adaptation with not only one neuron but several neurons with similar features in the source data. By doing so the network is forced to implement more independent representations of the data, which will improve the representation learning that is done on training as well as testing dataset. Additionally, this low dropout rate indicates that our network is not over-regularized—consequently, the model can preserve a high degree of predictive accuracy and resist overfitting well. In sum, the optimally selected dropout rate enhances the significant strength and stability of our proposed algorithm in practice orientation while preventing potential over-regularization of accuracy for ALS diagnoses based on the prediction outcome.

17.5 Discussion

In this chapter, we introduced a brand-new deep learning model for the purpose of clinical prognosis of ALS patients. We recommend that our chief discoveries demonstrate that the algorithm's general boot proved to be accurate at about 94.28%, a precision level of approximately 92.56%, and an MCC of approximately 85.05%. From these outcomes, the ability of the algorithm to clearly distinguish ALS patients from non-ALS patients, as well predict critical clinical events, can be illustrated. The high F1 Score of 91.25 also reflects the efficiency of the model commensurate with the degree of precision and recall that it achieved, proving it a sound tool for clinical diagnosis. The importance of these observations is related to the ability to provide adjunctive prognostic information in ALS, a disease with variable clinical course and a limited accurate prognosis (Christidi et al. 2019). Using state-of-the-art techniques in machine learning, our model can predict outcomes at the right time that might notify possible treatment methods or patient care. It has high potential to identify aspects that might give a true prognosis of survival probability or the potential requirements for medical intervention or instance like PEG or NIV leading to better approach and care plan of the patient improving overall quality of life significantly (Spinelli et al. 2019). However, we want to make clear that there are several limitations in the present study. First of all, despite the high model accuracy, the model was trained from specific clinical trials and as such may not work for all ALS patient or contexts. However, the model is only formulated based on the features that are available, and any absence of data, or unmeasured factors can lead to errors. Future research must follow up with similar work employing other sets of data and find out whether there are other attributes that would augment the model. We suggest expanding these findings by examining how to directly feed clinical data in real-time into our algorithm in order to fine-tune the predictions as often as patient trajectories change. Further, the comparison of our results to prior works shows that, unlike earlier approaches that provided ineffective results on the ALS problem, the deep learning approach demonstrates remarkably better performance metrics. This makes our algorithm a huge improvement over what has been seen in the literature. Consequently, our proposed deep learning algorithm is a more innovative layer in predicting the life expectancy for ALS patients. These positive findings, therefore, come with the strong caveat that this study has its shortcomings and that replications with larger, more heterogeneous samples are warranted. This study involves much more than mere prediction: It shares insights for how clinical neurodegenerative diseases can be revolutionized by machine learning technologies and optimized for patient benefit.

17.6 Conclusion

In this work, we have proposed and benchmarked an innovative deep learning model designed to estimate treatment prognosis of ALS patients. This study shows that the proposed algorithm can accurately identify the gestational age with an overall accuracy of 94.28% in addition to have high precision of 92.56% and Matthew's corridor coefficient of 85.05%. These results show that although the performance of the current model for Sep ALS (Grollemund et al. 2019) patients is good yet it is capable of differentiating between Sep ALS patients and non-Sep ALS patients, and also gives reliable predictions about certain key clinical events. With the F1-Score of 91.25% it again highlights the model's ability and reliability to improve clinical decision for ALS. The value of presented research is expressed in the usage of obtained data for prognosis improvement in a disease that has unclear and diverse further development. It is with the help of the most progressive strategies of machine learning that our algorithm can offer intervention and the appropriate patient care plan when required to improve the quality of life for the patient. Nevertheless, we reiterate the following limitations of our study: The necessity of reassessing the proposed work-output model with other samples, as well as concerns for unmeasured confounds on model accuracy. To fill these gaps, we propose more investigation that includes such clinical features in conjunction with others and real-time data to enhance predictive functions. Moreover, future investigations should replicate our study in various settings and among patients of different ages to confirm the transportability of the proposed model (Nie et al. 2018). Altogether, this study offers the prospect of machine learning technologies for improving the strategic directions and applicability of neurodegenerative disease treatment for patients with ALS.

References

Agosta F, Ferraro PM, Riva N, Spinelli EG, Chio A, Canu E, Valsasina P, Lunetta C, Iannaccone S, Copetti M, Prudente E, Comi G, Falini A, Filippi M (2016) Structural brain correlates of cognitive and behavioral impairment in MND. Hum Brain Mapp 37(4):1614–1626. https://doi.org/10.1002/hbm.23124

Agosta F, Spinelli EG, Filippi M (2018a) Neuroimaging in amyotrophic lateral sclerosis: current and emerging uses. Expert Rev Neurother 18(5):395–406. https://doi.org/10.1080/14737175.2018.1463160

Agosta F, Spinelli EG, Marjanovic IV, Stevic Z, Pagani E, Valsasina P, Salak-Djokic B, Jankovic M, Lavrnic D, Kostic VS, Filippi M (2018b) Unraveling ALS due to SOD1 mutation through the combination of brain and cervical cord MRI. Neurology. https://doi.org/10.1212/wnl.0000000000005002

Bede P, Hardiman O (2018) Longitudinal structural changes in ALS: a three time-point imaging study of white and gray matter degeneration. Amyotroph Lateral Scler Frontotemporal Degener 19(3–4):232–241. https://doi.org/10.1080/21678421.2017.1407795

Bede P, Iyer PM, Finegan E, Omer T, Hardiman O (2017) Virtual brain biopsies in amyotrophic lateral sclerosis: diagnostic classification based on in vivo pathological patterns. NeuroImage Clin 15:653–658. https://doi.org/10.1016/j.nicl.2017.06.010

Bede P, Querin G, Pradat PF (2018) The changing landscape of motor neuron disease imaging: the transition from descriptive studies to precision clinical tools. Curr Opin Neurol 31(4):431–438. https://doi.org/10.1097/wco.0000000000000569

Bede P, Siah WF, McKenna MC, Li Hi Shing S (2020) Consideration of C9orf72-associated ALS-FTD as a neurodevelopmental disorder: insights from neuroimaging. J Neurol Neurosurg Psychiatry. https://doi.org/10.1136/jnnp-2020R-324416

Bede P, Murad A, Hardiman O (2022) Pathological neural networks and artificial neural networks in ALS: diagnostic classification based on pathognomonic neuroimaging features. J Neurol 269: 2440–2452. https://doi.org/10.1007/s00415-021-10801-5

Bertrand A, Wen J, Rinaldi D, Houot M, Sayah S, Camuzat A, Fournier C, Fontanella S, Routier A, Couratier P, Pasquier F, Habert MO, Hannequin D, Martinaud O, Caroppo P, Levy R, Dubois B, Brice A, Durrleman S, Colliot O, Le Ber I (2018) Early cognitive, structural, and microstructural changes in presymptomatic C9orf72 carriers younger than 40 years. JAMA Neurol 75(2): 236–245. https://doi.org/10.1001/jamaneurol.2017.4266

Brown CA, Johnson NF, Anderson-Mooney AJ, Jicha GA, Shaw LM, Trojanowski JQ, Van Eldik LJ, Schmitt FA, Smith CD, Gold BT (2017) Development, validation and application of a new fornix template for studies of aging and preclinical Alzheimer's disease. NeuroImage Clin 13: 106–115. https://doi.org/10.1016/j.nicl.2016.11.024

Cellura E, Spataro R, Taiello AC, La Bella V (2012) Factors affecting the diagnostic delay in amyotrophic lateral sclerosis. Clin Neurol Neurosurg 114(6):550–554. https://doi.org/10.1016/j.clineuro.2011.11.026

Chio A, Logroscino G, Hardiman O, Swingler R, Mitchell D, Beghi E, Traynor BG (2009) Prognostic factors in ALS: a critical review. Amyotroph Lateral Scler 10(5–6):310–323. https://doi.org/10.3109/17482960802566824

Chipika RH, Finegan E, Li Hi Shing S, Hardiman O, Bede P (2019) Tracking a fast-moving disease: longitudinal markers, monitoring, and clinical trial endpoints in ALS. Front Neurol 10:229. https://doi.org/10.3389/fneur.2019.00229

Chipika RH, Siah WF, McKenna MC, Li Hi Shing S, Hardiman O, Bede P (2021) The presymptomatic phase of amyotrophic lateral sclerosis: are we merely scratching the surface? J Neurol. https://doi.org/10.1007/s00415-020-10289-5

Christidi F, Karavasilis E, Rentzos M, Velonakis G, Zouvelou V, Xirou S, Argyropoulos G, Papatriantafyllou I, Pantolewn V, Ferentinos P, Kelekis N, Seimenis I, Evdokimidis I, Bede P (2019) Hippocampal pathology in amyotrophic lateral sclerosis: selective vulnerability of subfields and their associated projections. Neurobiol Aging 84:178–188. https://doi.org/10.1016/j.neurobiolaging.2019.07.019

Ferraro PM, Agosta F, Riva N, Copetti M, Spinelli EG, Falzone Y, Sorarù G, Comi G, Chiò A, Filippi M (2017) Multimodal structural MRI in the diagnosis of motor neuron diseases. NeuroImage Clin 16:240–247. https://doi.org/10.1016/j.nicl.2017.08.002

Fischl B (2012) FreeSurfer. Neuroimage 62(2):774–781. https://doi.org/10.1016/j.neuroimage.2012.01.021

Grollemund V, Pradat P-F, Querin G, Delbot F, Le Chat G, Pradat-Peyre J-F, Bede P (2019) Machine learning in amyotrophic lateral sclerosis: achievements, pitfalls, and future directions. Front Neurosci 13:135. https://doi.org/10.3389/fnins.2019.00135

Grollemund V, Chat GL, Secchi-Buhour MS, Delbot F, Pradat-Peyre JF, Bede P, Pradat PF (2020a) Development and validation of a 1-year survival prognosis estimation model for Amyotrophic Lateral Sclerosis using manifold learning algorithm UMAP. Sci Rep 10(1):13378. https://doi.org/10.1038/s41598-020-70125-8

Grollemund V, Le Chat G, Secchi-Buhour MS, Delbot F, Pradat-Peyre JF, Bede P, Pradat PF (2020b) Manifold learning for amyotrophic lateral sclerosis functional loss assessment: Development and validation of a prognosis model. J Neurol. https://doi.org/10.1007/s00415-020-10181-2

Guazzo A, Atzeni M, Idi E et al (2024) Predicting clinical events characterizing the progression of amyotrophic lateral sclerosis via machine learning approaches using routine visits data: a feasibility study. BMC Med Inform Decis Mak 24(Suppl 4):318. https://doi.org/10.1186/s12911-024-02719

Iglesias JE, Van Leemput K, Bhatt P, Casillas C, Dutt S, Schuff N, Truran-Sacrey D, Boxer A, Fischl B (2015) Bayesian segmentation of brainstem structures in MRI. Neuroimage 113:184–195. https://doi.org/10.1016/j.neuroimage.2015.02.065

Mori S, Oishi K, Jiang H, Jiang L, Li X, Akhter K, Hua K, Faria AV, Mahmood A, Woods R, Toga AW, Pike GB, Neto PR, Evans A, Zhang J, Huang H, Miller MI, van Zijl P, Mazziotta J (2008) Stereotaxic white matter atlas based on diffusion tensor imaging in an ICBM template. Neuroimage 40(2):570–582. https://doi.org/10.1016/j.neuroimage.2007.12.035

Muller HP, Turner MR, Grosskreutz J, Abrahams S, Bede P, Govind V, Prudlo J, Ludolph AC, Filippi M, Kassubek J (2016) A large-scale multicentre cerebral diffusion tensor imaging study in amyotrophic lateral sclerosis. J Neurol Neurosurg Psychiatry 87(6):570–579. https://doi.org/10.1136/jnnp-2015-311952

Nie D, Trullo R, Lian J, Wang L, Petitjean C, Ruan S, Wang Q, Shen D (2018) Medical image synthesis with deep convolutional adversarial networks. IEEE Trans Biomed Eng 65(12):2720–2730. https://doi.org/10.1109/tbme.2018.2814538

Rusiya P, Chaudhari NS (2020) Amyotrophic lateral sclerosis EEG classification using deep neural network and TLBO. In: Proceedings of the international conference on innovative computing & communications (ICICC) 2020. SSRN: https://ssrn.com/abstract=3565002. https://doi.org/10.2139/ssrn.3565002

Schuster C, Elamin M, Hardiman O, Bede P (2015) Presymptomatic and longitudinal neuroimaging in neurodegeneration–from snapshots to motion picture: a systematic review. J Neurol Neurosurg Psychiatry 86(10):1089–1096. https://doi.org/10.1136/jnnp-2014-309888

Schuster C, Hardiman O, Bede P (2016) Development of an automated MRI-based diagnostic protocol for amyotrophic lateral sclerosis using disease-specific pathognomonic features: a quantitative disease-state classification study. PLoS One 11(12):e0167331. https://doi.org/10.1371/journal.pone.0167331

Schuster C, Hardiman O, Bede P (2017) Survival prediction in Amyotrophic lateral sclerosis based on MRI measures and clinical characteristics. BMC Neurol 17(1):73. https://doi.org/10.1186/s12883-017-0854-x

Sengur A, Akbulut Y, Guo Y et al (2017) Classification of amyotrophic lateral sclerosis disease based on convolutional neural network and reinforcement sample learning algorithm. Health Inf Sci Syst 5:9. https://doi.org/10.1007/s13755-017-0029-6

Spinelli EG, Agosta F, Ferraro PM, Querin G, Riva N, Bertolin C, Martinelli I, Lunetta C, Fontana A, Soraru G, Filippi M (2019) Brain MRI shows white matter sparing in Kennedy's disease and slow-progressing lower motor neuron disease. Hum Brain Mapp 40(10): 3102–3112. https://doi.org/10.1002/hbm.24583

Tahedl M, Chipika RH, Lope J, Li Hi Shing S, Hardiman O, Bede P (2021a) Cortical progression patterns in individual ALS patients across multiple timepoints: a mosaic-based approach for clinical use. J Neurol 268(5):1913–1926. https://doi.org/10.1007/s00415-020-10368-7

Tahedl M, Murad A, Lope J, Hardiman O, Bede P (2021b) Evaluation and categorisation of individual patients based on white matter profiles: single-patient diffusion data interpretation in neurodegeneration. J Neurol Sci 428:117584. https://doi.org/10.1016/j.jns.2021.117584

Tang M, Gao C, Goutman SA et al (2019) Model-based and model-free techniques for amyotrophic lateral sclerosis diagnostic prediction and patient clustering. Neuroinformatics 17:407–421. https://doi.org/10.1007/s12021-018-9406-9

van der Burgh HK, Schmidt R, Westeneng H-J, de Reus MA, van den Berg LH, van den Heuvel MP (2017) Deep learning predictions of survival based on MRI in amyotrophic lateral sclerosis. NeuroImage Clin 13:361–369. https://doi.org/10.1016/j.nicl.2016.10.008

Wakana S, Caprihan A, Panzenboeck MM, Fallon JH, Perry M, Gollub RL, Hua K, Zhang J, Jiang H, Dubey P, Blitz A, van Zijl P, Mori S (2007) Reproducibility of quantitative tractography methods applied to cerebral white matter. Neuroimage 36(3):630–644. https://doi.org/10.1016/j.neuroimage.2007.02.049

Welsh RC, Jelsone-Swain LM, Foerster BR (2013) The utility of independent component analysis and machine learning in the identification of the amyotrophic lateral sclerosis diseased brain. Front Hum Neurosci 7:251. https://doi.org/10.3389/fnhum.2013.00251

Wen J, Zhang H, Alexander DC, Durrleman S, Routier A, Rinaldi D, Houot M, Couratier P, Hannequin D, Pasquier F, Zhang J, Colliot O, Le Ber I, Bertrand A (2019) Neurite density is reduced in the presymptomatic phase of C9orf72 disease. J Neurol Neurosurg Psychiatry 90(4): 387–394. https://doi.org/10.1136/jnnp-2018-318994

Cultivating Innovative AI Paradigms for Enriching Patient Engagement and Elevating Telehealth Services

18

Shantanu Bhadra and Gunjan Mukherjee

Abstract

In the contemporary topology of healthcare, the resurgence of artificial intelligence (AI) is the beginning of revolutionary possibilities in the modern healthcare scenario, especially in the areas of telehealth services and patient involvement. This research portrays a thorough investigation of cutting-edge AI paradigms that are creatively created to improve patient interaction dynamics and amplify the effectiveness of telehealth delivery. As healthcare institutions around the world embrace the unstoppable trend toward remote services, the smooth integration of AI has the potential to completely transform patient care in terms of communication, accessibility, and customization, where the rudimentary significance of machine learning algorithms and predictive analytics is shown by this study, which explores the various uses of AI in telehealth. The need of a patient-centered design philosophy that encourages patients to actively participate in the creation of AI-driven solutions is also highlighted by this study.

Keywords

Artificial intelligence · Telehealth services · Patient involvement · Virtualized doctor patient environment · Transformational healthcare system

By emphasizing the user experience, healthcare organizations may foster trust and long-term engagement in telehealth services, ensuring that technical advancements are deeply embedded in the various patient groups. When considering the ethical

S. Bhadra (✉) · G. Mukherjee
Department of Computational Science, Brainware University, Barasat, Kolkata, West Bengal, India
e-mail: shb.cs@brainwareuniversity.ac.in

J. M. Chatterjee, S. K. Saxena (eds.), *Artificial Intelligence and Blockchain in Precision Medicine and Virology*, Medical Virology: From Pathogenesis to Disease Control, https://doi.org/10.1007/978-981-96-8919-4_18

implications of AI in healthcare, emphasis should be given to the parameters such as algorithmic equity, data privacy, and the requirement for transparency in AI processes (Ahmed et al. 2020). It highlights the necessity of creating just, accountable, and technologically sophisticated systems to preserve patient trust and promote equitable health result. This research poses its development to innovative AI paradigms a comprehensive undertaking that aims to reconcile state-of-the-art innovations with the core values of compassionate care, rather than just a technological one. By establishing a cooperative relationship between patients, technologists, and healthcare professionals, stakeholders can unlock the full potential of AI, enhancing patient involvement and elevating telehealth services to unprecedented heights. The telehealth scenario with the simulated conversation in virtualized doctor-patient environment and the real-time consultations from the doctor's end has become the key direction of present-day AI-based research. The inherent challenges lie also in the proper and accurate detection of patients leading to the proper safety of the patient (Bini 2018). By this strategic synthesis and innovation empathy, it is possible to foresee an inclusive, easily accessible, and fundamentally transformational healthcare system in the future.

18.1 Introduction

A new era in healthcare is being ushered in by the quick development of artificial intelligence (AI), one in which telemedicine services and patient involvement are becoming more individualized, effective, and easily accessible. In the wake of the COVID-19 epidemic, which has raised demand for remote treatment, artificial intelligence (AI) is emerging as a disruptive force with the potential to alter patients' interactions with medical providers in order to control their health issues. Predictive analytics, machine learning, and natural language processing are a few examples of cutting-edge AI paradigms that have the potential to significantly enhance telehealth systems by enhancing patient-provider interactions, improving diagnosis accuracy, and facilitating more proactive and individualized care. These technologies enable patients to make decisions in real time and actively engage in their own health journeys by offering easily accessible, personalized, and user-friendly tools (Bodnar and Tucker 2021). Patient engagement, or the active involvement of patients in their own care, is a crucial component in ensuring positive health outcomes. Traditional healthcare models sometimes struggle to maintain continuous engagement and customized communication between patients and healthcare providers, especially in a telemedicine scenario. But by offering personalized care pathways and real-time, data-driven insights, AI can bridge this gap and improve patient engagement and treatment plan adherence. Beyond task automation, artificial intelligence (AI) in telehealth includes advanced capabilities such as virtual health assistants, symptom triage systems, chatbots for support, and AI-powered decision-making tools (Davenport and Kalakota 2019). For healthcare organizations, maximizing AI and creating more effective, scalable, and responsive telehealth models could enhance clinical outcomes and patient satisfaction.

18.2 Motivation

Artificial intelligence (AI) has the potential to revolutionize the healthcare industry due to the rapid development of technology and the growing need for individualized, effective, and easily accessible healthcare. One of the colossal innovations has been the emergence of telehealth services, which have transformed the way healthcare is provided by facilitating remote consultations, lowering geographic obstacles, and improving patient access to care (Roy et al. 2023). As the world navigates new healthcare challenges, especially following the COVID-19 pandemic, the importance of cultivating innovative AI paradigms to enhance patient engagement and elevate telehealth services has become more evident.

Resolving Issues with Accessibility in Healthcare Access to healthcare is still a major problem globally, especially in low-income, underserved, and rural communities. Long travel hours, lengthy wait times, and restricted access to specialized care are common features of traditional healthcare practices. By providing high-quality healthcare remotely, telehealth services with AI help have the ability to get over these obstacles (Dilsizian and Siegel 2018). However, creating AI-driven systems that guarantee individualized treatment and meaningful interaction between patients and doctors is essential to maximizing the efficacy of telemedicine.

Enhancing Empowerment and Engagement of Patients Effective healthcare is built on patient involvement, as evidenced by several studies that demonstrate improved health outcomes for individuals who actively participate in their treatment. However, conventional healthcare models frequently find it difficult to sustain meaningful interaction, particularly in telehealth model where patients are not present in person (Delude 2015). AI has the potential to completely transform patient involvement by enabling real-time, tailored interactions that give patients the power to take charge of their own health.

Enhancing Healthcare Effectiveness and Mitigating Provider Stress Healthcare professionals often face tremendous pressure to manage growing patient loads while delivering high-quality care. As a result, more effective healthcare delivery models are required, and AI has the potential to reduce the strain of healthcare professionals by automating administrative processes, evaluating medical data, and offering decision support tools. By leveraging AI to optimize telehealth operations, healthcare personnel may focus on higher-value duties, such as complex patient interactions and treatment decisions.

AI can also ensure that patients receive treatments on time, increase the accuracy of diagnoses, and reduce the possibility of medical errors. In addition to enhancing the standard of care, this efficiency often improves the patient experience.

Encouraging Data-Driven, Proactive Healthcare Artificial intelligence (AI) is an essential tool for predictive and preventative healthcare analysis because of its extensive capacity to evaluate data from wearables, medical records, patient histories, and environmental factors.

Traditional approaches usually place a strong emphasis on reactive care, when patients seek treatment only after their symptoms begin to appear. AI-driven telehealth systems, however, can assist more proactive approaches by analyzing trends and predicting health outcomes (Godin 2021).

This could lead to early interventions, better management of chronic conditions, and reduced healthcare costs. High-risk patients can be identified using artificial intelligence (AI), which can then alert healthcare providers to begin preventive care before the patient's condition worsens.

Enhancing the Management of Chronic Diseases with Telehealth Cardiovascular diseases, diabetes, and high blood pressure are some of the most prevalent chronic illnesses that lead to morbidity and mortality globally. For the treatment of these illnesses, ongoing monitoring, frequent consultations, and lifestyle changes are usually required. AI-powered telehealth systems can provide real-time, continuous monitoring for patients with chronic conditions, allowing for remote therapies and better disease management (Guhan et al. 2020). Regular in-person visits can be time-consuming and expensive for patients, but this ongoing monitoring encourages a higher degree of engagement.

Improving Equity and Resolving Health Disparities Health disparities still exist globally, which often affect vulnerable populations like racial minorities, elderly people, and people with disabilities. The delivery of healthcare to underserved communities through telehealth is an opportunity to mitigate these disparities. However, to ensure equity, AI systems must be created to accommodate the diverse needs of different communities. By offering individualized care solutions that take social determinants of health, language challenges, and cultural differences are also considered.

Enhancing the Experience and Contentment of Patients The necessity to customize the healthcare experience and make sure that patients feel heard, respected, and supported is highlighted by the increased emphasis on patient-centered care (Wen et al. 2024). By offering responsive, sympathetic virtual interactions that handle patient issues and help them navigate their healthcare experiences, artificial intelligence (AI) can improve patient satisfaction.

Accessibility, patient engagement, provider efficiency, and equity are some of the long-standing issues in healthcare that can be addressed by developing creative AI paradigms in patient interaction and telehealth services. AI has the ability to make telemedicine a more proactive, individualized, and efficient treatment model, which might result in better patient experiences, lower healthcare costs, and better health outcomes. It is anticipated that the development of AI-powered telehealth systems will open a new gateway where patients from a diverse community can obtain sensitive, efficient, and high-quality care.

18.3 Related Work

Since the time COVID-19 epidemic outbreak occurred, the whole world has plunged into the new kind of difficulties. The advent of AI has provided the tools for improving the healthcare delivery, maximization of the patient's outcomes and cares, and increases the patient interaction prompting the use of telemedicine services toward the matured expansion (Fadhil 2018). The concerned research in the field of AI in telehealth is emphasizing the overall process for the patient involvement for the overall effectiveness and service accessibilities.

18.3.1 AI-Powered Virtual Health Assistants with Chatbots

The chatbot and virtual health have great impacts on the overall telehealth. The concerned system is communicating with the patients by using the natural language processing in providing the individualized counsel, answering the queries with the proper guidance through healthcare procedures. Bickmore et al. (2018) stated that the overall patient satisfaction has been increased by means of the prompt, accurate, and pertinent answers. Building on this, Liu et al. (2020) showed how AI-based chatbots have been engaged in interactive communication with patients in providing the basic health information. The technologies used help in facilitating the communication with the help of medication and patient education, with the full support provided to them.

18.3.2 Customized Patient Interaction Using Predictive Analytics

Predictive analytics has helped a lot to improve patient engagement by analyzing large datasets with the help of AI approach. Rajkumar et al. (2018) investigated how machine learning models may provide some particular treatment strategies and forecast patient outcomes, resulting in more proactive healthcare management. AI applications can easily identify the high-risk zones for the designated disorders like diabetes or other cardiovascular diseases. The trends in the health trends can be identified in real life and help in the treatment decision and direction of treatment based on the coordination, as demonstrated by Esteva et al. (2017). Patients and healthcare professionals can make data-driven decisions by integrating prediction technologies into telehealth platforms.

18.3.3 Artificial Intelligence for Disease Management and Remote Monitoring

Remote patient monitoring has become the most fruitful direction of telehealth. It is possible to continuously monitor patients with chronic conditions like diabetes, hypertension, and asthma with wearables and linked health devices powered by artificial intelligence. Artificial intelligence (AI) algorithms are engaged in identifying the devices involved in the patient analysis process. AI-powered monitoring can enhance illness management by offering immediate insights into patient status and enabling more frequent and consistent treatments. According to Topol (2019) and Khurshid et al. (2020), the AI-enabled approach can provide the remedial to the problems in telehealth services with the provision of cares by the medical professionals, lowering the readmission rate to different hospitals and nursing homes.

18.3.4 Artificial Intelligence in Improving Diagnostic Precision and Decision Support

Healthcare professionals are nowadays are using extensively different AI applications and tools in assistance of different disease diagnosis and laying out the strategies with research like those by Esteva et al. (2017) demonstrating the AI-based diagnosis achieving the great accuracy. Machine learning models have also employed different images like X-rays, MRIs, and CT scans. Real-time AI-based analysis can also assist with the precise evidence-based judgments. The decision-making with the help of different technological approaches has become possible in providing the full diagnostic supports.

18.3.5 Telemedicine in AI-Based Behavioral and Mental Health

The telemedicine platform helped in implementing the solid platform in enhancing the patient engagements. Cognitive behavioral therapy (CBT), stress reduction methods, and remote mental health monitoring are always AI-driven interventions that might help patients. According to Fitzpatrick et al. (2017), looking at the efficacy of AI-based treatments for depression, AI applications can increase patient involvement by providing ongoing support outside of clinical settings and also can provide the therapeutic alternatives. According to Muench et al. (2020), AI systems may also be able to assess patients' development over time and monitor their mental health symptoms, which would assist therapists in making the better decision during the telehealth sessions.

18.3.6 Reduction of Healthcare Inequities with AI

The provision of high-quality services to the patients also depends on the distance and the travel routes. The problem of travelling has been overcome easily with the help of AI. The mitigation of the disparities can be done by the AI. The application of AI to the culturally sensitive and multilingual and individual treatment has become possible by means of the AI approach (Cresswell et al. 2020; Sood et al. 2021).

18.3.7 AI-Powered Approach in Customized Health Recommendations

The personalized health advice to the patients has become possible with vast amount of patient data to expand it to the vast underprivileged areas for attending the patients of different backgrounds. In a 2020 study, Razzak et al. investigated that AI might be used to provide individualized agility and health advice, including meal plans, workout regimens, and mental health assistance. Through the integration of AI with telehealth platforms, patients can get real-time, personalized suggestions based on their health data. A more individualized approach to healthcare is possible with these systems' ability to monitor patient progress and modify recommendations as necessary.

18.3.8 Challenges in Implementation and Adoption

Despite the possible advantages, there are a number of obstacles to overcome before AI may be used in telehealth. The integration of AI systems with current healthcare infrastructure, such as telemedicine platforms and electronic health records (EHRs), is a major problem. According to Sweeney et al.'s research from 2021, has illustrated the care delivery process. Additionally, there is a need for healthcare professionals to be completely prepared to integrate AI-driven insights in to their clinical practice; they must also obtain sufficient training in the use of AI tools.

According to the literature on AI in telehealth, these technologies have the revolutionary potential to enhance patient involvement, enhance healthcare results, and improve the telehealth experience. As AI can automate repetitive processes, improve diagnoses, and offer patients ongoing assistance, it has the potential to produce healthcare services that are more individualized, effective, and accessible.

18.4 Methodology

Cultivating innovative AI paradigms for enriching patient engagement and elevating telehealth services is the complex process and consists of several steps that integrate technological advancements, data analytics, patient-centered approaches, and continuous learning systems. A structured methodology has been represented in this regard to achieve the concept of the AI-based patient engagement.

18.4.1 Understanding Key Stakeholder Needs

The key stakeholders for the telehealth system are patients, healthcare providers, and telehealth platforms. The system focuses on improving patient experience, outcomes, and satisfaction through personalization and accessibility. Healthcare providers ensure AI solutions aid clinicians in delivering efficient, accurate care while reducing administrative burdens. Telehealth platforms leverage AI to enhance platform usability, patient interaction, and clinical effectiveness.

18.4.2 Data Integration and Analysis

This part of the work is very important as it helps gathering the diverse data sources, such as patient demographics, medical histories, wearable device data, and telehealth interactions. The collected data has been standardized, anonymized, and stored securely in compliance with HIPAA or other regulatory standards. The AI model has been implemented to analyze data in real time to offer personalized care insights, suggest interventions, or track health trends.

18.4.3 Personalized AI Models for Patient Engagement

Patient engagement is very important step and offers the good level of understanding and interpretation so that the problems of patients can easily be tracked and the course of action can be taken. NLP is used to facilitate seamless, conversational interactions between patient and AI systems, whether through chatbots or voice assistants. The developed AI tools easily predict health trends and outcomes, providing proactive health advice tailored to the individual patient's needs. AI has been used to provide tailored nudges to patients for maintaining engagement in their health journey, whether it's medication reminders, exercise encouragement, or diet adjustments. The sentiment analysis is a tool to gauge patient emotions during telehealth consultations and offer insights to clinicians for improving patient care.

18.4.4 Improving Telehealth Service Efficiency

Enhancing the technological infrastructure, such as fast internet and intuitive platforms, is necessary to increase the efficiency of telehealth services and guarantee seamless consultations. It also entails maximizing accessibility by setting up appointments, being available around the clock, and installing telehealth kiosks in underprivileged regions. Processes can be streamlined through data integration with

AI for triage and diagnosis and Electronic Health Records (EHR). Appropriate training helps healthcare providers, and education and follow-up care improve patient engagement. Automating administrative chores is possible, and data analytics and ongoing feedback aid in service improvement. A safe, convenient, and successful telehealth experience depends on maintaining privacy and regulatory compliance.

18.4.5 Enhancing Clinical Decision Support

The machine learning algorithm has been used to analyze diagnostic images, lab results, and patient histories, supporting healthcare professionals in making quicker, more accurate decisions. The AI systems act as the decision support system that recommend treatment plans based on a patient's unique characteristics, history, and evidence-based guidelines, improving clinical decision-making. The risk prediction model predicts potential health risks such as heart disease, diabetes, or mental health crises, allowing healthcare providers to intervene early.

18.4.6 Human-Centered Design and User Experience

The interface design is the important step toward ensuring the AI tools and its application, whether for patients or healthcare providers, are user-friendly and intuitive to encourage adoption. AI platforms are accessible to all, including those with disabilities, by using voice recognition, visual cues, and text-to-speech features. The global or diverse populations ensure that the AI system is capable of supporting multiple languages and cultural nuances in patient interaction.

18.4.7 Continuous Learning and Feedback Loops

Real-time learning system enables AI models to learn and adapt based on real-time feedback from patients and clinicians. The system should improve over time by recognizing patterns in patient behavior and treatment outcomes. The monitoring systems have been set up to monitor the performance of AI tools, ensuring that they remain accurate, unbiased, and effective in improving health outcomes. Patient and provider feedback system regularly collect feedback from patients and healthcare providers to evaluate AI's effectiveness, usability, and areas for improvement.

18.4.8 Collaboration and Interdisciplinary Integration

The collaborative process establishes teams that include AI experts, clinicians, data scientists, behavioral psychologists, and healthcare administrators to ensure that AI solutions are clinically relevant, technically robust, and patient-centric. The partnerships and the collaboration with technology companies and startups specializing in AI help to stay at the forefront of innovation and bring new capabilities into telehealth solutions.

18.4.9 Ethical Considerations and Equity

Bias mitigation ensures that AI models are developed and tested for fairness, preventing bias in patient care or access to services. Transparency and accountability implement transparency in AI decision-making processes, allowing patients and clinicians to understand how AI-generated recommendations are made. The concept of equitable access design in AI tools ensures all patients, regardless of socioeconomic status or location, have equitable access to high-quality telehealth services.

18.4.10 Regulatory and Compliance Alignment

The adherence to standards ensures that AI applications comply with health regulations like HIPAA (US), GDPR (EU), and others relevant to patient privacy and data security. The AI governance implements governance frameworks that outline how AI is used, monitored, and adjusted, ensuring that all stakeholders are aligned and that ethical concerns are addressed.

The outcome of such concept has been multifaceted in the enhanced patient satisfaction where the accessibility and personalization leading to the more positive and intense experiences. The timely treatement preocess can ensure the betterment of the health cpondition overall. The reduction of clinician's workload and the prompt and streamline service delivery can pprovide the smooth system flow. The scalability future of the AI helps in managing the increased demands, covering the rural and undeserved areas.

18.5 Case Study

In the following section, two related case studies have been presented.

18.5.1 AI-Driven Virtual Health Assistant for Chronic Disease Management

18.5.1.1 Objective

The improvement of patient treatment planning for the case of chronic disease with AI-powered system.

Implementation Procedure

The healthcare system can use the AI-enabled healthcare system for providing the personalized healthcare services, medication remainders, and other lifestyle directions.

The use of NLP helps patients to be engaged in the two-way communication with the health expert systems and parallelly applies the machine learning algorithms to analyze the health records and to extract different patient behavior.

Benefits of the Systems

The statistical data gathering in this respect shows considerable improvement in the results. The patient adherence to the system has increased by 30%. The patient engagements with the healthcare assistants have increased to 85% with regularized interaction pattern. There has been an improvement in the resource optimization with sharp falls in in-person follow-ups and appointments. NLP enabled process for sentiment analysis in support of the mental health.

18.5.1.2 Objective

The enablement of the mental health of patients of diverse categories with the help of communication patterns.

Implementation Procedure

The mental patients can suffer from different types illness like the diseased people cause the concept of NLP to interpret signs of the depressions or any other anxieties and stress. The implementation of sentiment analysis algorithms concerned to different mental health patterns and proper recommendation systems for the purpose of therapy sessions or other kind of self-helps.

Benefits of the System The statistical data gathering and its verification propound some good results of improvements. The early detection of mental health issues has been increased by 25%. Patients expressed the higher trusts for the empathetic and responsive nature. The effective communication pattern can provide the more targeted and effective care according to the behavioral pattern of patients. The result of the NLP-based analysis of the patient notes has been provided in Fig. 18.1. The analysis shows the frequencies of the patient terms (Fig. 18.2).

The growth of the telemedicine is very prominently observed in the recent era all over the world. Different statistical estimation has been carried out on the available data to find out the growth dynamics of the trends. The sudden growth pattern of the telemedicine usage among different people has been influenced heavily by several factors. The chief contribution comes from the factors like live video, live telephone, email, different health apps, texture message, and picture videos. The digital literacy among people across the globe has mitigated the problem of remote medicine distribution. Figure 18.3 shows the statistical analysis of global telehealth data.

The percentage fluctuation of data has become prominent from the presentation with the live phone-based result estimated at the maximum value. The result of

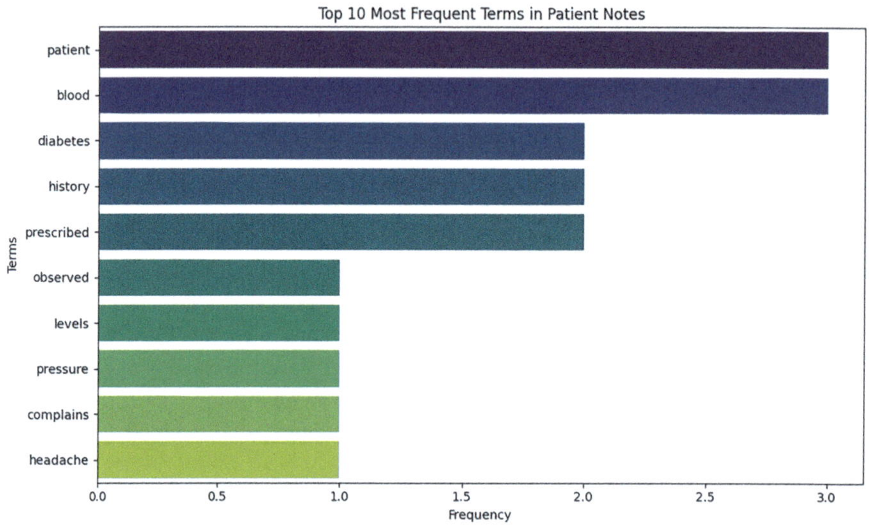

Fig. 18.1 Variation of patient notes with respect to the frequency values

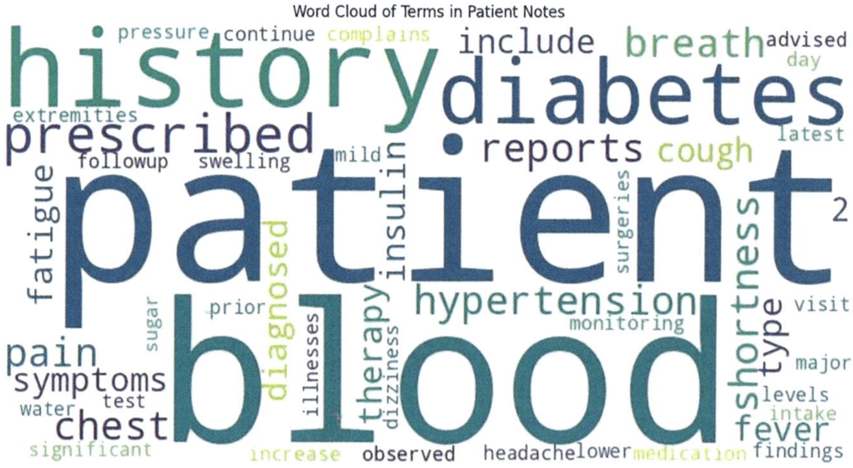

Fig. 18.2 Cloud version of the patient notes

comparative analysis for the live phone has been computed at the maximum level in the years 2018 and 2019.

The snippet of health system data has been presented in Table 18.1, where the detailed matrix of data has been represented in tabular form.

It is obvious from the tabular view that satisfaction level of telemedicine application has been projected with variable satisfaction level with the projected outcome. The good mix of outcome value presents the success rate of the satisfaction values.

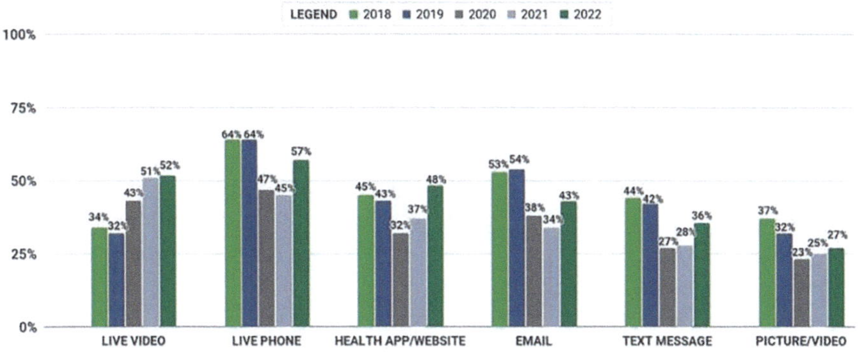

Fig. 18.3 Statistical display of telehealth data year-wise. (https://rockhealth.com/insights/consumer-adoption-of-digital-health-in-2022-moving-at-the-speed-of-trust/)

18.6 Conclusion

The rapid evolution of artificial intelligence (AI) presents unprecedented opportunities to enhance patient engagement and revolutionize telehealth services. This study underscores some important derivations. AI acts as a catalyst for the patient-centered care, where different AI aspects like AI-driven solutions, natural language processing, predictive analysis, and personalized recommendation engines have demonstrated their ability to improve patient communication, foster trust, and enable tailored care. By leveraging AI to deliver personalized experiences, telehealth platforms can achieve higher patient satisfaction and adherence to treatment plans. The seamless integration of AI into telehealth ecosystems significantly enhances service delivery. From real-time virtual assistants and symptom-checking tools to remote monitoring and risk stratification, AI offers robust support in addressing patient needs efficiently while reducing the burden on healthcare providers. Despite its potential application toward many different aspects of human lives in the modern civilized world, the deployment of AI in telehealth faces challenges, including data privacy concerns, algorithmic bias, and the need for robust regulatory frameworks. These issues call for transparent governance, stakeholder collaboration, and continuous refinement of AI algorithms to ensure ethical and equitable outcomes. The work culture trends thereof follow some future prospects, where the future of telehealth lies in the fusion of AI with other emerging technologies, such as the Internet of Things (IoT) and blockchain, to build comprehensive, secure, and adaptive healthcare solutions. Additionally, fostering interdisciplinary research and innovation will drive the development of novel AI paradigms that further optimize patient engagement and telehealth efficiency. The possibility of patient empowerment through AI has taken the flexible turn where the AI has taken the potential to shift paradigm of care from provider-centric to patient-centric by empowering individuals with tools to make informed decisions. Education, transparency, and intuitive design are crucial for ensuring that patients can effectively use AI-enabled

Table 18.1 The age- and gender-wise health condition with the prescribed satisfaction level and outcome

Age	Gender	Health condition	Session purpose	Session duration (min)	Satisfaction level	Preferred device	Internet speed (Mbps)	Outcome
78	Male	General checkup	Consultation	41	3	Tablet	25.69562787	Neutral
18	Other	Diabetes	Follow-up	38	5	Smartphone	48.2515913	Negative
38	Male	Respiratory issues	Consultation	42	5	Tablet	5.202591974	Neutral
70	Other	Respiratory issues	Consultation	53	1	Tablet	2.387673985	Positive
72	Other	Diabetes	Consultation	13	5	PC	40.60751756	Neutral
70	Female	Respiratory issues	Therapy	19	2	Tablet	35.08960266	Positive
66	Other	Diabetes	Follow-up	43	3	Tablet	49.52520885	Neutral

telehealth services. The prospects of the work open up new horizon toward more advancement of the robustness of the system.

18.7 Future Scope

Future developments in artificial intelligence (AI) will have a significant impact on telehealth services. The delivery of healthcare could be completely transformed by AI as technology advances, especially in terms of increasing patient interaction and telehealth service efficiency. Personalized patient care will be a key area of emphasis. AI is capable of producing highly individualized treatment plans by analyzing enormous volumes of patient data, such as genetic information, lifestyle choices, and medical histories. This will enable healthcare providers to meet the requirements of each individual and improve patient outcomes. AI-powered virtual assistants will be used to schedule appointments, issue reminders, and offer real-time support to aid patients in navigating their healthcare experiences. AI in preventive care will utilize predictive analytics to find trends in patient data and anticipate health issues. This will reduce long-term healthcare costs and enable early intervention for potential issues such as mental health issues, diabetes, or cardiovascular illnesses. AI will be integrated with remote monitoring systems to provide further continuous, real-time surveillance of patient health data. AI may be able to evaluate wearable data to identify trends and alert medical personnel to any anomalies, enabling timely interventions. AI-powered chatbots will offer a dependable and convenient channel of contact to patients in the mental health field who need emotional assistance. Moreover, AI will improve the efficiency of administrative tasks and free healthcare practitioners to focus more on patient care by utilizing robotic process automation (RPA) for tasks like scheduling, billing, and data management. In the impending years, AI will play a significant role in enhancing the efficiency, usability, and customization of telehealth services, which will make healthcare more patient-centered and cost-effective.

References

Ahmed MU, Ruskin HJ, Mahdi AE (2020) Machine learning for healthcare: advanced technologies and applications. Artif Intell Med 104:101842. https://doi.org/10.1016/j.artmed.2020.101842

Bickmore TW, Trinh H, Olafsson S, O'Leary TK, Asadi R, Rickles NM, & Cruz R (2018) Patient and consumer safety risks when using conversational assistants for medical information: An observational study of Siri, Alexa, and Google Assistant. Journal of Medical Internet Research, 20(9), e11510. https://doi.org/10.2196/11510

Bini SA (2018) Artificial intelligence, machine learning, deep learning, and cognitive computing: what do these terms mean, and how will they impact healthcare? J Arthroplast 33(8): 2358–2361. https://doi.org/10.1016/j.arth.2018.02.067

Bodnar T, Tucker R (2021) Predictive analytics in healthcare: the role of AI in patient engagement. Health Informatics J 27(1):111–123. https://doi.org/10.1177/1460458221101085

Cresswell K, Williams R, Sheikh A (2020) Developing and applying a formative evaluation framework for health information technology implementations: qualitative investigation. Journal of Medical Internet Research 22(6):e15068

Davenport T, Kalakota R (2019) The potential for AI in healthcare. Future Healthcare J 6(2):94–98. https://doi.org/10.7861/futurehosp.6-2-94

Delude C (2015) Deep phenotyping: the details of disease. Nature 527(7576):S14–S15. https://doi.org/10.1038/527S14a

Dilsizian SE, Siegel EL (2018) Artificial intelligence in medicine and healthcare: introduction and overview. J Nucl Med 59(9):1–6. https://doi.org/10.2967/jnumed.118.217432

Esteva A, Kuprel B, Novoa RA, Ko J, Swetter SM, Blau HM, Thrun S (2017) Dermatologist-level classification of skin cancer with deep neural networks. Nature 542(7639):115–118. https://doi.org/10.1038/nature21056

Fadhil A (2018) Beyond patient monitoring: conversational agents' role in telemedicine and healthcare support for home-living elderly individuals. arXiv preprint arXiv:1803.06000. https://arxiv.org/abs/1803.06000

Fitzpatrick KK, Darcy A, Vierhile M (2017) Delivering cognitive behavior therapy to young adults with symptoms of depression and anxiety using a fully automated conversational agent (Woebot): a randomized controlled trial. JMIR Mental Health 4(2):e7785

Godin G (2021) AI in primary care: a focus on patient engagement. J Prim Care Community Health 12:2150132721100242. https://doi.org/10.1177/21501327211002424

Guhan P, Awasthi N, McDonald K, Bussell K, Manocha D, Reeves G, Bera A (2020) Developing an effective and automated patient engagement estimator for telehealth: a machine learning approach. arXiv preprint arXiv:2011.08690. https://arxiv.org/abs/2011.08690

Khurshid A (2020) Applying blockchain technology to address the crisis of trust during the COVID-19 pandemic. JMIR Medical Informatics 8(9):e20477

Liu C, Jiang J, Xiong C, Yang Y, Ye J (2020) Towards building an intelligent chatbot for customer service: Learning to respond at the appropriate time. In Proceedings of the 26th ACM SIGKDD international conference on Knowledge Discovery & Data Mining (pp. 3377–3385)

Muench DE, Olsson A, Ferchen K, Pham G, Serafin RA, Chutipongtanate S, Grimes HL (2020) Mouse models of neutropenia reveal progenitor-stage-specific defects. Nature 582(7810):109–114

Rajkumar A, Oren E, Chen K, Dai AM, Hajaj N, Liu PJ, & Dean J (2018) Scalable and accurate deep learning for electronic health records. npj Digital Medicine, 1(18). https://doi.org/10.1038/s41746-018-0029

Roy K, Khandelwal V, Goswami R, Dolbir N, Malekar J, Sheth A (2023) Demo alleviate: demonstrating artificial intelligence enabled virtual assistance for telehealth: the mental health case. arXiv preprint arXiv:2304.00025. https://arxiv.org/abs/2304.00025

Sood M, Sharma S, Sood I, Sharma K, Kaushik A (2021) Emerging evidence on multisystem inflammatory syndrome in children associated with SARS-CoV-2 infection: a systematic review with meta-analysis. SN Comprehensive Clinical Medicine 3(1):38–47

Topol EJ (2019) High-performance medicine: the convergence of human and artificial intelligence. Nature Medicine 25(1):44–56

Wen B, Norel R, Liu J, Stappenbeck T, Zulkernine F, Chen H (2024) Leveraging large language models for patient engagement: the power of conversational AI in digital health. arXiv preprint arXiv:2406.13659. https://arxiv.org/abs/2406.13659

Ethical and Legal Considerations in Leveraging AI and Blockchain for Equitable Healthcare Access for Disabled Populations

19

Shashwata Sahu, Navonita Mallick, Ramesh Chandra Sethi, and Prativa Panda

Abstract

Artificial intelligence (AI) and blockchain have revolutionized healthcare in many aspects, such as precise diagnosis, tailored treatments, and streamlined data management. However, their adoption raises some profound ethical and legal questions, particularly around ensuring that marginalized groups like those with disabilities have equal access to the technologies. This chapter explores the possibilities for how such technologies might meet the health needs of people with disabilities while also implicating the challenges that may emerge. This chapter addresses the ethical-legal concerns about equity for individuals with different abilities through AI and blockchain. The research is meant to shed light on barriers to and opportunities for equitably deploying these technologies. This study employs a doctrinal approach consisting of a literature review, case identifications, and policy papers to analyze current legal frameworks, policies, and ethical guidelines. The interview touches on issues such as AI algorithmic bias and the problem of consent and data ownership about blockchain. It also evaluates how combining AI and blockchain could force health empowerment and informed consent among patients. The findings indicate that such technologies benefit disabled people's access to health promotion; these physical and ethical concerns of fairness and data protection need addressing. Enhancing existing laws is necessary to protect the rights of people with

S. Sahu (✉) · N. Mallick · P. Panda
School of Law, KIIT-DU, Bhubaneswar, India

R. C. Sethi
Department of Social Sciences and Humanities, School of Law, Kalinga Institute of Industrial Technology (KIIT), Deemed to be University, Bhubaneswar, India

J. M. Chatterjee, S. K. Saxena (eds.), *Artificial Intelligence and Blockchain in Precision Medicine and Virology*, Medical Virology: From Pathogenesis to Disease Control, https://doi.org/10.1007/978-981-96-8919-4_19

disabilities. Policies to equalize access to care must be driven by strong policies and careful design.

Keywords

AI in healthcare · Blockchain · Disabled populations · Ethical considerations · Legal frameworks

19.1 Introduction

The measure of a society is how it treats its most vulnerable members.
—Mahatma Gandhi

Without a doubt, among all the technologies of the twenty-first century, artificial intelligence (AI) and blockchain are the two at the forefront of transformation, and if they time together and combine, they could turn healthcare on its head. It has whisked up precision diagnostics, tailor-made therapy plans, resource allocation, and patient care, with phenomenal improvements in outcomes (Tagde et al. 2021). Blockchain enables secure data management, increases transparency, and speeds up consent processes, resulting in better healthcare systems. This combination would allow for inclusive healthcare solutions catering to diverse populations (Li et al. 2024). However, these improvements mean little if they are not equitable; disabled groups remain disproportionately underserved in the healthcare landscape. Many disabled people may already have trouble getting quality healthcare due to physical inaccessibility, personalized treatment, and data privacy issues. According to WHO, over 15% of humanity lives with some disability, and the majority of them are discriminated against and neglected in the provision of health services (Clemente et al. 2022). Technological solutions such as AI and blockchain would perpetuate these biases and become a part of the biased decision-making processes or a design that would overlook the healthcare needs of differently-abled populations if they are not designed inclusively. It is essential to explore how these technologies can be leveraged ethically and legally to advance equity in healthcare and overcome these challenges.

The ethical and legal issues around AI and blockchain in healthcare must be addressed to ensure that these technologies can benefit all people, including those with disabilities. This is especially pronounced in the case of AI algorithms that replicate the bias present in the training data and subsequently make discriminatory decisions. Blockchain is a fascinating, decentralized, and secure data technology (Farhud and Zokaei 2021). However, when personal health information is at stake, problems with data ownership and consent arise. These critical points of discussion as this study attempts to address both of these gaps by examining ways to responsibly implement these technologies and ensure disabled populations are not left behind in the digital health transformation. This is mainly related to SDG 3, which provides healthy lives and promotes well-being for all ages, and SDG 10, which reduces inequalities. Disabled populations are among the most hard-pressed demographics, and their inclusion in these technologies will help them reach many global goals (Guégan et al. 2018). Thus, by describing actionable insights to develop the ethical and legal frameworks that make this equitable access possible, this study can guide policymakers, healthcare providers, and health technology developers to bridge these gaps between people's assets and their health. Even though AI and blockchain have rapidly penetrated healthcare, unique frameworks addressing the needs of this subset population are sorely lacking. Health and disability studies clash here, with a few studies noting a disconnect between the hype and questioned applications in healthcare. At the same time, other scholarship remains primarily technical advancements or general ethical questions without zeroing in on how these technologies and ethical instructions layer over existing, disability-focused healthcare practices (Alowais et al. 2023). We also observed that legal safeguards for disabled persons in AI-based healthcare systems are insufficient, especially in developing economies. This study fills some of that gap by examining the equity barriers and providing an integrated ethical & legal framework.

These problems have been underscored in recent reporting. A recent WHO report states that over 1 billion people around the world live with some form of disability, and a significant proportion of them have no access to primary healthcare services. According to the United Nations Development Programme (UNDP), thanks to systemic barriers such as cost, physical inaccessibility, and discriminatory practices, disabled people face significant and unequal health disparities (WHO 2023). The Rights of Persons with Disabilities Act of 2016 in India mandates equal access to health care but continues to be insufficiently implemented, particularly regarding technology-driven systems where access to technology programs is required. According to a McKinsey report, the global AI healthcare market decided to overlook disabled populations, although they expect disabled populations to constitute $187 billion by 2030 collectively (Osman and Sun 2023). Similarly, in Deloitte's 2024 survey on blockchain in healthcare, these challenges around data ownership and consent management disproportionately impact underserved populations, including those with disabilities (Deloitte 2024). Thus, these insights highlight the importance of having ethical and legal frameworks to inform the use of AI and blockchain in healthcare systems and include marginalized communities of people with disabilities.

19.2 Research Objectives

The primary aim of this chapter is to focus on the ethical and legal issues concerning the specific needs of people with disabilities who use blockchain technologies to maintain their health status and information about their health. It aims to study how these technologies can enhance access to and equity in health care and the potential ethical challenges such as bias in AI algorithms, privacy, and informed consent issues. The study also evaluates the adequacy of existing legal frameworks and the need for new policies to ensure disabled individuals' rights are safeguarded when healthcare systems utilize AI. Thus, the research envisions possible barriers to equitable service provision and recommends an ethical and legal corporate healthcare delivery system to ascertain lucidness among underprivileged segments of the population.

19.3 Research Methodology

This chapter takes a doctrinal approach and considers existing laws, rules, and ethical principles, where relevant, addressing the use of AI and blockchain in the healthcare sector. Through literature review, case studies, and qualitative analysis of legal documents and policy papers, the research documents the applications of these technologies, the risks, and the legal frameworks designed to protect vulnerable groups.

19.4 Conceptualization and Theoretical Framework

Through diagnosis, treatment, and data management, AI and blockchain are changing the face of healthcare. This ensures quality data and delivers solutions tailor-made through AI and secure data storage through consent transparency and decentralized control of the blockchain. Together, they address significant inclusion and data ownership issues, empowering people with disabilities. Aligned with ethical structures surrounding fair, autonomous, and beneficent practices in applying these technologies, underpinned by rights-based legal structures, the intersection of AI and blockchain can help fill the accessibility gap by incorporating these aspects in developing equitable, inclusive healthcare infrastructures.

19.4.1 AI in Healthcare

Since AI was discovered, it has kept upgrading healthcare, creating something that the twenty-second century can only dream of. AI applications include precision medicine, diagnostics, treatment planning, and predictive analytics. AI-based systems use complex algorithms that exploit large volumes of medical data, including a patient's medical history, genetic information, and clinical imaging, to produce high-accuracy diagnoses and personalized options for treatment (Johnson et al. 2021). Machine learning algorithms can, for instance, predict the course of a disease or identify patterns in medical imagery with unmatched precision. But these advances also pose challenges. A core issue is the diversity of AI systems. Many AI training datasets use data from models that do not include a diverse cross section of the population, particularly people with disabilities. When these individuals have no voice, institutional decisions are made with bias, disenfranchizing marginalized populations. AI-assisted diagnostic devices could fail to recognize specific diseases common within marginalized populations or misinterpret medical data because the algorithms on which the diagnostic devices are based incorporate little or no information about particular disabilities (Khalifa and Albadawy 2024). Jargon aims to combine precision medicine with digital empowerment flows and invites us to the potential of tailoring our healthcare to our precise needs. AI has also been embraced by precision medicine, in which personal genetic, environmental, and lifestyle characteristics are recognized to be of excellent value in disabled populations with complex multiple medical needs (Shah and Heshmat 2024). Moreover, with faster and more accurate evaluations, AI-based diagnostics have the potential to fill gaps in access to healthcare. However, we must address AI systems' lack of inclusiveness to enjoy those benefits. To demonstrate how to use AI ethically in medicine, we need to deploy big algorithms trained in ways that can minimize bias and promote fairness.

19.4.2 Blockchain in Healthcare

The medical world needs blockchain technology coverage for better data security, consent management, and transparency challenges. Fundamentally, blockchain represents distributed ledger technology that enables the recording of transactions among the participants in a decentralized network by allowing them to establish a secure, unalterable ledger. This technology is decentralized and is very appropriate for health information management to avoid a central governing authority. Some of the main applications of blockchain in healthcare are related to ensuring data consistency (Haleem et al. 2021). Cryptographic mechanisms used in blockchain ensure that patients' records are immovable and thus unable to be breached. This is an essential factor for disabled patients, who are often restricted from accessing health services because of institutional barriers and are more vulnerable to breaches of privacy compared to others. Additionally, blockchain technology encourages efficient management by using smart contracts and self-executing contracts within the structure of the blockchain. The execution of these contracts will enable patients to gain more control over their medical data, thus leading to greater openness and trust in the healthcare system (Andrew et al. 2023). Another characteristic of the blockchain technology is its ability to produce immutable logs. This may prove helpful in continued care delivery to at-risk populations as it affords an all-inclusive permanent record of medical engagements. Blockchain technology comes with several challenges, such as integration into traditional systems, scalability issues, and literacy needed to operate these platforms (Singh et al. 2023). Nevertheless, these are surmountable challenges that have opened promising new opportunities to empower people with disabilities through blockchain, either by securing data or by improving their access and control over their care.

19.4.3 Ethical and Legal Theories

Integrating new technologies like AI and blockchain into the health sector opens new ethical and legal questions in health care to resolve, primarily including fairness, autonomy, and beneficence. Moral principles act as guidelines when resolving such issues, allowing for new technology that reflects human values (Fig. 19.1).

In addition, equity in healthcare technology must be developed impartially and equitably. With AI, that means developing algorithms that are cognizant that different populations—people with disabilities—will use and engage with tech and that we are not erratically replicating exclusion. Blockchain is a little too transparent technology for its implementation to block certain groups from access due to technical or infrastructural limits, as it sits on the bottom end of this spectrum (Gurevich et al. 2023). Autonomy, the moral referencing of the ability to make choices regarding their care, is another moral agent to emulate. Blockchain makes it possible to create consent management systems that point out who has limited access to their data and which data they can access. Similarly, both AI systems need to have the ability to augment patient autonomy instead of impinging upon it by providing information that fosters decision-making while not compelling those who wish to avoid such action. The second principle is beneficence—the duty of health care providers and technologists to do good at all costs (Funer and Wiesing 2024). Such a principle is especially relevant for disabled groups who are continuously subjected to systemic neglect. This should be in tandem with deploying AI and blockchain technologies that better the accessibility and outcomes of healthcare for such groups (Fig. 19.2).

From the legal dimension, a vital role of a human rights-based framework is the protection of interests as the healthcare system begins to be increasingly influenced by technology. This will ensure autonomy and privacy rights for individuals in a society where informed consent and fair access to healthcare services are ensured.

Fig. 19.1 Ethical and legal considerations in the integration of AI and blockchain in healthcare

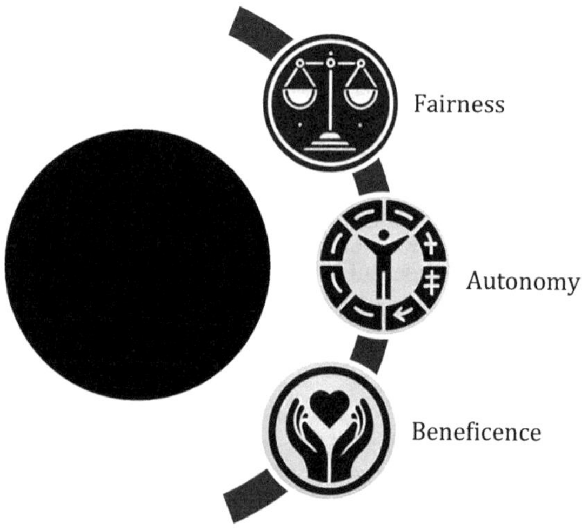

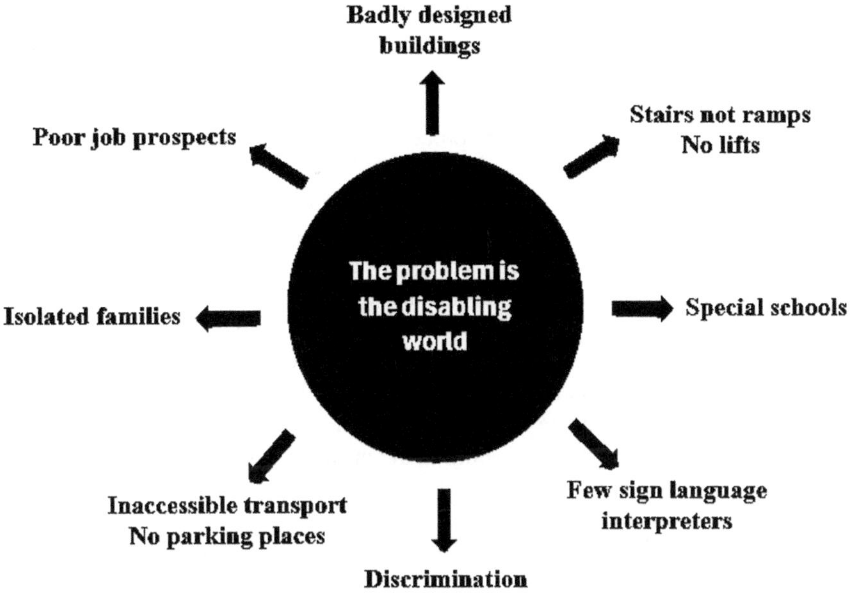

Fig. 19.2 Right-based approach

Current legal frameworks, such as the General Data Protection Regulation (GDPR) in the European Union and the Rights of Persons with Disabilities Act passed in 2016 in India, could provide an excellent framework for addressing many of the issues associated with these issues (Krupiy and Scheinin 2023). Such frameworks tend to have limited features explicitly designed for AI and blockchain, and many must introduce additional regulations to address their distinct challenges.

19.4.4 Intersection of AI and Blockchain

The convergence of both technologies has enormous transformative potential to help tackle health inequalities, particularly for disabled communities. Thus, blockchain can enhance AI's work by ensuring that processed data is immutable and verifiable. Together, these technologies can tap into a strong ecosystem for transforming access to and egalitarianism in healthcare. Thus, this intersection entails theoretically creating AI-engineered diagnostic systems with blockchain-postulated authorization approaches (Bhumichai et al. 2024). AI will analyze patient data and offer personalized treatment recommendations, for example. From a blockchain perspective, patients could still have control over how and in what cases their data are being used. This integration addresses many efficiency and privacy concerns often associated with tech-based healthcare solutions.

In addition, merging AI with a blockchain framework opens up additional inclusiveness. Blockchain's transparency can provide a way to solve biases in AI algorithms. This holds particularly true in healthcare due to the disparate impact biased algorithms can have on treatment outcomes. Utilizing AI and blockchain at the same time can serve to analyze the situation and prevent bad applications of AI, on the one hand, and hold the models accountable and transparent, on the other. In addition, by streamlining healthcare processes with AI and blockchain, administrative burdens and inefficiencies can be reduced. In operational terms, blockchain can potentially automate the soft side of insurance claims processing via clever contract mechanisms, while AI will process patient histories to predict their needs (Bathula et al. 2024). These can be useful to disabled people facing bureaucratic hurdles to access medical facilities. Many hurdles are yet to be overcome before the entire potential of AI is matched with that of blockchain. Widespread adoption also faces challenges, including the inability to scale due to interoperability across systems, the digital divide, and regulatory gaps. To address these problems, policymakers, technologists, and healthcare practitioners must develop frameworks that value diversity and inclusion while focusing on the needs of people with disabilities.

19.5 Ethical Considerations in Leveraging AI and Blockchain in Healthcare

As a result of the combined AI and blockchain technology, significant innovations are developing in healthcare, including high-precision rank-enabled diagnostics, effective data management, and delivery of personalized healthcare to patients. Although they have been beneficial in implementing those technologies, some ethical issues arise, especially as they are integrated into healthcare systems for marginalized people, such as people with disabilities. Therefore, this raises much debate on issues including algorithmic bias in AI, the call for transparency and accountability in the AI systems, and data ownership and privacy, among others. These must be addressed for fair and just health solutions to be produced.

19.5.1 Bias in AI Algorithms

AI models are trained with data and operate based on the data. But for these systems to be, this data must be high quality and representative. The most relevant ethical risk with AI in healthcare is using discriminatory models and algorithms, leading to discriminatory healthcare delivery. Bias is the easiest thing to train into systems when training datasets are unrepresentative and include data from underrepresented populations, such as disabled people (J. Rhem 2023). For example, an AI model could be developed based on data from non-disabled people. In such situations, the system can misdiagnose or mismanage conditions characteristic of that community. Such biases can increase already existing inequalities in healthcare delivery and end up delivering less complete or adequate care to disabled populations. There is a need to recognize several types of discriminatory health outcomes. Such tests with the involvement of AI might not be very effectively utilized for determining health conditions that are pretty prevalent in a given society. AI-based recommendations on treatment can also be discriminatory against people with disabilities because there is very little data available on such populations. Such impacts are severe and skewed and manifest in late diagnoses, wrong treatments, and a loss of confidence in the technology-based healthcare system (Vela et al. 2022). Several measures can be used to reduce bias. First, the development of AI systems should be made on varied and representative datasets that include information about both people with disabilities and from marginalized communities. A deep data collection will result in algorithms that can address the complete healthcare spectrum. Secondly, the auditing and continuous monitoring of AI systems can be done to rectify decisions by biases (Hanna et al. 2024). Such audits must be multi-professional and include input from healthcare professionals, technologists, and ethicists to conduct comprehensive assessments. Finally, making their algorithm development and decision-making transparent builds trust, responsibility, and the following essential concerns.

19.5.2 Transparency and Accountability

AI and blockchain in healthcare are presented with core ethical values of transparency and accountability. However, these technologies' complexity and often opacity make it challenging to realize such principles. AI algorithms, especially in deep learning, work as "black boxes," so their reasoning is opaque even to those who create them. This general lack of clarity can lead to particular difficulties in health care, where the stakes of medical decision-making are high. For example, patients and healthcare providers could struggle to comprehend how the recommendations provided by an AI system reach a specific treatment plan (Marey et al. 2024). Such vagueness may undermine trust and lead to resistance to the uptake of AI-driven healthcare solutions. For instance, as described above, E-voting through blockchain is transparent in data transactions; however, technical complexities are inherent in the blockchain itself and may not be comprehensible to non-technical domain users. Patients do not always understand how their data is stored, shared, or used, which can lead to concerns regarding misuse and exploitation. Various steps can be taken to solve these issues. For the first step, AI developers must emphasize explainability in their models, and as a result, algorithms should be able to present clear and easily interpretable reasons for their decisions. It also means designing easy-to-navigate interfaces that make the information accessible to patients and healthcare providers (Pereira et al. 2023). The second area will be making data flows via blockchain easier to comprehend by visualizing the transaction/action or simplifying the explanation of the data execution. Accountability systems must finally be implemented so that the developers and organizations deploying these technologies are held responsible for the ethical consequences. This may involve third-party oversight, certification processes, and adherence to ethical guidelines.

19.5.3 Data Ownership and Privacy

The ethical issues surrounding AI and blockchain in healthcare approaches are crucial. Such technologies rely on a pool of patient data, with AI learning from the pool while blockchain securely manages data. However, this data collection, storage, and use present serious ethical challenges considering informed consent, data security, and ownership. Informed consent is fundamental to ethical healthcare and empowers patients to determine how their data should be used. Nonetheless, realistic consent to AI and blockchain can be pretty tricky (Shinde et al. 2024). The patients cannot see how their data will be processed by AI algorithms or stored by blockchain systems. This ignorance can lead to patients unwittingly agreeing to practices with their information they'd never otherwise accept. Their data may, for example, be employed for commercial gain without their full awareness. Another key issue is data security. Blockchain is also not vulnerable. It is common to talk about immutable and tamper-proof data, but it is also impenetrable. Due to mismanagement of the private key, patient data may be compromised by cyberattacks, unauthorized access, and privacy breaches. Data breaches, coupled with the systemic barriers that disabled individuals already face while traversing healthcare, can, in some cases, exacerbate the outcomes, further marginalizing disabled individuals (Saeed et al. 2022). Here are some potential solutions to these ethical challenges. Overall, blockchain makes it easy for patients to manage their data. Firstly, blockchain systems offer consent management features that allow patients to determine who has access to their data and under what conditions. Any options users gain must be easily manageable, giving patients more control over their decision-making regarding their data. GPH and data protection help Strategies to support providers' interest in better data protection. Strategies to protect providers with data protection are, firstly, everyone can access data only after they log in to the servers, programs, and appliances; secondly, everyone should create a config from doctor to administrator; and thirdly, everyone must be identified before entering the system (Porsdam Mann et al. 2021). This should be followed by creating legal frameworks that define rights to data ownership and ensure that patients remain in control of their data.

19.6 Legal Framework and Gaps

Integrating AI and blockchain into healthcare systems provides new pathways for enhanced access and efficiency. But these technologies also raise highly herculean legal challenges, particularly in ensuring that vulnerable populations, such as people with disabilities, are treated equitably. They can help influence the legal frameworks critical to privacy and data protection, and that must be followed to respect academic guidelines to fight discrimination and uphold ethical principles in technology deployment. In this connection, this chapter first examines the existing legal framework and the deficient legal protection.

19.6.1 Current Legal Frameworks

Countries have unique laws meant to safeguard individuals' rights and their data and keep them private. Below is a summary of crucial rules governing the use of AI and blockchain in health care:

Table 19.1 summarizes the legal frameworks adopted across various countries concerning AI and blockchain functionalities in the healthcare sector. Although these legal frameworks provide necessary controls that govern data-related practices and ethical issues regarding data protection, privacy, and consent management, there are some significant weaknesses, especially for disabled populations, because of the risks and challenges these technologies present to these groups. For example, regulations like GDPR and HIPAA are majorly concerned with the security aspects of data but fail to account for AI biases or the decentralized nature of blockchain (Ducato 2020). This necessitates a comprehensive, unbiased review of all these technologies to facilitate their ethical and equitable deployment in various healthcare systems.

19.6.2 Gaps in Legal Protection

There are such legal systems, though the present state of law has many gaps that could be filled to deal with abstract legal and ethical issues behind AI and blockchain in care for populations of people with disabilities. The present gaps are only limited provisions due to algorithmic bias, the decentralized nature of blockchain, and a lack of a compact, right-based approach (Fig. 19.3).

So far, no laws do a good job of addressing bias in AI systems. As mentioned above, for instance, the GDPR increases transparency regarding data processing. However, data algorithmic audits cannot be required to detect or mitigate bias. This is concerning for the healthcare spectrum, as AI algorithms biased toward people living with disabilities could lead to further unequal treatment. Disabled people, for example, require adequate healthcare to address their specific needs. Partisan algorithms may not be programmed to address their needs, meaning they get an inconclusive diagnosis or inadequate medical services (Alvarez et al. 2024). To avoid those risks, stringent regulations, regular audits, and accountability from AI developers are necessary. By its very nature, blockchain technology presents its legal compliance challenges. Due to its decentralized architecture, data is not stored in any single location, creating multiple data storage nodes, which hinders the identification of any one party that can be held accountable. GDPR and HIPAA assume centralized data control, which does not align with how blockchain is structured. One of the examples is the GDPR's "right to be forgotten," which contradicts blockchain immutability, where data, once stored, cannot be wiped. This creates a legal gray area that existing frameworks cannot handle. The rights of disabled people, who might depend on blockchain to securely manage their data for healthcare purposes, are not enshrined in this legally binding framework. They are at particular risk (Han et al. 2023). Current legal safeguards in India for persons with disabilities, such as

Table 19.1 International regulations governing AI and blockchain applications in healthcare

Country/ region	Law/regulation	Key features	Relevance to AI and blockchain in healthcare
European Union (EU)	General Data Protection Regulation (GDPR)	Ensures data protection, privacy, and transparency; emphasizes informed consent and data portability	Applicable to AI and blockchain systems, requiring accountability and data security compliance
United States (USA)	Health Insurance Portability and Accountability Act (HIPAA)	Protects sensitive patient health information (PHI); mandates confidentiality and security measures	Provides data protection standards but does not fully address blockchain or AI-specific challenges
India	Data Protection And Digital Privacy Act (DPDPA), 2023	Regulates data protection, security, and user consent in digital systems; introduces penalties for violations	It provides a framework for blockchain and AI in digital healthcare but needs a better AI-specific focus
India	Rights of Persons with Disabilities Act (RPWD), 2016	Guarantees equal access to healthcare and rights for disabled individuals	It addresses accessibility but lacks specific provisions for AI and blockchain in healthcare
Singapore	Personal Data Protection Act (PDPA)	Focuses on data security and user consent for personal data collection and processing	Relevant to blockchain's decentralized data storage and AI'S data-driven operations
China	Personal Information Protection Law (PIPL)	Strict data protection law focuses on personal data collection, storage, and cross-border data sharing	Addresses healthcare data privacy but lacks AI-specific algorithmic transparency requirements
United Kingdom (UK)	Data Protection Act (DPA) 2018	It complements GDPR and focuses on data security, transparency, and user rights over personal data	Addresses blockchain's transparency and AI'S data-driven healthcare applications
Australia	Privacy Act 1988 (Amended in 2020)	Ensures privacy and security of personal health information; introduces mandatory data breach reporting	Relevant to AI and blockchain for maintaining transparency and accountability in healthcare data
Canada	Personal Information Protection and Electronic Documents Act (PIPEDA)	Regulates data collection, consent, and disclosure in commercial and healthcare settings	It provides a framework for blockchain's consent management but is limited in addressing AI biases
South Africa	Protection of Personal Information Act (POPIA)	Emphasizes privacy, security, and accountability in personal data processing	Valid for blockchain's secure data transactions and AI'S privacy-sensitive healthcare operations
Brazil	General Data Protection Law (LGPD)	Protects personal data and mandates transparency and accountability in data use	Encourages data security and consent, applicable to AI and blockchain in healthcare systems

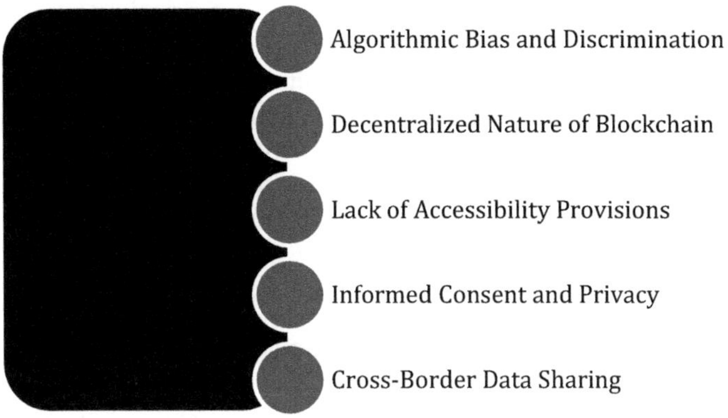

Fig. 19.3 Major gaps in legal protection

the RPWD Act of 2016, primarily emphasize accessibility in the physical realm, often neglecting accessibility in digital and technological spaces. Many systems, such as AI and blockchain, require some level of technical literacy or access to specific high-end devices, which, due to their disability, are unable to do and, therefore, unable to benefit from the advantages presented (Jain and Jain 2024). Modern legislation must catch up to the needs of our evolving world with inclusivity provisions for digital accessibility standards so that these technical solutions serve populations with disabilities as well. Such informed consent is a pillar of healthcare ethics, but adequately addressing the complexities of consent in the context of AI and blockchain systems goes beyond the current laws. Many programs trained on data through algorithms have indecipherable bits, which most people find challenging to empathize with as they do not know how their data will be used. Likewise, although blockchain ensures data security, it frequently has technical consent management systems that non-technical individuals do not interpret easily. These issues disproportionately impact disabled individuals who are already at a disadvantage when it comes to accessing healthcare (Dankar et al. 2019). Legal mechanisms must prioritize unburdening consent processes and ensure that patients know the implications of their decisions. Because of the decentralized nature of the blockchain, it enables cross-border data sharing, which can be instrumental in any global healthcare initiative. However, this also poses jurisdictional questions because such data may be subject to one and, at the same time, multiple and conflicting legal regimes. For example, suppose we have a blockchain system that stores patient data across nodes in various countries. In that case, for that system to comply with the law, it must adhere to the data protection laws and regulations for each such jurisdiction (Taherdoost 2023). Current frameworks such as GDPR are ineffective in resolving these tensions, and patients are left to negotiate a cumbersome legal framework that leaves them open to privacy violations.

19.7 Case Studies

In healthcare, transformative AI and blockchain technologies are possible solutions for disabled populations. These groups face unique challenges regarding receiving quick resolutions to their issues. This section has analyzed how these technologies work in practical, real-world scenarios and how they effectively resolve issues like accuracy in diagnosis, secure data transmission, and the scalability of healthcare systems.

19.7.1 AI in Diagnostics for Disabled Populations

AI has revolutionized diagnostics as it can diagnose disabilities and rare diseases that typically require specific knowledge and facilities early on. For instance, AI-based deep learning tools are used to analyze genetic and medical image data to diagnose congenital disorders or developmental disabilities in their children. One interesting example is Google's DeepMind, which developed a criterion model for interpreting retinal scans. For instance, the technology has already been applied to the diagnosis of early diabetic retinopathy; this is a potentially blinding condition that is diagnosed in many patients who receive a diagnosis of diabetes (Hasani et al. 2022). Nevertheless, if this problem is approached early enough, timely intervention could reduce the prospects of severe disability. Similarly, AI models that are being developed to identify behavioral signs suggestive of ASD can be helpful in the early diagnosis of the disorder by doctors. Nevertheless, prompt and precise diagnostic assessments can potentially improve health outcomes and facilitate an enhanced quality of life for individuals with disabilities (Okoye et al. 2023). It may be imperative to ensure the representation of various demographics or disabilities that could be omitted from training datasets to mitigate biases associated with the comprehensive application of these systems.

19.7.2 Blockchain for Consent Management

If implemented in patient data management, it could enable data security, transparency, and consent. There may be much potential in the most essential and promising use case of empowering patients, especially people with disabilities, in control of their health data, access, and the terms on which access is permitted. It has already proven to be a practical concept successfully practised through MIT's MedRec project. An example of a blockchain project for this is MedRec, which stores patient medical history on an immutable blockchain but allows access permissions to the patient (Cumyn et al. 2023). This will provide data security and empower patients as they have more control over their healthcare decisions. For disabled people who generally have complex medical histories and engage frequently with multiple service providers, this type of system decreases the possibilities of miscommunication and problems with data management. Another area in which blockchain might solve an issue is informed and correct consent. The smart contracts allow for encoding conditions under which an entity can use the patient data, ensuring that all data usage aspects meet the necessary ethical and legal requirements. (Marbouh et al. 2022). For example, a cognitively disabled patient may want to have a trusted guardian or health care provider as the custodian of her consent settings, in which case the data will be used according to her desires.

19.7.3 Integrated AI and Blockchain Solutions

This powerful combination creates a seamless ecosystem of integrity, transparency, and inclusivity inherent in the systemic nature of security and privacy in blockchain that addresses all shortcomings of healthcare solutions. These technologies are integrated to deliver personalized healthcare alongside privacy and data transparency. Use of blockchain in the healthcare industry due to this integration, IBM Watson Health is working with blockchain platforms for clinical trial management. Used to mine through patient information to find potential patients and match them to relevant clinical trials, blockchain keeps the information shared between parties secure and tamper-proof (Maslove et al. 2018). That's particularly useful for disabled individuals, who have historically been underrepresented in clinical studies. Concerns such as these ensure, in combination with the system, that their data will be treated ethically and, more importantly, that they will be equally represented in research that may help their conditions. Another example is the e-Health system in Estonia, where blockchain technology is used to store electronic medical records of patients in a secure manner while AI algorithms mine that data and share back predictions. This has opened the door for personalized treatment plans and improved patient outcomes. These systems allow seamless access between disability patients and all providers of healthcare services while minimizing duplication of effort and relieving the administrative burden. AI and blockchain are also exploring synergies in remote healthcare delivery. One such example is AI-based telemedicine platforms that use blockchain technology to verify the identity of patients and healthcare providers, thus securing their communication (Maher et al. 2023). These systems are particularly effective for disabled individuals, as those existing in remote or underserved places can use quality healthcare without traveling a long distance.

19.8 Findings and Recommendations

There is some hope concerning implementing AI and blockchain technology in the health system to improve access and outcomes for disabled individuals. However, this research reveals that critical ethical and legal gaps need to be filled if this potential is to be realized. The importance of this study lies in sunscreen innovation. Still, one of the challenges is the potential bias in AI systems, particularly in cases where the datasets used for building training do not represent all varieties of divergence found in disabled corporations. That leads to algorithms that miss the nuance required to properly assess and serve the medical needs of these groups, particularly worsening health disparities. Data design without the breadth of human assumptions can also lead to incorrect or strict diagnostics, leading to people receiving incorrect treatments or sub-optimal treatment in the first place, and that is where traditional analytics is also in a dire state (Olawade et al. 2025). Blockchain provides a fantastic insight into the federation of data security and consent. While blockchain technology offers unmatchable transparency and immutable data storage, its implementation and representation in public exits are not user-centered enough to provide room for people with disabilities. This population may also find complicated interfaces or technical terminology in a complex system, making it less accessible.

It also points out lapses in the regulation of these technologies. Existing laws, such as the General Data Protection Regulation (GDPR) and the Health Insurance Portability and Accountability Act (HIPAA), protect against data protection and privacy. However, they may not fully encompass the idiosyncrasies of AI algorithms or the decentralized nature of blockchain. Similarly, laws like India's Rights of Persons with Disabilities Act, 2016, where the focus is on physical accessibility, fail to recognize the existence of digital and technological barriers that disabled populations face (Edemekong et al. 2025). Nonetheless, these challenges underscore the burgeoning transformative potential of inclusive tech to impact healthcare equity positively. When inclusive, they further personalize diagnostics and treatment plans relevant to a disabled population. The data privacy and permission administering features of blockchain give patients the option to manage their privacy in a way that ensures the decentralized ownership of their health.

Here are the recommendations to bridge the identified gaps to promote the equitable deployment of AI and blockchain technologies in healthcare:

1. Inclusive Algorithm Design: From step one, algorithm design should still be within focus for AI developers. This entails building heterogeneous datasets such as persons with disabilities so that biases do not exist inclusively. There is a need for routine audits that may detect and rectify discriminatory patterns within the decision-making process. Implementation of explainable artificial intelligence, XAI, is also in need whereby patients and health providers can understand why the AI-based decision was reached, thereby providing for both trustworthiness and accountability (Schwartz et al. 2022).
2. Build strong legal frameworks: Existing legal frameworks must evolve to reflect the unique issues thrown up by AI and blockchain. Governments and regulatory bodies need to enact laws demanding that there is documentation of algorithmic fairness and that makers of advanced systems publish their training data in addition to the methodology used. In the case of blockchain, regulation should focus on data ownership, cross-border data flow, immutability, and balancing transparency with privacy. Based on the skill context, some existing laws, such as the GDPR, can be modified to contain clauses addressing the role of AI and blockchain in healthcare while appropriately protecting disabled populations.
3. Policy Recommendations for Ethical Principles in Technology Governance: Integrating ethical principles such as fairness, autonomy, and beneficence into

the ongoing policies governing the deployment of AI and blockchain. However, whatever new technologies the future brings must be user-friendly and accessible to disabled people, and policymakers should set standards to make that happen. Blockchain platforms might, for instance, provide disability, simple assistive technologies, or other types of disability advantages. Then, policies should encourage the broad development of healthcare technologies that specifically address the needs of marginalized groups, e.g., grants and subsidies for inclusive innovation.

4. Involves Collaboration Between Stakeholders: There is more to the successful integration of these technologies than just design; it is also a matter of collaboration between various stakeholders, including technologists, healthcare providers, policymakers, and people with disabilities themselves. Creating multi-disciplinary committees to regulate the ethical application of AI and blockchain would ensure that varied perspectives are accounted for. This approach should include the participation of disabled people in the design and testing stages of such technologies to meet their needs.
5. Capacity Building and Education: We also need to educate healthcare providers and patients about the benefits and risks of these technologies. Along with the training, awareness programs for patients regarding AI or blockchain tools to be used by healthcare professionals will create understanding and trust for the healthcare facility provider. Thus, the whole process will promote the acceptance of AI or blockchain in the healthcare sector. For disabled populations, such initiatives need to be made considering their requirements and using means for communication or formats that are easily accessible.

This research is primarily doctrinal and based on literature and case analysis to identify where ethical and legal gaps can exist. Although this method can glean critical theoretical insights, it does so without data to support its conclusions as they play out in practice. Future studies should collect primary data from those with disabilities and their healthcare providers to examine first-hand perspectives and experiences.

19.9 Conclusion

Innovation is the ability to see change as an opportunity – not a threat.
—Steve Jobs

Specifically, this study provides an example of how technologies, such as artificial intelligence (AI) and blockchain, that are disruptive in their very nature can disrupt healthcare and ensure healthy lives, particularly for people with disabilities. The main findings suggest profound ethical and legal deficiencies that impede the equitable implementation of these technologies. Some significant challenges revolved around AI algorithm bias, blockchain accessibility, and the inadequacy of the legal frameworks (Junaid et al. 2022). Despite these barriers, the study demonstrates the potential of inclusive technologies to enhance health equity by providing personalized diagnostics, enabling data management safety, minimizing data access constraints, and enabling informed consent.

Ultimately, ethical and legal considerations related to data ascension should be superior to decisions that impact data use. In this regard, the advantages of data ascension should be accessible to all without aggravating the present inequalities. Fairness, autonomy, and beneficence are basic values that can guide the development and use of AI and blockchain systems. This would indicate that laws and regulatory regimes should be updated as new questions and issues emerge regarding technological progress, like algorithmic bias, data ownership, and cross-jurisdictional data transfer. When ethics are built into enforceable legal frameworks, these technologies can be used in a way that is clear, fair, and responsive to societal needs (Gold Nmesoma Okorie et al. 2024). Closing the accessibility gap for people with disabilities requires inclusive and transparent methodologies in the endeavor. This objective calls for the dedication of policymakers, technology experts, healthcare professionals, and individuals with disabilities. This is essential to fair healthcare systems that ensure justice by creating equal AI algorithms, doing work with the simplistic UI of Blockchain, and protecting these vulnerable groups through law. While these insights are purely theoretical, my empirical work will test these insights in future research. Working with disabled people in participatory research and investigating region-specific issues will be invaluable in shaping inclusive technology (Bekbolatova et al. 2024). Moreover, understanding the long-term social and economic implications of AI and blockchain in healthcare can feed into the long-term refinement of policies and practices. Overcoming these obstacles will allow the global healthcare ecosystem to embrace these technologies while addressing the challenges associated with equity, inclusiveness, and ethical governance.

References

Alowais SA, Alghamdi SS, Alsuhebany N, Alqahtani T, Alshaya AI, Almohareb SN, Aldairem A, Alrashed M, Bin Saleh K, Badreldin HA, Al Yami MS, Al Harbi S, Albekairy AM (2023) Revolutionizing healthcare: the role of artificial intelligence in clinical practice. BMC Med Educ 23(1):689. https://doi.org/10.1186/s12909-023-04698-z

Alvarez JM, Colmenarejo AB, Elobaid A, Fabbrizzi S, Fahimi M, Ferrara A, Ghodsi S, Mougan C, Papageorgiou I, Reyero P, Russo M, Scott KM, State L, Zhao X, Ruggieri S (2024) Policy advice and best practices on bias and fairness in AI. Ethics Inf Technol 26(2):31. https://doi.org/10.1007/s10676-024-09746-w

Andrew J, Isravel DP, Sagayam KM, Bhushan B, Sei Y, Eunice J (2023) Blockchain for healthcare systems: architecture, security challenges, trends and future directions. J Netw Comput Appl 215:103633. https://doi.org/10.1016/j.jnca.2023.103633

Bathula A, Gupta SK, Merugu S, Saba L, Khanna NN, Laird JR, Sanagala SS, Singh R, Garg D, Fouda MM, Suri JS (2024) Blockchain, artificial intelligence, and healthcare: the tripod of future—a narrative review. Artif Intell Rev 57(9):238. https://doi.org/10.1007/s10462-024-10873-5

Bekbolatova M, Mayer J, Ong CW, Toma M (2024) Transformative potential of AI in healthcare: definitions, applications, and navigating the ethical landscape and public perspectives. Healthcare 12(2):125. https://doi.org/10.3390/healthcare12020125

Bhumichai D, Smiliotopoulos C, Benton R, Kambourakis G, Damopoulos D (2024) The convergence of artificial intelligence and Blockchain: the State of play and the road ahead. Information 15(5):268. https://doi.org/10.3390/info15050268

Clemente KAP, Silva SVD, Vieira GI, Bortoli MCD, Toma TS, Ramos VD, Brito CMMD (2022) Barreiras ao acesso das pessoas com deficiência aos serviços de saúde: Uma revisão de escopo. Rev Saude Publica 56:64. https://doi.org/10.11606/s1518-8787.2022056003893

Cumyn A, Ménard J-F, Barton A, Dault R, Lévesque F, Ethier J-F (2023) Patients' and members of the public's wishes regarding transparency in the context of secondary use of health data: scoping review. J Med Internet Res 25:e45002. https://doi.org/10.2196/45002

Dankar FK, Gergely M, Dankar SK (2019) Informed consent in biomedical research. Comput Struct Biotechnol J 17:463–474. https://doi.org/10.1016/j.csbj.2019.03.010

Deloitte (2024) Deloitte's first Disability Inclusion @ Work 2024 survey reveals that workplace accessibility is a significant challenge for many | Deloitte Global. https://www.deloitte.com/global/en/about/press-room/deloittes-first-disability-inclusion-work-2024.html

Ducato R (2020) Data protection, scientific research, and the role of information. Comput Law Secur Rev 37:105412. https://doi.org/10.1016/j.clsr.2020.105412

Edemekong PF, Annamaraju P, Afzal M, Haydel MJ (2025) Health insurance portability and accountability act (HIPAA) compliance. In: StatPearls. StatPearls Publishing. http://www.ncbi.nlm.nih.gov/books/NBK500019/

Farhud DD, Zokaei S (2021) Ethical issues of artificial intelligence in medicine and healthcare. Iranian J Public Health 50. https://doi.org/10.18502/ijph.v50i11.7600

Funer F, Wiesing U (2024) Physician's autonomy in the face of AI support: walking the ethical tightrope. Front Med 11:1324963. https://doi.org/10.3389/fmed.2024.1324963

Guégan J-F, Suzán G, Kati-Coulibaly S, Bonpamgue DN, Moatti J-P, Guégan J-F, Suzán G, Kati-Coulibaly S, Bonpamgue DN, Moatti J-P (2018) Sustainable development goal #3, "health and well-being", and the need for more integrative thinking. Veterinaria México OA 5(2). https://doi.org/10.21753/vmoa.5.2.443

Gurevich E, El Hassan B, El Morr C (2023) Equity within AI systems: what can health leaders expect? Healthc Manage Forum 36(2):119–124. https://doi.org/10.1177/08404704221125368

Haleem A, Javaid M, Singh RP, Suman R, Rab S (2021) Blockchain technology applications in healthcare: an overview. Int J Intellig Netw 2:130–139. https://doi.org/10.1016/j.ijin.2021.09.005

Han H, Shiwakoti RK, Jarvis R, Mordi C, Botchie D (2023) Accounting and auditing with blockchain technology and artificial intelligence: a literature review. Int J Account Inf Syst 48:100598. https://doi.org/10.1016/j.accinf.2022.100598

Hanna M, Pantanowitz L, Jackson B, Palmer O, Visweswaran S, Pantanowitz J, Deebajah M, Rashidi H (2024) Ethical and bias considerations in artificial intelligence (AI)/machine learning. Mod Pathol 38:100686. https://doi.org/10.1016/j.modpat.2024.100686

Hasani N, Farhadi F, Morris MA, Nikpanah M, Rahmim A, Xu Y, Pariser A, Collins MT, Summers RM, Jones E, Siegel E, Saboury B (2022) Artificial intelligence in medical imaging and its impact on the rare disease community: threats, challenges and opportunities. PET Clin 17(1):13–29. https://doi.org/10.1016/j.cpet.2021.09.009

Jain S, Jain M (2024) Revisiting the conceptual terrains of the right to accessibility in India: the role of judicial enforcement. Laws 13(4):54. https://doi.org/10.3390/laws13040054

Johnson KB, Wei W, Weeraratne D, Frisse ME, Misulis K, Rhee K, Zhao J, Snowdon JL (2021) Precision medicine, AI, and the future of personalized health care. Clin Transl Sci 14(1):86–93. https://doi.org/10.1111/cts.12884

Junaid SB, Imam AA, Balogun AO, De Silva LC, Surakat YA, Kumar G, Abdulkarim M, Shuaibu AN, Garba A, Sahalu Y, Mohammed A, Mohammed TY, Abdulkadir BA, Abba AA, Kakumi NAI, Mahamad S (2022) Recent advancements in emerging technologies for healthcare management systems: a survey. Healthcare 10(10):1940. https://doi.org/10.3390/healthcare10101940

Khalifa M, Albadawy M (2024) AI in diagnostic imaging: Revolutionising accuracy and efficiency. Comput Meth Progr Biomed Update 5:100146. https://doi.org/10.1016/j.cmpbup.2024.100146

Krupiy T, Scheinin M (2023) Disability discrimination in the digital realm: how the ICRPD applies to artificial intelligence decision-making processes and helps in determining the state of international human rights law. Hum Rights Law Rev 23(3):ngad019. https://doi.org/10.1093/hrlr/ngad019

Li K, Sai AR, Urovi V (2024) Do you need a blockchain in healthcare data sharing? A tertiary review. Explorat Digital Health Technol:101–123. https://doi.org/10.37349/edht.2024.00014

Maher M, Khan I, Prikshat V (2023) Monetisation of digital health data through a GDPR-compliant and blockchain enabled digital health data marketplace: a proposal to enhance patient's engagement with health data repositories. Int J Informat Manag Data Insights 3(1):100159. https://doi.org/10.1016/j.jjimei.2023.100159

Marbouh D, Simsekler MCE, Salah K, Jayaraman R, Ellahham S (2022) Blockchain for patient safety: use cases, opportunities and open challenges. Data 7(12):182. https://doi.org/10.3390/data7120182

Marey A, Arjmand P, Alerab ADS, Eslami MJ, Saad AM, Sanchez N, Umair M (2024) Explainability, transparency and black box challenges of AI in radiology: impact on patient care in cardiovascular radiology. Egypt J Radiol Nucl Med 55(1):183. https://doi.org/10.1186/s43055-024-01356-2

Maslove DM, Klein J, Brohman K, Martin P (2018) Using Blockchain technology to manage clinical trials data: a proof-of-concept study. JMIR Med Inform 6(4):e11949. https://doi.org/10.2196/11949

Okorie GN, Udeh CA, Adaga EM, DaraOjimba OD, Oriekhoe OI (2024) Ethical considerations in data collection and analysis: a review: investigating ethical practices and challenges in modern data collection and analysis. Int J Appl Res Soc Sci 6(1):1–22. https://doi.org/10.51594/ijarss.v6i1.688

Okoye C, Obialo-Ibeawuchi CM, Obajeun OA, Sarwar S, Tawfik C, Waleed MS, Wasim AU, Mohamoud I, Afolayan AY, Mbaezue RN (2023) Early diagnosis of autism spectrum disorder: a review and analysis of the risks and benefits. Cureus 15:e43226. https://doi.org/10.7759/cureus.43226

Olawade DB, Bolarinwa OA, Adebisi YA, Shongwe S (2025) The role of artificial intelligence in enhancing healthcare for people with disabilities. Soc Sci Med 364:117560. https://doi.org/10.1016/j.socscimed.2024.117560

Osman A, Sun S (2023) Lack of disability data impedes health care equity | McKinsey. McKinsey Health Institute. https://www.mckinsey.com/mhi/our-insights/the-missing-billion-lack-of-disability-data-impedes-healthcare-equity

Pereira BMB, Torres JM, Sobral PM, Moreira RS, Soares CPDA, Pereira I (2023) Blockchain-based electronic voting: a secure and transparent solution. Cryptography 7(2):27. https://doi.org/10.3390/cryptography7020027

Porsdam Mann S, Savulescu J, Ravaud P, Benchoufi M (2021) Blockchain, consent and prosent for medical research. J Med Ethics 47(4):244–250. https://doi.org/10.1136/medethics-2019-105963

Rhem JA (2023) Ethical use of data in AI applications. In: Ethics—scientific research, ethical issues, artificial intelligence and education [working title]. IntechOpen. https://doi.org/10.5772/intechopen.1001597

Saeed H, Malik H, Bashir U, Ahmad A, Riaz S, Ilyas M, Bukhari WA, Khan MIA (2022) Blockchain technology in healthcare: a systematic review. PLoS One 17(4):e0266462. https://doi.org/10.1371/journal.pone.0266462

Schwartz R, Vassilev A, Greene K, Perine L, Burt A, Hall P (2022) Towards a standard for identifying and managing bias in artificial intelligence (No. NIST SP 1270; p. NIST SP 1270). National Institute of Standards and Technology (U.S.). https://doi.org/10.6028/NIST.SP.1270

Shah H, Heshmat S (2024) Precision medicine and AI: revolutionizing personalized healthcare. Www.Cognizant.Com. https://www.cognizant.com/se/en/insights/blog/articles/precision-medicine-and-ai%2D%2Drevolutionizing-personalized-healthcare

Shinde R, Patil S, Kotecha K, Potdar V, Selvachandran G, Abraham A (2024) Securing AI -based healthcare systems using blockchain technology: a state-of-the-art systematic literature review and future research directions. Trans Emerg Telecommun Technol 35(1):e4884. https://doi.org/10.1002/ett.4884

Singh Y, Jabbar MA, Kumar Shandilya S, Vovk O, Hnatiuk Y (2023) Exploring applications of blockchain in healthcare: road map and future directions. Front Public Health 11:1229386. https://doi.org/10.3389/fpubh.2023.1229386

Tagde P, Tagde S, Bhattacharya T, Tagde P, Chopra H, Akter R, Kaushik D, Rahman MH (2021) Blockchain and artificial intelligence technology in e-Health. Environ Sci Pollut Res 28(38): 52810–52831. https://doi.org/10.1007/s11356-021-16223-0

Taherdoost H (2023) Privacy and security of Blockchain in healthcare: applications, challenges, and future perspectives. Sci 5(4):41. https://doi.org/10.3390/sci5040041

Vela MB, Erondu AI, Smith NA, Peek ME, Woodruff JN, Chin MH (2022) Eliminating explicit and implicit biases in health care: evidence and research needs. Annu Rev Public Health 43(1): 477–501. https://doi.org/10.1146/annurev-publhealth-052620-103528

WHO (2023) Global report on health equity for persons with disabilities. https://www.who.int/teams/noncommunicable-diseases/sensory-functions-disability-and-rehabilitation/global-report-on-health-equity-for-persons-with-disabilities

Analysing the Impact and Challenges of Deep Learning Models in Virology

20

Niti Patil, Shweta Agrawal, and Sukrati Agrawal

Abstract

Virology has seen significant advancement in virus detection, understanding, and containment due to advancements in traditional and computational methods. The use of deep learning models made significant improvements in prediction of accuracy, data analytics efficiency, and biological system pattern analysis. This chapter discusses the challenges and contributions of deep learning methods in virology, starting with an introduction to the field and discussing applications during the coronavirus pandemic. Key bottlenecks include data availability, interpretability, and ethical considerations. This chapter concludes with recommendations for the effective application of deep learning in virology. It emphasizes the importance of collaboration between virologists and AI experts to harness its potential to protect humanity against viral diseases.

Keywords

Virology · Deep learning · AI in virology · Computational virology · Model · Interpretability · Ethical considerations in AI · Coronavirus · AI for health

20.1 Introduction

Virology is a branch of microbiology that is concerned with the study of viruses, their structure, classification, genetic regulation, and interactions with the host viruses are acellular microscopes that fold DNA or RNA encapsulated in a protein

N. Patil · S. Agrawal (✉) · S. Agrawal
Department of Computer Science and Engineering, Indore Institute of Science and Technology, Indore, MP, India

J. M. Chatterjee, S. K. Saxena (eds.), *Artificial Intelligence and Blockchain in Precision Medicine and Virology*, Medical Virology: From Pathogenesis to Disease Control, https://doi.org/10.1007/978-981-96-8919-4_20

coat in a protective capsid. They lack basic biological characteristics such as respiration, metabolism, or growth, but simply reproduce within the host cells and exhibit life-like properties. These two classify them as living and unconscious of them, making them unique in the biological world.

Virology study has greatly expanded in importance with the critical role viruses play in both causing and addressing diseases. They are often seen as harmful causes responsible for illnesses like influenza, hepatitis, AIDS, and even certain cancers; viruses also possess beneficial traits that are harnessed in medical and biotechnological applications. For instance, viruses are important tools in gene therapy, where they are modified to deliver therapeutic genes to cells, and in vaccinology, where they serve as the foundation for developing vaccines against infectious diseases.

Viruses are extremely diverse, including humans, animals, plants, bacteria, and even parasites. Viruses are classified by their genetic material and by the mode they spread. Some viruses, like DNA viruses, rely on DNA polymerase for replication, with bacteriophages and animal viruses being well-known examples. RNA viruses have very high mutation rates, making it difficult to develop vaccines against diseases such as influenza and smallpox. Technological advancements, especially in the field of computational sciences, have revolutionized virology.

Since viruses have always been instrumental in causing diseases, but also curing them, the field of virology has risen to become increasingly important. Even as harmful pathogens responsible for diseases ranging from AIDS to hepatitis, influenza, and some cancers, viruses possess favourable characteristics that are used advantageously in biotechnological and medicinal applications. For instance, viruses play an important role in vaccinology—they form the foundation upon which vaccines are developed to treat particular diseases, and in gene therapy, where the virus is used for transferring therapeutic genes into cells.

The relevance of virology has lately been underscored by global challenges such as the COVID-19 epidemic. Virology has continued to widen its scope, provide innovative solutions in the combating of emerging viral threats, and deepen understanding of mechanisms behind viruses. Its input is pivotal in responding to medical challenges while working on advancements in biotechnology and therefore offering new therapeutic strategies. It is the cornerstone of both modern science and health research (Wang et al. 2020).

This chapter gives an in-depth review and analysis of the relation between virology and deep learning by deeply and carefully examining their intersections. This chapter begins with an introduction to virology giving a foundation for comprehending the biology and clinical features of viruses. After that, it explains how deep learning involves advanced virological applications and research. This is followed by a comprehensive study of different deep learning models and their uses in virology. This chapter also assesses each model's limitations and difficulties. Furthermore, a case study explaining the contribution of deep learning against COVID-19 is provided. At the end, this chapter provides future aspects and insights into collaborating deep learning technology into virological research (Fig. 20.1).

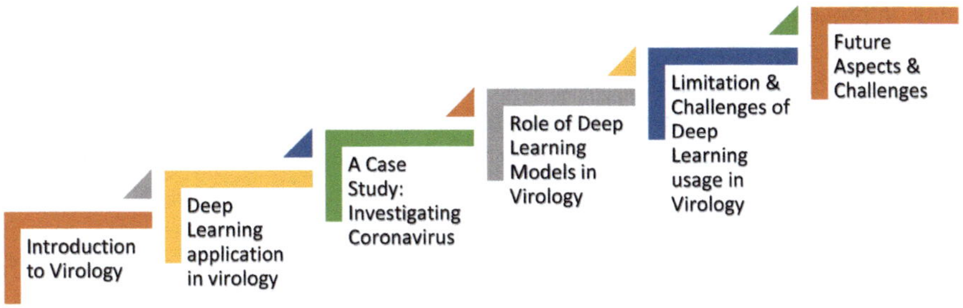

Fig. 20.1 Outline of chapter

20.2 Overview of Deep Learning

Deep learning is a powerful and most popular tool used to analyse and interpret large datasets, especially in the field of virology. It does great in dealing with complex biological data. Nowadays, a variety of applications in the medical sector are using deep learning models like recurrent neural networks (RNNs), transformer architectures, and convolutional neural networks (CNNs), which are widely used in image recognition, natural language processing (NLP), and sequence analysis. Some of the applications of deep learning in virology are as follows:

- Tracking evolution and mutations through the analysis of viral genomes.
- Virus detection through high-precision techniques that make use of genomic data and imaging.
- Insight into the way viruses interact with their hosts.
- Vaccine development accelerates and improves diagnostic methods.
- Using epidemiological models predicting outbreaks.

Deep learning, therefore, is a subset of AI whereby a computer system can learn from large amounts of data. It is somewhat parallel to how humans think, that is, through a series of virtual "neurons" in stacked layers. Such technological supremacy has become most useful in the field of virology, where scientists study viruses and their influence.

Deep learning aids virologists in making sense of large, complex data, such as genetic material recovered from viruses or patterns of virus that spread during outbreaks. Unlike traditional methods, which tend to require much manual work, DL can automatically find hidden patterns within the data, making it faster and easier to understand how viruses behave, forecast probable outbreaks, and develop new treatments or vaccines.

For instance, by examining the nature of viruses, in terms of genetics, DL would determine mutations potentially creating new varieties like influenza and COVID-19 (Han et al. 2023). Moreover, DL helps to design specific antiviral drugs by discovering molecules that potentially can stop reproduction in a given virus. Secondly, DL aids in processing microimages from the electron microscope to further increase the identification accuracy of a virus.

The efficiency of virology is improved by combining deep learning with traditional laboratory techniques. This integration enables researchers to respond to viral threats more quickly and develop better ways to prevent or treat viral infections (Jin et al. 2023).

20.3 Applications of Deep Learning in Virology

20.3.1 Viral Genome Analysis

CNNs and RNNs have been used to categorize viral genomes and predict their potential as pathogens. In addition to predicting mutations and identifying viral sequences, deep learning helps us comprehend evolution. Deep learning methods like transformers and RNNs could help handle larger genomic datasets. Models like VirHunter can identify new viruses from metagenomics data successfully. As an example, a research paper (Jin et al. 2023) explored how DL monitors viral development and vaccine updates by tracking genomic variants.

20.3.2 Virus Detection and Classification

In the class of deep neural networks, convolutional neural networks (CNNs) are among the best classifiers for identifying and classifying viruses. In addition, AM684 proposed DL-based methods for utilizing radiological data to detect COVID-19 infections. Further development of pathogen discovery also relies on virus typing and models such as DETIRE for identifying new viruses.

20.3.3 Diagnostics

The revolutionization in diagnostics due to deep learning can be visualized with applications mostly related to images. As a simple example, the author in research paper (Wang et al. 2020) has built DL for assessment on chest CT for diagnosis and prediction of the outcomes of COVID-19. Deep learning models to predict the potential diagnosis at the earliest moment (Alzubaidi et al. 2021). Furthermore, RT-PCR data also checks models which assesses the amount of viral load in infected people and determines infection's level.

20.3.4 Vaccine and Drug Development

Deep learning accelerates the discovery of vaccines and drugs by identifying immunogenic regions in viral genomes and testing antiviral compounds. The role of DL in identifying potential SARS-CoV-2 vaccine targets is highlighted in paper (Wu et al. 2023). New pharmaceutical molecules have also been designed through reinforcement learning algorithms and GANs. Drug repurposing models screen existing molecules for antiviral activity.

20.3.5 Epidemiology and Outbreak Management

To facilitate proactive public health measures, the predictions are made by models regarding viral outbreaks. In the paper (Shoeibi et al. 2024) examined the use of deep learning to forecast the prevalence of COVID-19. Deep learning has played a crucial role in monitoring the dissemination of illnesses such as COVID-19 and influenza. It tracks mutations in real time, aiding in the evaluation of the effects of emerging variants.

20.3.6 Metagenomic Analysis

Deep learning analyses vast amounts of metagenomic data to find viruses in a variety of settings, including human microbiomes, soil, and water. This ability advances our knowledge of the ecological roles and diversity of viruses.

20.4 Impact of Deep Learning in Virology

Deep learning is used in artificial intelligence (AI) and has been recently evolved to be of significant importance to virology as it provides deep analysis of biological data and helps in problem-solving ranging from understanding the viruses, their action, and their effects on late human health (Joshi et al. 2021). Below are the key roles of deep learning in virology:

- Handling Complex Data: Imaging data, epidemiology records, and viral genomes are all extensive and complex. Deep learning shines in processing such data to find patterns and insights.
- Speed and Efficiency: DL models automatically perform tasks, such as locating mutations or determining diseases, which would take humans significantly longer to complete.
- Accuracy: Because deep learning minimizes human error, research and diagnostic outcomes are more trustworthy.
- Flexibility: DL models could be optimized toward a variety of tasks, such as tracing evolutionary changes in the virus and leading to discovery.
- Predictive Power: Through historical data, DL is able to predict patterns, such as the spread of a virus or the appearance of mutations, which helps in taking preventative measures (Fig. 20.2).

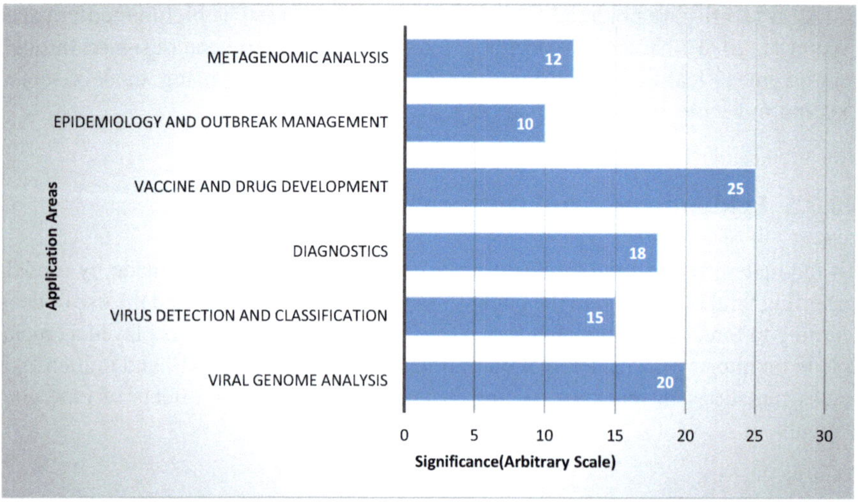

Fig. 20.2 Bar chart illustrating the various applications of deep learning in virology

Here is a chart representing the relational significance of deep learning and virology applications on an arbitrary scale. As shown by the chart, viral genome analysis has higher significance as deep learning helps in the analysis and identification of large genomic datasets and new viruses and predicting mutations that are most crucial components for understanding and improvement in viral evolution and vaccine development. Virus diagnostics find it most significant due to the role of DL in analysing scans, like chest CT scans, and RT-PCR data for prior and accurate diagnostics. Virus detection and classification have moderate significance among all areas of virology, as CNNs and other DL models are widely used for virus identification and detection through imaging and radiological datasets, as seen during the COVID-19 pandemic. Among all the application areas studied, vaccine and drug development has the highest significance, as DL accelerates the discovery of vaccines and drugs by identifying immunogenic regions and testing antiviral compounds. This played a vital role in SARS-CoV-2 vaccine development. Last but not the least, epidemiology, outbreak management, and metagenomic analysis have lower significance relative to others because its focus is on predicting outbreak patterns and mutation tracking, which, while vital, often relies on integration with other disciplines like public health. As DL scrutinizes metagenomic data to reveal viruses in diverse environments, it helps us understand their ecological roles.

Note: The significance values (arbitrary scale) reflect the estimated contribution of these applications based on criteria such as real-world impact in the context of the COVID-19 pandemic, research activity, and potential benefits. These values serve to underscore where deep learning makes the most difference and thus inform prioritization in virology research.

20.5 Case Study: Investigating Impact of Deep Learning in Coronavirus Prediction, Handling, and Maintenance

Improved diagnostic accuracy, prognostic assessments, and treatment protocols made deep learning of utmost significance for the expansion of coronavirus study research, such as SARS-CoV-2. Summaries of pivotal publication research studies clearly elaborating upon the contributions are seen in the references given below in a diagnostic role.

- An innovative DL methodology for the COVID-19 diagnosis and prognosis has been developed utilizing chest CT scans. The lung segmentation and diagnostic analysis module constitutes the proposed model. Excellent accuracy and clinical decision-making were made possible by diagnosis using COVID-19Net and dense segmentation using DenseNet121-FPN (Wang et al. 2020).
- Deep learning approaches for early screening of COVID-19 have been investigated. The research could identify several very accurate DL models but point out that for practical applications, for improving their dependability, clinical validation and standardization are required (Alzubaidi et al. 2021).
- New deep learning methods for COVID-19 in terms of diagnosis, prognosis, and managing therapy have been explored. This research showed how the DL models would be able to analyse imaging data and predict patient's outcome, making it possible for more personalized care (Jin et al. 2023).
- DL applications in lesion localization, medical imaging, and severity assessment for COVID-19 diagnosis were considered. To make the findings precise, the paper considered preprocessing methods and architectures like ResNet (Wu et al. 2023).
- Applying DL models to drug discovery and medical imaging has been focused. Reinforcement learning and transfer learning have been applied to identify COVID-19 treatment candidates and vaccination targets (Bharati et al. 2021).
- DL methods for the prediction of COVID-19 epidemic prevalence have been explored. The research focused on innovative forecasting models that include social and environmental factors (Shoeibi et al. 2024).
- DL models for COVID-19 detection have been classified using radiological information like CT scans and X-rays. Infection diagnosis, both supervised and unsupervised methods, was incorporated into the models (Ouchicha Archana et al. 2023).
- An extensive review of DL applications in the COVID-19 pandemic has been presented, such as drug development, epidemiological modelling, and diagnostics; in this study, the flexibility and the versatility of DL approaches are also emphasized (Staudemeyer and Morris 2019).
- Methods for diagnosing COVID-19 using ML and DL were systematically reviewed. The study addressed issues with the clinical deployment of the models and examined their performance (Salama et al. 2024).

- Several DL approaches have been explored to identify variants such as Omicron, and the challenges with COVID-19 testing methods have been discussed (Khan et al. 2024).

Together, these experiments demonstrate how deep learning significantly enhances our understanding of, ability to recognize, and power to manage COVID-19 with more effective responses to the pandemic.

20.6 Deep Learning Architectures Commonly Used in Virology

Deep learning has become important in microbiology because it allows scientists to model, test, and predict complex biological processes. Listed below are some of the most common deep learning algorithms used in microbiology, along with some examples of their use.

20.6.1 Convolutional Neural Networks (CNNs)

Convolutional neural networks (CNNs) are a class of deep learning algorithms, which are used to process structured network data, such as images, video frames, and audio signals. CNNs are highly efficient in tasks like pattern recognition, feature extraction, and image processing. Computer vision has been adapted and is used in areas such as image classification, object recognition, facial recognition, and self-driving vehicles.

CNN draws inspiration from the human visual system, in which the brain interprets information by recognizing features, such as edges, shapes, and textures, and generates significantly improved understanding of the features the hierarchical models are identified by (Irmak 2020).

20.6.1.1 Applications of CNN in Virology

Convolutional neural networks (CNNs) have been widely used in virology, leading to remarkable advances in virus analysis, classification, and identification. The following are notable applications of CNNs in virology, each supported by relevant research studies:

COVID-19 Detection from Chest X-Ray Images

CNNs are used to analyse viral proteins and predict the virulence of influenza A viruses, contributing to the understanding of the virus. Irmak (2020) demonstrated two new CNN algorithms to classify chest X-ray images. One of the models achieves an accuracy of 98.92% in discriminating between positive and negative cases. The other model is able to classify the images into three groups named: COVID-19, normal, and pneumonia, and achieve an average accuracy of 98.27%.

Virulence Prediction of Influenza A Viruses

The application of convolutional neural networks (CNNs) is made in predicting influenza A viruses based on their virulence factors involving the analysis of viral protein sequences. Lee et al. (Yifan et al. 2022) introduced VirPreNet, which is a CNN-weighted ensemble, to predict influenza A virus outbreaks. It analyses all eight bacterial protein sequences and has proven to have a better prediction rate than traditional methods.

Virus Detection Using Nanoparticles and Deep Neural Networks

The combination of nanoparticle technology with CNN contributes to accurate and rapid bacterial detection, which greatly improves the efficiency of bacterial detection methods.

In paper(Irmak 2020) author described a nanoparticle-based smartphone system which is capable of detecting viruses, can identify viruses on a microchip, and can label them with nanoparticles, utilizing CNN-based deep neural networks to analyse the data with high sensitivity and specificity.

Classification of Influenza Hemagglutinin Protein Sequences

CNNs contribute to the understanding of infectious diseases by classifying influenza hemagglutinin proteins for host specificity. Research Analysis: Chrysostomou et al. (2021) developed a CNN-based model to classify the hemagglutinin protein sequences of influenza A viruses. With greater accuracy than that of previous studies, it determines whether the virus can infect human, bird, or rodent hosts.

Automated Detection of Intact Adenovirus from TEM Imaging

CNNs aid virus characterization by assisting in the automation of intact adenovirus identification from transmission electron microscopy (TEM) images. Rukundo et al. (2023) designed a CNN-based tool that automates segmentation and detection of intact adenoviruses in TEM images. This model appears to recognize complete viruses as intact and all debris, shattered particles and artefacts as non-intact, thereby increasing virus detection performance.

20.6.1.2 Challenges and Limitations of Using CNN in Virology

- Data availability and quality: Limited viral datasets and inappropriate labelling prevent accurate model training. Small or imbalanced datasets can lead to over-accuracy.
- Generalization and overfitting: CNNs struggle to generalize to new viruses and run the risk of overfitting the training data. This prevents the detection of new bacteria.
- Translation power and translators: CNN often creates "black boxes", making their decisions difficult to understand. This includes trust and regulatory approval in clinical practice.
- Computational cost and resources: Training of CNNs requires a large amount of computing power and resources, making them expensive and resource-intensive, especially for large datasets.

20.6.2 Recurrent Neural Networks (RNNs) and Long Short-Term Memory Networks (LSTMs)

Recurrent neural network (RNN) is a type of artificial neural network designed specifically for processing sequential data. RNNs are well-suited to tasks involving time-series data, natural language processing, speech recognition, and more because they have a feedback loop that allows information to persist, unlike traditional feedforward neural networks. RNNs' major feature is their capacity to record temporal dependencies by preserving a hidden state that holds details about previous time steps.

An RNN uses the hidden state from the previous time step to process each input at a time step. RNNs can model sequences based on its design, which records information about previous inputs and uses it to forecast future inputs (Sherstinsky 2020).

The gradient issue vanishes during backpropagation, which makes it challenging for the model to pick up long-term dependencies. RNN models have this drawback, particularly when executing lengthy processes. To overcome the drawbacks of conventional RNNs, particularly the vanishing gradient issue, RNNs have been modified. The study in paper (Zhao et al. 2023) suggested that LSTM provides superior memory, allowing the network to better store long-term expectations.

LSTMs consist of memory cells that store information over time, and gates such as input gates, forget gates, and output gates, controlling the flow of information into and out of the memory. This is what makes LSTMs choose which information is relevant to be kept and which to discard, thus maintaining relevant information for long sequences.

Key components of LSTMs:

- Input Gate: It decides the amount of new input information that is to be stored in the memory cell.
- Forget Gate: This determines which information present in the memory cell needs to be forgotten or removed.
- Output Gate: This decides the amount of information in the memory cell to be produced at each time step.
- LSTM has been a very dominating architecture for sequential tasks as it has the ability to learn long-term dependencies and does not suffer from the vanishing gradient problem.

20.6.2.1 Applications of RNN and LSTM in virology

RNNs and LSTMs are nicely perfect for obligations that contain sequences of data, which includes viral genome analysis, time-collection information in epidemiology, or the taking a look at of viral evolution. Below are a number of the important approaches wherein RNNs and LSTMs may be applied in the field of virology.

Viral Genome Sequence Analysis

Predicting nucleotides or amino acids in a viral genome sequence with RNN or LSTM helps in understanding the viral genome shape and identifying ability mutations, but can also expect the outcomes of adjustments on the genome. Through training RNNs or LSTMs on big datasets of viral genomes, these fashions may want to predict how mutations at particular positions in the genome might affect viral replication, infectivity, or drug resistance. This aligns with the need to monitor viral mutations, as proven in preceding studies on viral sequence prediction (Sherstinsky 2020).

Viral Evolution and Phylogenetic Tree Construction
By tracing the distinctive adjustments in viruses' genomes acquired over the years that give rise to newer virallines, RNNs and LSTMs can be the reason version the virus's evolutionary dynamic. These forms of models are finally compared with the previous statistics to offer forecasted predictions as to how new variations of their genetic strands will emerge because of the capability of viral strains' evolution through time. LSTMs may be applied to the development of phylogenetic bushes seeing that they examine temporal patterns in viral sequence records. This is important for monitoring the evolution of viral strains and their relationships to every different. This is consistent with the techniques discussed by (Staudemeyer and Morris 2019), evaluating the sequential facts.

Epidemiological Modelling and Disease Forecasting
RNNs and LSTMs can expect the spread of viral illnesses over time. Through modelling time-series records, as an instance, the wide variety of cases or deaths, these fashions can be capable of expecting the trajectory of a deadly disease or pandemic, permitting public health authorities to plan and reply successfully. LSTMs are used to analyse actual-time epidemiological data-contamination charges, vaccination coverage, public health interventions, and tuning of the patterns in which way the virus is spreading, expecting how cases will be unfolded in the future. The method used for the early detection and forecasting of COVID-19 changed into comparable (Shoeibi et al. 2024).

Viral Pathogenesis and Host Immune Response
LSTMs can be utilized to version how viral proteins interplay with the host's immune gadget. This kind of modelling gives insight into the dynamics of a viral contamination's development and into the way that the immune system responds to infections. LSTM fashions can be skilled on medical facts, including viral load, immune markers, and affected person demographics, and are expecting affected person effects in viral infections. This is beneficial for figuring out people at high threat of intense disease and guiding treatment choices. Similar tactics had been explored by Sherstinsky (2020) for viral contamination predictions.

Drug Discovery and Development
LSTMs networks' potential may be used for the prediction of the potential goals of drugs using analysis of viral genome sequences and identification of genes or proteins involved in the cycle of virus's existence. The predictions help enhance the efficacy of drug discovery projects using healing dealers and focus on significant

viral proteins centred for their functioning. Just by their chemical properties and current ways of movement, LSTMs could be designed to express the efficacy of antiviral drugs. To predict new drugs to be developed, LSTMs will come in handy with existing datasets of drug interactions with viruses. Drug interactions with viruses (Staudemeyer and Morris 2019), which have explored the role that in-depth knowledge plays in the discovery of drugs.

Host–Virus Interaction Prediction

RNNs and LSTMs can predict how viral particles input host cells by learning from viral series records and receptor binding patterns. This is of crucial importance to understand the tropism of viruses, i.e. the cells or tissues that are infected through the virus, and in the development of targeted antiviral therapies. LSTMs may be used to take a look at the temporal dynamics of how viruses evolve to steer clear of the host immune machine. Through modelling the genetic changes in viral proteins that interact with host immune receptors, LSTMs might be beneficial for figuring out specific viral mutations with related immune evasion.

Time-Series Data in Viral Infections

RNNs and LSTMs are in particular effective at analysing time-series information, including viral load measurements, affected person monitoring statistics, and the progression of disorder signs and symptoms over the years. These fashions may be used to song the dynamics of viral infections in individual patients or populations.

Synthetic Data Generation for Viral Research

RNNs and LSTMs may also be implemented in producing artificial viral genome sequences to mimic mutations that occur in the virus affecting the structural and purposeful shape. It permits scientists to perceive what sort of mutation does have a massive impact on viral pastime, helping researchers in making vaccine and antiviral medicine advancements. In cases in which real medical facts are scarce or incomplete, RNNs and LSTMs may be used to generate synthetic patient information, inclusively.

20.6.2.2 Challenges and Limitations of Using RNNs and LSTMs in Virology

- Data Requirements: RNNs and LSTMs require large amounts of high-quality sequential data for training. In virology, this may be limited, especially for rare viral strains or early-stage infections.
- Complexity: Both RNNs and LSTMs are computationally intensive, particularly when dealing with long sequences or large datasets. This can make training and inference slow and resource demanding.
- Interpretability: Like many deep learning models, RNNs and LSTMs are often considered black-box models, making it challenging to interpret how the model arrives at specific predictions.

20.6.3 Transformer Architectures

Transformer architectures, provided inside the paper Attention is All You Need with the aid of Vaswani (2017), have dominated the field of deep mastering, notably in natural language processing (NLP). Self-attention mechanisms are the hallmark innovation of the transformers: self-attention can be thought of as the opportunity for fashions to weigh words (or tokens) of any distance from every other inside the sentence. This allows transformers to process sequential information effectively, hence making them a powerful resource in many applications, including ones beyond traditional NLP, like genomics and virology.

In a transformer version, the input records are fed through layers of self-attention observed by feed-forward networks. The architecture includes encoder and decoder layers, with attention mechanisms applied at every layer to compute the relationships among special elements of the input series.

20.6.3.1 Applications of Transformer Architectures in Virology

Analysis of intricate biological data, including viral genomes, protein sequences, and viral activity in the host body, is necessary for virology, the study of viruses and viral diseases. Transformer models may be used to increase research and our knowledge of viral processes in a variety of virology-related fields. Transformer architectures are useful in the following important areas:

Predicting Protein Structure

Instructions for viral replication are encoded by a chain of nucleotides (DNA or RNA) that is called a viral genome. Transformers are useful in the analysis of viral genomes because they would be able to look at very long genetic statistical sequences and perhaps find patterns or relationships in sequences of nucleotides. Transformer models can be fine-tuned to extract motifs (positive sequences of biological significance), predict viral behaviours, and verify alterations in viral genomes. This is especially true for BERT (bidirectional encoder representations from transformers) and GPT (generative pretrained transformer), for example, predicting the possible effects of positive mutations on the viral genome, such as alterations in treatment resistance or infectivity (Zhao et al. 2023).

Protein Structure Prediction

Since proteins are crucial for viral replication, immune device evasion, and host interplay, they are very important for understanding how viruses work. Protein structure prediction from amino acid sequences is a challenging task. Transformation designs can be adapted to predict secondary and tertiary structures of proteins by mimicking the interactions among amino acids in a protein set. The deep learning version of AlphaFold has helped augment the prediction of protein structure amazingly. The viruses' proteins, with mutations, might trade in their structure and, therefore their characteristic. This too can be predicted using transformers, especially attention-based models (Shoeibi et al. 2024).

Phylogenetic Analysis and the Evolution of Viruses

Mutations and recombination events in the viral genome are what propel viral evolution. In virology, it is far critical to recognize how viruses trade, adapt to their hosts, and become proof against remedies (which include antiviral medicinal drugs or vaccinations). By studying numerous series alignments of viral genomes or proteins, transformer models can be used to take a look at phylogenetic trees. They may be used to predict the origins of novel viral strains, spot evolutionary traits, and forecast capability in future viral improvement. Transformers may be used, for instance, to discover genetic variations in RNA viruses (consisting of coronaviruses or the flu) that could bring about new traces to wonderful characteristics (Skandarani et al. 2023).

Predicting Viral Protein–Host Interactions

Understanding how viral proteins interact with host cell machinery is important for developing antiviral medications and vaccines.

The troublesome interactions between host and viral proteins can be represented using transformers. Models in conjunction with TAPE (transformer for protein sequences) or ViT (vision transformer) can be used to predict how viral proteins would interact with host cell receptors in order to get insight into the process of viral infection.

Identifying the viral proteins required for infection and thus useful for targeting will help identify potential healing targets (Chen et al. 2018).

Drug Discovery and Screening

The process of identifying substances that can interact with certain viral targets is known as drug discovery. This may be accomplished by examining enormous chemical libraries, identifying compounds that show promise, then using transformer models to predict how effectively those compounds will function against viral proteins.

Transformer-based models for chemical and molecular data, such as ChemBERTa and MolBERT, may be used to predict how a certain chemical might react with viral proteins, leading to the discovery of new antiviral medications (Eraslan et al. 2019).

Viruses Pathogenesis Prediction

In general, virology research aims to understand how viruses infect hosts. This includes studying the impact of viruses on different tissues of the host, causing host symptoms and evade the host's immune system. The transducers can be used to process large datasets on gene expression and host–virus interactions. To find indicators of disease development or to predict the course of virus infection, for example, it is possible to train a transformer model to predict which host components are required for virus replication and which viral proteins are involved in evading the immune system (Alzubaidi et al. 2021).

Literature Review and Text Mining

Virology makes use of published research, clinical trial, and extensive epidemiological data. Text mining is useful in extracting useful insights from scientific and medical literature. It uses transformer models such as BERT and GPT. These models are useful for literature reviews. By filtering through published research, trend analysis finds research on new viruses or summarizing information about a specific strain of virus such as COVID-19 (Zitnik et al. 2018).

20.6.3.2 Challenges and Limitations of Using Transformer Architectures in Virology

- Model Size and Training Cost: Transformers have huge model sizes with billions of parameters, which demand huge computational resources for training and storage and hence are costly to develop and maintain (Zhao et al. 2023).
- Interpretability and Transparency: These models are often regarded as "black boxes", which makes it difficult to explain how they arrive at specific decisions, a challenge in fields requiring transparent reasoning, such as healthcare or finance (Skandarani et al. 2023).
- Bias and Ethical Issues: Transformers are prone to inherit and amplify the biases present in the training data, which could lead to outputs that are biased or even dangerous, especially for sensitive applications (Shoeibi et al. 2024).
- Data Efficiency and Fine-Tuning: Transformer models generally require significant amounts of data to be pre-trained and fine-tuned, making them resource-hungry and not very suitable for limited amounts of data (Zitnik et al. 2018).
- Handling Long Sequences: Although it performs well on medium-length sequences, transformers struggle with very long input sequences because of memory and computational bottlenecks, especially in tasks involving long documents or complex contexts (Chen et al. 2018).

20.6.4 Generative Adversarial Networks (GANs)

The beauty of machine learning models became provided through Goodfellow et al. 2014 Both generator and discriminator neural networks make up a GAN and are simultaneously taught using an approach known as "adversarial schooling".

Generator (G): This community creates synthetic information that mimics actual facts, such as text, photos, or even sound. As it learns, the generator's initial output of random noise is transformed into greater sensible statistics.

Discriminator (D): With the use of the true statistics distribution, this community verifies the information that is being produced from the generator as either valid or false. The discriminator's goal is to decide between valid and false information.

The networks are engaged in a sport during which the generator tries to fool the discriminator via producing files, which are much like actual data. The discriminator does work to extend its capacity to categorize fake information. Until such a time when the generator generates data so realistic that the discriminator cannot reliably and consistently make the distinction among them, this learning process is iterated (Staudemeyer and Morris 2019).

20.6.4.1 Applications of Generative Adversarial Networks (GANs) in Virology

In domains like text generation and picture synthesis, GANs have demonstrated notable effectiveness in producing realistic data. Their uses in virology can be extensive and significant, particularly in fields where the creation of synthetic biological data is required for study, medication development, and comprehension of viral activity. The process of generating synthetic snap shots or genomic data that complement the available datasets is called synthetic records generation. Antiviral drug discovery includes two things: candidate compound creation and simulating drug-goal interactions. Enhancement of low-resolution virus photographs for analysis is referred to as viral imaging. The following are some important virology domains where GANs might be useful:

Synthesis of Viral Genomes and Modelling of Mutations

Researchers can also attain artificial records for a whole lot of viral lines by way of using GANs to generate synthetic viral genomes. These artificial genomes may be used to mimic viral outbreaks, inspect how viruses exchange and increase through the years, and comprehend how various mutations might affect a deadly disease's conduct. GANs can be taught to simulate and forecast the evolution of viral mutations. Knowing how viruses like influenza, HIV, or coronaviruses quickly trade to keep away from immune responses or expand remedy resistance may be very helpful in this regard. For instance, GANs would possibly produce viable viral genome modifications and forecast the results of those mutations on viral replication, infectivity, or resistance to antiviral medicines (Staudemeyer and Morris 2019).

Creation of Protein Structures

GANs have the ability to create useful protein structures from amino acid sequences because viral proteins are crucial for infection and immunological machine evasion. This topic is essential to virology. By generating synthetic data on protein systems and interactions, GANs can be used to model the interactions between viral proteins and host proteins, perhaps identifying new therapeutic targets. By providing plausible 3D systems of viral proteins, GANs may also improve vaccines and capsules more quickly in situations when experimental methods are expensive or time-consuming (Zhao et al. 2023).

Artificial Information for Models of Viral Disease

Clinical results have proven that the number of viruses has stimulated patient data, disease spread, and virus infection models are examples of synthetic datasets that can be created using GANs. These models can be used to predict the spread of disease and maximize public health initiatives. By generating synthetic data on the progression of viral infections in host organisms, GANs can help researchers model the pathogenesis of various viruses. This may help understand how different viruses cause disease and predict patient outcomes based on genetic changes (Shoeibi et al. 2024).

Virtual Screening and Drug Discovery
GANs can produce new drug-like compounds that could interact with viral targets. Virtual screening may then be used to assess the possible effectiveness of these medicines in treating specific viral illnesses. The compound generation by GANs with certain properties, such as the affinity to viral proteins, can help to find potential candidates. Development of Drug-Resistant Strain Molecules: GANs can help develop drugs that target specific viral variants by mimicking drug-resistant viral strains (Skandarani et al. 2023).

Medical Imaging and Image Synthesis
GANs are valuable resources for creating medical images, including MRI, CT, or histology images that mimic the impacts of viral infections on tissues. In situations where annotated medical images are sparse, such synthetic images can be leveraged to train diagnostic algorithms and expand databases.

GANs could aid the development of more accurate diagnostic instruments also by generating virtual images of viral infections. This can train models to identify viral lesions or markers in scientific scans that have been contaminated by viruses like COVID-19, HIV, or hepatitis (Kingma and Welling 2013).

Viral Evolution Simulation in Population Research
Simulation of Virus Evolution: ANs can be used to simulate how a viral population changes over time in viral population dynamics. Artificial viral populations can be created by scientists to study how viruses adapt to different environments or how a strain of a virus spreads throughout a population. Gene Drift and Shift Simulation: GANs can be used to simulate the genetic drift and shifts of viruses, which have a major contribution in the creation of viruses such as influenza. Besides, this further aids in a better development of vaccines as well as a better prediction of stress (Eraslan et al. 2019).

Facilitating the Design of Viral Vaccines
GANs can also be used to create vaccine adjuvants, which are compounds that toughen the immune device's reaction to an antigen. GANs can recommend novel adjuvants that might improve the effectiveness of vaccinations in opposition to viruses through gaining knowledge from the adjuvant information that is already had (Zhang et al., 2023).

By training GANs on databases of viral protein structures, researchers can generate novel antigens that could be used to enhance vaccines. It may be studied whether those synthetic antigens can trigger an immune response (Shi et al. 2024).

Antiviral Resistance Prediction and Analysis

GANs can assist in forecasting how viruses may additionally grow to be immune to antiviral medications or vaccinations by generating artificial information on viral genomes, proteins, and mutations. This makes it viable to plan countermeasures earlier. GANs can assist in looking forward to the appearance of drug-resistant viral mutations that can manual the advent of next-era antiviral remedies (Ouchicha Archana et al. 2023).

20.6.4.2 Challenges and Limitations of Using Generative Adversarial Networks (GANs) in Virology

- Data Scarcity and Quality: GANs require large amounts of high-quality, annotated data for training, which can be difficult to obtain in virology due to the limited availability of labelled viral datasets.
- Model Stability and Training Issues: Training GANs is notoriously unstable, often plagued by problems such as mode collapse (where the generator produces limited outputs) or failure to converge, which makes it challenging to produce reliable models for viral data.
- Interpretability and Transparency: GANs are usually viewed as black-box models, making it challenging to understand how they generate results. This lack of transparency is a concern when trying to understand viral behaviours or mutations.
- Generalization to Novel Viruses: GANs are prone to overfitting on the specific datasets they were trained on and therefore may fail to generalize to new, unseen viruses, thus limiting their ability to model rare or emerging viral strains (Albery et al. 2021).
- Ethical Issues in Generation: The ability of GANs to produce realistic viral data gives rise to several ethical issues related to misuse, such as the creation of harmful viral mutations or producing misleading data for research or public health purposes.

20.6.5 Auto-encoders

A class of neural networks called auto-encoders based on artificial work are used for unsupervised learning purposes, especially for tasks of dimensionality reduction, feature extraction, and data compression. They consist of two main parts: an encoder compresses input data into lower-dimensional space, and a decoder reconstructs original data from this compressed form. The model is trained to minimize the difference between the original input and the reconstructed output; the most common choice of loss function would be mean squared error.

Auto-encoders are widely used for anomaly detection, denoising, and generative modelling, among other applications. They can learn efficient representations of data, and this has particularly made them applicable in image processing, natural language processing, and medical diagnostics. Although auto-encoders are quite simple, they have been highly effective in representing complex patterns in the data, thereby often forming the basis for advanced models such as VAEs and GANs (Zitnik et al. 2018).

In numerous domains, which include virology, auto-encoders are being utilized increasingly for tasks which include function extraction, anomaly detection, and data discount. Some packages of auto-encoders inside the research and comprehension of viruses and viral illnesses are listed underneath.

20.6.5.1 Applications of Auto-encoders in Virology

Analysis of Viral Genomic Data

Viral genomes are made from prolonged nucleotide sequences. By reducing the dimensionality of those sequences, auto-encoders can facilitate the identification of crucial patterns or functions in viral genomes without affecting important information (Wang et al. 2020). This can help researchers more correctly and compactly look into the genetic composition of viruses. Researchers can extract vital components of viral genomes which are important for comprehending how mutations arise and the way they affect viral behaviours via compressing the information by using the auto-encoders. This can be useful for monitoring viral stress evolution and forecasting the consequences of mutations (Alzubaidi et al. 2021).

Predicting the Structure of Proteins

When blended with convolutional layers, auto-encoders can also be used to expect or rebuild protein systems from amino acid sequences (convolutional auto-encoders). This use is especially useful in virology, in which investigating viral replication, developing vaccines, and designing drugs—all depend upon an understanding of the shape of viral proteins (Chen et al. 2018).

Auto-encoders may learn to map viral proteins into a decrease-dimensional space by using schooling on large datasets of recognised protein sequences. This makes it easier to spot styles and correlations that are important for comprehending how proteins work.

Classification and Identification of Viruses

Auto-encoders are great at anomaly detection, which means they are extremely useful for the detection of novel or rare viral strains. Training an auto-encoder on a set of known viral genomes might cause it to learn a dense representation of commonalities of viral sequences. It will have trouble reconstructing a new sequence significantly different from those in the training set, and so the auto-encoder will identify this as an anomaly. This may be helpful for the early detection of the emergence of novel or emerging viral strains.

In addition to the classification above, auto-encoders can also group viral strains based on their protein structures or genomic sequences. Once an auto-encoder learns a compressed representation of the viral data, it can be used with a classifier to classify viruses into groups (Shoeibi et al. 2024).

Modelling Virus–Host Interactions

Auto-encoders can learn compressed representations of complex biological information, including host cell protein–virus protein interactions. This may help scientists identify critical relationships that impact pathogenicity, immunological evasion, or viral replication. Treatment strategies, including antiviral drugs and vaccines, are based on such relationships.

Understanding the infection and pathogenicity of different virus strains, this depends on the ability of the auto-encoder to characterize both viral proteins and host cell receptors and predict how strongly viral proteins bind to host cell receptors (Wu et al. 2023).

Drug Discovery and Virtual Screening

One of the types of auto-encoders that may be used to layout new drug-like compounds which could probably interact with viral targets is variational auto-encoders (VAEs). These fashions are examined from a database of regarded antiviral tablets and can create new molecules that resemble successful tablets in terms of their residences. Then, their capability efficacy towards some viral infections can be determined.

Auto-encoder may be able to expect possibilities of drug resistance because of evolution in viral genomics. For this reason, auto-encoders may be able to teach identifying changes that likely effect the responsiveness of the viruses to drugs even as encoding each type of viral sequences and existing interplay among recognized capsules (Staudemeyer and Morris 2019).

Denoising and Pre-processing of Bio-Data

Because of technological constraints or experimental mistakes, biological datasets—like gene expression data or protein sequences—are frequently noisy. Before doing more analysis, denoising auto-encoders may be employed to eliminate this noise and enhance the data quality. In high-throughput sequencing investigations, where mistakes might result in inaccurate findings, this is especially crucial.

The raw data output from next-generation sequencing may contain unreliable and noisy data. Hence, preprocessing or cleaning up of NGS data can enhance the accuracy of following studies such as viral genome assembly, gene expression analysis, and mutation discovery (Alzubaidi et al. 2021).

Epidemiological Modelling and Prediction

Using auto-encoders, the epidemiological pattern of virus transmission in the population can be modelled. It is possible to recognize patterns hidden within the data that are not easily identifiable based on learning representations from epidemiological data, such as the rate of infection and spread over time, which may lead to guiding public health actions or predicting future epidemics.

Viral spread simulation: Auto-encoders may also be combined with other models of machine learning to simulate how viruses are spread among populations. This is helpful in determining the causes behind some diseases' spreading rapidly and to assess the potential impact of containment measures such as vaccination or social distancing (Khan et al. 2024).

20.6.5.2 Challenges and Limitations of Using Auto-encoders in Virology

- Data Scarcity and Noise: Good-quality virological datasets are scarce because of ethical issues and experimental costs, and noise in data can also impact the model's accuracy (Skandarani et al. 2023).
- Poor Interpretability: Auto-encoders are "black-box" models, and hence it is hard to biologically interpret the features that are extracted (Chen et al. 2018).
- Overfitting Risks: Auto-encoders tend to overfit on small datasets and learn to encode noise rather than meaningful patterns (Eraslan et al. 2019).
- Computational Requirements: Training deep auto-encoders requires a lot of resources, which may be a limitation for smaller research facilities (Hinton and Salakhutdinov 2006).

20.6.6 Graph Neural Networks (GNNs)

This is a type of deep learning model working on graph-structured data which is known as graph neural networks (GNNs). It has been seen that in some domains, especially in virology, because the interactions in biology are basically graph structured, they have found great promise. Complex inter-linking among the nodes that present entities such as proteins, viruses, or cells, and interconnecting edges describing interactions among the above-mentioned entities may be represented by GNNs.

20.6.6.1 Applications of Graph Neural Networks in Virology

Virus–Host Interaction Modelling

GNNs may be applied to version complicated organic networks, such as host cell proteins and virus proteins interactions. Graphs are typically used to symbolize these interactions where proteins are nodes and their interactions are edges. GNNs can expect new interactions by means of schooling on such graphs, that's vital to apprehend viral pathophysiology and replica (Liu et al. 2024).

GNNs can discover proteins within the host cellular that are liable to contamination by means of viruses. Identification of important host–virus interplay networks may be used to perceive new objectives for antiviral drugs (Yifan et al. 2022).

Protein Structure Prediction and Drug Discovery

Proteins, especially viral proteins, can be represented as graphs where amino acids are nodes and chemical bonds or physical closeness are edges. Predicting protein folding enables GNNs to learn structure-function correlations, which is crucial for understanding viral protein structures and developing drugs that block them (Li et al. 2022).

GNNs are used for the prediction of interaction between viral proteins and drug-like compounds. The binding affinities of GNNs can predict the possible inhibitors of viral proteins by considering molecules and proteins as graphs (Liu et al. 2024).

Virus Classification and Variant Detection
Genomic sequences can be represented as graphs, where edges represent structural connections or proximity and nucleotides are nodes. Other than predicting how sequence variation might create novel viral strains or modify pathogenicity, GNNs can be trained to classify viral genomes based on structure (Zhang et al. 2024). By applying GNNs to genomic graphs, new mutations or changes in the viral genome may be discovered, such as altering transmissibility, evading immunology, or resisting treatment (Jin et al. 2023).

Epidemiological Modelling and Prediction of Spread
Virus transmission network simulation: Using humans as nodes and edges as transmission links, GNN can model virus transmission networks. By revealing how the virus spreads throughout society. This information can be used to evaluate containment strategies and predict future outbreaks. Prediction of epidemic patterns: GNN models the dynamics of epidemics by classifying patterns in network data. These models can be used to identify groups that need special attention—high risk areas and hyper-spreading events (Zhang et al. 2024).

Antiviral Drug Resistance Prediction
GNNs can predict how alterations to the viral genome might lead to drug resistance against antiviral drugs. GNNs rely on graph representations of viral genomes and known drug interactions to predict how mutations might alter the structure of the virus and treatment susceptibility. GNNs can help model the resistance pathways that viruses take by understanding the connections between different mutations and how they affect drug binding or viral reproduction. (Zhang et al. 2023; Wang et al. 2024).

Synthetic Biology and Vaccine Development
With GNNs, it is possible to create protein sequences resembling synthetic viral genomes or actual viruses in their structure. It is then possible to create novel virus strains in the lab with controlled environments and build and test these to eventually create vaccines. GNNs can be helpful in the discovery of viral epitopes by studying interactions between viral proteins and host immunological receptors. The interaction is represented as graphs, and the GNN will predict which epitopes are most likely to provoke an immune response and will be useful in the construction of vaccines. (Koca et al. 2022; Joshi et al. 2021).

Viral–Host Interaction Network Analysis
This can help in understanding the mechanisms of viral pathogenesis by analysing the interconnection networks between the viruses and the host proteins working together for the pursuit of infection. In this approach, by analysing such networks,

GNNs help in predicting the function of some connections in the process of viral pathogenesis and locating possible therapeutic targets. By analysing viral–host networks with GNNs, scientists can find important host characteristics that viruses take advantage of during infection. This information is crucial for comprehending the processes underlying pathogenicity (Chrysostomou et al. 2021; Rukundo et al. 2023).

20.6.6.2 Challenges and Limitations of Using Graph Neural Network (GNN) in Virology

- Sparse and Noisy Data: The protein interaction networks and drug resistance data in virology are generally sparse or noisy, thus complicating the task of learning a reliable GNN (Zhang et al. 2024; Chrysostomou et al. 2021).
- High Dimensionality: Biological data inherently has a very high dimension, and any attempt to fit the data to the GNN might lead to information loss. The training of GNNs on large biological datasets requires tremendous computational resources that might be inaccessible to most virology labs (Joshi et al. 2021).
- Heterogeneity of Viral Data and Overfitting Risks: Virology deals with diverse data types, including genomic, proteomic, and epidemiological data. The integration of these heterogeneous data sources into a unified GNN framework is not straightforward. Overfitting is a concern, especially when training on limited data with complex graph structures (Rukundo et al. 2023).
- Dynamic Interactions: Many virological processes, such as viral mutations and host–virus interactions, are dynamic over time, requiring spatiotemporal graph models that are computationally intensive (Wang et al. 2024).
- Model Interpretability: GNNs have often been viewed as "black-box" models, making it tough to interpret the biological meaning of learned features, especially in critical applications such as drug resistance prediction (Zhang et al. 2023).
- Scale-up of Big Graphs: Epidemiological models and protein interaction networks contain millions of nodes and edges, which poses a challenge to scaling-up GNN architectures (Zhang et al. 2024).
- Domain-Specific Bias: GNNs can inherit biases from training data, which leads to poor generalization across different viral strains or host populations (Chrysostomou et al. 2021).
- Ethical and Privacy Concerns: Using patient-derived datasets for training GNNs raises ethical concerns related to privacy and data sharing, which limits data availability (Koca et al. 2022).

20.7 Comparative Study

A comparative analysis of different deep learning architectures in virology helps to establish the pros and cons of various approaches for some specific goals, like virus classification and detection or finding new treatments. An organized comparison of different important architectural can be found below (Table 20.1 and Fig. 20.3):

Table 20.1 Comparative study of different deep learning architectures in virology

Architecture	Best for	Strengths	Limitations	Example model
CNNs	Genomic patterns and image analysis	High accuracy of spatial features	Lack of good sequential modelling	Deepvirfinder
RNNs/ LSTMs	Sequence modelling and temporal assignments	Capture of long-term dependencies	Vanishing gradient problems	Deepviral
Transformers	NLP tasks and long-range dependencies	Scalable attention mechanism	Computationally expensive	Progen
GANs	AI and image super-resolution	Good data generation	Training instability	GAN for TEM imaging
Auto-encoders	Dimensionality reduction, anomalies	Unsupervised learning	Limited pattern complexity	DeepVirome
GNNs	Relational data and interaction	Relational modelling that works	Sensitive to graph inputs	DeepGraphGO

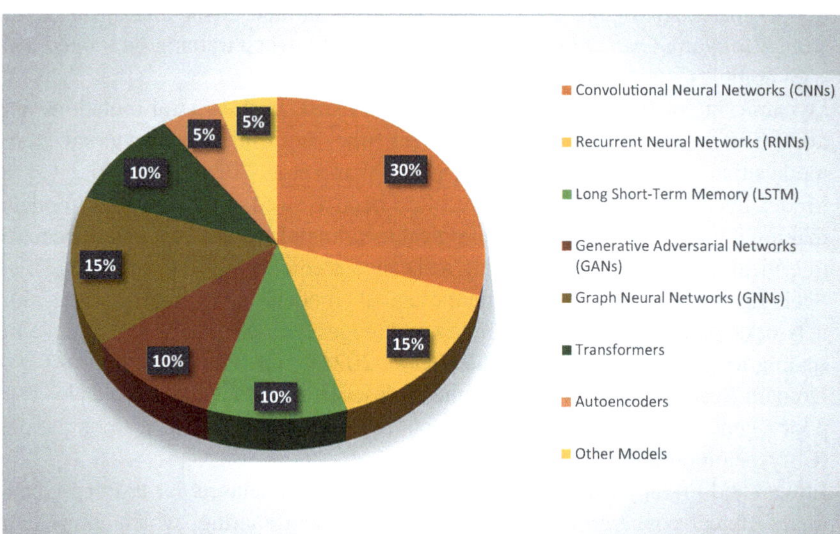

Fig. 20.3 Chart representing contribution of various deep learning models in virology

This illustrates the relative contributions of several deep learning models used in virology research. Each model has critical applications from viral detection and genomic analysis to drug discovery and epidemiological modelling. Percentages refer to relative contributions of each model toward tackling these key challenges in the field.

20.8 Future Work and Discussion

Future research should focus on the following:

- Collaboration in delivering standardized datasets.
- Techniques of explainable AI (XAI).
- Reducing computational costs with edge computing.
- Enhancing virologists' collaboration and data sharing.

In contemporary virology, DL has emerged as a core part that brings new insights and answers to problems like COVID-19. DL makes faster and more accurate diagnosis, drug development, and control of an outbreak by merging genetic data, imaging, and epidemiological models. More contributions are expected from this evolving discipline towards virology and public health.

References

Albery GF, Becker DJ, Brierley L, Brook CE, Christofferson RC, Cohen LE et al (2021) The science of the host–virus network. Nat Microbiol 6(12):1483–1492

Alzubaidi M, Zubaydi HD, Bin-Salem AA, Abd-Alrazaq AA, Ahmed A, Househ M (2021) Role of deep learning in early detection of COVID-19: scoping review. Comput Methods Programs Biomed Update 1:100025

Bharati S, Podder P, Mondal M, Prasath VB (2021) Medical imaging with deep learning for COVID-19 diagnosis: a comprehensive review. arXiv preprint arXiv:2107.09602

Chen Y, Liu X, Li Y, Quan C, Zheng L, Huang K (2018) Lung cancer therapy targeting histone methylation: opportunities and challenges. Comput Struct Biotechnol J 16:211–223

Chrysostomou C, Alexandrou F, Nicolaou MA, Seker H (2021) Classification of influenza hemagglutinin protein sequences using convolutional neural networks. In: 2021 43rd Annual International Conference of the IEEE Engineering in Medicine & Biology Society (EMBC). IEEE, pp 1682–1685

Eraslan G, Avsec Ž, Gagneur J, Theis FJ (2019) Deep learning: new computational modelling techniques for genomics. Nat Rev Genet 20(7):389–403

Goodfellow I, Pouget-Abadie J, Mirza M, Xu B, Warde-Farley D, Ozair S. et al (2014) Generative adversarial nets. Advances in neural information processing systems

Han W, Chen N, Xu X, Sahil A, Zhou J, Li Z et al (2023) Predicting the antigenic evolution of SARS-COV-2 with deep learning. Nat Commun 14(1):3478

Hinton GE, Salakhutdinov RR (2006) Reducing the dimensionality of data with neural networks. Science 313(5786):504–507

Irmak E (2020) A novel deep convolutional neural network model for COVID-19 disease detection. In: 2020 Medical Technologies Congress (TIPTEKNO). IEEE, pp 1–4

Jin S, Liu G, Bai Q (2023) Deep learning in COVID-19 diagnosis, prognosis and treatment selection. Mathematics 11(6):1279

Joshi RC, Kaushik M, Dutta MK, Srivastava A, Choudhary N (2021) VirLeafNet: automatic analysis and viral disease diagnosis using deep learning in Vigna mungo plant. In Advances in Ecological Informatics

Khan A, Khan SH, Saif M, Batool A, Sohail A, Waleed Khan M (2024) A survey of deep learning techniques for the analysis of COVID-19 and their usability for detecting omicron. J Exp Theor Artif Intellig 36(8):1779–1821

Kingma DP, Welling M (2013) arXiv preprint arXiv:1312.6114.

Koca MB, Nourani E, Abbasoğlu F, Karadeniz İ, Sevilgen FE (2022) Graph convolutional network based virus-human protein-protein interaction prediction for novel viruses. Comput Biol Chem 101:107755

Li J, Wu YN, Zhang S, Kang XP, Jiang T (2022) Deep learning based on biologically interpretable genome representation predicts two types of human adaptation of SARS-CoV-2 variants. Brief Bioinform 23(3):bbac036

Liu Z, Wan G, Prakash BA, Lau MS, Jin W (2024) A review of graph neural networks in epidemic modeling. In Proceedings of the 30th ACM SIGKDD Conference on Knowledge Discovery and Data Mining, pp 6577–6587

Ouchicha Archana K, Kaur A, Gulzar Y, Hamid Y, Mir MS, Soomro AB (2023) Deep learning models/techniques for COVID-19 detection: a survey. Front Appl Math Stat 9:1303714

Rukundo O, Behanova A, De Feo R, Ronkko S, Oja J, Tohka J (2023) Convolutional neural networks for automatic detection of intact adenovirus from TEM imaging with debris, broken and artefacts particles. arXiv preprint arXiv:2310.19630

Salama GM, Mohamed A, Abd-Ellah MK (2024) Machine learning and deep learning covid-19 diagnosis system: key achievements, lessons learned, and a transfer learning algorithm. Soft Comput 28:13715–13742

Sherstinsky A (2020) Fundamentals of recurrent neural network (RNN) and long short-term memory (LSTM) network. Phys D Nonlin Phenom 404:132306. Li, W., Deng, X., Shao, H., & Wang, X. (2021). Deep Learning Applications for COVID-19 Analysis: A State-of-the-Art Survey. CMES-Computer Modeling in Engineering & Sciences, 129(1)

Shi X, Tao Y, Lin SC (2024) Deep neural network-based prediction of B-Cell epitopes for SARS-CoV and SARS-CoV-2: enhancing vaccine design through machine learning. arXiv preprint arXiv:2412.00109.

Shoeibi A, Khodatars M, Jafari M, Ghassemi N, Sadeghi D, Moridian P et al (2024) Automated detection and forecasting of covid-19 using deep learning techniques: a review. Neurocomputing 577:127317

Skandarani Y, Jodoin PM, Lalande A (2023) Gans for medical image synthesis: an empirical study. J Imaging 9(3):69

Staudemeyer RC, Morris ER (2019) Understanding LSTM–a tutorial into long short-term memory recurrent neural networks. arXiv preprint arXiv:1909.09586

Vaswani A (2017) Attention is all you need. Advances in Neural Information Processing Systems

Wang S, Zha Y, Li W, Wu Q, Li X, Niu M et al (2020) A fully automatic deep learning system for COVID-19 diagnostic and prognostic analysis. Europ Respirat J 56(2):2000775

Wang B, Wang T, Du X, Li J, Wang J, Wu P (2024) Microbe-drug association prediction model based on graph convolution and attention networks. Sci Rep 14(1):22327

Wu S, Ren Y, Yang A, Chen X, Pu X, He J et al (2023) Deep learning and medical imaging for COVID-19 diagnosis: a comprehensive survey. arXiv preprint arXiv:2302.06611

Yifan W, Gao M, Zeng M, Zhang J, Li M (2022) BridgeDPI: a novel Graph Neural Network for predicting drug–protein interactions. Bioinformatics 38(9):2571–2578. https://doi.org/10.1093/bioinformatics/btac155

Zhang P, Hu X, Li G, Deng L (2023) AntiViralDL: computational antiviral drug repurposing using graph neural network and self-supervised learning. IEEE J Biomed Health Inform. https://doi.org/10.1109/JBHI.2023.3328337

Zhang Z, Veerapaneni R, Ayoola M, Das AR, Chen Z, Nanduri B, Ramkumar M (2024) Leveraging graph neural networks for MIC prediction in antimicrobial resistance studies. In: 2024 46th Annual International Conference of the IEEE Engineering in Medicine and Biology Society (EMBC). IEEE, pp 1–4

Zhao J, Sun J, Shuai SC, Zhao Q, Shuai J (2023) Predicting potential interactions between lncRNAs and proteins via combined graph auto-encoder methods. Brief Bioinform 24(1):bbac527

Zitnik M, Agrawal M, Leskovec J (2018) Modeling polypharmacy side effects with graph convolutional networks. Bioinformatics 34(13):i457–i466

GPSR Compliance

The European Union's (EU) General Product Safety Regulation (GPSR) is a set of rules that requires consumer products to be safe and our obligations to ensure this.

If you have any concerns about our products, you can contact us on ProductSafety@springernature.com

In case Publisher is established outside the EU, the EU authorized representative is:

Springer Nature Customer Service Center GmbH
Europaplatz 3
69115 Heidelberg, Germany

Batch number: 10725820

Printed by Printforce, the Netherlands